# Preserving the Legacy

# Introduction to Environmental Technology

## Ann Boyce

## JOHN WILEY & SONS, INC.

New York   Chichester   Weinheim   Brisbane   Singapore   Toronto

This book is printed on acid-free paper. ☉

Technical Illustrations: *Glen McKenzie*
*Doug Millhoff, Zogmorph Designs*

Photo Research: *Stuart Kenter Associates*

This material is based on work supported by the National Science Foundation under Grant No. DUE94-54521. Any opinions, findings, and conclusions or recommendations expressed in this material are those of the author(s) and do not necessarily reflect those of the National Science Foundation.

**Library of Congress Cataloging-in-Publication Data:**

Available upon request

Printed in the United States of America

10  9  8  7  6  5  4

# DEDICATION

For my students, who have been my greatest teachers.

# Table of Contents

# Preface

*Introduction to Environmental Technology* is a text designed to provide the college student with an introductory but comprehensive overview of topics relating to the broad field called environmental technology. The book has been organized to provide a logical flow of information and allow for teaching flexibility. It has been my endeavor to provide a text that will maximize the learning experience.

The focus of this textbook is on two general approaches to environmental management, *regulations* and *technology*. During the last few decades increasing numbers of laws have been enacted in an effort to stop and reverse environmental impacts of a previous era when there were few such laws. Technologies for cleanup and control of environmental pollutants have also undergone a phenomenal growth period, driven in many cases by the new legal requirements.

*Technology* is the name we generally give to the practical use of scientific discoveries to solve everyday problems. Technology and science are therefore closely associated and environmental technology is a field of applied science. *Science* is the search for explanations about the workings of our world. It is based on the principle of cause and effect, which is another way of saying that there is a logical reason or cause for every change in our environment.

Scientists study the factors that cause pollution of water and air, such as how various chemicals react in the atmosphere. Engineers use this information to develop technological solutions by designing more effective pollution control equipment. Technicians understand the basic scientific principles at work in pollution control equipment in order to operate, monitor, sample, or regulate it.

Each chapter in this book will introduce you to applicable science concepts to better appreciate the technology aspects that are covered. Discussions involve biology, chemistry and physics, which are the three main branches of science. Biology is the study of living things. Chemistry is the study of the composition, structure, properties, and reactions of matter. Physics is the study of how matter and energy are related. The coverage of these scientific disciplines is very basic since this course assumes that no prior science courses have been taken.

The overall goal of *Introduction to Environmental Technology* is to develop your awareness of the many facets of science, technology, and public policy that are involved in environmental management and protection. Environmental technology is a relatively new field that also has interesting and exciting connections with other non-science academic disciplines such as political science and sociology. These interrelationships have been noted in the text to the extent possible. You are encouraged to use the information in this text in an integrated fashion with knowledge gained in other classes when analyzing issues.

The following are specific goals I had in mind while writing this text:

1) Acquire the vocabulary or "jargon" of this discipline.

2) Understand the many facets and interdisciplinary aspects of environmental technology.

3) Kindle interest in the topic for responsible citizenship.

4) Provide a framework of basic information that can be built upon using critical thinking skills.

5) Provide a basis for classroom discussion.

6) Gain an appreciation for the need to study the sciences.

7) Avoid oversimplification through recognition of the complexity of the decision-making and public policy aspects of this field.

8) Appreciate the importance of environmental protection to current and future inhabitants of the earth.

9) Recognize career opportunities and options.

I hope this text provides the foundation for you to meet these goals and that, through the topics presented here, you consider a career that puts you in a position to further the strides being made to protect the environment. I have found my career in environmental technology to be continually challenging and rewarding. I hope that you too will find your niche within this worthwhile field and will join me in a lifetime adventure of far-reaching importance.

## Acknowledgments

I wish to thank the many people who helped make this book a reality.

■ The Partnership for Environmental Technology (PETE), and INTELECOM Intelligent Telecommunications, for making the *Preserving the Legacy* Project, and this book, possible.

■ The National Science Foundation for the foresight to support Advanced Technology Education.

■ Howard Guyer, Academic Team Leader for the *Preserving the Legacy* Project, for his indefatigable efforts on behalf of the Project, and for the countless hours spent editing the text and rewriting a number of the Science & Technology sections.

■ Amber Taylor and Jeff Gurican, student researchers, for their invaluable assistance.

■ Sarah Bremner, for her skill in coordinating the work of reviewers, photo researchers, and graphic artists.

■ Sally Beaty, for her leadership in directing the *Preserving the Legacy* Project, and the following members of the National Academic Council for their encouragement and technical expertise in the reviewing of the manuscript: Doug Feil, Kirkwood Community College; A.J. Silva, Eastern Idaho Technical College; Eldon Enger, Delta College; Dave Boon, Front Range Community College; Doug Nelson, SUNY Morrisville; Bill Engel, University of Florida; Ray Seitz, Columbia Basin College; and Stephen Onstot, Fullerton College.

■ Judy Sullivan, for her creative text design, attention to detail, and transformation of the manuscript into a student-friendly text.

■ Larry Fanucchi, Division Chair, Applied Science and Technology, Bakersfield College, and the Administration at Bakersfield College for supporting me in this project.

■ My family and friends whose belief in me kept me going.

Ann Boyce
Bakersfield College, 1996

# How Did We Get To This Point?

## ■ Chapter Objectives

After completing this chapter, you will be able to:

1. **Define**, in the broad sense, hazardous materials, and explain how they differ from hazardous wastes.

2. **Describe** major historical events that gave rise to the environmental movement, and identify changes in public policy caused by the movement.

3. **List** the major federal agencies involved in environmental, health, and safety regulation and their areas of authority.

4. **Describe** future challenges for the EPA.

5. **Explain** changes in environmental policies and regulations since the 1970s.

6. **List** the types of impacts that environmental regulations have on the various segments of society.

## ■ Terms and Concepts to Remember

Asbestos
Atom
Atomic Mass
Atomic Number
Atomic Theory
Bonded
Carrying Capacity
Chlorofluorocarbons (CFCs)
Cleanup
Compounds
Contaminant
Department of Transportation
 (DOT)
Disposal
Drinking Water Standards
Electrons
Elements
Enforcement
Environmental Medium or
 Compartment
Environmental Protection
 Agency (EPA)
Environmental Technology
 (ET)
Episode (Pollution)
Groundwater

Hazardous Materials
Hazardous Wastes
Inorganic Compounds
Isotopes
Media
Molecule
Montreal Protocol
Neutrons
Nucleus
Occupational Safety and
 Health Administration
 (OSHA)
Oil Spill
Organic Compounds
Pollutant
Pollution
Protons
Sewage
Superfund
Surface Water
Sustainable Development
Underground Storage Tank
 (UST)
Waste
Wetlands

## ■ Chapter Introduction

This chapter explores the origins, history, and impact of the modern environmental movement. Environmental consciousness, expressed in a variety of ways, has traditionally played a role in shaping public policy. Since the 1970s, however, environmental awareness has grown significantly, a reflection of the general public's interest in environmental issues. A number of key events and the activities of several individuals and organizations were especially instrumental in broadening and changing the scope of the environmental movement. These changes resulted in new legislation designed to preserve and protect the environment. Environmental health is now perceived as both a national and a global priority.

The enactment of numerous laws dealing with environmental issues since the 1970s created demand for new governmental agencies to set standards and enforce compliance with regulations. Because of the wide scope of the legislation, nearly all sectors of society today feel their impact. As a result, career opportunities for students in many environmental fields have increased significantly and are examined in this chapter.

# 1-1 History and Origins

## ■ Overview and Introduction

### Concepts

- ET is a field that applies regulations and scientific principles to protect public health and the environment.

- Hazardous materials can cause health or environmental damage if improperly managed.

- DOT, EPA, and OSHA are the primary federal agencies involved in hazardous materials regulation.

- Adopting a "systems approach" to environmental regulation will help prevent cross-media transfer of pollution.

- Preventing, rather than repairing environmental damage, is currently the main focus of the EPA.

**Figure 1-1:** Over the past several decades, we have become more aware of the impact our actions have on the environment.

**Figure 1-2:** The environmental compartments.

**Environmental Technology (ET)** refers to the knowledge and skills that are necessary to manage, work with, and control hazardous materials and **pollutants**. These skills, based on applied science principles, are designed to reduce human health and ecological risks while being fully in compliance with governmental regulations. The field of ET is relatively new. The general public has shown increasing interest and political influence in the last three decades in areas relating to public health and the environment. Greater public involvement led the government to enact more laws related to protection of public health and the preservation of the environment. This change in the direction of public policy has resulted in the 1990s being heralded as the "decade of the environment."

**Hazardous materials** is a general term used to describe substances that may either cause or increase the occurrence of death or irreversible illness in the population, or pose a hazard to human health or the environment when improperly treated, stored, transported, disposed of, or otherwise managed. In this general sense, the term can include hazardous wastes, hazardous substances, toxic materials, hazardous chemicals, radioactive substances, and infectious or medical wastes. A **Hazardous waste** is a hazardous material for which there is no further use. These terms also have very specific legal definitions as they apply to various legal mandates. While we will be using the more general definitions at this point in the text, legal definitions must be used when official determinations are made. Later chapters will provide more information on legal definitions of many of these terms.

The increased numbers of laws and public awareness of the potential harm that hazardous materials can pose has resulted in growth within the field of ET. This growth has created excellent career opportunities for students. Environmental professionals and industry analysts see some staggering challenges and opportunities in the years to come for those involved in this broad, many-faceted, and dynamic field. Career opportunities are expected to increase in a variety of support, informational, and affected industrial segments. These opportunities are discussed in more detail later in this unit.

As in most fields, ET uses a specialized vocabulary, including a multitude of acronyms used for describing various activities, laws, and technologies. Acquiring the fundamentals of this new vocabulary is one of the goals of this introductory course. Without this vocabulary, it would be impossible to understand journals, articles, speakers, and many of the foundational concepts necessary to progress to more specialized classes. To assist in reaching this goal, each unit will present a list of terms and concepts discussed in that chapter that you will need to master.

Take for instance, the terms **environmental medium** or **environmental compartment** that are described in Box 1.1 These terms refer to the part of the environment, typically air, water, soil, and biota (living things), that **contaminants** are transmitted through or carried by. A spill of a hazardous material that evaporates easily could, for instance, affect all four of the compartments. The spill would immediately contaminate the air and soil, then it could move

# Box 1.1

# An EPA Administrator's Overview

*In late 1988, the Environmental Protection Agency published a document entitled "Environmental Progress and Challenge: EPA's Update." The Administrator of the EPA at the time, Lee M. Thomas, provided an overview statement in that document that is reproduced in part below. This statement touches on many of the areas that will be dealt with in this course, and as such provides a good introduction to the field.*

For the past 30 years, the American people have been involved in a great social movement known broadly as "environmentalism." We have been concerned with the quality of the air we breathe, the water we drink, and the land on which we live and work. This concern has focused on the wise use of our natural resources and the preservation almost exclusively in terms of economic growth and prosperity. Calvin Coolidge immortalized this focus when he observed that the "business" of America was just that – business. Our nation was blessed with seemingly endless resources, hard-working people, and unlimited opportunity.

Our natural resources were exploited indiscriminately. Waterways served as industrial pollution sinks; skies dispersed smoke from factories and power plants; and the land proved to be a cheap and convenient place to dump industrial and urban wastes. Industrial growth during the period following World War II was unparalleled in the history of the nation.

We enjoyed a prosperity never before known. Unfortunately, we also were accumulating an environmental debt of staggering proportions.

By the late 1960s, Americans began to recognize an emerging crisis. We had witnessed serious environmental degradation in every medium. The air in many industrial cities was deemed unhealthy; Lake Erie lay on its deathbed; the Cuyahoga River erupted in flames; pesticides like DDT took their toll on wildlife. Over the next decade and a half Congress passed a series of far-reaching laws that prescribed needed changes in the way the nation conducted its business. Slowly, a process was built that today incorporates environmental considerations into the basic decision-making of government and industry.

Our society is now more aware of the environment and the need to protect it. While economic growth and prosperity are still important goals, opinion polls show overwhelming public support for **pollution** controls and a pronounced willingness to pay for them. This latter point, perhaps the ultimate measure of commitment, has been borne out over the last two decades as Americans spent billions of dollars for cleaner air, water, and land.

Because many of the solutions of the past decades merely transferred pollutants from the water to the air or from the air to the land, we must adopt a more integrated or "systems" approach to environmental protection. We can no longer think simply of clean air or clean water; we must work for a clean environment.

We need to work harder to prevent environmental problems by reducing the amount of **wastes** from our homes and from industry. We need to recycle more waste. Future waste management should prevent **disposal** problems by reducing the amount of waste as a first step.

We also must reassess our notion of "environmental safety." We now know that many of the things we do and chemicals we need to sustain our modern lifestyle pose some risk to people and to our ecosystem. We have to develop better ways to assess these risks and make the choices which balance the benefits with the risks. This may be one of our toughest missions in the next decade.

Increased public understanding of environmental problems, risk, and solutions will be even more critical to our success than in the past. The problems we deal with today are not primarily the large smokestack concerns of the 1970s. As these major sources of pollution are being controlled, Americans will have to recognize that our individual actions in our homes, the products we buy, and how we choose to relax all can affect the quality of our environment. We will have to face choices in our daily lives to balance the risks of these actions with the benefits. The EPA will have to play a large role in educating and involving the public in its decisions.

The responsibility for implementing our nation's environmental laws also is changing. Unlike the majority of issues in the 1970s, centralized pollution control efforts will not effectively address all of the major problems. State, local, and Indian tribal governments are now playing an ever more significant role in environmental protection. We need to discover how the EPA can best assist and encourage an even stronger future role for these players.

Finally, we now recognize through problems such as acid rain, global warming, and pollution of our oceans that the quality of the environment in America is also dependent on how the rest of the world treats our planet – just as our actions affect

## Box 1.1

other nations. One of our biggest challenges will be to assist and educate other nations and actively negotiate multinational agreements to protect the global environment.

### EPA's Future Agenda

Our agenda should include several key components. It should take into account the problem of cross-media transfer of pollution. It should, to the extent possible, reflect priorities that are set on the basis of risk. It should encourage full involvement of the public in making tough choices in the future. It should promote an effective role for federal, state, local, and American Indian tribal governments. It should recognize the global nature of some issues. Finally, and perhaps most critically, it should strive to prevent pollution by reducing the amount of waste we produce, by recycling what we can, and by making other sound management approaches to avoid more costly cleanups in future years.

### A Systems Approach

The environment is an integrated system. There is no such place as "away" where we can throw things. For example, when we remove pollutants from the air, we often inadvertently transfer them to the water or to the land. If we simply transfer the pollution, it likely will come to rest at the point of least regulation. The point of least regulation, however, may not be the point of least risk.

To meet the challenges of the 1990s and the next century, a more systematic approach to protecting the environment must be taken, one involving a coordinated strategy by the EPA and other government programs for achieving the maximum affordable reduction of the most significant risks.

### Reducing Risk

We cannot eliminate all toxic chemicals from the environment. Given the tens of thousands of chemicals used today, basic decisions must be made as to which ones should be controlled, to what levels, and at what cost. This requires an assessment of the risks to human health, welfare, and the environment that are posed by different pollutants in different locations. It requires decisions on how to reduce the most significant risks, taking into consideration the benefits and the costs, as well as other public concerns. It requires the establishment of priorities through risk-based decision-making.

The EPA's challenge is to improve the way risk-based decisions are made in the future. To do this, we will need better risk information and better science.

### Public Education and Involvement

The EPA must share with the public its knowledge of the scope and severity of the environmental challenges before us. With improved information and an understanding of the reality of risk in an industrial society, the public can do a better job helping us establish our national environmental directions.

The EPA must also help people to understand that they are part of the pollution problem and its solution. The attention of the last 15 years has been on controlling large sources of pollution such as smokestack industries and municipal sewage disposal facilities. As we strive for further environmental improvements in the 1990s, the focus of the nation's pollution control efforts must also include small sources that are widespread throughout the country. For example, we need to pay greater attention to how each of us can recycle more of our household waste, properly use and dispose of lawn chemicals, and drive and maintain our cars to reduce air pollution. We cannot pass to someone else the responsibility for controlling pollution. Each of us must recognize the part that we play in environmental protection and become "part of the solution."

Further progress, therefore, will require lifestyle changes by the American people. All of us – individual homeowners, farmers, shopkeepers, automobile drivers – will have to make tough choices between convenience, the costs of goods and services, and a cleaner environment. The EPA's job should grow from primarily "the enforcer" to include greater emphasis on helping citizens make informed choices in their daily lives.

### State/EPA Partnership

The EPA and the states need to continue to strive toward a proper balance in our working relationship. The challenges that face us now are not as amenable to centralized "command and control" approaches as past problems. The EPA must further recognize the increasing capabilities and responsibilities of states – including Indian tribal governments and local governments – in protecting the environment.

The EPA must provide better technical support to state governments as they assume more responsibility for environmental program funding and management. This means more training and technical assistance to help states assess and address risks. It means improving data management so that

# Box 1.1

states and the EPA have access to each other's information, and creating other ways for all of us to share our expertise. A true partnership also requires the EPA to be more sensitive to the separate needs of the states. Risks are not distributed uniformly nationwide; the EPA must look for new ways to address priorities that differ around the country and recognize the growing role of state, local, and Indian tribal governments in shaping their own environmental agendas.

Although state and federal roles are changing, the EPA will continue to carry out its statutory responsibility for setting standards and ensuring compliance with them. Our challenge is to get this job done while also giving states enough flexibility to solve important local problems that are not national priorities.

## Environmental Outlook: Global

In the largest sense, the earth is a single, integrated ecosystem shared by all the people living on it. As the world's population and economy continue to grow, that ecosystem is being strained in a number of ways. Chlorofluorocarbons are threatening the stratospheric ozone layer; global emissions of carbon dioxide are contributing to a gradual warming of the earth's atmosphere; species of flora and fauna are being lost worldwide at an accelerating rate. These kinds of changes to the global environment are of especially serious concern because they have the potential to affect the quality of life of literally everyone on earth.

Moreover, in a number of places natural resources shared by neighboring nations are being degraded. For example, acid rain is harming aquatic ecosystems in the northeastern United States, southeastern Canada, and Europe. Shared water bodies like the Mediterranean Sea and the Gulf of Mexico are being polluted by the combined economic activities of the different countries that border them.

But whether these emerging environmental problems are global or regional, our response to them will entail international cooperation. Because pollution does not stop at international political boundaries, we are going to have to find new ways of cooperating with the community of nations to find shared solutions to shared problems.

International cooperation in this area will be complicated by the need to factor in the special economic circumstances of developing nations. Poorer countries often find it difficult to sustain their natural resource base in the face of immediate demands for food, fuel, and jobs. They have fewer resources to invest in waste disposal and pollution

control. But as was proved by the recently signed **Montreal Protocol** to protect the stratosphere, nations are capable of resolving their economic and political differences in the interest of protecting a shared environment. We will have ample opportunity to put that lesson to work in the years ahead.

## Preventing Future Environmental Problems

Environmental protection will be a never-ending battle against contaminated "hot spots" unless we take steps now to prevent them from developing. We are placing more emphasis on pollution prevention in all EPA programs. We need to do more of it, and so do state and local agencies as well as the private sector.

We as a nation can do many things to prevent environmental problems from developing in the first place. We should reduce the amount of waste from our homes and industries before that waste becomes a disposal problem. Municipal recycling and industrial waste reduction ease the economic and environmental burden of waste disposal. Source reduction and recycling should become the centerpiece of a progressive national waste management strategy.

We will continue to restrict the use of toxic chemicals in places where they might enter drinking water supplies or endanger fish and wildlife. We will continue to ban the disposal of untreated wastes on the land, where it threatens human health or the environment. We will continue to identify sensitive **wetlands** and restrict harmful development before irreparable damage is done. We will continue to promote better farming practices that will prevent agricultural chemicals from contaminating **ground** and **surface waters**. In short, we will do a better job of planning to prevent future problems. By doing so, we avoid costly cleanup later and we avoid the loss of irreplaceable resources. If we take precautions today, we will be making an important investment in a safer and cleaner environment tomorrow.

We have begun to do long-range planning at the EPA. We are trying to determine what environmental results we want to see in the not-too-distant future and the best strategies we can employ to achieve them. To do this well requires tremendous vision in establishing goals, taking into consideration the tough choices we and the American public need to make between environmental and other social goals.

It requires creativity and hard-nosed realism in designing systematic, coordinated program strategies, and it requires dedication to follow through. This is EPA's ultimate challenge.

through the soil to reach ground or **surface water** supplies which could be ingested by humans or animals.

Although more than a dozen federal agencies deal with various specialized aspects of environmental protection, there are three primary federal agencies that have the responsibility of enforcing the broadest and most encompassing laws relating to public health and the environment. These three agencies are the **Environmental Protection Agency (EPA)**, the **Occupational Safety and Health Administration (OSHA)**, and the **Department of Transportation (DOT)**.

The U.S. EPA was founded as an independent agency by President Richard Nixon in 1970. The EPA's mandate is to ensure compliance with laws and develop regulations designed to protect the environment. Its primary responsibilities include protection of natural resources, soil, water and air compartments through control and reduction of air and water pollution by radioactive materials, pesticides, and other hazardous materials. More recently, the EPA has become involved in global environmental issues such as global warming and the depletion of the ozone layer. Lee Thomas, the administrator of the EPA during the Bush administration, described the challenges confronting the agency in 1988 in an excerpt reprinted for you in Box 1.1. When he was elected President, Bill Clinton appointed Carol Browner to head the EPA. In a May 1993 publication the EPA was listed as employing 20,000 people in a wide variety of technical jobs. The administrative offices of the EPA are located at 401 M St., SW, Washington, D.C. 20460.

Since the DOT is the federal department that develops and enforces regulations dealing with all aspects of transportation, it also regulates the transportation of hazardous materials. This department employs nearly 70,000 persons. The DOT's administrative offices are located at 400 7th St., SW, Washington, D.C. 20590.

As an agency of the federal Department of Labor, OSHA develops and enforces regulations pertaining to worker health and safety. OSHA employs 18,000 persons and is jointly headed by the Assistant Secretary of Occupational Safety and Health and the Secretary of Labor. Administrative offices are located at 200 Constitution Ave., NW, Washington, D.C. 20210.

## ■ Historical Perspectives

### Concepts

- Thoreau and Muir were early champions of conservation.

- Human activities have had a significant impact on the environment and other living things.

- *Silent Spring* heralded in the modern environmental movement.

- The Montreal Protocol was an agreement by 24 nations to limit CFCs.

■ The Love Canal crisis raised public awareness about the dangers of indiscriminate dumping of hazardous waste.

Every generation of Americans has been confronted with questions about the relationship between human beings and their environment. Concern about the preservation of and reverence for the environment can be seen in the works of American writers as diverse as Thomas Jefferson and Ralph Waldo Emerson. The ideas of Henry David Thoreau, the American philosopher, naturalist, and essayist, particularly influenced policy makers during the nineteenth century. Thoreau lamented the waste and destructive living conditions in the commercial cities of the eastern seaboard so much that in 1845 he decided to remove himself altogether from civilized society. He lived for 26 months in a simple cabin at Walden Pond, a wilderness area two miles outside of Concord, Massachusetts. There he kept a journal in which he recorded in detail his observations on nature which he later used to write his famous work, *Walden*. His philosophy on nature concluded that wilderness was vital to human existence and saw it neither as an enemy to be conquered nor as a resource to be exploited. His appeals for "national preserves in which the bear and the panther, and some even of the hunter race may still exist, and not be civilized off the face of the earth – not for idle sport or food, but for inspiration mid our own true recreation" elicited a positive response from the American people and contributed to the establishment of the first national park in Yellowstone, Wyoming in 1872.

Figure 1-3: Henry David Thoreau.

"We can never have enough of nature. We must be refreshed by the sight of inexhaustible vigor, vast and titanic features, the sea-coast with its wrecks, the wilderness with its living and its decaying trees, the thunder-cloud, and the rain which lasts three weeks and produces freshets. We need to witness our own limits transgressed, and some life pasturing freely where we never wander."

From: Walden, "Spring" (1854)

In the period of American westward expansion following the Civil War, scientists and policy makers grappled less with the philosophical meaning of wilderness than they did with the political and economic realities of the growing industrial and agricultural society. A central figure among scientists and naturalists in the West who addressed these issues was the conservationist John Muir. A student of geology and a skilled amateur botanist, Muir combined an interest in scientific and practical knowledge with a reverence for nature similar to Thoreau's. Today his nature writings are still of interest, but he is best known as the first president of the Sierra Club, an organization founded in 1892 with the mission of conserving and making accessible the mountain regions of the Pacific Coast for the purposes of exploration and enjoyment. The Sierra Club drew its early membership from scholars and dedicated outdoors enthusiasts, the majority of whom lived in Northern California. However, because of the national popularity of Muir's writings, the club's membership grew from its original 27 members to 160 within its first year. It now has over 630,000 members. From its inception, when it contributed to the creation of six national parks including Yosemite, the Sierra Club established itself as a force in shaping U.S. conservation policies. The scope of the club's activities has now widened to include protection of the earth's scenic and ecological resources and addressing the issues of the urban environment, overpopulation, the use of DDT, offshore drilling, and disposal of hazardous wastes.

1914, the year of John Muir's death, was a turning point in the general public's environmental consciousness. Martha, the last living passenger pigeon, one of America's species of wild birds, died in the Cincinnati Zoo. She was the last of a species that had been abundant in the eastern part of North America. In 1648, settlers in the Massachusetts Bay Colony wrote of avoiding starvation by eating these birds. In the early 1800s, it was estimated that there were over 5 billion of these pigeons, considered a low cost and endless source of food. The passenger pigeon accounted for one-third of America's bird population. John James Audubon, the painter and naturalist, described a flight of passenger pigeons as a "torrent of life" that took three days to pass overhead. Professional pigeon hunters supplied markets with vast numbers of the birds. One poultry dealer in New York received a shipment of 18,000 in one day. Human activities resulting in the over-exploitation of this species led to its extinction. Martha became a powerful symbol and reminder of how even a very successful species is no match for humans if it is indiscriminately hunted or deprived of the space and conditions it needs for survival.

"It is change, continuing change, inevitable change, that is the dominant factor in society today. No sensible decision can be made any longer without taking into account not only the world as it is, but the world as it will be. . . . This, in turn, means that our statesmen, our businessmen, our everyman must take on a science fictional way of thinking."

Isaac Asimov from "My Own View" (published in The Encyclopedia of Science Fiction, 1978)

The post World War II and Korean War years witnessed rapid economic growth, an explosion of new technologies, and the rise of an automobile-based urban culture, each of which had significant environmental impact. It was a new era of abundance and consumption symbolized especially by petrochemical products, including plastics, pesticides, additives to food and fuels, detergents, solvents, and abrasives. During this period, air and water pollution were the two most significant forms of environmental degradation generating public concern. By the 1960s the polluting of Boston Harbor, Santa Monica Bay, and the darkening of the Los Angeles skyline were highly visible symbols of pollution.

The single work often credited with starting the modern environmental movement is *Silent Spring*, written in 1962 by Rachel Carson. A former biology teacher and researcher with the Bureau of U.S. Fisheries, Carson wrote about the widespread use of pesticides, particularly DDT. As the title *Silent Spring* indicates, her book documented the disappearance of songbirds. Carson attributed their reduction in number to the disruption of their reproductive processes from the accumulation of pesticides such as DDT in their bodies. What distinguished Carson's book was her argument that public health and the environment were inseparable, and that the particular issue of pesticide use was a public issue to be dealt with at the level of policy and not decided as a technical issue to be determined solely by industry and government officials. Americans, alerted by Carson's frightening message, helped shape public policy that eventually led to the banning or severe restriction of a number of pesticides, including DDT. Despite its popular success, *Silent Spring* was not without its critics. The book was sharply attacked by reviewers, scientists, pesticide "experts" and members of the chemical industry.

Rachel Carson was an inspiration to others interested in studying and writing about environmental issues. The success of her book paved the way for growth in public concern for improving the quality of our environment. College students became more interested in the field, and colleges and universities offered more courses in ecology. Other books published in the late 1960s and early 1970s took advantage of this new interest and understanding of environmental issues. Most notably, *The Population Bomb*, written by Paul Erlich in 1968; *The Closing Circle* by Barry Commoner, 1971; and *Limits of Growth*, written by a research team from Massachusetts Institute of Technology in 1972.

Figure 1-4: Rachel Carson.

---

"The 'control of nature' is a phrase conceived in arrogance, born of the Neanderthal age of biology and the inconvenience of man."

From: Silent Spring, Chapter 17

---

The environmental movement of the late 1960s distinguished itself with a new level of activism that culminated in the celebration of Earth Day on April 22, 1970. Earth Day began as the idea of Wisconsin senator Gaylord Nelson and Maine senator Edmund Muskie for a "National Teach-In on the Crisis of the Environment." As word spread throughout colleges and universities and to other environmental organizations, each group saw an opportunity to participate in a dialogue that they hoped would help define a national environmental agenda. When the day was done, it was estimated that over 10 million people had participated in Earth Day events at locations across the country.

The first highly publicized environmental crisis that captured the nation's interest occurred during the late 1970s in a small community called Love Canal near Niagara Falls, New York. Originally a 16-acre waterway built by William T. Love in 1895 as part of a venture for a model city, the abandoned project became the site for casual dumping of highly toxic industrial chemicals over several decades by a nearby Hooker Chemicals plant. The site was also used for the disposal of municipal wastes. When filled, the pit was covered over and later purchased for one dollar for the building of a school for the new residential community being developed in the area.

Residents in the Love Canal area began to notice foul-smelling and noxious chemicals in their yards, on the school grounds, and seeping into their basements and suspected them of causing health problems in the community. Under the outspoken leadership of Lois

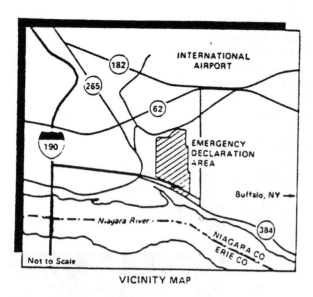

Figure 1-5: Love Canal emergency declaration area.

Gibbs, a local housewife and mother, Love Canal residents initiated a community health survey, staged protests, and challenged politicians and EPA officials. The Love Canal Homeowner's Association demanded that the government purchase homes on or adjacent to the dump site because of the extensive damage caused by toxic contamination. Initially, government agencies did not know how to handle the case since no one at the time had experience in hazardous waste contamination matters on this scale. Ultimately, the residents of Love Canal prevailed. In 1978, the New York Commissioner of Health issued a finding that "hazardous chemical wastes at the 'Love Canal Chemical Waste Landfill' constituted a public nuisance and an extremely serious threat and danger to the health, safety, and welfare of residents nearby." Pregnant women and children under two years of age living closest to the contaminated site were advised to move from the area temporarily, and all residents were told to avoid using their basements and eating homegrown produce.

The area was declared a federal disaster area in 1978 by President Jimmy Carter, an unprecedented event for a hazardous waste site. By declaring the area a disaster, federal aid became available for both the relocation of residents and for the containment of wastes. New York State also provided funds for the purchase of homes in the vicinity. Two years later, President Carter again declared a federal emergency which made funds available to "permanently relocate more than 500 families from a 232-acre area surrounding the Love Canal." (New York State Commissioner of Health, "Love Canal, Emergency Declaration Area – Decision on Habitability," September, 1988). Ultimately, 2,500 people were evacuated from their homes, and an elementary school and 230 homes were torn down. All of these contaminated materials were added to the original landfill and buried under a clay and plastic cap which now covers 40 acres.

The Love Canal crisis alerted the nation to the problem of hazardous waste disposal and many of the ways unregulated disposal threatens human health and the environment. This **episode** led to the enactment of a federal program in 1980 called **CERCLA (Comprehensive Environmental Response, Compensation, and Liability Act)**, better known as "Superfund," designed to provide moneys to clean up areas like Love Canal. CERCLA established a waste management system which included mechanisms for identifying sites, identifying responsible parties, and for selecting cleanup technologies.

Another environmental tragedy that caused public outcry occurred on March 24, 1989, when the Exxon oil tanker, *Valdez*, loaded with over one and a quarter million tons of crude oil, ran aground on Bligh Reef off the coast of Prince William Sound in Alaska. Most estimates indicate that about one-quarter of its load of oil was released. The effects of this massive **oil spill** to fragile shoreline and marine wildlife were brought to living rooms worldwide via live news coverage. This disaster resulted in huge cleanup costs and changes in the regulation of oil transport.

Many of the environmental problems that we have today are the result of activities that were once perfectly acceptable and even desirable but are now considered unacceptable due to new knowledge of their effects on our environment. When Hooker Chemical filled Love Canal, it was acting legally and following the common practices used to dispose of hazardous waste. DDT was considered a miracle chemical and was applied liberally to residential areas and often directly to humans to control mosquitoes and other insects. Another example is the use of a group of substances known as **chlorofluorocarbons** (CFCs). In the 1930s they were considered a miraculous discovery because of their non-toxic and non-combustible properties. They were widely used for many years as aerosol propellants and refrigerants. Then, in 1976, the National Academy of Sciences reported a connection between CFCs and the depletion of the earth's protective ozone layer. In 1987, 24 nations signed an agreement to reduce by 50 percent the most widely used and most detrimental of the CFCs. This agreement, known as the Montreal Protocol, represented the beginning of efforts toward world cooperation to protect the environment. In 1992, a revision to the Protocol called for total elimination of CFCs in industrial nations by 1996. We look back at these practices and activities and wonder how past generations could have caused such great harm to humans and to the environment. What actions of ours will be viewed similarly in 50 or 100 years?

## ■ Checking Your Understanding (1-1)

1. Explain what hazardous materials are and how they differ from hazardous waste.

2. List the three main federal agencies involved in hazardous materials regulation and enforcement and describe the aspects of environmental health and safety each is responsible for.

3. List the environmental compartments and explain why they are called "compartments."

4. What did Lee Thomas recommend as the "first step" in reducing environmental waste?

5. On environmental issues, what is the EPA's position on attitudes and activities of other nations?

6. Explain the statement, "The point of least regulation may not be the point of least risk."

7. List three things that Lee Thomas says we can all do to "become part of the solution" to the environmental problems we face.

8. Describe the two actions that should "become the centerpiece of a progressive national waste management strategy."

9. Describe the types of writings that influenced the public and sparked interest in conservation and protection of the environment.

10. What is the significance of the Montreal Protocol?

11. How do heavily publicized environmental tragedies affect public policy?

# 1-2 Population and Growth

## ■ Global and National Issues

### Concepts

■ Environmental concerns are global concerns.

■ Much environmental damage results from activities once considered acceptable.

■ Today, noncompliance with environmental laws carries stiff fines and penalties.

Although we tend to think of environmental issues as problems that we can solve by addressing individual sources of pollution (i.e. factories), the truth is that environmental concerns are complex issues with global dimensions. On June 13, 1992, the twentieth anniversary of the first United Nations Conference on the Environment, the Earth Summit, was held in Rio de Janeiro, Brazil. Never before had so many heads of state gathered together for a common purpose. By their attendance at the conference, the world's leaders demonstrated their commitment to preserving the planet's natural resources. The goal of the Earth Summit was to demonstrate that unity of purpose and global cooperation would benefit all the countries of the world.

Both national and international economic development play a significant role in the degradation of the environment. Industrialized nations today know their wealth and prosperity have come at the cost of the environment. The concern has shifted from "How do we clean up the mess?" to "How can we prevent future messes like this?" Industrialized countries as well as countries with developing economies are beginning to seek solutions beyond technology. These nations now encourage alternative ways of conceptualizing environmental safety that allow them to continue to make economic prosperity a first priority. In the U.S., economic structures are gradually changing to accommodate the need to sustain the natural environment.

Some critics of government regulation say that the pendulum is beginning to swing too far in the direction of environmental protection and away from enterprise and economic progress. Many businesses that are unable to produce the quantity and quality of products they once did claim that environmental regulations make it impossible to compete in a global marketplace. Placing so many environmental regulations on businesses and industries is due in part to the fact that decision makers want to avoid repeating past errors. Much of the environmental damage that has occurred has been the result of activities that were acceptable practices at one time, but are now known to be environmentally hazardous. For example, we now know that one quart of used motor oil, if disposed of improperly, can potentially contaminate thousands of gallons of fresh water. It may travel downward through soil and find its way into groundwater. The practice of pouring used oil into the sewer is now illegal, but how many millions of quarts had been disposed of in just that manner prior to our understanding of their impact? As our knowledge base increases and improves, laws change to reflect the new knowledge and improve our management of hazardous materials.

In addition to polluting activities that were once acceptable, many people have seen environmental devastation caused by willful and negligent actions. People want their children and grandchildren to enjoy the same basic things that they have taken for granted: clean air, water, land, and a life free of the health hazards associated with environmental pollution. Therefore, economic productivity is no longer considered the ultimate measure of success. Many laws have been enacted with severe penalties for those involved in activities that cause environmental degradation. Noncompliance with environmental laws can result not only in penalties, but negative publicity, imprisonment, and lawsuits.

## ■ Population and Sustainable Development

### Concepts

- A global goal is to provide sustainable development and economic prosperity.

- Many global problems are interconnected.

- Industrialized nations use a disproportionate amount of the world's resources.

There has always been ample economic incentive not to consider environmental issues when making business decisions. Most developing countries have based the entire value of natural resources on the costs of their extraction and distribution. This has resulted in little or no incentive to develop long-range plans for sustainable resource use. The focus on short-term economic gain has been viewed as more important than the long-term environmental loss. As imprudent and non-sustainable practices are implemented, the resource being used will likely be depleted – permanently! The failure to consider the full environmental costs has led to over consumption and environmental degradation. Clear-cutting of the tropical rain forests is a good example of this narrowly-focused perspective.

In the past, governments of nations with developing economies have sold rain forest land to developers who use it for timber and for cattle grazing. The proceeds from the land sale and subsequent export tariffs on timber and beef provide vital income needed for their economic growth and political stability. However, after the forests are clear-cut and cattle have grazed, the land becomes depleted in a very short period of time. The soil left behind is not rich enough to support agriculture, and it is left to erode. Being susceptible to floods, the soil fills nearby rivers with silt and causes a variety of other environmental problems. Devoid of its necessary components, the land can no longer carry out its original function and purpose.

Tropical rain forests play a vital role in the life of our planet. Their immense value comes not from the soil itself, but from the canopy of vegetation overhead that contributes to the balance of moisture, carbon dioxide, and temperature for the entire planet. Among the thousands of species of plants and animals that inhabit these forests are resources scientists have shown to be invaluable in the development of new medicines.

Environmentalists have influenced governments, agencies, corporations and industries to invest in projects to teach indigenous people how to replant and reforest their land. Governments have been given money to train individuals in enforcement activities and to punish those who practice clear-cutting in the rain forests. With these projects, the vast resources from the forests themselves can be harvested and sold so that not only the forests are sustained, but so are the communities who inhabit them.

We are beginning to understand that many of the global problems we face, such as poverty, population growth, industrial development, and the destruction of the environment, are quite closely interconnected. We cannot solve any one of these problems without looking at the big picture and addressing them all. The development of the earth to provide a basic level of comfort for all people and the protection of the global environment are two sides of the same coin – human survival.

The human population of the earth is increasing at a staggering rate. The current world population is five times that which it was in 1800. The population has doubled in the last 40 years from 2.5 billion to 5 billion and is expected to reach nearly 6.5 billion in the year 2000. Ninety percent of this growth will occur in the world's poorest countries.

At present and projected growth rates, the world's population will reach 10 billion by 2030 and will approach 30 billion by the end of the twenty-first century. These levels correspond closely to estimates by the U.S. National Academy of Sciences of the maximum **carrying capacity** of the entire planet. Carrying capacity refers to the ability of a region to support its inhabitants without degrading available resources.

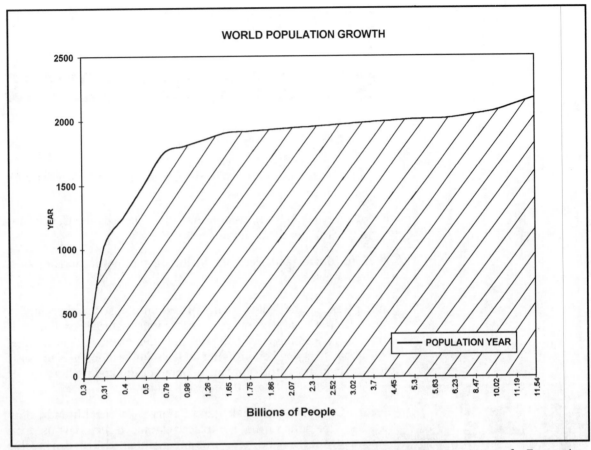

**Figure 1-6:** World population growth.  (Source: United States Population Division, Department for Economic and Social Information and Policy Analysis)

A major factor affecting this dramatic increase in population involves longer life expectancies through improved medical practices and better nutrition. Population directly relates to a multitude of environmental quality issues. As population increases, there are fewer resources to go around, putting greater strain on existing food, water, and energy resources, not to mention the environmental impact of the various resultant wastes.

The U.S. population grows by approximately 2.2 million each year, making it one of the fastest growing industrialized countries. Although this country is rich in natural resources, at our present rates of consumption we are using them up at a faster rate than we are able to replace them. Because of our high standard of living, our population also consumes a disproportionate amount of the world's resources. For example, the U.S. population represents approximately 5 percent of the world population but uses 25 percent of the world's energy resources.

More recently, nations have gathered to discuss the environmental impact of the projected global population increase.  At an international conference sponsored by the EPA and the International Association for Clean Technology (IACT) held in Washington, D.C. in the summer of 1990, the goal repeated again and again was that of **sustainable development**. Sustainable development refers to the management of resources in such a way as to enable us to meet current needs without jeopardizing the ability of the earth's future inhabitants to meet theirs.

## ■ Curriculum and Career Opportunities

### Concepts

■ Most industries require skilled, trained personnel to assist with compliance.

■ Educational institutions offer Associate Degrees and Certificates in Environmental Technology.

■ Core courses and electives, including communications classes, are required.

In the wake of the requirements placed on industry and society to minimize the negative impacts of hazardous materials on the environment, a need has developed for individuals who are knowledgeable about safety and environmental laws and technologies to assist industry with compliance. Some of the areas where career opportunities exist are described below.

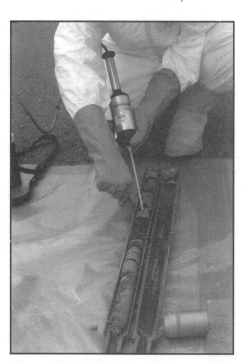

**Figure 1-7:** Opportunities for employment in the field of environmental technology continue to increase.

**Occupational Health and Safety** The health and safety of persons in the workplace relating to materials used on site needs to be monitored and maintained. This can involve training and supervision in handling hazardous materials or the use of personal protective gear.

**Regulatory Agencies** Federal, state, and local agencies need workers to enforce hazardous materials and environmental laws.

**Industry** Opportunities exist in companies – small businesses or large corporations – that process hazardous materials or that produce hazardous waste as a by-product. There is a growing demand for qualified persons knowledgeable about the laws and regulations.

**Transportation** Hazardous materials must be moved from source to user by road, rail, air, or water. Hazardous wastes must be transported to treatment or disposal sites. These positions require an understanding of regulations and an ability to maintain documents and other important information required by law.

**Public Service** Hazardous materials emergencies are typically handled by public service agencies such as fire, police, and rescue squads.

**Environmental Consulting Firms** These firms provide specialized information as a service to businesses and industries needing environmental advice.

**Environmental Cleanup Contractor**
The cleanup of contaminated sites is one of the fastest growing areas of ET work. This work may involve heavy equipment, complicated machinery, or a chemical/biological process.

**Education/Training** Training requirements mandated by law for handlers of hazardous materials continue to increase. Many companies do their own training in-house and need knowledgeable persons to perform this function.

**Banking** Acquiring and financing property is far riskier than it was in the past. Records have to be searched and the property assessed for potential contamination.

**Environmental Laboratories** These laboratories evaluate soil and water samples to determine levels of contaminants.

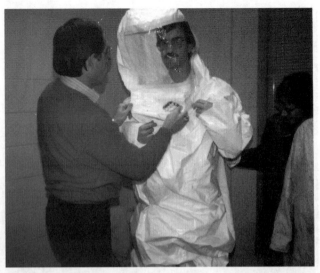

**Figure 1-8:** Students "suit up" in a typical Environmental Technology classroom.

**Insurance** This is a growing field with a broad range of aspects pertaining to ET such as liability for accidents, spills, emergencies, and site cleanup.

In order to meet the growing needs of industry and the regulatory community, colleges and universities throughout the nation have partnered with industry and governmental agencies in order to ascertain specific needs and to develop curricula in ET. The curricula that has resulted is typically competency-based and developed in cooperation with industry advisory groups.

Surveys conducted throughout the nation indicate that completion of a Certificate or Associate Degree will allow students to realize many career opportunities in the field of ET. In most community colleges, the ET Associate Degree Program is a two-year program. Some colleges also have Certificate Programs that are typically one-year programs.

Typical course requirements for ET Certificate and Degree programs include:
- – Introduction to Environmental Technology
- – Principles of Inorganic, Organic, and Biochemistry
- – Environmental Science or Ecology
- – Human Biology
- – Hazardous Waste Generation, Reduction, & Treatment
- – Basic Toxicology
- – Hazardous Waste Management Applications
- – Safety & Emergency Response
- – Hazardous Materials Management Applications
- – Basic Industrial Hygiene

The course names and credits may vary slightly in different areas of the country, but several of these courses will comprise the core curriculum. Other courses may be required given the specific needs

# Science & Technology

## Misplaced Matter

It has been said that "dirt is only misplaced matter." In a similar way, it could be said that waste is no longer useful matter. Today scientists use the term "matter" to describe all the things around us that have mass and occupy space. Prior to 450 B.C. it was believed that all matter was composed of the four basic elements: fire, air, water and earth. The Greek philosopher, Empedocles, challenged that belief. He reasoned that if you cut a piece of copper repeatedly, you would eventually reach the smallest piece of copper that still retained all of the chemical and physical properties of that substance. He called this piece the **atom**, and described it as being absolutely small, full, incompressible, without pores, and homogeneous. Unfortunately Aristotle (360 B.C.) preferred the previous explanation, and the word atom did not appear in print again for the next 2,000 years.

The dawn of the modern atomic theory is credited to the English schoolmaster John Dalton. After reviewing data from numerous investigators, he re-introduced the atom concept to explain their experimental results. Nearly 100 years later, after much experimentation, J. J. Thomson, Ernest Rutherford and Niels Bohr proposed theories that led to refinements in our understanding of the composition and internal structure of atoms.

The **Atomic Theory** proposes that atoms are composed of a **nucleus**, surrounded by a cloud of **electrons**. The electrons have a negative charge. The particles that make up the nucleus are the positively charged **protons** and the uncharged **neutrons**. Although protons and neutrons are much smaller in diameter than electrons, both protons and neutrons weigh nearly 2,000 times more than the electron.

A carbon atom – All carbon atoms have six protons in their nucleus.

Most of the space occupied by an atom is occupied by the much lighter electrons, but nearly all of its mass is located in the nucleus. It is the combined sum of the proton's and neutron's mass that determines the atom's **atomic mass**.

Atoms are very small. Billions and billions can exist on a pinhead. Atoms, however, appear to be mostly empty space. It has been calculated that if the nucleus were enlarged to the size of a baseball, the diameter of the atom would be approximately 2.5 miles. According to the Atomic Theory, every-

thing we are familiar with – solids, liquids and gases – are made up of these miniature solar systems.

Solids, liquids and gases that cannot be further chemically broken into simpler substances are called **elements**. Only a few elements (examples: copper, gold, silver, sulfur) are found in nature in a pure uncombined form; the remainder must be separated and purified through some type of chemical processing. Of the 109 or more elements known today, only 89 are naturally occurring. The remainder are man-made using elaborate atom-smashing devices. Each element can be identified by either its symbol or **atomic number**, which corresponds to the number of protons in its nucleus. The nucleus of all atoms of the same element contain the same number of protons, but they may differ in the number of neutrons resulting in **isotopes**.

When two or more elements bond together, they form chemical **compounds**. Chemical compounds can be subdivided into two general types – **inorganic** and **organic**. Inorganic compounds are typically composed of only a few atoms. For example, oxygen and silicon, two of the most abundant elements in the earth's crust, form silicon dioxide ($SiO_2$), the compound present in quartz. Originally, organic compounds were considered to be only those compounds found in living or previously living materials. Now the definition has been expanded to include all compounds containing the element carbon.

Organic compounds range from very simple to those that contain hundreds of atoms. When two or more atoms within a compound are found to be sharing the electrons, as they do in most organic compounds, then the group is called a **molecule**. In the simplest and most abundant organic compound, methane ($CH_4$), the four hydrogen atoms are **bonded** (sharing electrons) with one carbon atom forming the methane molecule.

Stop for a moment and look around, then consider that all this diversity is the result of just a few elements and how they have been connected. Whether each of these substances is useful or a waste may be just a matter of its location.

of a region or area. Electives may also be restricted to certain types of classes. Industry advisors have noted how important communication skills, both oral and written, are to the ET practitioner. For this reason, many programs require specific communications credits. Some examples of the types of courses that may be found on an electives list for ET students are:

- Introduction to Small Business Computers
- Speech Communication
- Technical Writing
- American Government
- Business and Social Responsibility
- Fundamentals of Public Administration
- Organizational Communications

## ■ Checking Your Understanding (1-2)

1. Describe how skill and knowledge about environmental and hazardous material matters could be used by a person working in the following industries:

   - real estate
   - education
   - construction
   - transportation

2. List two major factors involved in longer human life expectancies.

3. Describe ways an increasing human population negatively impacts the environment.

4. Describe an imaginary sustainable society.

5. Explain how economic factors affect the future security of tropical rain forests.

6. Discuss the factors a business owner might have to consider when deciding how to comply with environmental regulations affecting his/her business.

7. Describe the relationship between the terms atom, molecule, nucleus, and element.

# 1-3 Impacts

## ■ The Changing Nature of Environmental Regulations

### Concepts

- Both the scope and strictness of environmental regulations has increased over the years.

- Small businesses, farms, and municipalities are being affected by environmental regulations at a higher level than in the past.

- There are many local, state, and federal environmental regulations.

- The EPA has programs to support regulatory compliance.

The driving force of ET is the multitude of environmental regulations dealing with hazardous materials and environmental management. It is important for the student, therefore, to understand the impacts of these laws as they touch all sectors of society – agriculture, business, government, and individuals. As these regulations have evolved, their impacts upon society have increased.

Over the years, the regulatory mission of federal, state, and local agencies has broadened considerably, as has their reach into society. New regulations pertaining to hazardous waste and material management and occupational health and safety for employees affect hundreds of thousands of farms, businesses, and municipalities.

In 1970, the most pressing environmental problems seemed obvious, and the perpetrators easy to identify. Soot and smoke from automobiles and smokestacks were fouling our air, and sewage and wastewaters from municipal and industrial outfalls were contaminating our rivers and streams. Air and water pollution regulations have done a great deal to abate these more visible forms of pollution. The air in most of our cities today is far cleaner and healthier. Thousands of miles of rivers and streams, and thousands of acres of lakes have been restored for fishing and swimming. These accomplishments are especially impressive when seen in the context of the economic expansion and population growth that occurred during the same period. There are 25 percent more people in the United States now than 20 years ago, and our gross national product has increased by over 60 percent.

But the job is far from finished. While traditional environmental programs continue, we are now confronted with new challenges. Environmental dangers from the use of toxic substances and the disposal of hazardous wastes are now a major concern. More and more contaminants are being discovered in our drinking water. This has caused the enactment of more stringent **drinking water stand-**

ards and the development of programs to protect our groundwater from pesticides and leaking **underground storage tanks (UST)**.

These new environmental programs present regulatory problems that in many ways are similar to traditional pollution control programs. Traditionally, one of the most difficult determinations to be made in any regulatory scheme is how stringent standards should be and what technologies are available to meet those standards. New programs, however, present the added challenge of assuring compliance by the ever-increasing number of small sources that now must be regulated. Over the past 20 years, for example, the EPA has developed complex permit programs that govern the discharge of contaminants by large facilities. Now, the EPA must establish programs that govern the activities of hundreds of thousands of small sources, namely small businesses.

The new environmental programs not only bring many new groups into the regulated community, but also affect groups that were previously less regulated. Farming provides a striking example of the broadened scope of the new environmental programs. For example, farmers were initially only affected by restrictions on pesticides. Now, they must comply with a greater number of pesticide restrictions and with new regulations covering surface water runoff, the storage and handling of pesticides and other toxic substances, and the disposal of hazardous wastes. These new regulations are impacting over 2,000,000 farms in the United States today.

Municipalities provide another example of the broad scope of the new environmental programs. The EPA's regulations cover over 35,000 municipal governments in the United States and many other separate governmental jurisdictions including water supply districts and school systems. Although local governments provide a broad range of services, the only significant environmental regulations that affected them in 1970 were those pertaining to wastewater collection and treatment systems. Now, municipalities find that several of their services are governed by environmental regulations. Two services commonly provided by local governments – waste disposal and drinking water supply – are the primary focus of new environmental programs. In both of these areas, several new environmental regulations are bringing about significant changes. In addition, municipalities are required to form local planning committees to coordinate plans for dealing with hazardous substance emergencies. Even local school systems, hardly considered polluters in the conventional sense, now find that they must comply with new regulatory programs covering radon contamination and the removal of **asbestos** from buildings.

In today's complex environmental setting, the major enforcement missions of federal, state, and local agencies cover:

- air and water pollution
- radiation
- protection of drinking water
- transportation, treatment, storage, and disposal of hazardous wastes
- use of toxic and hazardous chemicals in the workplace
- pesticides
- transportation of hazardous materials

– citizens' and employees' right to know about chemical health and safety hazards
– the storage of substances in underground tanks

Not only do these environmental regulations touch every sector of American life, but each sector finds itself covered by several different kinds of environmental regulations. The EPA conducted studies on the impact of environmental regulations on small business, agriculture, and municipalities. The conclusions of these studies were published in September of 1988. The following is a summary of the highlights.

Eighty-five federal environmental regulations having the potential for large and far-reaching impacts were studied. It is important to note that only regulations under the jurisdiction of the EPA, and not DOT, OSHA, or other state/local regulations, were used for this analysis.

The study projected that by 1996, municipalities would have to charge an additional $100 per year to each household to meet the additional regulatory burden. The individual environmental regulations that account for the largest potential increases in expenses in small communities are the drinking water and sewage treatment requirements. The costs of solid waste disposal, asbestos removal in schools, and underground storage tank regulations, when totaled, also account for a significant portion of the costs borne by smaller communities.

The **Regulatory Flexibility Act (RFA)** is a federal law that requires each agency to analyze how its regulations affect the ability of small entities to invent, produce, and compete. Under the provisions of this law, agencies are required to balance the burdens imposed by regulations against their benefits, and propose alternatives to regulations that create economic disparities between different-sized entities.

The law states that "small business and other small organizations have long shouldered the same burden of federal regulations as their larger competitors and with fewer resources." These groups include small businesses, small governments, and small not-for-profit organizations. The RFA was enacted in 1980 "to ensure that regulators do not burden small units disproportionately by imposing uniform regulations on all entities regardless of size."

It would seem that the EPA's 85 regulations would overwhelm any small business, but the actual impacts vary greatly. Some businesses will be affected adversely by the regulations, but others – particularly those that provide pollution control products or services – will find that their businesses grow.

An examination of statistics provided by the U.S. Small Business Administration (SBA) reveals that 70 percent of the 3.5 million small businesses in the United States are in sectors of the economy that produce minimal pollution – wholesale and retail trade, finance, and services. Most of these businesses will not be affected directly by any of the 85 EPA regulations. Small businesses that contribute to environmental problems will incur additional costs to comply with the regulations, however, and in some industries the costs may be high.

The small business sector study examined the impacts in nine industries judged likely to be adversely affected by environmental regulations. The study found that costs may be high for three of the industries – electroplating, wood preserving, and pesticide formulating and packaging. If costs prove to be as high as estimated and cannot be passed on to consumers, some businesses in these industries may be forced to discontinue part of their operations or to close. Businesses such as gas stations, trucking firms, and farm supply stores with leaking underground storage tanks may face corrective action costs beyond their financial means.

The EPA regulations that appear to be most often responsible for high costs in the industries studied are those covering the:

- handling and reporting of toxic chemicals
- handling, treatment, and disposal of hazardous wastes
- operation of underground storage tanks.

Although cost estimates are available for only some of the regulations, indications are that they will affect a large number of firms, entailing costs in the $5,000 to $10,000 range. Although these costs may be managed easily by businesses of moderate size, they present difficulties for those with fewer than 20 employees. It is these very small companies that comprise the majority of U.S. businesses.

The objective of the agriculture sector study was to examine the effect of EPA actions on the financial condition of farms in the United States. Because of the complexity of this sector and the many uncertainties, this study had to limit its focus to a few "representative" farm types.

For livestock and major field crops, three specific farm types were examined: 1) an Illinois corn/soybean farm, 2) a Mississippi cotton/soybean farm, and 3) a Kansas cattle/wheat farm. On each farm, loss in income attributed to environmental regulations ranged from three percent to "substantial" (no percentage given). The Kansas cattle/wheat farm was already in vulnerable financial condition and was predicted to go out of business one year earlier than it otherwise would have.

Specialty crops studied were apples, tomatoes, potatoes, peas, caneberries (e.g. raspberries, blackberries, etc.), and peanuts. This study examined changes in net returns per acre. In the least costly scenario, the decrease ranged from 1 percent to 8 percent. Under the most costly regulatory scenario, however, losses increased substantially, particularly for apple producers in New York and Michigan, where predicted losses were 60 per-

**Figure 1-9:** EPA regulations impact agricultural practices.

cent and 84 percent, respectively. Large differences in the impact of EPA regulations on crops grown in different regions occurred because some of the proposed restrictions involve pesticides that are used in some regions and not in others.

Even though the results of this study must be considered preliminary, these figures show that EPA actions could create economic problems for some farms. In response to the findings of these sector studies, several areas for potential policy initiatives have been suggested.

The EPA has implemented a number of activities to help support small communities' compliance with environmental regulations. These include establishing better lines of communication among the EPA, community leaders and citizens, and extending technical and financial assistance programs. Public partnerships might be promoted, for example, to allow two or more communities to share expertise, purchase services and goods in larger volumes for discounts, or to raise capital in larger, more cost-effective blocks. Two or more communities might create a joint venture for a particular purpose such as construction of a water supply system, or might work with private companies to assist in providing environmental services to help reduce costs.

Because the new environmental programs cut across many industries and affect thousands of small businesses, new compliance strategies may be needed to supplement the EPA's traditional enforcement efforts. Many programs are available to help small businesses learn about and comply with environmental regulations. These include educational programs, preparing standardized responses to paperwork and other requirements, helping to expand environmental services, and fostering new technologies. All of these programs can be developed with the cooperation of governmental agencies and industry trade associations.

The EPA and other agencies have an important responsibility and challenge ahead in trying to reduce environmental risks associated with the activities of small businesses, municipalities, and agriculture in a cost-effective manner. Businesses and the public have a vested interest in the outcome of these decisions. Toward this end, they must take seriously their responsibility to provide input in the process of regulation.

## ■ Checking Your Understanding (1-3)

1. Explain why the EPA is now concentrating more effort on small sources of pollution, like farms and small businesses, instead of on large facilities and industries.

2. Describe what progress, if any, has been made in environmental clean-up efforts since 1970.

3. List the municipal services that have become the primary focus of new environmental programs.

4. Describe some of the policy considerations that might assist municipalities, small businesses, and agricultural businesses in their efforts to comply with environmental regulations.

# Governmental Processes

**2**

## ■ Chapter Objectives

After completing this chapter, you will be able to:

1. **Describe** the governmental and administrative processes that give rise to environmental laws and regulations at the state and federal levels.

2. **Compare** the responsibility and jurisdiction of federal and state agencies involved with environmental protection and workplace health and safety.

3. **Define** terms that are associated with environmental laws and regulations and their creation.

4. **Describe** the kinds of action and authority available to local jurisdictions concerning land use planning, air quality policies, hazardous materials ordinances, building and fire codes, and other significant elements.

## ■ Terms and Concepts to Remember

Abatement
Administrative Order
Air Quality Standards
Asbestos Hazard Emergency
    Response Act (AHERA)
Atomic Energy Act (AEA)
Bill
Case Law
Clean Air Act (CAA)
Clean Water Act (CWA)
Code of Federal Regulations
    (CFR)
Comprehensive Environmental
    Response, Compensation,
    and Liability Act (CERCLA)
Conversion Factor
Degradation
Department of Energy (DOE)
English System
Environmental Impact
    Statement (EIS)
Federal Insecticide, Fungicide,
    and Rodenticide Act (FIFRA)
Felony
Freedom of Information Act of
    1966 (FOIA)
Hazardous Materials
    Transportation Act (HMTA)
Initiative Process
Inspection Warrant
Joint and Several Liability
Labeling
Lobbyist

Marine Protection, Research,
    and Sanctuaries Act
Measuring System
Metric System
Misdemeanor
Monitoring
National Environmental Policy
    Act (NEPA)
NEPA
Occupational Safety and Health
    Act (OSHA)
Ordinances
Paragraph
Part
Precedents
Privacy Act of 1974
Private Law
Public Law
Resource Conservation and
    Recovery Act (RCRA)
Safe Drinking Water Act
    (SDWA)
Section
Slip Law
Statutes
Statutory Law
Technology
Title
Toxic Substance
Toxic Substances Control Act
    (TSCA)
Unit
Water Quality Standards Public
    Law

## ■ Chapter Introduction

In the last chapter, we discussed how increased environmental consciousness has changed the way society operates. We noted that these new attitudes about the environment have also given rise to greater restrictions on the way we live, especially in the way we do business. Developments in science, technology, and industrialization have been echoed by corresponding governmental activities such as the numbers of laws enacted.

This chapter will address the legal processes and mechanisms through which laws and regulations are created. Practitioners in the field of Environmental Technology need to know how to locate, read and understand numerous environmental laws and regulations that are the result of these specific government processes.

# 2-1 Lawmaking

## ■ What are Laws?

### Concepts

- Laws are broad legal standards.
- Public and private are the main divisions of law.
- Case law differs from statutory law.

Laws, also called **statutes** (and sometimes acts), are the legal standards that govern society. If you think about the laws you are familiar with, you will notice that many (if not most) are restraining; for example, speed limits, drinking age, building occupancy limits. Because ours is a large, complex, industrialized society, we are governed by many laws. Laws are enacted with the goal of creating a better society. In rare cases, however, laws are repealed when they are found either unenforceable, not producing the desired effect, or if they are shown to violate other existing rights or protections under the law. For example, in 1918 Congress ratified the 18th Amendment to the Constitution which prohibited the manufacture, transportation, and sale of alcoholic beverages. The Liquor Laws, as they were known, were very difficult to enforce and had the negative effect of contributing to an illegal black market, bootlegging, creating an environment for the spread of organized crime, and generally failing to meet social reform objectives. Therefore, in 1933, Congress passes the 21st Amendment which repealed the 18th.

There are two main divisions of law: **Public Law** (often called Criminal Law) and **Private Law** (often called Civil Law). Public Law includes:

- laws describing the basic rules of our system of government,
- criminal acts against society,
- the operation and establishment of administrative agencies, and
- court cases.

Private Law is designed to settle disputes between parties, such as a property settlement in a divorce action. The government's role in private law is to act as "judge" when one party claims that another party "injured" their person, reputation, or property; or that a party failed to carry out some obligation they were legally required to fulfill.

One of the oldest systems of law was the ancient Sumerian Code of Hammurabi, which dates to approximately 1900 B.C. This system of criminal and civil law defined over 280 legal standards ranging from property transactions to wrongdoing in the handling of public affairs

**Figure 2-1:** Code of Hammurabi.

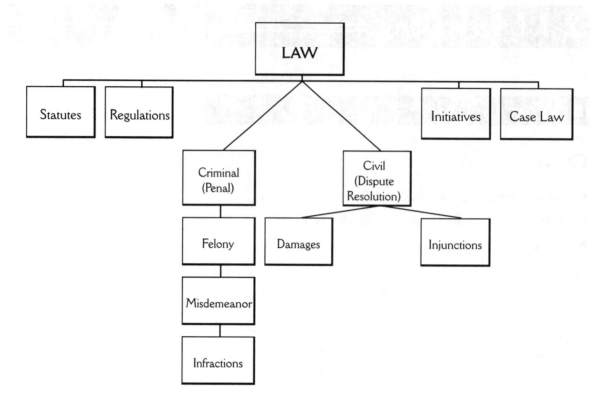

**Figure 2-2:** Divisions of public and private law.

to theft. It set up the penalties for certain crimes, which seem severe by modern standards. Theft, for instance, was punishable by death. Our criminal law system also prescribes punishments for violation of laws. For instance, fines and prison time are set forth in law for acts such as illegal dumping of hazardous waste.

While it may appear that laws are rigid, specific, consistent requirements, this is not the case. Laws are written to cover a broad range of activities and are not meant to address every potential situation. For this reason, they must be interpreted. You have probably heard reference to something being a "gray area" of the law. You have also probably noticed in reading the newspaper that different courts can interpret the same requirement of a law in different ways. Thus the laws that regulate our behavior are an interwoven body of court decisions, statutes, regulations, policies, and procedures that are continually changing.

Two other terms are important when talking about law: statutory law and case law. **Statutory law** is composed of written laws, the exact wording of which has been approved by a federal, state, or local legislative body. **Case law** consists of the written decisions made by courts on cases with related topics. These earlier decisions are called **precedents** and are used as patterns for decisions on similar cases that arise later.

## ■ Who Enacts Laws?

### Concepts

■ Executive Branch proposes, approves, disapproves, and enforces laws.

■ Legislative Branch enacts laws.

■ Judicial Branch examines and interprets laws.

The basic legal framework of our governmental system is set forth in the U.S. Constitution. States also have constitutions that provide their legal basis. The first three articles of the U.S. Constitution set up three separate branches of the Federal government – the executive, legislative, and judicial.

The Executive Branch consists of the Chief Executive, or President, and governmental bodies such as administrative agencies, that report to the Chief Executive. This branch of government proposes laws, approves laws, can veto laws passed by the legislative branch, and enforces the existing laws. The Executive Branch has the responsibility of setting up regulatory and administrative agencies to accomplish enforcement. For example, the U.S. Environmental Protection Agency was established in 1970 by President Richard Nixon to enforce environmental laws. The Chief Executive selects individuals to head each department or agency. The President's nominees are presented to the Congress which has the right to accept or decline the individual. When he began his term, President Bill Clinton nominated Carol Browner to head the EPA. Congress then accepted the nomination, and she was appointed Administrator of

**Figure 2-3:** Branches of the Federal Government.

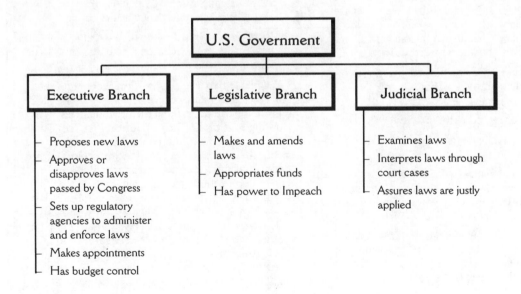

the EPA. The Cabinet, composed of the heads of governmental departments such as the State Department, the Department of the Interior, the Department of Justice, etc., serves as the Chief Executive's closest advisors. Their function is to advise the President and to establish policy. Discussions have taken place to elevate the position of Administrator of the EPA to a Cabinet-level position.

The Legislative Branch of the federal government is responsible for making laws. It is composed of two representative bodies, the Senate and the House of Representatives, collectively known as Congress. Each state is limited to two senators, while the number of representatives sent to the House of Representatives depends on the population of the state. Each state also has a legislative body. All states except Nebraska have a two-house structure similar to that of the federal government. Local governments, such as cities or counties, may have a council or board that performs the legislative function. Legislators in either house can propose new legislation. Each house of Congress must vote on and pass a piece of legislation before it is put before the Chief Executive to be either signed into law or vetoed. A veto from the President, however, can be overridden by a sufficiently large majority of legislators. Laws passed by federal and state legislatures are called statutes. A law passed by a local unit of government is called an **ordinance**.

## Box 2.1

### California's Proposition 65

Proposition 65, called the **Safe Drinking Water and Toxics Enforcement Act of 1986**, was an initiative passed overwhelmingly by the voters of California. It was prepared by a number of environmental organizations including the Environmental Defense Fund, Natural Resources Defense Council, and the Sierra Club. This law has a number of provisions that impact consumers, government, and businesses. Proposition 65:

■ Requires the Governor to designate a state agency to implement its provisions

■ Requires publishing of a list of chemicals causing cancer or reproductive toxicity

■ Allows development of regulations to implement the law

■ Prohibits businesses from discharging or releasing a chemical on the cancer and/or reproductive toxin list where it can potentially find its way to a source of drinking water

■ Prohibits any person in the course of doing business from knowingly and intentionally exposing any individual to a listed chemical without giving them prior warning

■ Provides for a reward or bounty of 25 percent of the penalty collected to a person who brings a successful enforcement action

■ Requires government employees to report within 72 hours any discharge that is likely to cause substantial injury to public health or safety

■ Increases (mostly doubles) the maximum penalties for serious violations of hazardous waste laws already on the books

Opposed to the proposition were the state's governor and many corporations in the chemical, petroleum, and manufacturing industries. According to the state's Fair Political Practices Commission, nearly 7 million dollars were spent on the proposition with large contributions coming from individual supporters as well as contributions from major corporations to oppose it.

In the years since its passage, a number of lawsuits have been settled. Most recently, in April of 1994, a record $1 million was paid by a major paint company for failure to make adequate warning on their paint products containing a toxic substance, toluene. Another manufacturer of a paint stripper paid penalties of $2,500 per day of exposure.

The Judicial Branch is made up of a system of courts of law; the U.S. Supreme Court is the highest court and its decisions are often the basis for new laws or changes in established laws. Any law that is passed by the Legislative and Executive Branches can be reviewed or examined by the Supreme Court if an argument is made that it violates the U.S. Constitution or the constitution of one of the states. The Court also interprets laws when there are "gray areas" or there are differences of opinion in lower court rulings.

Federal laws are enacted by Congress. These are the "supreme laws of the land" and no state can enact laws that are less strict than these. States, however, do have the right to enact laws that are more restrictive than federal laws. Many states have done so in the areas of environmental protection and workplace health and safety. The federal government can authorize a state to enforce its own law instead of a similar federal law if the state demonstrates that its law is at least as or more restrictive than the federal law.

In many states voters can take lawmaking into their own hands through the **initiative process** or through referendums. An initiative allows people to place local or state measures directly on the ballot in a general election. Before an initiative can be placed on the ballot as a proposition, it must first have a minimum level of support from the public. Support is usually demonstrated by obtaining a sufficient number of signatures of registered voters. Through a similar petition process, referendums allow the public to vote on any law already passed by the state legislature.

---

# Box 2.2

## Dave Willis – Dealing with Governmental Regulators

*David Willis has extensive experience in administration and operations management in a variety of environments. His particular emphasis is in turning unproductive organizations and agencies around. His background involves relations with industry groups/organizations, as well as the administration of regulations and procedures governing industries and agencies. He is currently the Process and Planning Director for the District of Columbia Government, Washington, D.C. Prior to this he was Deputy Director of the Hazardous Waste Management Program in the California Department of Health Services. He is a Littauer Fellow of Harvard University and continues his research on regulatory agency activities. He was chosen for this interview on dealing with governmental regulators because of his strong interest in "covering an area that is largely a mystery, even to those within government."*

*Q: What is all the frustration about governmental regulations?*

A: People go into private enterprise because they want to be their own bosses and make their own decisions. So what happens? In some of the worst ways imaginable, they end up working for the government. It is government which makes many of their most basic decisions, or fails to make them, or does so in ways that add costs and frustrations to the private enterprise.

*Q: What kinds of complaints do you get from the private sector about having to deal with government?*

A: The constant complaint of those in the private sector who have to deal with government – either in following its rules or getting its permission to take some action – is that bureaucrats aren't "real world." They don't set timetables or keep to them when they do set them. Their work always seems to cost more and produce less than promised. And they don't seem to care about the impact of their decisions on the private sector or the economy as a whole. They

## Box 2.2

just seem to want what they think is a perfect solution in an imperfect world. They have never had to meet a payroll, and they don't care about or even completely recognize the concept of the "bottom line."

**Q: Do you think these complaints are legitimate?**

A: There is more than some truth in these charges. On the whole, however, most bureaucrats are dedicated public servants who want to make reasonable, rational, and cost-effective decisions. And they would like to do it in as short a time frame as possible.

I know what you're going to ask me next. Then why don't they do it? Why don't they just recognize that time is money and that we can't have a perfect, risk-free world? Why can't they just make decisions and get on with it?

**Q: The questions did occur to me!**

A: When asked these questions by those in the private sector whom we regulate, we're sometimes not even sure of the forces that cause us to act counter to our stated desires. But, these institutional forces exist, nonetheless – and that is what makes it so difficult to get decisions out of government regulators.

---

There are few, if any, rewards in government for taking risks.

---

If you work with, and need decisions by, government regulators – and plan to be successful in obtaining these decisions – it is necessary for you to understand the imperatives and values that drive people in government. First, by making a decision at the behest of those being regulated, the bureaucrat is becoming accountable for the final outcome of the actions he or she is approving. That is, if something ultimately goes wrong, those in government who allowed the activity to go forward will be at least partially blamed for the failure. Secondly, there are few if any rewards in government for taking risks.

**Q: How does this all happen?**

A: Once government determines that an identified problem in society must be addressed, and once it identifies the type of solution needed, a governmental agency is assigned the responsibility of assuring that this problem will be avoided in the future. This is where the problems start. Any future failures or problems that originate from the actions

of the entity the government now regulates will be seen as the government's fault. For didn't it approve the actions, or at least the rules that guided the actions? The more that government regulates, the more responsible it becomes; and the more risk it takes in that regulation, the more vulnerable it becomes if the risk backfires. The moral here, of course, is that by taking little or no risk in its decisions, personal risk and accountability are also avoided.

A government agency is assigned the responsibility for creating rules and processes to be followed, to enforce those newly created laws, and to review proposed actions to assure the absence of the problem in the future. Public employees are required to peer into the future and say that if particular tasks are accomplished, things will be fine. In other words, they are being asked to prove a negative: that something will not happen.

**Q: Is that really possible?**

A: Short of surveying the entire universe of possibilities, that is impossible to prove. Therefore, there is always some unforeseen and unintended consequence that will probably occur once an action is taken. As a result, government decision makers are being called upon to take some level of risk each time they make a positive decision – or for that matter, any decision.

---

In the absence of clear rewards, government can often be described as not organized to succeed, but organized not to fail.

---

**Q: But is it so bad to take risks?**

A: To better understand, if not appreciate, this shyness about risk-taking, it may be important to review how government is organized. The absence of a bottom line, for example, has a very real impact on government. Without it there is no convenient way to measure return for risk. There are no monetary rewards to be weighed against the risk taken and no monetary rewards for taking risks. In the absence of clear rewards, government can often be described as not organized to succeed, but organized not to fail. If you look at government, where are the rewards and how are they apportioned? At the political level, it's very similar to business in a way that is negative to both – short- rather than long-term payoffs are the solutions of choice. In addressing these recurring bottom lines or re-elections, the

## Box 2.2

thrust is on real or perceived failures. The whole issue revolves around what is wrong, or the absence of things being wrong, rather than future benefits to be derived from risk-taking.

**Q: So are you saying that a successful bureaucrat doesn't "rock the boat?"**

A: It is the absence of problems generally, not the specific positive benefit derived, that is the test in the world of politics. This adverse impact is compounded by the reward system of government. In the interest of fairness, and to eliminate both real and perceived problems with government hiring and pay practices, the pay and promotion of government employees is very structured. Everyone at a certain level gets the same pay with little or no differential based on performance. While there are some bonus programs, they are not really designed to reward risks. This means that, unlike businesses which can have a variety of rewards that at least balance the risks, no such system exists in government.

**Q: So what happens in a government that is not rewarded for taking risks but stands to be punished for failure?**

A: Obviously, it avoids making decisions when there is a risk. One way to accomplish this is to distribute responsibility for making a final decision. Given our pluralistic society, and the fact that our government was organized to diffuse authority and power, and therefore, responsibility – and given the incremental way we tend to address most of our problems – spreading responsibility is quite simple. In fact, it is often difficult not to do so.

---

It is also easy to shift the blame for failure to act to the requester

---

Very seldom does sufficient authority rest in one person or entity. Many points of authority are usually involved in establishing the laws, regulations, processes, allocating resources, hiring people, etc. The result is that no one can be held completely accountable for the end results. In this confusion you can almost always avoid responsibility.

**Q: Is this the only problem?**

A: Given that we are also charged with foretelling the future – things will be fine if you follow our rules – it is also easy to shift the blame for failure to act to the requester. Since to prove a negative – i.e.,

something will not happen in the future – it would take all the information about the future in order to prove or disprove the premise (this is why it's impossible), a request for more information can always be justified. Yes, this is in part a euphemism for nit-picking. But regardless of whether it is seen as nit-picking or a legitimate need for more information, the fact remains that when asked to prove a negative, the request for more – time, study, information – can be supported. A classic example in this area is the Federal Food and Drug Administration (FDA). A major complaint against the FDA is that it takes too long to approve a drug. However, it will neither earn money from the drug's sale nor benefit from taking the drug. But the FDA will be blamed for the adverse effects of taking the drug. The agency will, therefore, attempt as exhaustive a study of the specific universe as possible. It will go down every street, byway, and blind alley conceivable in trying to develop enough information to safely predict the future. The result: approval seems to take forever. Yet in doing an exhaustive review, the FDA has carried out its assignment in a way that minimizes risks both to society and to the agency.

What also must be recognized is that the reviewers of any actions or proposed actions finally taken by government will always either apply 20-20 hindsight or will demand as much perfection as possible. In fact, in government agencies – or any other bureaucracy – the further removed you are from the actual occurrence (but still have review or approval responsibility) the more you can retreat into perfection and process. It is often to satisfy these later reviews that we in government ask for more and more information prior to recommending an action. In doing so, the bureaucrat hopes to ensure that there is very little risk in the decision required.

---

Applying due process without clearly defined objectives and with little recourse for those "caught" within due process has produced volumes of wonderful anecdotes about the supposed "stupidity" of bureaucrats.

---

**Q: Isn't this wholly ineffective?**

A: The concept of retreating into process is one of the principal ways government avoids taking clear-cut actions and thereby minimizes risk. For all the good that "due process" brings us in protecting

# Box 2.2

rights and assuring fairness and consistency of treatment, it is also a major cause of frustrating and often outright stupid government action. Applying due process without clearly defined objectives and with little recourse for those "caught" within due process has produced volumes of wonderful anecdotes about the supposed "stupidity" of bureaucrats. In some instances bureaucrats take actions that produce results which are almost inconceivable, but as long as the rules are followed, those making the decisions are protected. Someone else made the rules and to break them would be a risk, which if it turned out badly they would have clear responsibility. As there are few rewards for assuming risk, public employees don't take them. Why on earth should they? If all goes OK, someone else reaps the rewards and if it goes wrong they get the blame. The fact that so many bureaucrats do take risks in these circumstances is an occasion for admiration, not condemnation, of government employees.

*Q: So, how do you suggest we best deal with regulators?*

A: It needs to be pointed out that successful government programs do happen. Risks are taken and great benefits derived. Many of these cases, such as success in war, are generated by crises when the risks of not taking immediate action clearly outweigh the risks of taking action. But what about non-crisis oriented action?

In short, how can you get to the positive decision in the uncharted wilderness of a new area such as toxics, where the chance of making a mistake is greatly increased by the lack of certain knowledge and agreed-upon processes that can be followed? To do so, you must find out what the decision makers' comfort barrier is and what it will take to overcome that comfort barrier. By comfort barrier, I mean that level of confidence in the pertinent information that makes him or her "comfortable" with making a decision. Remember, unless it's an easy, safe decision, all the institutional pressures work against persons in government making a decision.

---

**Unless it's an easy, safe decision, all the institutional pressures work against persons in government making a decision.**

---

Knowing the laws, rules and regulations is not enough. In fact, throwing them in the face of a hesitant decision maker, as adversarial trained lawyers are apt to do, will just make him or her more prone to find another missing piece of information. As stated earlier, given that the decision is at least in part based on foretelling the future, there are always reasons to ask for more information. Also regardless of how complete and tightly written the laws, rules, and regulations are, there are always subjective decision points along the way, and it is the decision maker who is the final arbiter of these subjective decisions. Therefore, he or she will usually win an adversarial argument no matter how informed and talented the opponent is.

---

**The environment in which the decisions get made is the major determinant in overcoming or understanding the decision maker's comfort barrier.**

---

Another point to remember is that the bureaucrat's values are not the same as yours. Having a strong argument based upon your values and your risks versus your benefits is just not a real factor in the decision process. The government decision maker, even in an administration apparently sympathetic to business, is driven by a different view of the world and priorities. Therefore, you should make sure you understand the values that the decision maker brings to the process, since they guide the bureaucrat's way of making decisions within the latitude available in the laws and rules. Just remember: it is you who needs the positive decision if you are to function as you wish. Therefore, you have more to gain than does the government decision maker from making sure that your communications are productive and not negative.

Another of the things you must know to understand a government worker's comfort barrier is the environment in which he or she makes decisions. Where do these people fit in the hierarchy? Does their boss and his or her superiors support risk taking? Does the boss support their decisions even if wrong, or does he or she leave them hanging out to dry if it goes bad? How many others must they satisfy? Are there other government agencies or levels that must agree to the decision? Are the decision makers closely scrutinized by the public, press, and/or legislature? What is their track record? Have they been successful in the past so that their credibility is such that it can stand a failure now and then?

The environment in which the decisions get made is the major determinant in overcoming or understanding the decision makers' comfort bar-

## Box 2.2

rier. If the management is risk-oriented, if the agency has a lot of credibility and a good track record, if the subject matter is not a major current interest to the legislature, media, or public, and if the decision does not need concurrence by other agencies or levels of government, then the comfort barrier will probably be quite low: a positive decision can be quickly reached without tons of information. If the decision maker is new on the job and still uncertain about things, then the comfort barrier goes up. If the subject matter is controversial, if the management has a reputation for self-protection rather than staff protection, if others are involved or will be in a position to second guess and criticize, all of these and more can raise the comfort barrier.

---

Very seldom will you get the response, "Yep, you're right, here's your decision. Have a nice day."

---

Next, you should not get turned off by nitpicking. In the first place, he or she may consider it important, and if it overcomes a comfort barrier it is by definition important. Secondly, if it is nitpicking, it is often small and therefore should be easier to satisfy than to fight over. Also, it is really in your interest to satisfy these extra demands since they often have the effect of convincing the public that the government is really doing its job. This gives the decision maker more room to maneuver in the long run thereby making it easier to make positive decisions. This also has the effect over time of lowering the public's comfort threshold and thereby lessening the chance for more restrictive laws or pressures on the decision makers.

Q: *What about going to a higher, more experienced level?*

A: Of course, one of the things you can always do is resort to that time-honored tradition of going over their heads. Write or call the boss or the governor or your local legislator and complain about these intransigent bureaucrats who are standing in the way of progress and frustrating the will of the people in their inept interpretations of the laws. It might even get you resistance, and the time that would otherwise be spent working on your problem will be spent defending their actions. Very seldom will you get the response, "Yep, you're right, here's your decision. Have a nice day."

---

Chances are what you will get is a lot of sound and fury, but in the end you will still be asked to satisfy the expert's comfort barrier.

---

Much more often what you get is a bureaucracy that will go to great lengths to prove its case. And even if you win this time, what is your credibility going to be next time? Will you be greeted with open arms as one who we can be "comfortable" working with? Don't bet on it. Also your chance of getting a reversal of the present position is low since the bosses all the way up the line are faced with the same imperatives working against the decision as are those you are dealing with. Chances are what you will get is a lot of sound and fury, but in the end you will still be asked to satisfy the expert's comfort barrier. All you will probably accomplish is to add more time and difficulty to the process.

Lastly, there is a point that should not have to be said, but since the situations occur with some regularity, needs to be pointed out. Avoid turning in sloppy work! Whether it's in the interest of saving time, money, or just out of poor standards, it is surprising how much poorly-conceived and poorly-executed work gets presented as a basis for government to make decisions. When a bureaucrat is asked to take a risk, sloppy and incomplete work are poor ways to engender confidence and comfort.

---

Get strict definitions of such words as "adequate," "effective," and "sufficient."

---

Q: *Any other hints?*

A: In addition to knowing the laws, rules, and regulations, and recognizing that the bureaucrat's comfort barrier has to be met, you should also attempt to negotiate early on certain rules that the decision makers will follow in making their decisions. All too often we tend to treat the decision process in the same way the Supreme Court deals with pornography: We can't define what is enough information but will know the appropriate level when we see it. This is a cop-out and should not be accepted without attempting to negotiate a more positive set of agreed-upon standards. In other words, get government to define in as specific a way as possible just what the "comfort level" is and what

## Box 2.2

specific products or actions on your part will satisfy that comfort barrier. Get strict definitions of such words as "adequate," "effective," and "sufficient." Don't accept generalizations or protestations that we can't tell what is needed until we study your submittals, etc. It may be true that, upon review, more questions will come up, but we should have some fairly clear idea of what we need. Keep at it. Don't be adversarial, but be tenacious. Keep asking questions and requesting definitions.

If at all possible, get the bureaucracy to state its needs in the form of standards to be met or goals to be accomplished, rather than detailed processes to be followed. If you can accomplish this, you will have done both yourself and the bureaucracy a favor. For if you can both agree on standards and/or goals and leave the ways and means or process to you, you will have succeeded in eliminating the major impediment to bureaucratic action. If it is only the standard or goal that government must approve, then the only responsibility that the bureaucrat would have is if the standards or goals proved insufficient, then they could be raised, but the decision makers would no longer be responsible for guessing that this or that way of operating will or will not be safe. That now becomes the risk-taking business's responsibility.

I recognize that what I am proposing is close to heresy in many government circles. The idea that we pre-approve actions and focus on process rather than outcome is the way we have all been taught. How can we be sure that everyone is doing the same thing and that favoritism and independent action is not taking place unless we insist on certain processes being followed? Well, we can't. If it is more important that government control processes rather than standards and goals, or if we feel that the only way we can prevent harm is to assure that all actions are either pre-approved or follow the same process, then we are stuck with the slow, deliberate and frustrating system. Only by changing the basic paradigm will we change the way government works.

Along the same lines, more accountability must be placed on the major responsible policy makers – the legislators – if real change is to take place. The legislature should change its common approach of spelling out in detail, or directing the bureaucracy to spell out in detail, the processes to be followed. Rather, they should concentrate on setting goals, standards, and ways to measure success or failure in meeting them. In this way, they would continue to place responsibility where it belongs.

*Q: How would business feel about this?*

A:   Of course, there is a down side for business in all of this. If government is no longer to be blamed for actions, then business will have to take responsibility, and suffer probably more dire consequences than they do now. However, they also have a return and reward for taking this risk more quickly and cheaply. By rearranging the playing field, we can do much to improve business, government, and society as a whole.

## ■ Getting From an Idea to a Law

### Concepts

■ A bill is a proposal for a law introduced by a legislator.

■ Bills must pass through both houses of Congress and go to the President before becoming law.

A law starts as an idea when someone identifies a need to control or change society's activities in some way. However, the fact that an idea is born does not guarantee it will successfully progress over the many hurdles required for it to become a law. In its formative stages, a proposed law is called a bill. A **bill** is a proposal to change, amend, repeal, or add to existing law.

Any member of Congress, whether from the House of Representatives or the Senate, can introduce a bill. "Joint Resolutions" are submitted by both houses simultaneously. The ideas for bills can come from citizens, congressional members and their staff, legislative committees, the President, governmental agencies, special interest groups, lobbyists, trade associations and the like. After introduction, a bill is referred to an appropriate policy committee that may hold hearings or assign it to a subcommittee. After receiving recommendations and conducting public hearings, the committee meets to take action on the bill. A committee's response can include: take no action, rewrite the bill completely, amend the bill as written, or leave the bill as is. A written report from the committee will often accompany the bill as it is placed on the appropriate house's calendar. Finally, it will be debated and called to vote.

Once the bill has been passed in one house, it is then sent to the other house where it goes through essentially the same process. If there are major differences in the bill passed by each house, a conference may be called. The bill is then assigned to a special conference committee made up of members of both houses. This committee attempts to reconcile differences and negotiate a compromise bill that is acceptable to both houses. Each house can then accept or reject the conference report.

When both houses have approved a bill, the final version signed by the Speaker of the House and the President of the Senate goes to the President for final approval. The President may sign the bill into law, let it become a law without signing it, or veto the bill.

Public comment plays a very important role in this process. The public may attend congressional committee hearings and testify as to the desirability and workability of the proposal. Many changes are made throughout the process due to this type of input.

**Lobbyists**, who got their name in the early 1800's when they milled around the lobbies of public buildings trying to get the attention of lawmakers, represent special interests and they aggressively pursue lawmakers in the hopes of influencing legislation that will benefit their interests. Lobbying is an activity that has as many critics as it does supporters; it has long been an integral part of the legislative process and is an activity that is protected by the First Amendment as a right of free speech. All major industries employ lobbyists and so do most interest groups including environmental groups.

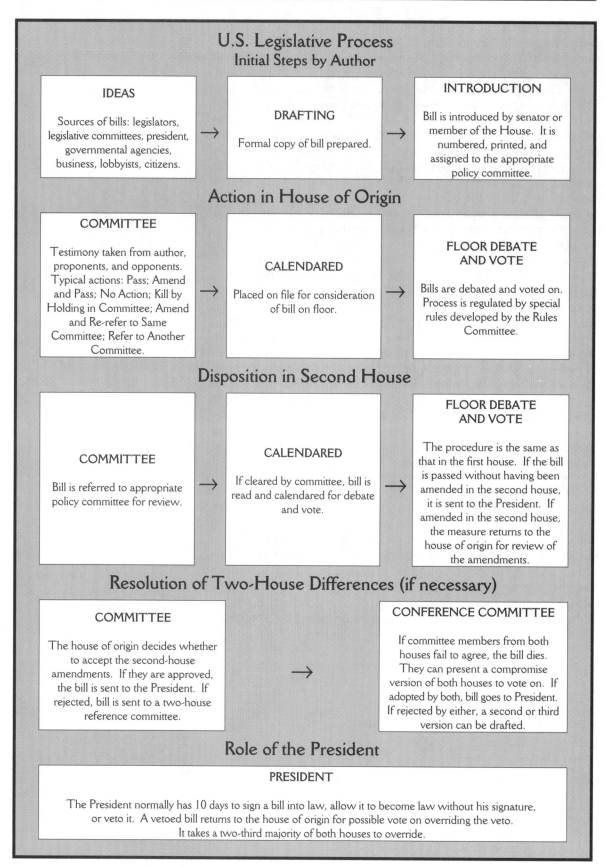

## U.S. Legislative Process
### Initial Steps by Author

**IDEAS**

Sources of bills: legislators, legislative committees, president, governmental agencies, business, lobbyists, citizens.

→

**DRAFTING**

Formal copy of bill prepared.

→

**INTRODUCTION**

Bill is introduced by senator or member of the House. It is numbered, printed, and assigned to the appropriate policy committee.

### Action in House of Origin

**COMMITTEE**

Testimony taken from author, proponents, and opponents. Typical actions: Pass; Amend and Pass; No Action; Kill by Holding in Committee; Amend and Re-refer to Same Committee; Refer to Another Committee.

→

**CALENDARED**

Placed on file for consideration of bill on floor.

→

**FLOOR DEBATE AND VOTE**

Bills are debated and voted on. Process is regulated by special rules developed by the Rules Committee.

### Disposition in Second House

**COMMITTEE**

Bill is referred to appropriate policy committee for review.

→

**CALENDARED**

If cleared by committee, bill is read and calendared for debate and vote.

→

**FLOOR DEBATE AND VOTE**

The procedure is the same as that in the first house. If the bill is passed without having been amended in the second house, it is sent to the President. If amended in the second house, the measure returns to the house of origin for review of the amendments.

### Resolution of Two-House Differences (if necessary)

**COMMITTEE**

The house of origin decides whether to accept the second-house amendments. If they are approved, the bill is sent to the President. If rejected, bill is sent to a two-house reference committee.

→

**CONFERENCE COMMITTEE**

If committee members from both houses fail to agree, the bill dies. They can present a compromise version of both houses to vote on. If adopted by both, bill goes to President. If rejected by either, a second or third version can be drafted.

### Role of the President

**PRESIDENT**

The President normally has 10 days to sign a bill into law, allow it to become law without his signature, or veto it. A vetoed bill returns to the house of origin for possible vote on overriding the veto. It takes a two-third majority of both houses to override.

**Figure 2-4:** U.S. legislative process.

## ■ Where to Find Federal Laws

### Concepts

■ Federal laws are found in the United States Code (USC)

■ Laws are not the same as regulations

Congressional statutes are found in the U.S. Code (USC). The first official publication of a statute is in a form generally known as a **slip law**. In this form, each law is published separately as an unbound pamphlet. Slip laws can be obtained by writing to:

> Superintendent of Documents
> Government Printing Office
> Washington, D.C. 20402

Each year, all public laws of permanent interest are gathered and inserted into the U.S. Code (USC) and laws with amended sections are updated. Laws on similar subjects appear together in the U.S. Code. Statutes, or laws, should not be confused with regulations. Federal laws set forth only the broad outlines of a national policy. The power to specify the exact conditions under which the objectives of a federal law will be carried out is delegated to independent agencies of the federal government or the states. These regulations are based on the implementing agency's interpretation of the law and set specific compliance requirements.

Updated copies of the statutory codes are published annually. Government repository libraries throughout the country keep government documents and codes on hand for the public's use. The Office of the Federal Register publishes a list of libraries that report having these documents. Legal libraries within county or state facilities may also be a good source for obtaining these documents.

## ■ Checking Your Understanding (2-1)

1. List three other names for laws.

2. What is the difference between a bill and a law?

3. How do Public Laws and Private Laws differ?

4. How do case law and statutory law differ?

5. What are precedents and how do they fit into the legal process?

6. Describe the functions of the three branches of the U.S. Government.

7. List three ways, either directly or indirectly, citizens can be involved in the lawmaking process.

# 2-2 Rulemaking

## ■ What Are Rules?

### Concepts

■ Rules, or regulations, are developed by regulatory agencies.

■ Regulations, based on law, define more specific requirements for compliance.

■ Rulemaking must follow a prescribed process.

As we have seen, the Legislative Branch can make or change laws and the Executive Branch assigns agencies to enforce laws. Agencies, when handed a law passed by Congress, begin what is known as the rulemaking process. Rulemaking, or developing regulations, serves to "put meat on the bones of the law." For example, a law may say, "The waters of this state shall be protected from pollution and other sources of degradation." This statement is a broad, encompassing requirement. The state agency assigned to enforce this law can develop regulations that further define how to achieve this goal. It may, for instance, require that all underground tanks storing hazardous materials be regularly monitored for leaks and tested for tightness once each year. This would, indeed, help achieve the goal of protecting groundwaters from leaking underground tanks even though the original law did not specify how to do it.

The main federal agencies and departments of interest to students learning about environmental technology include:

■ Environmental Protection Agency (EPA)

■ Department of Transportation (DOT)
  - Coast Guard
  - Research and Special Programs Administration

■ Department of Labor
  - Occupational Safety and Health Administration
  - Mine Safety and Health Administration

■ Nuclear Regulatory Commission (NRC)

■ Department of the Interior
  - Fish and Wildlife Service
  - Land Management Bureau

■ Department of Energy

These agencies develop regulations based on laws they are assigned to enforce. The development of regulations must follow a prescribed system. On the federal level, this is called the Federal

Register System and is designed to keep the public informed at each step of the rulemaking process.

It is important for the public and industry to be aware of rulemaking activities. Since most laws outline only broad, general provisions, agencies often interpret them differently than the businesses that are affected by the law. Proposed regulations have been altered by the persuasive activities of people willing to provide testimony and technical information to regulators.

## ■ The Federal Register System

### Concepts

■ Rules are published and announced using the Federal Register System.

■ There are two major publications making up the system:
- the Federal Register (FR) and
- the Code of Federal Regulations (CFR).

In the 1930s, the passage of The Federal Register Act established a uniform system for publishing agency regulations. Because Congress was giving increasing authority to federal departments and agencies to develop rules, a plan for disseminating information about these agencies' activities needed to be developed. The Federal Register Act required the agencies to:

- File rulemaking documents with the Office of the Federal Register
- Allow for public inspection of the rulemaking documents

The Office of the Federal Register was required to:
- Publish rulemaking documents in the Federal Register (FR)
- Permanently codify (place in numerical arrangement) rules in the **Code of Federal Regulations (CFR)**

The Federal Register is published daily, Monday through Friday (except on official holidays) by the:

Office of the Federal Register
National Archives and Records Administration
Washington, D.C. 20408

The Federal Register is distributed solely by the:

Superintendent of Documents
U.S. Government Printing Office
Washington, D.C. 20402

1. General Provisions
2. (Reserved)
3. The President
4. Accounts
5. Administrative Personnel
6. (Reserved)
7. Agriculture
8. Aliens and Nationality
9. Animals and Animal Products
10. Energy
11. Federal Elections
12. Banks and Banking
13. Business Credit and Assistance
14. Aeronautics and Space
15. Commerce and Foreign Trade
16. Commercial Practices
17. Commodity and Securities Exchanges
18. Conservation of Power and Water Resources
19. Customs Duties
20. Employees' Benefits
21. Food and Drugs
22. Foreign Relations
23. Highways
24. Housing and Urban Development
25. Indians
26. Internal Revenue
27. Alcohol, Tobacco Products and Firearms
28. Judicial Administration
29. Labor
30. Mineral Resources
31. Money and Finance: Treasury
32. National Defense
33. Navigation and Navigable Waters
34. Education
35. Panama Canal
36. Parks, Forests, and Public Property
37. Patents, Trademarks, and Copyrights
38. Pensions, Bonuses, and Veterans' Relief
39. Postal Service
40. Protection of Environment
41. Public Contracts and Property Management
42. Public Health
43. Public Lands: Interior
44. Emergency Management and Assistance
45. Public Welfare
46. Shipping
47. Telecommunication
48. Federal Acquisition Regulations System
49. Transportation
50. Wildlife and Fisheries

**Figure 2-5:** List of CFR titles.

Publication of a rule in the Federal Register has the following legal effects:

– It is the official notice of the existence of a document.
– It becomes the true copy of the original document.
– It indicates the date of the regulation's issuance.
– It provides evidence that is acceptable in a court of law.

In the 1940s, changes to the Federal Register Act increased the scope of public participation in the rulemaking process. These changes made public comment a requirement of the process. Public comment can take the form of "written data, views, or arguments with or without opportunity for oral presentation . . ." (U.S. Code Section 553(c)). To facilitate this, a requirement was imposed that no regulation could become effective unless at least 30 days had passed from its date of publication, with only a few exceptions. This was to allow time for public input. Agencies were also required to publish statements of organization and procedural rules.

The daily Federal Register can be purchased in paper format, microfiche, or magnetic tape. Subscription information can be found after the frontispiece in the paper version.

Proposed rules and final rules are published in the Federal Register as well as notices of proposed rulemaking and meeting notices. Rules are cited by volume and page number. For example, 58 FR 4462 in the Federal Register shown to the left contains the final rules established for confined space entry by the Occupational Safety and Health Administration. The cover of the Federal Register indicates the date of issue as well as the volume and number. Since 1936, a new volume begins each year.

A Table of Contents in each issue lists rules, proposed rules, and notices in order by governmental agency. The following is an sample listing from the contents page.

**Figure 2-6:** Cover of the daily Federal Register (FR). (See Figure 2-9 for a sample page from this document.)

## Table of Contents

| | Page |
|---|---|
| Explanation | v |
| **Title 40:** | |
| Chapter I—Environmental Protection Agency | 3 |
| **Finding Aids:** | |
| Material Approved for Incorporation by Reference | 223 |
| Table of CFR Titles and Chapters | 225 |
| Alphabetical List of Agencies Appearing in the CFR | 241 |
| Table of OMB Control Numbers | 251 |
| List of CFR Sections Affected | 271 |

**Figure 2-7:** Table of contents of the Federal Register.

A second publication of the Federal Register System is the Code of Federal Regulations. CFR's are revised annually. Each year all the rules published in the daily Federal Register are compiled into the appropriate code book according to topic. There are 50 titles representing the areas that are regulated. They are listed in Figure 2-5.

Each rule in the Federal Register is preceded by supplementary information that presents the regulation's background, its intent, any issues that arose during comment periods and how they were addressed. Supplementary information, however, is left out when rules in the Federal Register are compiled in the Code of Federal Regulations. Although not included in the CFR, this background information remains admissible in court. When clarification about the meaning of a regulation is necessary, this supplemental material is a valuable source of information.

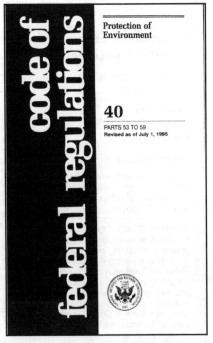

Figure 2-8: Cover of 40 CFR.

## ■ The Regulatory Numbering System

### *Concepts*

- Regulations are identified using a numbering system from broad to specific.

- The section (§) is the basic designation for a portion of a regulation.

The numbering system used for regulations is very similar for all types of regulations. Understanding this numbering system will help you to properly cite regulatory passages as well as locate them when you look for specific provisions. The following lists the major divisions from broad to specific:

**Title** – each title covers a broad area; e.g., 49 CFR is about transportation, 29 CFR is about labor which includes occupational health and safety, and 40 CFR is about protection of the environment (see Figure 2-5 for a complete list).

**Part** – a part is a body of regulations about the same topic within a title. These are designated by Arabic numbers. Part 260 in 40 CFR deals with definitions relating to hazardous waste; Part 261 is on hazardous waste identification. Subparts may be used, and if so will be designated by capital letters.

**Section** – the section, identified by the symbol §, is preceded by the number of the part followed by a period. For example, § 261.4(a)(1)(ii) of 40 CFR excludes domestic sewage from the definition of solid waste (which in turn means it can't be a hazardous waste, but we'll get to this topic later). The number 261 refers to the part, the 4 refers to the fourth section of that part, and the numbers and letters following refer to the paragraphs within the section. The section is the basic unit of CFR organization.

4462    Federal Register / Vol. 58, No. 9 / Thursday, January 14, 1993 / Rules and Regulations

**DEPARTMENT OF LABOR**

**Occupational Safety and Health Administration**

**29 CFR Part 1910**

[Docket No. S–019]

**RIN 1218–AA51**

**Permit-Required Confined Spaces**

**AGENCY:** Occupational Safety and Health Administration (OSHA), U.S. Department of Labor.
**ACTION:** Final rule.

**SUMMARY:** The Occupational Safety and Health Administration (OSHA) hereby promulgates safety requirements, including a permit system, for entry into those confined spaces, designated as permit-required confined spaces (permit spaces), which pose special dangers for entrants because their configurations hamper efforts to protect entrants from serious hazards, such as toxic, explosive or asphyxiating atmospheres. The new standard provides a comprehensive regulatory framework within which employers can effectively protect employees who work in permit spaces.

Few OSHA standards specifically address permit space hazards. These standards, in turn, provide only limited protection. OSHA has determined, based on its review of the rulemaking record, that the existing standards do not adequately protect workers in confined spaces from atmospheric, mechanical and other hazards. The Agency has also determined that the ongoing need for monitoring, testing and communication at workplaces which contain entry permit confined spaces can be satisfied only through the implementation of a comprehensive confined space entry program. OSHA anticipates that compliance with the provisions of this standard will effectively protect employees who work in permit-required confined spaces from injury or death.

**EFFECTIVE DATE:** This final rule will become effective on April 15, 1993.

**ADDRESSES:** In compliance with 28 U.S.C. 2112(a), the Agency designates for receipt of petitions for review of the standard, the Associate Solicitor for Occupational Safety and Health, Office of the Solicitor, Room S–4004, U.S. Department of Labor, 200 Constitution Avenue NW., Washington, DC 20210.

**FOR FURTHER INFORMATION CONTACT:** Mr. James F. Foster, U.S. Department of Labor, Occupational Safety and Health Administration, Office of Information and Consumer Affairs, Room N3647, Washington, DC 20210, (202) 523–8151.

**SUPPLEMENTARY INFORMATION:**

**I. Background**

Many workplaces contain spaces which are considered "confined" because their configurations hinder the activities of any employees who must enter, work in, and exit them. For example, employees who work in process vessels generally must squeeze in and out through narrow openings and perform their tasks while cramped or contorted. For the purposes of this rulemaking, OSHA is using the term "confined space" to describe such spaces. In addition, there are many instances where employees who work in confined spaces face increased risk of exposure to serious hazards. In some cases, confinement itself poses entrapment hazards. In other cases, confined space work keeps employees closer to hazards, such as asphyxiating atmospheres or the moving parts of a mixer, than they would be otherwise. For the purposes of this rulemaking, OSHA is using the term "permit-required confined space" (permit space) to describe those spaces which both meet the definition of "confined space" and pose health or safety hazards.

In its June 5, 1989 NPRM (54 FR 24080), OSHA determined, based on its review of accident data, that asphyxiation is the leading cause of death in confined spaces. The asphyxiations that have occurred in permit spaces have generally resulted from oxygen deficiency or from exposure to toxic atmospheres. In addition, there have been cases where employees who were working in water towers and bulk material hoppers slipped or fell into narrow, tapering, discharge pipes and died of asphyxiation due to compression of the torso. Also, employees working in silos have been asphyxiated as the result of engulfment in finely divided particulate matter (such as sawdust) that blocks the breathing passages.

The Agency has, in addition, documented confined space incidents in which victims were burned, ground-up by auger type conveyors, or crushed or battered by rotating or moving parts inside mixers. Failure to deenergize equipment inside the space prior to employee entry was a factor in many of those accidents. OSHA notes that the NPRM (54 FR 24080–24085) discussed the hazards which confront employees who enter permit spaces and the inadequacy of existing regulation in greater detail. Additionally, Section II of this preamble, *Hazards*, presents a detailed discussion of the hazards to which permit-space entrants have been exposed, demonstrating that this final

rule is reasonably necessary to protect affected employees from significant risks.

OSHA has determined, based on its review of the rulemaking record, including investigation reports covering "permit space" fatalities (Exhibits (Ex.) 10 through 13 and 16), that many employers have not appreciated the degree to which the conditions of permit space work can compound the risks of exposure to atmospheric or other serious hazards. Further, the elements of confinement, limited access, and restricted air flow, can result in hazardous conditions which would not arise in an open workplace. For example, vapors which might otherwise be released into the open air can generate a highly toxic or otherwise harmful atmosphere within a confined space. Unfortunately, in many cases, employees have died because employers improvised or followed "traditional methods" rather than following existing OSHA standards, recognized safe industry practice, or common sense. The Agency notes that, as documented in the NPRM, many of the employees who died in permit space incidents were would-be rescuers who were not properly trained or equipped.

In addition, OSHA believes that, as noted in the NPRM (54 FR 24098), the failure to take proper precautions for permit space entry operations has resulted in fatalities, as opposed to injuries, more frequently than would be predicted using the applicable Bureau of Labor Statistics models. The Agency notes that, by their very nature and configuration, many permit spaces contain atmospheres which, unless adequate precautions are taken, are immediately dangerous to life and health (IDLH). For example, many confined spaces are poorly ventilated— a condition that is favorable to the creation of an oxygen deficient atmosphere and to the accumulation of toxic gases. Furthermore, by definition, a confined space is not designed for continuous employee occupancy; hence little consideration has been given to the preservation of human life within the confined space when employees need to enter it.

Accordingly, the Agency has determined that it is necessary to promulgate a comprehensive standard to require employers to take appropriate measures for the protection of any employee assigned to enter a permit space. OSHA believes this new standard will help eliminate confusion and misunderstanding by clearly stating employer responsibilities.

The record and determinations that are discussed in this final rule

**Figure 2-8:** Page from the daily Federal Register.

**Paragraph** – whenever further division of a section is necessary, as in the example above, the divisions are designated as paragraphs with the following order:

|  | Symbol |
|---|---|
| Paragraph | (a), (b), (c), etc. |
| Further Subdivisions | (1), (2), (3), etc. |
|  | (i), (ii), (iii), (iv), etc. |
|  | (A), (B), (C), etc. – note that sometimes lower case italic letters are used for this subdivision |
|  | (1), (2), (3), etc. |
|  | (i), (ii), (iii), etc. |

This "nesting" of parts, sections, paragraphs, and subparagraphs can be confusing. For instance, notice that the lower case letters start the first divisions of paragraphs. This runs contrary to the outline format with which we are most familiar. Sample excerpts from 40 CFR have been reproduced on the next page. Look over the organization of these excerpts to make sure you understand this basic numbering system.

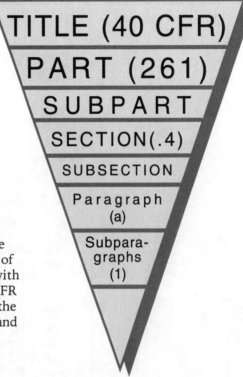

**Figure 2-9:** Hierarchy of regulatory numbering system.

## ■ Checking Your Understanding

1. Name two differences between laws and regulations.

2. How are proposed and final federal rules "announced?"

3. Given the following citation, 29 CFR 1910.120, identify the section, part and title.

4. How do the Federal Register and the Code of Federal Regulations differ?

5. List three federal agencies that develop regulations on environmental or health and safety subjects.

6. Why might risk-taking be perceived differently by regulators and the business community?

# 2-3 Enforcement

## ■ Monitoring Compliance

### Concepts

■ Inspections and review of reports are the most common ways to monitor compliance.

■ The Freedom of Information Act allows public access to government records.

Regulatory agencies are not only charged with developing regulations but with enforcing them as well. Not complying with environmental laws and regulations can result in serious consequences for individuals and companies. For instance, transportation, treatment, storage, disposal, or export of any hazardous waste where one knowingly places another person in imminent danger of death or serious bodily harm carries a possible penalty of up to $250,000 and/or 15 years in prison; up to $1 million if a corporation is involved!

Agencies are given broad enforcement powers to ensure that individuals and corporations obey regulations. In addition to regulations, agencies can develop policies, procedures, and guidance documents to regulate the public and help meet their own enforcement obligations. Any action taken by an agency must comply with the provisions of the statute they are designated to enforce. Remember, only a legislative body can make or amend laws. Agencies can only develop regulations to enforce laws.

The goal of any enforcement program is to ensure that the provisions and intent of the original law are met by all affected parties. EPA states that:

> This requires close monitoring of facility activities and quick legal action where noncompliance is detected. Facility inspections by Federal/State officials are the primary tool for monitoring compliance. When noncompliance is detected, legal action may follow.

The Compliance Evaluation Inspection (CEI) is the most common method used to determine whether or not a facility or person is complying with environmental regulations. These inspections may be routine (e.g., you will be inspected once a year), they can follow complaints or "tips" from a citizen or employee, or can result from referrals from other agencies. The inspection usually consists of a file review prior to going to the site, a site evaluation, a review of records at the site, and a written evaluation of the facility's compliance.

Inspections can be conducted with or without the cooperation of the owner or operator. Most regulations provide the enforcement agency with reasonable access to observe operations, examine permits and records, take samples, and question employees and others about the activity in question. If the owner or operator is uncooperative, access can be forced through the use of an **inspection warrant**. This is a legal document that the inspector obtains from a judge.

Employees of the regulatory agency typically make inspections and perform reviews of reports to assess compliance. Some agencies hire independent contractors to perform these functions. In either case, the inspector will have appropriate identification and must present it upon request. It is good practice for a knowledgeable representative of the company to accompany an inspector while making an inspection. Notes and observations made by the company will help to ensure future compliance and may come in handy during legal proceedings that may result from the visit.

Another type of inspection that is conducted is called a Case Development Inspection (CDI). This type of inspection is performed in support of an enforcement action when serious violations are known or suspected. The type of enforcement action dictates the type of activities that need to be performed under this type of inspection.

A second way that an agency assesses compliance is through a review of reports. Environmental regula-

**Figure 2-10:** Inspector making a compliance evaluation inspection.

tions often require that companies complete reports. These reports may go to federal or state agencies and be used to build a data base for the effectiveness of the regulations, track program progress, and/or determine specific compliance requirements. For example, if you ship hazardous waste to a disposal site, you must send a copy of a document called a Uniform Hazardous Waste Manifest to the state or federal EPA. This document details what the load is and where it is going. Further, if you don't receive acknowledgment from the disposal facility that they received your waste within a specified time, an "exception report" must also be submitted to the appropriate regulatory agency. Each of these reports is a requirement of the regulations and carries fines and penalties if not completed correctly.

Under the **Freedom of Information Act of 1966**, the public is entitled to access information gathered as a result of governmental agency functions. This can include inspection reports, other inspection documents, agency records, sample results, photographs, and maps acquired during inspection activities. There are certain exemptions under the act involving release of trade secrets, financial information, anything that can jeopardize national security, and any information that may compromise a court case. A complementary act, called the **Privacy Act of 1974,** requires that federal agencies provide individuals with information pertaining to them that is kept in governmental files. Under the act, the government must amend or correct these files if inaccurate.

## ■ Enforcement Actions

### Concepts

■ Regulatory agencies have three main options for enforcement actions.

■ The burden of proof differs for criminal versus civil enforcement actions.

Enforcement actions that are typically available to regulatory agencies include:
– Administrative Enforcement Actions
– Civil Actions
– Criminal Actions

Regulatory agencies usually have policies and procedures that guide regulators as to which of these options to pursue. This usually depends on the severity of the violation, the past history of the violator, and the level of intent shown by the violator. (See Box 2.3, for more information on this topic.)

An Administrative Enforcement Action involves only the enforcement agency and not the court system. This type of action can range from issuing an informal notice of noncompliance to the violator to a more formal **administrative order** with a public hearing. Administrative actions tend to be less complicated and labor intensive, and therefore, less expensive to perform for all parties involved. Informal letters to a violator are often called "notices of violation" (NOV) or "notices of deficiency" (NOD). These are usually used for relatively minor violations and may be satisfied by the correction of the problem(s) identified. With more severe violations, or if the violator does not make required corrections under an informal notice, an administrative order may be issued. Fines can sometimes be assessed as part of the administrative order. Orders can be used to force a facility to comply with specific regulations, to take corrective action, to perform monitoring, testing, and analysis, or to address a threat of harm to human health and the environment.

Civil Actions are formal lawsuits that must be resolved in the court system. Attorneys within the federal Department of Justice (DOJ) handle civil cases for federal agencies. At the state level, the Attorney General's Office (or similarly named office) takes these cases to court, and in cities or counties this may be done by the District Attorney. In Civil Actions, the government can force an

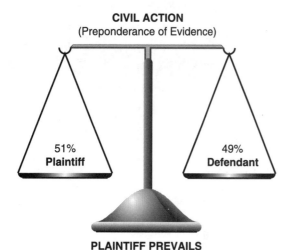

**CIVIL ACTION**
(Preponderance of Evidence)

51%
**Plaintiff**

49%
**Defendant**

**PLAINTIFF PREVAILS**

**Figure 2-11:** Burden of proof in criminal case ("winning" party only needs to demonstrate slightly

## Box 2.3

# An Official Enforcement Response Policy

The following is an excerpt of an official policy of the Department of Toxic Substances Control (the Department), California EPA, on the subject of enforcement actions:

### Purpose

This policy provides guidance to employees of the Department for determining the appropriate enforcement response — both administrative and judicial — following the discovery of violations of the Hazardous Waste Control Act or implementing regulations. The purpose of the Enforcement Response Policy is to provide a general framework for identifying and classifying violations and to ensure that timely, consistent and appropriate enforcement actions are taken by the Department.

### Policy Statement

It is the goal of the Department to pursue the attainment and maintenance of a high rate of compliance within the regulated community by maintaining a thorough inspection program and by taking timely, visible and appropriate enforcement actions against serious violators. It is the policy of the Department to enforce hazardous waste laws promptly, assuring that violators:

- are notified of violations
- comply with the pertinent violations
- are punished for the violations, as provided by law, and
- do not gain a substantial economic benefit from non-compliance.

To make enforcement response timely, the Department's policy is to complete inspection reports and reports of violation within 30 calendar days of the date of inspection, and to take enforcement action within 135 calendar days of the date that violations are documented or the date of inspection when no additional documentation of violations is necessary.

It is also Department policy to allocate enforcement resources in proportion to the significance of the violation, since violations vary considerably in severity and since enforcement resources are limited. By concentrating our efforts on the most serious violators, we will best strengthen our enforcement program.

Finally, it is the Department's policy that all cases where a formal enforcement action has been taken will be diligently pursued until such time as the violations have been resolved to the Department's satisfaction.

It is the intent of this policy to allow the flexibility to choose among the various appropriate enforcement options based upon circumstances of each particular case.

### Types of Violations

For purposes of determining the appropriate enforcement response, violations are divided into two broad categories: Class I violations and Class II violations. These terms have been selected to coincide with the EPA's terminology and thus simplify enforcement reporting to the federal agency.

Class I violations are defined as chronic violations by a recalcitrant party and/or significant deviations from statute and regulations, compliance order provisions or permit conditions, which are designed to:

a) assure that hazardous wastes are destined for and delivered to authorized treatment, storage or disposal facilities.

b) prevent releases of hazardous waste or constituents to the environment.

c) assure early detection of such releases.

d) assure adequate financial resources in case of releases or to undertake necessary actions at the time of the facility's closure.

Class I violations always require a formal enforcement action be taken. A formal enforcement action is defined as:

a) civil litigation

b) criminal litigation

c) administrative orders (both penalty and non-penalty)

Class II violations are any violations of the state's statutory or regulatory requirements that do not meet the definition listed above for Class I violations. Class II violations may be handled with an informal enforcement action such as a Report of Violation which contains a date-specific compliance schedule.

## Box 2.3

### Enforcement Options

Enforcement options can be viewed as a pyramid with the most stringent measures, which are used least frequently, at the top of the pyramid:

- Criminal action
- Civil action and penalty
- Corrective action order/ penalty
- Report of violation

The options also involve different agencies as the level of response rises. Reports of violation, corrective action orders, and penalties are all administrative actions taken by the Department. Civil and criminal actions require referral to the city attorney, district attorney, or Office of the Attorney General.

Exercise of the Department's enforcement discretion in choosing an enforcement response involves a variety of factors listed in more detail below. In general, however, the facts of the particular case, the interest and capacity of the various agencies to handle the case, and the priority of the violations must all be weighed.

Among the factors that should be considered are:

- extent of deviation from legal requirements
- potential for harm – real threat to humans and the environment
- firm's compliance history
- good faith efforts to comply
- deterrent effect on the violator and on the regulated community
- special, unusual or other mitigating factors

The weight given to various factors in particular cases will vary, depending on the circumstances. Nevertheless, the following chart provides guidance to select the appropriate option.
NOTE: Where there are multiple violations of varying degree of priority, all violations should be included in one enforcement action based on the highest priority violation.

### Criteria for Selecting Enforcement Response

- **Criminal Action**
  - One or more criminal acts.
- **Civil Action**
  - One or more criteria indicating Class I violations.
  - Need for injunctive relief.
  - Possible precedential issues.

  - Multi-jurisdictional issues.
  - Cases with statewide significance.
  - Slim likelihood of settlement.
  - Enforcement of existing Departmental administrative orders or agreements.
- **Administrative Penalty**
  - May be assessed without a corrective action order if compliance has already been achieved.
  - A penalty should generally be assessed for every violation unless the violation is included in an enforcement response to multiple violations and the violation, standing alone, would only result in the Report of Violation.

### Informal Administrative Action

- **Corrective Action Order**
  - Multiple Class II violations where compliance has not been achieved.
  - Class II violations where specific corrective action should be ordered by the Department.
  - One or more Class I violations, but none of acute life-threatening nature.
  - No issues are of a precedential nature.

### Criteria for Referral to Other Agencies

- **District Attorney/City Attorney**
  - Criminal cases.
  - Civil cases in counties and cities where the local prosecutor has significant interest in hazardous waste prosecutions.
  - Civil cases originating at the local level.
- **Attorney General**
  - Federal facilities.
  - Cases with statewide significance.
  - Cases requiring extensive coordination with other state agencies.
  - Cases involving major cleanup activity.
  - Cases with local prejudice (e.g. major employer in small county – DA conflict).
  - Cases in which the Department seeks to establish a judicial precedent.
  - Violations of enforcement of existing Departmental administrative orders or agreements.
  - Coordination of multi-jurisdictional cases.
- **Environmental Protection Agency**
  - It is more effective to coordinate the state violations with a federal enforcement action.
  - The state is unable to take appropriate action under existing authorities.

individual or company to comply with applicable regulations. They can suspend or revoke permits or authorizations, or require immediate action to remove an imminent hazard to human health or the environment. Large penalties can also be imposed. It is easier to establish guilt in civil cases because the "burden of proof" for the plaintiff, in this case the agency that brings the lawsuit, is preponderance of evidence. This means that the weight of evidence need only be slightly in favor of the plaintiff in order to win the case.

Criminal Actions can result in fines and/or imprisonment. This type of enforcement action is reserved for the most severe violations such as making a false statement on required reports or compliance documents, putting another person in imminent danger of death or serious bodily injury, or illegal disposal of hazardous waste. Within criminal actions, there are two basic divisions – **misdemeanors** and **felonies**. Less serious crimes are misdemeanors which are punishable by a fine and/or a maximum sentence of one year in jail. More serious crimes are felonies and are punishable by imprisonment for more than one year. Felonies such as murder or treason are capital offenses and may be punishable by death. Criminal cases, having more serious consequences for the convicted, require a higher level of assurance of approaching the "truth." To secure a conviction, a prosecutor must provide proof of guilt beyond a reasonable doubt. In contrast to civil cases, this does not equate to a slightly greater weight of evidence.

**CRIMINAL ACTION**
(Beyond a Reasonable Doubt)

51%
Plaintiff

49%
Defendant

**DEFENDANT PREVAILS**

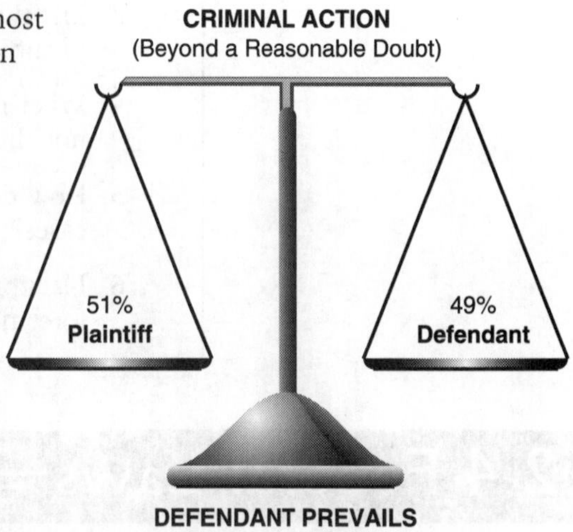

Figure 2-12 Burden of proof in criminal case (in this case the plaintiff must have greater weight of evidence in order to provide proof "beyond a reasonable doubt").

When determining who to file a lawsuit or a criminal case against, regulatory agencies have broad authority. Most environmental regulations make all parties that had anything to do with the illegal activity liable. Additionally, in some cases, a property owner can be liable even if he or she did not have knowledge of the illegal activity. This liability can extend to: the costs of clean-up, which can be enormous; the expenses associated with investigating the extent of contamination in soil and groundwater; any penalties or fines; damages associated with injury to persons or property; and expenses for legal help. The **Comprehensive Environmental Response, Compensation, and Liability Act of 1980 (CERCLA)** is a federal law that makes owners and operators liable for hazardous waste releases on their property, even if they are not responsible for the release, because the act imposes **joint and several liability**. This means that all potentially responsible parties, which include owners and operators, are responsible together and individually ("severally" is from the root word sever which means to separate, or break off). Ultimately, under this act, the EPA can sue and recover from all wrongdoers or from any one of the wrongdoers. For this reason it is often referred to as a "deep pocket" provision.

## ■ Checking Your Understanding (2-3)

1. Explain why a Case Development Inspection would not be considered a routine inspection.

2. What is the difference between the Freedom of Information Act and the Privacy Act?

3. List two activities performed by a regulator as part of a Compliance Evaluation Inspection.

4. What recourse does an inspector have if a facility owner does not allow access for inspection?

5. How does the burden of proof differ for civil versus criminal cases?

6. List three ways Administrative Actions differ from other enforcement actions.

# 2-4 Federal Laws

## ■ Synopses of Federal Laws

### Concepts

■ There are a number of laws related to specific areas of environmental health and safety

■ All laws except NEPA have assigned enforcement agencies

■ Scientists use the scientific method to find out more about our physical world

There are a number of key federal laws dealing with environmental protection and employee health and safety. The following is a quick overview:

**National Environmental Policy Act (NEPA)** – NEPA was signed into law on January 1, 1970. It authorized a broad policy requiring that environmental consequences be considered whenever the federal government engaged in any activities that could have a negative impact on the environment. It created the Council of Environmental Quality as advisor to the President.

NEPA "requires each federal agency to prepare a statement of environmental impact in advance of each major action, recommendation, or report on legislation that may significantly affect the quality of the human environment." The statement that accomplishes this is called an **Environmental Impact Statement** (EIS). The EIS aids the agency in its decision-making process of whether to grant or refuse the proposed action. *Enforcement Agency:* NONE; Enforced by citizen and other lawsuits.

**Resource Conservation and Recovery Act (RCRA)** – RCRA was enacted in 1976 to deal with both municipal and hazardous waste problems and to encourage resource recovery and recycling. It is administered by EPA's Office of Solid Waste and Emergency Response (OSWER).

RCRA requires states to develop and implement waste disposal plans. States are required to inventory all existing waste disposal sites and determine whether the sites are in sound condition.

To carry out RCRA's provisions for dealing with hazardous wastes, EPA has developed a national hazardous waste management system to monitor the movement of significant quantities of hazardous wastes from "cradle to grave." Under the system, hazardous waste generating facilities must identify the wastes they create and report the means of disposal. Transport of wastes is also regulated and tracked. *Enforcement Agency:* Environmental Protection Agency (EPA).

## Science & Technology

### Science and Lawmaking?

Many people's image of science and scientists come from the Saturday afternoon movie matinees they remember as children. These colorful and exciting "mad scientists" found in such memorable movies as "The Man With Two Brains" and "Journey to the Center of the Earth" certainly do not do justice to the formalized system of science. Science may be considered to be the development of scientific laws and principles through the use of systematized observations and experimentation. Applied science, also called **technology**, is the practical application of these scientific laws and principles.

So, what is the relationship between science and governmental law? One of the more whimsical take-offs on the Universal Law of Gravitation was recently seen on a bumper sticker stating, "Gravity, it's the law." Obviously, this is different from the law limiting speed to 65 mph. Observations and experimentation over many years have established that one can always count on gravity having measurable effects. But to get to this level of confidence, it took a lot of systematic measurements and observations by many people over a long period of time.

It is scientists then that attempt the ever-continuing challenge to find the answers about our environment. If we want to know whether a chemical can cause cancer or if emissions from a certain industry are likely to contribute to the formation of ozone, it takes scientific methods to find reliable answers.

With the type of long-term impacts environmental policy decisions can have, citizens should always question the basis for making them. It is all too easy to get swept up in the emotional aspects of many issues and overlook the valid science that would help solve the problem.

Unfortunately, the government and the public do not always look at scientific facts and conclusions before enacting laws seemingly based on scientific argument ƒ This is a source of frustration with many scientists because public policy decisions are often made, at best, with sketchy scientific information. Frequently the public is not willing to take chances with a consequence that may be irreversible.

Realistically, there is no simple, straightforward method that provides fail-safe answers to all our questions. We have found that we cannot assume a scientific theory is always valid. Good science takes time and usually needs to incorporate the use of statistical methods since not every

**Federal Clean Water Act (CWA)** – CWA was signed into law in 1972. The objective of the CWA is to restore and maintain the "chemical, physical, and biological integrity of the nation's waters." Its long-range goal is to eliminate the discharge of pollutants into navigable waters and to make national waters fishable and swimmable.

Achieving water quality standards requires that controls be placed on sources of pollution; therefore, municipal sewage systems and industrial discharges of pollution are subject to a number of requirements. Nationwide standards are established for each type of industry and for pollutants based upon the availability and economic feasibility of technology. *Enforcement Agency:* EPA.

## Box 2.4

## Lily Wong, U.S. EPA Regulator

*Lily Wong has worked at the USEPA since 1976. She has a B.S. degree in Biology of Natural Resources and a Masters of Public Health in Environmental Health, both obtained from the University of California at Berkeley. She is a native of San Francisco.*

**Q:  What type of work have you done with the EPA since 1976?**

A:   My first job was in the NPDES permit program under the Clean Water Act. From there I worked on permits for the discharge of dredged or fill material to U.S. waters, and wetlands protection. I then transferred to the Enforcement Section of the Hazardous Waste Management Division where I worked as an inspector and compliance officer. I am now Section Chief of that section for USEPA Region 9.

**Q:  What are your responsibilities and duties in your current position?**

A:   I am a first line supervisor and oversee a section that is responsible for enforcing the regulations at hazardous waste management facilities regulated by the Resource Conservation and Recovery Act ("RCRA"). My staff is responsible for conducting inspections and taking the appropriate follow-up enforcement actions to ensure that facilities comply with the law.

**Q:  How did you become interested in the environmental field?**

A:   I had always been interested in nature and the outdoors, and the environmental field seemed to be a logical next step for a professional career.

**Q:  What is most rewarding or satisfying about the work you do?**

A:   Instead of producing widgets, I believe that the work we do at the USEPA is important and matters.

In addition to this philosophical answer, what is most rewarding to me is to be able to do a job well.

**Q:  What are the more difficult or challenging aspects of your job?**

A:   My greatest challenge is to be more efficient at what I do because I find that there usually isn't sufficient time to do the job.

**Q:  How does working as a regulator differ from being in private industry?**

A:   At this point, I have only worked as a regulator and have not worked in the private sector. My guess is that profits would be the driving force for decision-making in private industry. As a regulator, compliance with the regulations is more important.

**Q:  What advice can you give to those seeking environmental careers?**

A:   Read and become knowledgeable on environmental issues so that you could better determine what areas may be of interest to you professionally. Try to get a job (summer or internship) in the environmental field so you can confirm your interest, and begin establishing references and contacts. A summer job or internship will also demonstrate to a prospective employer your commitment and interest in the environmental field.

**Q:  What best prepared you for the work you do?**

A:   I think it helped to have a keen interest in the environment and to have a logical and analytical approach to thinking and problem-solving.

**Safe Drinking Water Act (SDWA)** – The SDWA was passed by Congress in 1974 with the goal of establishing federal standards for drinking water quality, protecting underground sources of water, and setting up a system of state and federal cooperation to assure compliance with the law. The EPA administers this act.

Under the SDWA there are two types of drinking water standards. For substances that have an adverse effect on health, compliance is mandatory. For substances or conditions that affect color, taste, and smell of drinking water, SDWA provides guidelines.

The intent of SDWA is to fill the gap left by the enactment of the CWA two years earlier. While the CWA was passed to clean up and protect streams and other surface waters, the SDWA was intended to protect underground water sources. *Enforcement Agency:* EPA.

**Federal Clean Air Act (CAA)** – The CAA was enacted in 1970 and amended most recently in 1990 with the fundamental objective of protecting public health and welfare from harmful effects of air pollution. To achieve this goal, the act sets maximum acceptable levels for pollution in ambient (outdoor) air.

Under the CAA, it is the responsibility of state and local governments to control pollution sources so that air pollution is at or below levels necessary to comply with the federal air quality standards.

The CAA also provides for setting nationwide emission standards for a variety of air pollution sources such as automobiles. *Enforcement Agency:* EPA.

**Federal Insecticide, Fungicide, and Rodenticide Act (FIFRA)** – This law was originally enacted in 1947 but was significantly amended and strengthened in 1972, 1975, and 1978. Under FIFRA, all pesticides must be registered with the EPA before being marketed. On the basis of available scientific data concerning their safety and efficacy, the EPA reviews and approves labeling, directions for use, precautions, and warnings. From the given data, a decision is also made to determine whether the pesticide should have general or restricted usage. The EPA also sets maximum safe levels for pesticide residues in human and animal food. FIFRA requires EPA to develop and oversee applicator training programs. Another aspect of FIFRA involves research programs on pesticides and monitoring of pesticide levels in the environment. *Enforcement Agency:* EPA.

**Comprehensive Environmental Response, Compensation, and Liability Act (CERCLA), a.k.a. "Superfund"** – The superfund law was passed by Congress in 1980 and amended most recently in 1986 as the Superfund Amendments & Re-authorization Act (SARA). CERCLA provides a system for identifying and cleaning up chemical and hazardous materials released into any part of the environment. The law also provides authority for the EPA to collect the cost of cleaning up a release from the responsible party(s) and establishes a fund to pay for cleaning up environmental contamination where no responsible party can be found or where those responsible for the release will not or cannot pay for cleanup. The EPA can subsequently go to court to secure reimbursement to the fund by responsible parties who will not pay voluntarily.

Money for the various types of cleanups authorized under the law comes from fines and other penalties collected by the government from a tax imposed on chemical feedstocks, and from the U.S. Treasury. A separate fund established under the law is authorized to collect taxes imposed on active hazardous waste disposal sites to finance monitoring of the sites after they close. *Enforcement Agency:* EPA.

**Figure 2-13:** CERCLA regulates and provides funds for cleanup of chemical and hazardous material spills.

**Asbestos Hazard Emergency Response Act (AHERA)** – This act requires certain procedures to be followed for asbestos abatement in school buildings. It requires all asbestos inspectors to receive certification from the EPA and requires that asbestos abatement response plans be designed by accredited persons. It ties into OSHA worker health and safety programs by requiring the use of OSHA safe work practices in asbestos removal work. Extensive record keeping is required. *Enforcement Agency:* EPA.

**Toxic Substances Control Act (TSCA)** – Congress enacted TSCA in 1976. The act deals with two major kinds of problems. First, newly created chemicals or chemicals entering into commerce for the first time may do serious damage to humans and the environment before their potential danger is known. Second, many existing chemicals may require more stringent control.

To deal with the problem of new, potentially toxic substances, the act imposes a system of pre-market notification to the EPA if a company wishes to market a new chemical, or significantly expand uses of an existing chemical. The EPA determines whether there is sufficient information about the chemical to predict the health and environmental effects, and can require additional testing before the chemical is sold.

The EPA also requires testing of some chemicals already in commerce. If a chemical presents an unreasonable risk to health or the environment, and there is sufficient data to evaluate its toxicity, testing may be ordered resulting in possible restrictions or a ban. *Enforcement Agency:* EPA.

**Marine Protection, Research, and Sanctuaries Act (or Ocean Dumping Act)** – This act regulates what can be dumped into the ocean in order to protect the marine environment. It restricts allowed dumping to designated locations and strictly prohibits the dumping of materials such as radioactive and biological warfare substances. *Regulatory Agency:* EPA, U.S. Coast Guard conducts surveillance.

**Occupational Safety and Health Act (OSHA)** – The OSH Act was passed in 1970 to "assure so far as possible every working man and woman in the nation safe and healthful working conditions." To achieve this goal, the Act authorizes several functions such as encouraging safety and health programs in the workplace and en-couraging labor-management cooperation in health and safety is-sues. It also provides for research and training in occupational safety and health. The main objectives of OSHA, however, are the devel-opment of regulatory standards, record keeping of injuries and training, and enforcement programs. *Enforcement Agency:* Occupa-tional Safety & Health Administration (OSHA).

# Science & Technology

## Comparison of the Metric and English Measuring Systems

Any time you measure something, it results in both a number and a **unit**. For example, 3 apples, 3 feet, and 3 meters all have the same number, but different units. When writing down measure-ments, both the number and its units must always be recorded. A **measuring system** provides the units for expressing length, volume, weight and time. In America we use two systems, the **English system** for everyday business and the **metric sys-tem** when doing scientific work. The English sys-tem includes such familiar units as feet and inches, tons and pounds, and gallons and quarts. The metric system uses units like meters and centime-ters, kilograms and grams, and liters and milliliters to measure similar quantities. Both systems use the same units for expressing time. Today, each of these systems is used on part of the reporting forms; therefore, as an environmental technician you will need to develop the ability to work with either system.

The metric system has several features that makes it more practical and easier to use. One of these features is the relationship between the units within the system. In the English system we must memorize that 12 in = 1 ft, 3 ft = 1 yd, 5280 ft = 1 mile, 2000 lbs. = 1 ton, 16 oz = 1 lb., 4 qt = 1 gallon, etc. Did you ever think about how many of these **conversion factors** we must have memo-rized to be able to operate within our system? Here are three more useful conversion factors that allow you to move between the systems:

Metric to English Conversion Factors

| | |
|---|---|
| Length | 2.54 cm = 1 in |
| Volume | 1 liter = 1.06 quarts |
| Weight | 454 g = 1 lb. |

In the metric system there is one base unit for each kind of measurement; meter for lengths, gram for weights, and liters for volumes. Then prefixes are used with these base units to give both smaller and larger units. The meaning of each of these prefixes remains constant throughout all measurements. For example, 1 kilometer is 1,000 meters; 1 kilowatt is 1,000 watts; 1 kilogram is 1,000 grams; etc. The prefix kilo always means 1000 of the base unit. The other prefixes that are commonly used are:

| Accepted Abbreviation | Unit | Times the Base Unit |
|---|---|---|
| k | Kilo | 1,000. |
| h | Hecto | 100. |
| da | Deka | 10. |
| | Base | 1 |
| d | Deci | 0.1 |
| c | Centi | 0.01 |
| m | Milli | 0.001 |
| μ | Micro | 0.000001 |

Since these prefixes are based on the decimal system, like our money, you only have to move the decimal point when changing from one unit to another. For example, the dollar is the base unit for our money system. We have other subdivi-sions of money like dimes and pennies. If our money were in the metric system, we would call the dime a decidollar, and the penny (cent) a centidollar, and a one-thousand dollar bill would be a kilodollar bill!

If an item were priced as 1.50 dollars or an object was 1.50 meters long, then the following would be equivalent:

| | |
|---|---|
| 1.50 dollars | 1.50 meters |
| 15.0 decidollars | 15.0 decimeters |

**Atomic Energy Act (AEA)** – The original Atomic Energy Act was enacted in 1954 to provide controls over the possession, development, and use of radioactive materials. The act authorized the creation of the Atomic Energy Commission which was abolished by later amendments to the Act. Currently, various aspects of this act are enforced by two agencies, the **Department of Energy (DOE)** and the Nuclear Regulatory Commission (NRC). *Enforcement Agency:* DOE and NRC.

**Hazardous Materials Transportation Act (HMTA)** – The HMTA was passed in 1975 and authorized the Secretary of Transportation to establish criteria for handling and transporting hazardous materials. This act applies to shippers, carriers, and manufacturers. The Department of Transportation is the administering agency for this Act.

The purpose of this act is to regulate commerce by improving the protections afforded the public against risks associated with the transportation of hazardous material. *Enforcement Agency:* Department of Transportation (DOT).

## ■ Checking Your Understanding (2-4)

1. What federal law is most likely to be violated by a barge dumping radioactive waste in the Gulf of Mexico?

2. What federal law is most likely to be violated by setting up a burn dump?

3. What federal law is most likely to be violated by not assessing the effects to the environment of siting a new amusement theme park in Denver?

4. What federal law is most likely to be violated by a company that formulates a new pesticide and immediately puts it on the market without testing its harmful effects?

5. What federal law is most likely to regulate the allowed use of alternative fuels?

6. What federal law should protect employees planning to enter a confined space?

7. How many meters are there in 100 kilometers?

8. How many centimeters are there in 1 meter?

9. How many micrograms are there in 2 grams?

# Health Effects of Hazardous Materials

## ■ Chapter Objectives

After completing this chapter, you will be able to:

1. **Recognize** the health effects of toxic substances.

2. **Define** principal terms of toxicology.

3. **Describe and give examples** of the dose/response relationship.

4. **Rank** relative toxicity using lethal dose values.

5. **Describe** the routes by which toxicants can enter the body.

6. **Relate** variability of reaction and response to differences between individuals.

7. **Describe** major components of the risk assessment process.

8. **Give examples and discuss** the concepts of voluntary, involuntary and relative risk.

Introduction to Environmental Technology ■ 61

## ■ Terms and Concepts to Remember

Absorption
Accumulation
Acute Toxicity
Additive Effect
Air Contaminant
Antagonism
Benign Tumors
Cancer
Carcinogen
Carcinogenicity
Chemical Properties
Chemical Reaction
Chemical Substances
Chronic Toxicity
Container
Dose
Dose-Response Relationships
Epidemiology
Exposure
Exposure Assessment
Gas
Hazard Evaluation
Interaction
Latency Period
$LC_{50}$
$LD_{50}$
Liquid
Local Effect
Lowest Observable Adverse
  Effect Level (LOAEL)

Malignant Tumors
Mutagen
Mutagenicity
Mutations
No Observable Adverse Effect
  Level (NOAEL)
Pesticides
Physical Properties
Physical States
Potentiation
PPM/PPB
Risk Assessment
Risk Characterization
Risk Management
Safety Factor
Sensitizers
Solid
Solubility
Synergism
Systemic Effect
$TC_{50}$
$TD_{50}$
Teratogen
Teratogenicity
Toxic
Toxicant
Toxicity
Toxicology
Toxin
Volatile

## ■ Chapter Introduction

In the last chapter we learned how governmental processes bring about laws and regulations controlling environmental and health hazards. We also learned that the public plays a vital role in the lawmaking and rulemaking processes. The level of public concern over the use, storage, and transport of chemicals in our society has greatly increased in the last few decades. This concern stems from increased awareness about the harmful effects of chemicals on human life, animals, and environmental quality.

    This chapter explores the basic principles and terms used to describe the harmful potential of toxic substances. It also looks at how we as a society make decisions about how to manage these risks.

# 3-1 Basic Toxicology

## ■ What is Toxicology?

### Concepts

- Toxicology is the study of poisons.

- Toxicity is the relative ability to cause harmful effects.

- Health hazards are different from safety or physical hazards.

**Toxicology** is the study of poisons, which are substances that cause harmful effects to living things. These effects are called **toxic** effects and can range from minor irritation of mucus membranes to serious damage to the liver or other organs, cancer or death. A lethal effect (death) is the most severe toxic effect, but, as noted above, there are many potential levels of harm between it and perfect health. Toxicologists are scientists who study the types and severity of harmful effects caused by toxic substances. They also determine which body systems are targeted by certain poisons and how likely it is that a given effect will occur.

Toxicology, although considered a relatively new science, has deep roots in human culture. Even the earliest human communities had to be knowledgeable about the dangers of poisons in their environment for their very survival. People knew how to distinguish plants that were safe to eat from those that weren't. They knew that the bite of one spider caused death, while others caused illness, and some just a skin irritation. There are indigenous people today who still use their knowledge of the poisonous properties of plants and animals to enhance the effectiveness of their weapons for hunting and protection. Natives of the rain forest regions of Central and South America, for instance, use the secretions of the "poison arrow frog" on the tips of blow darts.

The word "toxic" comes from the Latin translation for a Greek word *toxikon* which means arrow poison. In Greek, a bow was a *toxin* and a drug, a *pharmacon*. The combined form for an "arrow drug" was *toxon pharmacon*. The Romans modified the first part of this combination to make toxicum, their equivalent for "poison." In English, toxicum became "toxin," and the science of poisons, "toxicology." Currently, we use the term **toxin** for poisons that are produced by living organisms. **Toxicant** is used for toxic substances that are manufactured by humans.

Philippus Aureolus Paracelsus (1493–1541), a Swiss physician who studied alchemy, a medieval chemical science that sought to change base metals into gold and to discover a universal cure for disease, is considered the father of toxicology. In his study of the effects of chemicals as cures for diseases, he observed that "All substances are poisons; there is none which is not a poison. The right

**Figure 3-1:** Poison hemlock, a member of the parsley family commonly found growing naturally as a weed, is deadly if any part is eaten. It was used extensively in ancient Greece to execute criminals. Probably its most famous victim was Socrates. Its alkaloids affect the nervous system causing paralysis, convulsions, and death.

**Figure 3-2:** Paracelsus.

dose differentiates a poison and a remedy." He introduced many chemicals, such as laudanum, mercury, sulfur, iron and arsenic, into use as medicines.

There are many tales in the literature and history of classical Greece and ancient Rome, Egypt, China, and in cultures of pre-Columbian Americas of the use of poisons by religious and political leaders. In ancient Greece and Rome, an ordinary person could acquire poisons in three strengths: those that acted quickly; those acting slower causing lingering illness; and those that had to be given to the victim over a period of time to achieve the desired effect. Among history's most famous victims of fatal poisonings are the philosopher Socrates who was condemned to death and drank a lethal potion of hemlock, and the Egyptian queen Cleopatra who died from a self-inflicted bite of an asp, a small venomous snake. One of the most notorious poisoners was the Queen of France, Catherine de Medici. She tested toxic concoctions, mainly involving arsenic, on a number of people and recorded the results of her studies in detail. She carefully recorded symptoms, the progression of disease, and the cause of death of each of her "test subjects."

Mattien Joseph Orfila (1787–1853) is considered the father of modern toxicology. As a professor at the University of Paris he was the first to explore the connection between the chemical nature of poisons and their effects on living things. He did this using scientific studies on laboratory animals such as dogs. Modern day toxicologists have built on these techniques to identify, quantify, and explain the harmful effects caused by toxic substances. In some cases they test how animals are affected by various doses of a chemical. They may also use microorganisms like bacteria to study how a chemical affects genetic material passed from generation to generation. Humans and other animals can be studied over a long time period to assess the affect of low concentrations of chemicals in their air or water supplies. These studies assess the toxicity of substances being studied.

**Toxicity** is defined as the relative ability of a substance to cause harmful effects to living things. Another way of saying this is the potential of a substance to cause harm. We have already learned from Paracelsus that all chemicals can cause harm, therefore, the toxicity of a chemical will tell us whether it will take a large or small quantity of the chemical to do its damage. As shown in Figure 3-4, if it takes a large quantity of a chemical to produce a harmful effect, we would consider that chemical to have very low toxicity. If it takes only a small amount, the chemical would be considered highly toxic.

Various factors play a role in determining the toxicity of a substance. The chemical makeup of the substance is important. Certain elements and combinations of elements have a tendency to be more toxic than others. Even the way the various atoms are arranged in the molecule can make

**Figure 3-3:** Catherine de Medici.

**Figure 3-4:** If death would occur in an adult who ingested either the amount shown of chemical "x," or the amount shown of chemical "z," it is obvious that "z" is far more toxic.

a difference. The ability of the chemical to find its way into the bloodstream, which is called **absorption**, is also important. Some chemicals are not as easily absorbed and therefore are less likely to cause harm. Lastly, some chemicals are readily changed within our bodies to less toxic substances which can be easily eliminated without causing harm.

Sometimes the term "hazardous" is incorrectly used to mean toxic. It should be kept in mind that toxicity is only part of the numerous hazards we are faced with when we work with chemicals. There are a number of serious hazards besides toxicity, such as the potential for fires, explosions, and corrosive burns. Toxic hazards are considered health hazards. Fires, explosions, violent chemical reactions, corrosive burns, and a host of other hazards are included in the much broader category of safety or physical hazards.

## ■ How Do Exposures Occur?

### Concepts

■ Dose and exposure have different meanings.

■ There are three critical components needed for an exposure to occur.

■ Inhalation, dermal absorption, ingestion, and injection are the primary routes of exposure.

■ Acute and chronic effects result from exposures that differ in time frame, number, and concentration.

Toxic chemicals cause damage by altering normal body functions such as breathing, heart beat, liver function, and kidney function. In order to have any effect, toxic chemicals must first come into contact with the body and be absorbed. For most toxicants to cause damage, a sufficient quantity must enter the body. Terms used to express quantities of toxicants coming into contact with and entering the body are dose and exposure. These terms are often loosely used to mean the same thing; however, technically they are not equivalent. **Dose** refers to the actual amount of chemical that enters and reacts with body systems to cause harm. **Exposure** is the amount of toxic chemical our body comes into contact with in the air we breathe, the food we eat, or the compound we spill on our arm.

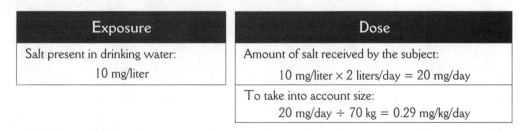

| Exposure | Dose |
|---|---|
| Salt present in drinking water: 10 mg/liter | Amount of salt received by the subject: 10 mg/liter × 2 liters/day = 20 mg/day |
| | To take into account size: 20 mg/day ÷ 70 kg = 0.29 mg/kg/day |

**Figure 3-5:** Difference between dose and exposure.

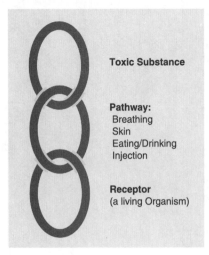

**Toxic Substance**

**Pathway:**
Breathing
Skin
Eating/Drinking
Injection

**Receptor**
(a living Organism)

**Figure 3-6:** Exposure chain.

For example, let's say that the water we drink contains 10 mg/liter of salt. Our exposure is to 10 mg/liter of salt. The dose, or actual amount received, must take into account the amount of water we drink and our weight. Our dose from this exposure would be 0.29 mg/kg per day. See Figure 3-5 to learn how this dose was determined. The subtle difference between the terms can be better understood when we think about the following relationships: the higher the concentration of the exposure, the larger the dose; also, the longer the exposure, the larger the dose. Exposure values are used to calculate dose.

Exposure can only take place when a toxic agent comes into contact with the body. The exposure chain in Figure 3-6 shows the three critical links required for this to occur. If any of the links of the chain are broken, an exposure cannot occur. In later chapters we'll see how personal protective clothing/equipment and containment can serve to break the exposure chain. The middle link, involving pathways, will be discussed in this section.

The path that a toxic agent uses to enter the body has a major influence on the severity and location of toxic effects. The four basic ways toxic chemicals can enter the body, referred to as routes of exposure, are:

– **Inhalation** – through the respiratory system
– **Dermal Absorption** – through the skin
– **Oral** – through the digestive tract
– **Injection** – enters directly into veins, muscles, etc.

Inhalation and injection are the two routes that provide the most rapid movement of toxic substances into internal systems. Inhalation, or breathing a substance into the lungs, is also the most common and serious type of exposure in the workplace. The air we breathe contains the oxygen necessary to sustain life. For this reason, the process of transferring oxygen from the air into the bloodstream is rapid and easily accomplished. Unfortunately, if we inhale air that also contains a toxic substance, it too can rapidly and easily find its way into our bloodstream where it can move to other body systems and cause toxic effects. The next section of this chapter will cover in greater detail the workings of major body systems and how they can be harmed by toxic substances.

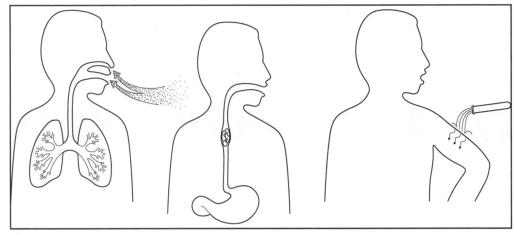

**Figure 3-7:** Three of the four routes of exposure.

Our skin is the largest organ in our body. It provides a protective barrier that serves to keep foreign substances out of our bodies. Unfortunately, this barrier is not sufficient to prevent the entry of many chemicals that can readily pass through the skin and enter the bloodstream. Some chemicals irritate the skin, causing itching and redness. Corrosive chemicals can burn the skin. Chemicals that can dissolve oils and greases can dissolve skin oils leaving the skin dry, cracked, and even more susceptible to the absorption of chemicals. As seen in Figure 3-8, absorption occurs more easily at body locations where the skin is thinner.

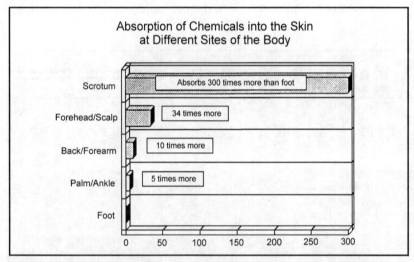

**Figure 3-8:** Comparison of absorption by skin at various sites of the body. (Source: Adaptation by OCAW/Labor Institute from E. Hodgson and P.E.

The eyes are particularly sensitive to harm by chemicals. Many chemicals can burn or irritate delicate eye tissue. Chemical contact with the eye should always be regarded as serious and proper action taken to reduce the chances for permanent eye injury. This is one of the easiest exposures to prevent by using appropriate eye protection such as goggles or face shields.

Ingestion of toxicants is not common in the workplace. It can, however, occur if basic personal hygiene activities are neglected. For instance, a worker biting his or her nails, smoking a cigarette, or eating lunch without washing after handling chemicals may be exposed. Mustaches can collect particles from the air that can then be ingested. Metal dusts such as lead or cadmium are easily ingested this way.

Although not as problematic in the workplace, ingestion of toxic substances is a serious problem at home with children. Some children tested in inner cities have been found to have elevated blood lead levels that are potentially harmful to the blood and central nervous system. The majority of the cases investigated involved sources in the child's immediate environment. An example is eating paint chips from older buildings that were painted with lead-based oil paints. Emissions from vehicles using leaded gasoline before it was phased out have resulted in lead accumulating in the soil next to major freeways. Soil may contain sufficient concentrations of lead to pose added risks to children who may play in and/or ingest it.

Medical facilities pose the greatest risk of injection by toxins and disease-causing organisms in the workplace. This is due to the types of instruments used, such as needles and scalpels. Outside the workplace, insects, scorpions, spiders, snakes, and other venomous animal bites or stings are another way that injections occur.

When looking at the types of harm caused by a toxic agent, both the long-term and short-term effects need to be evaluated. **Acute toxicity** is the result of a short-term exposure and causes effects that are felt at the time of exposure or shortly after. The Latin root *acu* means "sharp." An acute effect, then, is rapidly detected and is sharp

## Science & Technology

### PPM, PPB, and Percent

Many regulations describing limits for various contaminants use the designations **parts per million (ppm)** and **parts per billion (ppb)** when describing concentrations or dose. One part in a million parts is smaller than most people realize. It is one dollar in a million dollars or 1 cent out of 10,000 dollars.

One way to get a feel for ppm is to compare distances traveled. For example, think of a destination that is 16 miles away from your present location. Now, in your mind, divide that trip into a million parts. Each one of those parts is a ppm; did you guess that it amounted to *one inch*? If we were to take a ppb of that same trip we would travel 1/1000th of an inch! In fact, 17 inches is a ppb of the distance between the earth and the moon. Here are some other interesting comparisons:

PPM   – One half minute in a year
      – If you are 32 years old, 17 minutes of your life.
PPB   – 1 square foot in 36 square miles
      – 1 pinch of salt in 10 tons of potato chips
      – If you are 32 years old, a ppb would be one second out of your life

The basics of the metric system were covered in Chapter 2. Let's look at the relationships between metric units and ppm and ppb. In Figure 3-5 the dose was determined to be 0.29 mg/kg per day. We know that 1 mg represents .001 or 1/1000th of a gram. A kg is 1000 grams. We see then that mg/kg is the same as:

$$\frac{.001g}{1000g}$$

If we multiply the numerator and the denominator by 1,000 we have not changed the value of the fraction. It is the same as saying 1 in 10 is equivalent to 10 in 100.

$$\frac{.001g}{1000g} \times \frac{1000}{1000} = \frac{1}{1,000,000}$$

From this we can see that mg/kg is equivalent to parts per million. If we do the same type of comparison with ug/kg we will find that it is equivalent to parts per billion.

$$mg/kg = ppm$$
$$ug/kg = ppb$$

Although it would only be true for dilute aqueous solutions at 0°C, it is common practice to consider a mg/l as ppm and a ug/l as ppb.

Lastly, you should be able to convert between percentage and parts per million or parts per billion. Using the same methods we used before, let's find out how many ppm 1% is.

1% is the same as 1 in 100 so we'll start there:

$$\frac{1}{100} \times \frac{10,000}{10,000} = \frac{10,000}{1,000,000}$$

From this we can see that 1% is equivalent to 10,000 ppm. Therefore, air is made up of approximately 21% oxygen; this is the same as saying air is comprised of approximately 210,000 ppm of oxygen.

or intense. An example would be getting a whiff of a chlorine compound while putting bleach in your washing machine. You would probably immediately start coughing from the chlorine's irritation of your breathing passages and lungs.

**Chronic toxicity** is due to long-term exposure and causes effects that appear after months or years of exposure. The Greek root *chron* means "time," hence this is an effect that occurs over time. An example would be regular small doses of ethyl alcohol (found in alcoholic beverages) over a long period of time. These small doses in themselves might cause only minor acute effects (e.g. giddiness) but can lead to damage of the liver from chronic exposure. Most toxic effects do not cause permanent, irreversible damage, but some do. Even in those situations where effects are reversible, full recovery may take a long time. **Cancer**, which is a disease caused by the unregulated overgrowth of cells (tumors), is an example of a potential chronic effect from exposures to some chemicals. A lung disease like emphysema or nervous system damage caused by heavy metals, drugs or alcohol are other examples of chronic health effects. Figure 3-9 shows the basic differences between acute and chronic effects.

| Acute Effect | Chronic Effect |
|---|---|
| Occurs immediately or soon after exposure. | Occurs over time. |
| Often involves a high exposure which results in a large dose over a short period. | Often involves low exposures which result in small doses over a long period. |
| Often reversible (unless death occurs) after exposure stops. | Many effects are not reversible. |
| Can be minor or severe; for example, small amounts of chlorine can cause eye or throat irritation; larger amounts can be serious or even fatal. | Chronic effects are still unknown for many chemicals. For example, most chemicals have not been tested for cancer or reproductive effects. |
| Relationship between chemical exposure and symptoms is generally, although not always, obvious. | It may be difficult to establish the relationship between chemical exposure and illness because of the long delay. |
| Knowledge of these effects is often based on human exposure. | Knowledge of these effects is often based on animal studies. |

**Figure 3-9:** Differences between acute and chronic effects.

## ■ Relative Toxicity

### Concepts

■ The lower the amount of toxicant required to cause death, the more toxic a substance is.

■ $LD_{50}$ and $LC_{50}$ are the most common ways to compare toxicities of different chemicals.

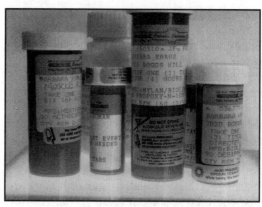

**Figure 3-10:** A dose involves both amount and time. For instance, you may need to take one tablet of decongestant every twelve hours.

■ Danger, warning, and caution are signal words required to describe the toxicity of certain substances such as pesticides.

While all chemicals must be handled carefully, some are so toxic that they require additional precautionary measures for safe use and handling. It is important, therefore, to be able to classify how toxic a substance is. To determine toxicity, laboratory experiments are conducted using various species of animals. Test animals can be exposed to the toxicant orally, dermally, by injection or inhalation. Lethality is the most common effect studied to determine toxicity.

We already learned that dose involves the concepts of quantity and time. This is really nothing new since, typically, when we think of dose we think of taking one decongestant tablet every twelve hours when we have a cold or one multi-vitamin tablet each day. The units of dose for substances administered orally or dermally are typically grams or milligrams of the substance being tested per kilogram of body weight of the test animal per unit of time. For doses administered by inhalation, the dose units are given as parts of the gas or vapor per million parts of air, or as milligrams or grams of airborne material per cubic meter of air per unit of time. For example, if we were conducting an experiment that required feeding rats 1 mg of a toxicant for every kilogram (2.2 pounds) of their body weight every six hours, the dose could be written 1 mg/kg/6 hrs or 4 mg/kg per day.

As the dose of a toxic substance is increased, the harmful effects are generally expected to increase. This is called the **Dose-Response Relationship**. It is a key concept in toxicology that is illustrated in the experimental results depicted in Figure 3-11. These results are loosely based on actual experimental data on the lethal effects of administering tetraethyl lead to rats orally (Sax, 1984). An effect is the biological change noted in the individual. Death, the effect in this experiment, is depicted in the figure by the rats that are "belly-up". A response refers to the percentage of a population demonstrating the harmful effect.

From this experiment we can see that as the dose was increased, the response (numbers affected) also increased. Note that none of the animals died when given a dose of 5 mg/kg of tetraethyl lead and then observed over a 24-hour period. (Oral lethality is usually determined using a 24-hour test and survivors are typically followed for 14 days.) However, 50 percent of them died at a dose of 36 mg/kg, and the entire population died with a dose of 75 mg/kg.

For tetraethyl lead, experimental evidence on a larger number of animals indicates that the lowest dose that causes a lethal effect is around 7.5 mg/kg. This dose is known as the **LOAEL** (**Lowest Observable Adverse Effect Level**). The dose that produces no harmful effect is known as the **NOAEL** (**No Observable Adverse Effect Level**). This means that for this substance the NOAEL is less than 7.5 mg/kg. This information coincides with the results indicated in Figure 3-11 where a dose of 5 mg/kg had no effect. This brings up the important concept of threshold. The threshold level is considered to be the lowest concentration that could produce a harmful effect. This level is different for each chemical. Variability among people exposed to a chemical also has an influence on its effects. Some people are more sensitive and therefore more susceptible to the harmful effects of certain chemicals than others. This concept of threshold is the premise on which standards for workplace exposure are based.

It should be obvious that in a workplace situation, we want no greater exposures than the threshold level. That way, we should suffer no ill effects from working with the chemical. Chapter 9 will explore these workplace standards further. Workplace standards for exposure are derived from experimental data with a **safety factor** added on. This might mean that although experimental results indicate that 1,000 mg/L of a particular chemical in air wouldn't cause any adverse effect, using a 1,000-fold safety factor would reduce the allowable concentration to 1 mg/L.

Lethal Dose 50 ($LD_{50}$) is the most common way to report dose-response. The "50" stands for 50 percent of the test population. This means that it is the dose resulting in the death of 50 percent of the test animals in a given time. Figure 3-11 indicates that the $LD_{50}$ for tetraethyl lead is 36 mg/kg. $LD_{50}$ is used with dermal and oral toxicity. Sometimes the words "dermal" or "oral" will be found next

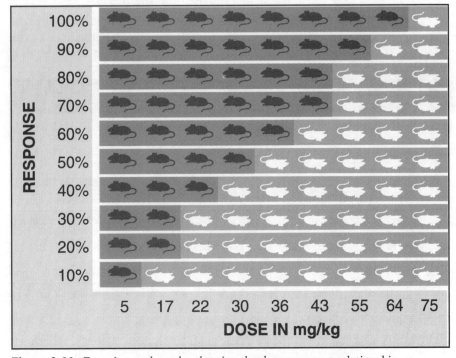

**Figure 3-11:** Experimental results showing the dose-response relationship.

## Lethal Dose 50 of Various Venoms Injected into Mice

| | |
|---|---|
| Australian Death Adder Snake Venom | 338 ug/kg |
| Costa Rican Snake, Agkistrodon Bilineatus Bilineatus Venom | 1322 ug/kg |
| Naja Naja Cobragift Venom (Sea Snake) | 180 ug/kg |
| Midget Faded Rattlesnake Venom | 3200 ug/kg |
| Scorpion Venom | 880 ug/kg |
| Lactrodectus Spider Venom (same genus as Black Widow) | 8500 ug/kg |

## Rat Oral Lethal Dose 50s for Various Common Substances

| | |
|---|---|
| Aflatoxin $B_1$ | 7 mg/kg |
| Aspirin | 1,000 mg/kg |
| Caffeine | 192 mg/kg |
| DDT | 113 mg/kg |
| Ethyl Alcohol | 14,000 mg/kg |
| Nicotine | 53 mg/kg |
| Sodium Chloride (common salt) | 3,000 mg/kg |

**Figure 3-12:** $LD_{50}$ Values. (Source: Sax, 1984 and EPA Training Manual)

to the subscripted 50 to identify the route of exposure (e.g. $LD_{50}$ dermal). The equivalent of $LD_{50}$ for toxicity from inhalation is $LC_{50}$ which stands for **Lethal Concentration 50 percent**.

Lethal dose and lethal concentration values are used extensively to compare the toxicity of various chemicals. Since the experimental tests used to determine these values have specific times associated with them, these doses are presented in shorthand without indicating time. Figure 3-12 shows the $LD_{50}$'s for several common and not so common toxic substances.

Dose-response can be reported in a variety of ways other than $LD_{50}$ and $LC_{50}$. Some of these tests involve looking for effects other than death. They can include $TC_{50}$ which is **Toxic Concentration 50 percent**. This gives us the concentration of inhaled air needed to produce an observed toxic effect in 50 percent of the test animals in a given time period. $TD_{50}$ is the **Toxic Dosage 50 percent**. This is the dosage by any route other than inhalation that produces an observed toxic effect in 50 percent of the test animals in a given time period.

**Pesticides**, chemicals used to kill pests such as insects, weeds, and other toxic substances, are required by law to be ranked and labeled according to their toxicity. The EPA requires the signal words Danger, Warning, and Caution to be used on containers of pesticides. Lethal doses determine which label is needed. Figure 3-13 shows the Lethal Dose ranges for the four pesticide categories. Remember that the smaller the amount of toxicant per kilogram of body weight, the more toxic the substance is.

Given the $LD_{50}$ numbers on this chart, a substance with the signal word "Danger" would only require a dose ranging from a taste to a teaspoonful to kill an adult weighing 150 pounds. A moderately

| Toxicity Category | | | | |
|---|---|---|---|---|
| Hazard | I | II | III | IV |
| Oral Lethal Dose 50 | ≤50 mg/kg Danger/Poison | 50–500 mg/kg | 500–5,000 mg/kg | > 5,000 mg/kg |
| Dermal Lethal Dose 50 | ≤0.2 g/kg | 0.2–2 g/kg | 2–20 g/kg | > 20 g/kg |
| Inhalation Lethal Concentration 50 | ≤0.2 mg/l | 0.2–2.0 mg/l | 2–20 g/l | > 20 g/l |
| Eye Irritation | Corrosive; Corneal opacity persists > 21 days | Corneal opacity persists < 21 days; irritation persists ≥ 7 days | No corneal opacity; irritation persists < 7 days | None |
| Skin Irritation (Draize Score | Corrosive (5.0) | Severe @ 72 hrs. (3.0–4.9) | Moderate @ 72 hrs. (2.0–2.9) | Mild @ 72 hrs. (1.0–1.9) |
| Signal Word | Danger (Highly Toxic) | Warning (Moderately Toxic) | Caution (Low-Order Toxicity) | Caution (Low Toxicity) |

**Figure 3-13:** Toxic ratings for pesticides. (Source: EPA Pesticide Regulations)

toxic substance with a "Warning" signal would require a teaspoon to a tablespoon to have the same result. Substances labeled with "Caution" would require an ounce to more than a pint. Figure 3-14 shows a broader scheme for looking at relative toxicity.

Figures 3-13 and 3-14 are helpful for comparing the toxicities of chemicals. It is important to remember, however, that these only assess acute lethal effects. Lethal dose values tell us nothing about

| Hodge-Sterner Toxicity Table | | |
|---|---|---|
| Experimental LD$_{50}$ Dose | Degree of Toxicity | Probable Lethal Dose for a 70 kg (154 lb.) Man |
| < 1.0 mg/kg | Dangerously Toxic | A Taste (a drop) |
| 1 – 50 mg/kg | Seriously Toxic | A Teaspoonful (5 ml) |
| 50 – 500 mg/kg | Highly Toxic | An Ounce (30 ml) |
| 0.5 – 5 gm/kg | Moderately Toxic | A Pint (500 ml) |
| 5 – 15 gm/kg | Slightly Toxic | A Quart (1 liter) |
| > 15 gm/kg | Extremely Low Toxicity | More than a Quart |

**Figure 3-14:** Commonly used toxicity rating system. (Source: Thomas J. Haley contributed chapter in Irving Sax, *Dangerous Properties of Industrial Materials*, Sixth Edition, 1984)

a toxicant's potential for harmful effects other than death, or for chronic or delayed effects that may ultimately result in death.

The following section will describe some of the other harmful effects of toxicants by exploring how the function of various systems in the body can be damaged. Some systems in the body can detoxify chemicals which means to change them chemically into a less toxic form which the body can eliminate. If chemical doses are received at a rate faster than our bodies can detoxify and eliminate them, then they have a greater chance to accumulate and do damage. Some chemicals, like ammonia and formaldehyde, leave the body quickly and do not accumulate at all. Other chemicals, such as lead or polychlorinated biphenyls (PCBs), are stored in the body for long periods of time. Storage sites can vary from bone, liver and kidneys, to fat. Some substances, like asbestos fibers, are virtually impossible for the body to eliminate.

## ■ Checking Your Understanding (3-1)

1. Based on the definitions given in this section, give two reasons why a corrosive burn would not be considered a toxic effect.

2. List three factors that would affect what the actual dose would be to a person who was exposed to an air contaminant.

3. Describe the main routes of exposure and give one example of an exposure (or potential exposure) to a toxic substance you have had for each route.

4. Explain how an exposure is prevented by breaking any of the three links of the exposure chain.

5. List three ways chronic and acute exposures differ.

6. Give an example of a chronic exposure and an acute exposure you have experienced in your life.

7. What is the difference between LOAEL and NOAEL?

8. When is $LC_{50}$ used instead of $LD_{50}$?

9. Which of the venoms listed in Figure 3-12 is most toxic?

10. Rank the toxicity of the common substances listed in Figure 3-12 using the Hodge-Sterner Toxicity Table.

11. What signal word indicates the highest toxicity on pesticide containers?

12. Discuss the limitations of using $LD_{50}$ alone to decide how harmful a chemical might be.

# 3-2 Toxic Effects

## ■ Types of Effects

### Concepts

■ Local and systemic are terms used to describe where damage occurs.

■ Latency period, reactions, and interactions complicate predicting the effects of toxins.

■ Sensitizers can cause allergic reactions.

■ The three basic physical states are solid, liquid, gas.

If a toxicant causes damage at the point where it first comes into contact with the body, it is considered to have produced a **local effect**. An example would be inhaling a chemical that irritates the nasal and breathing passages causing irritation, soreness and reddening of the passageway. Some of the common points of initial contact are: eyes, nose, throat, lungs, and skin. Toxicants also enter the body and travel in the bloodstream to reach vital organs such as the liver, kidneys, heart, or critical systems such as the nervous system (including the brain), and reproductive systems. Damage caused in this way is considered a **systemic effect**.

A toxicant can cause either local or systemic effects, or it can cause both. Consider ammonia gas. If you were to quickly inhale this gas, it would irritate your nose, throat, and lungs. Virtually no ammonia would pass from the lungs to the bloodstream and we would receive only a local effect (don't consider this less serious). An organic solvent like benzene, however, would, if spilled on skin, irritate the skin and also be absorbed through the skin to cause internal systemic damage. It is important to note that local effects are usually apparent and provide a warning that exposure has occurred. However, not all chemical exposures result in a local effect. Systemic effects, on the other hand, may occur without being felt or sensed.

The longer a person is exposed to a chemical, the more likely he or she is to be affected by it. The dose is also important in determining the likelihood of an effect. Chronic exposures, however, are particularly dangerous because some chemicals build up in the body in a process called **accumulation**, and because this type of exposure does not give the body a chance to repair itself.

Some effects take a long time to manifest themselves after the initial damaging exposure. The term **latency period** is used to describe this delay between the beginning of exposure and the resultant harmful effect(s). Chronic effects such as cancer have very long latency periods. For some chemicals the latency period can be 30 to 40 years (see Figure 3-15). This makes determining the cause of an illness especially difficult.

Predicting the effects of a single toxicant is often difficult since we are all exposed to multiple chemicals. Chemicals can combine with toxicants and alter their behavior. This is called an **interaction**. Another process called a **reaction** can occur when chemicals combine and form a new substance. This new substance may be even more toxic than either of the original chemicals. A good example is the mixing of household bleach and drain cleaner which react and result in two new chemicals: highly toxic chlorine gas and hydrochloric acid.

Unless there is data to the contrary, most health and safety regulations assume an **additive effect** in the interactions of chemicals. This means that the combined effect of two or more chemicals is equal to the sum of the effect of each agent alone. The op-

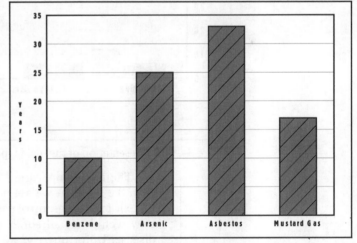

**Figure 3-15:** Average number of years after exposure for cancer first to appear. (Source: Levy and Wegman, 1983)

posite of this would be **antagonism** which is subtractive; one substance interferes with the action of another. An example of this would be the combined ingestion of methanol (wood alcohol) and ethanol (grain alcohol). Ethanol reduces the damage that would have been caused by the ingestion of methanol alone.

**Potentiation** is another type of interaction. In this case, one substance increases the effect of another substance causing an effect that it would not have had by itself. An example is acetone, which does not damage liver by itself but can increase the ability of carbon tetrachloride to do so.

Chemicals can also interact within the body to result in a health effect that is different from the effects of either chemical alone. **Synergism** is an interaction of two or more chemicals where their total effect is greater than the sum of their individual effects. An example is exposure to both asbestos and to one pack of cigarettes a day. It is estimated that each one of these exposures increases the risk of lung cancer about six times as compared to a person who does not receive the exposure. An exposure to both, however, may increase one's risk of lung cancer by 90 times! Unfortunately, there is not a lot of information to be found on the synergistic and other interactive effects of combinations of chemicals like this. See Figure 3-16 for another way to look at these interactions.

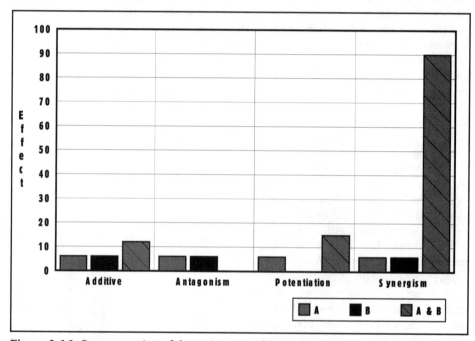

**Figure 3-16:** Representation of the main types of interactions between chemicals.

The last complication in predicting harmful effects has to do with the fact that we are not alike in our sensitivity to the damage of toxicants. If a group of people were all exposed to a toxic chemical, they would not all suffer the same harm. Not all would develop disease. Unfortunately, we can't know which one of us is more or less sensitive to a particular toxicant. Various factors play a role in how vulnerable we are. These include age, sex, inherited traits, our diet, state of health, use of medication, drugs, alcohol, and pregnancy. These factors determine whether a person will experience the harmful effects of a chemical at a lower or higher dose than other people. Figure 3-17 provides an example involving asbestos workers. It shows the percentage of workers exposed to asbestos that die of the three major asbestos-related diseases. Note that not all asbestos workers get these diseases.

Another interesting difference is that some people become allergic to certain chemicals. A very low dose often triggers an allergic response in much the same way that the small amount of venom from bee stings does to those who are allergic to it. Not all chemicals are prone to initiate allergic responses; those that do are called **sensitizers** and should be handled with special caution. An example of a strong sensitizer is toluene diisocyanate, a chemical used to make molded foam products such as dashboards. Some workers handling it became so sensitized that the small amounts of it released from dashboards inside of new vehicles could cause them to have severe allergic reactions.

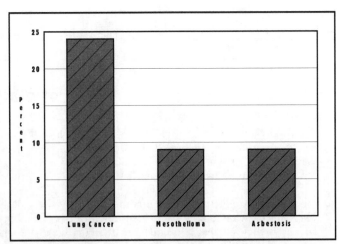

**Figure 3-17:** Percentage of asbestos workers that die of three major asbestos-related diseases.

Some chemicals are considered mild sensitizers and can cause an allergic reaction in certain people. Formaldehyde is strongly irritating and causes tearing of the eyes and a sore throat in anyone exposed to high enough levels. It also occasionally results in sensitization of an individual, which can result in an allergic reaction such as dermatitis at low levels. Most people, however, will never become sensitized to this chemical.

There are many ways that poisons work but they all involve changing how our body functions. One specific organ system can be affected or a number of systems can be targeted. Poisons can speed up or slow down various vital activities such as breathing or the beating of the heart, sometimes stopping these functions completely. Figure 3-18 shows the main body systems and the common symptoms of exposure to toxicants for each.

Historically, the major regulatory emphasis has been directed toward the potential for toxic substances to cause cancer. Now there is a new awareness that other toxic effects may be as widespread and severe, therefore needing to be regulated as well. The rest of this section surveys the body systems that are affected by toxic chemicals and provides information on non-cancer toxic effects.

| System | Affected Organs/Body Parts | Common Symptoms |
|---|---|---|
| Respiratory | Nose, trachea, lungs | Irritation, coughing, choking, tight chest |
| Gastrointestinal | Stomach, intestines | Nausea, vomiting, diarrhea |
| Renal | Kidney | Back pain, change in amount of urination, discolored urine |
| Neurological | Brain, spinal chord, behavior | Headache, dizziness, confusion, depression, coma, convulsions |
| Hematological | Blood | Tiredness, weakness |
| Dermatological | Skin, eyes | Rashes, itching, redness, swelling |
| Reproductive | Ovary, testes, fetus | Infertility, miscarriage, birth defects |

**Figure 3-18:** Body systems affected by toxic agents. (Division of Agricultural Sciences, University of California, *Toxicology: The Science of Poisons*, 1981)

# Science & Technology

## States of Matter

The three main forms, or **physical states** in which we find substances are: **solid, liquid**, and **gas**. Within each of these states additional descriptive terms are commonly used to classify substances.

The physical state of a toxicant is important to predict its most likely route of entry into the body. The same toxicant may be found in different forms and each may present a unique hazard. For instance, lead solder as it is purchased in its solid form is not likely to present a toxic hazard. It has no ready means to enter the body. However, heating it up to melt it for soldering turns it into a liquid which can be spilled onto the skin. If heated to very high temperatures, small particles called fumes are released into the air and are dangerous if inhaled.

■ Solid – An example is a stone. Most solids, unless very small, are not likely to be absorbed into the body. Some types of solids that can enter the body include:

– *Dust* – small solid particles dispersed in the air. Dusts can be created through grinding and other processes that pulverize solids. Inhaling larger particles of dusts generally causes no problem because our bodies can expel them. Smaller particles, however, are a real concern. Large concentrations of organic dusts in the air can even be explosive. An example is dust from grain which has been the cause of numerous silo explosions.

– *Fume* – small, fine solid particles in air formed when solid materials such as metals are heated to very high temperatures causing them to evaporate, then become solid again as they cool in the air. These very small particulates are easily inhaled and can be toxic.

– *Fiber* – a solid particle that has a length at least three times its width. The size and configuration of the fiber determines the degree of hazard it poses. Smaller fibers that are barbed, such as asbestos, can lodge in the sensitive lung tissue and cause serious damage, including cancer. Larger fibers are more readily trapped in the respiratory tract where they can be expelled before reaching the lungs.

■ Liquid – a material that does not retain its form but takes on the form of its container and has a specific volume. Many hazardous substances are found as liquids at room temperature (considered 68-70° F). Some liquids can cause local effects if poured or splashed on skin. Some pass through the skin to enter the bloodstream and target various body systems. Liquids evaporate, forming vapors, which can be inhaled.

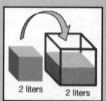

– *Mist* – liquid particles of varying sizes that are produced by the agitation or spraying of liquids. Mists can be inhaled or absorbed by the skin. Pesticide application is an example of a process that produces mist. Other industrial processes such as painting also result in the production of mists.

■ Gas – a material made up of atoms or molecules that move about freely; gases have no specific shape or volume. Some gases like acetylene are flammable; others are explosive, and/or toxic. Gases are not always easily detectable because they do not always cause local effects. Also they may have no color or odor. An example is carbon monoxide, an asphyxiant which causes uncon-

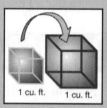

sciousness from lack of oxygen in the blood.

– *Vapor* – the gas form of a substance that is found primarily as a liquid at room temperature and pressure. Many chemicals found in the workplace evaporate readily, giving off vapors. Vapors can be irritating, causing local effects to the respiratory system and eyes. They can also enter the bloodstream through inhalation and then move to other parts of the body.

# ■ The Respiratory System

## Concepts

■ Breathing provides for necessary gas exchange.

■ Inhalation of toxicants can result in quick absorption into the bloodstream or lodging in the respiratory system.

■ The respiratory system's natural defenses are not sufficient to prevent harm by many toxic substances.

The respiratory and cardiovascular systems are responsible for transferring and transporting oxygen into and throughout the body. The air we breathe contains the oxygen needed to sustain life. Without oxygen a person can only live for a few minutes. Contrast this with water, which we can live without for days, and food, which we can live without for weeks. Every time we breathe, an exchange takes place. We take oxygen into our body and we expel a major waste product, carbon dioxide. In order to supply the oxygen needs of our body, we take in approximately 10,000 to 20,000 liters of air each day. Oxygen from the air passes into the bloodstream where it is supplied to the cells of all body tissues. If this oxygen transfer does not occur continuously, or if its effectiveness is altered, serious harm results.

Air is a mixture of gases. Its composition at sea level is approximately 78% nitrogen, 21% oxygen, and 1% trace gases including carbon dioxide, water vapor, and rare gases like argon. Unfortunately, the air we breathe can also contain contaminants which may be toxic. Examples are gases or vapors from the evaporation of liquids such as gasoline or formaldehyde. Breathing in chemicals can cause a variety of harmful effects, both local and systemic.

Inhalation is the most critical route of entry for most workers handling toxic chemicals because of: 1) the ability of gases and vapors to be taken quickly into the lungs and be absorbed into the bloodstream, and 2) the ability of some toxic agents to accumulate in the respiratory system itself. It is important to understand the gas exchange process and the physiology of the respiratory system so we can better understand how harmful effects can occur from inhaling toxic agents.

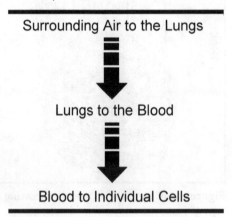

Surrounding Air to the Lungs

Lungs to the Blood

Blood to Individual Cells

**Figure 3-19:** The three levels of gas exchange involved in supplying the body with oxygen.

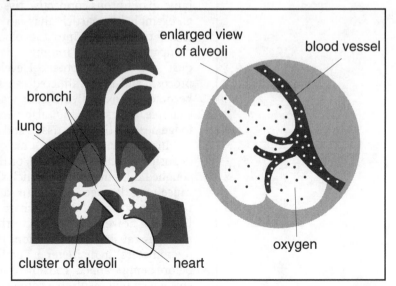

bronchi

lung

enlarged view of alveoli

blood vessel

cluster of alveoli

heart

oxygen

**Figure 3-20:** The respiratory system. Air enters the breathing passages through the nose and moves to the lungs. When it reaches the alveoli, it

The air we breathe enters the body through the nose and the mouth and is warmed and humidified before entering the lungs. After moving through many branches of the airways called bronchi, the air finally reaches the area where gas exchange takes place. This final destination in the lungs is comprised of about 300 million tiny air sacs called alveoli. Amazingly, if we were to lay out all the air sacs in an adult, they would cover a surface about the size of a singles tennis court! Oxygen and other chemicals that may be in the air are transferred into the bloodstream from the alveoli.

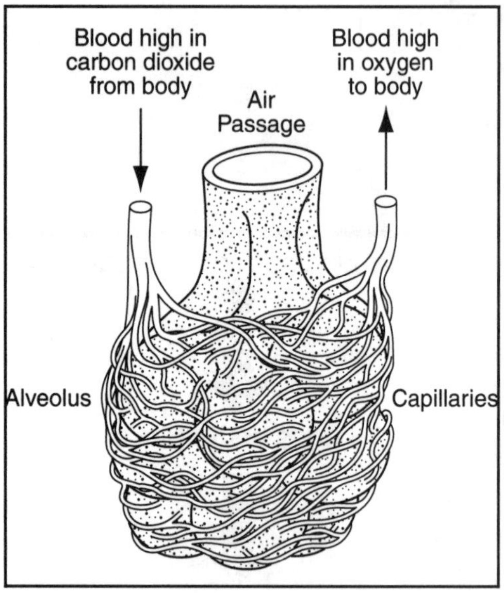

The large numbers of alveoli and their thin walls make oxygen transfer rapid and easy. Thin membraned capillaries bring blood in contact with the air sacs whose walls are only one cell in thickness. This is approximately one-millionth of an inch thick. These very thin membranes allow gases to pass back and forth through them with ease. This enables oxygen from the air we breathe to readily enter the bloodstream where it is carried to all parts of the body by the hemoglobin in the red blood cells. It also allows carbon dioxide to move from the bloodstream to the lungs to be exhaled. Breathing, then, is the process of taking in air and exhaling carbon dioxide.

**Figure 3-21:** Detail of an alveolus showing how oxygen and carbon dioxide are

The defense mechanisms used by the respiratory system to protect itself from injury include the ability to filter large particles and absorb gases before they reach the lungs. Trapping foreign particles is the job of the mucus. Cilia, which are whip-like hair cells lining the upper respiratory tract, sweep mucus and captured particles upwards to the esophagus where they are swallowed. Extra mucus is secreted if there is serious irritation or inflammation of the air passages.

Most particulates finding their way into the lungs are attacked and digested by specialized scavenger cells. Unfortunately, these defenses are not sufficient for all assaults to the respiratory system. Some particles cause an abnormal buildup of fibrous connective tissue which is called fibrosis. Silica is an example of a particle that can cause fibrosis, which hampers the transfer of oxygen into the blood. This is a serious problem in occupations such as mining, quarrying, and pottery glazing. Coal dust also causes fibrosis. Development of fibrosis is a gradual process, and those afflicted can suffer many years before damage becomes so extensive that it leads to death. Certain fibers – for example, asbestos – can also penetrate the available defenses and lodge deep in the lung tissue and cause permanent scars.

Inhaling dusts or mists may result in harmful particles being deposited in the bronchi and/or the alveoli. Larger particles may be coughed up, but smaller particles tend to stay in the lungs and can cause lung damage. Only gases and very small particles are able to reach the part of the lung where gas exchange occurs. Larger particulates, however, can irritate the lining of the upper respiratory passages causing inflammation. Particles less than 10 microns (10 micrometers) in diameter generally penetrate further into the lungs before being trapped, causing bronchitis (inflammation of the bronchi) which will result in a soreness of the chest and coughing. Some small particles can even reach the alveoli. Common contaminants such as carbon particles from smoke, vehicle exhaust, and tobacco smoke have various sized particles, some of which can find their way deep into the lungs.

Other low-level, long-term exposures can result in scarring of the lung or cause chronic bronchitis or emphysema. Dust particles can be the cause of some chronic lung injuries. Several outdoor air pollutants such as sulfur dioxide make breathing difficult for people whose respiratory systems have been compromised or weakened by other injuries or illnesses such as asthma. Tobacco smoke is known to cause cancer in humans and also increases the incidence of chronic bronchitis and emphysema (Figure 3-22). Recent evidence implicates second-hand smoke in increased chronic lung problems for family members of smokers.

Some common gases that penetrate alveoli include sulfur dioxide and nitrogen oxides. Both are emitted as products of combustion in processes such as burning fuel. These compounds react with the water vapor in our lungs causing the formation of acids. Ammonia and chlorine gas can also damage sensitive lung tissue. These gases dissolve in the mucus of the lungs and produce caustic solutions. These corrosive materials can easily damage fragile air sacs. Injured lung tissue allows liquids to move from capillaries into the alveoli resulting in pulmonary edema, which is the filling of the alveoli with water. This is an extremely dangerous situation since fluid-filled air sacs are unable to perform the vital oxygen transfer they are designed to provide. A person with extensive lung damage can literally drown in his or her own fluids.

Obviously, the potential effects from inhaling toxic chemicals vary greatly. Some chemicals, when breathed in, cause minor nose or throat irritation. Some exposures can result in chemical bronchitis leading to discomfort, coughing, or chest pain. With other exposures, there may be no obvious warning symptoms, but they can quickly cause death. Our sense of smell does not provide a reliable warning of exposure. Some toxic chemicals have no odor and cannot be detected at any concentration. Odorless carbon monoxide (CO), for example, has caused many deaths in the home and workplace. Other chemicals cannot be detected until highly dangerous quantities are present. For example, if you smell phosgene gas, you're already in serious danger. Our sense of smell can become desensitized to some smells. This means that we may no longer be able to smell a chemical we were initially able to smell; dangerous hydrogen sulfide ($H_2S$) is an example (Figure 3-23).

Figure 3-22: Lung with emphysema.

| Can be smelled at | 0.1–5.0 ppm |
|---|---|
| Eye irritation at | 10 ppm |
| Headaches at | 15 ppm |
| Convulsions/death at | 300 ppm |

Figure 3-23: Concentrations of $H_2S$ cause various effects.

# ■ The Cardiovascular System

## Concepts

■ The cardiovascular system carries needed substances to cells and carries away waste products.

■ Some chemicals interfere with hemoglobin's ability to carry large amounts of oxygen.

The cardiovascular system, comprised of the heart and blood vessels, is responsible for delivering oxygen and nutrients to all parts of the body. The heart and brain are particularly sensitive to lack of oxygen. They must receive adequate amounts of oxygenated blood in order to carry out their vital functions. Lack of blood to these organs will result in death in minutes. The average adult body contains about six quarts (5.7 liters) of blood. Every day, the heart pumps nearly 2,000 gallons (7,568 liters) of blood through almost 60,000 miles (100,000 km) of tubing in our bodies. This activity is automatically controlled by the central nervous system.

The heart is a muscle that pumps blood throughout the body through two separate circuits. The right side of the heart pumps blood directly into the lungs. In the lungs the blood picks up a new supply of oxygen and releases the carbon dioxide it picked up from body tissues. It then travels to the left side of the heart and from there is pumped to the brain, skin, intestines, liver and the other organs of the body.

The blood vessels that circulate blood and gases throughout the body are branching tubes that get smaller and smaller as they get closer to the areas where transfer occurs. The smallest blood vessels are called capillaries. They have very thin walls that allow oxygen, nutrients, and other substances in the blood to easily and quickly enter the tissues. They also pick up waste products and transport them to the lungs or kidneys for removal from the body.

Blood is a complex solution that transports all the substances that the body's organs and tissues use and eliminate. The liquid part of blood, called plasma, is composed of a variety of cells and substances. Blood transports proteins, vitamins, fats, minerals, and hormones which are necessary for the proper growth and functioning of body tissues. It also transports the wastes produced by cells, which are then filtered out when the blood passes through the kidneys. In addition, blood carries white blood cells that fight against infections, red blood cells that carry oxygen, and platelets that initiate the blood-clotting process.

The red blood cells are filled with an iron-containing protein called hemoglobin. As blood passes through the lungs, hemoglobin picks up large quantities of oxygen, carries it to the rest of the body, and releases it where it is needed. Some chemicals, however, interfere with the ability of hemoglobin to pick up oxygen. Carbon monoxide (CO) is a good example. It is probably the most insidious and dangerous of the common gases that can harm us. Hemoglobin has a much greater affinity for the CO than it does for oxygen, about 300 times more. Therefore, even if there is sufficient oxygen in the

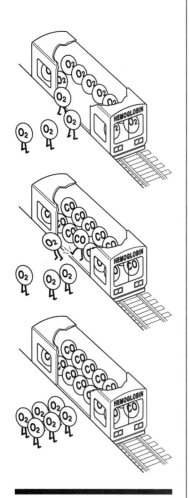

**Figure 3-24:** Carbon monoxide interference with oxygen transport.

air we breathe, hemoglobin will elect to transport CO instead. As the blood travels through the lungs, hemoglobin will pick up any CO that we have inhaled rather than oxygen that is also present. When this happens, vital organs are starved for oxygen. This condition is called chemical asphyxiation. If we displace about half of the oxygen in our blood, death will result. Other chemicals can cause red blood cells to rupture. When a cell ruptures, hemoglobin – a large molecule – is released into the bloodstream and damages the kidneys.

## ■ The Digestion and Filtration Systems

### Concepts

- When we ingest food or other substances, most absorption into the bloodstream takes place in the small intestine.

- Relatively little absorption takes place in the stomach.

- The liver functions as the chemical processing plant of the body.

- The kidneys maintain proper balance of body fluids and filter out waste products.

In addition to oxygen, all living things require food and water to survive. Food provides the raw materials for maintenance and repair and for the supply of energy. At various stages of digestion, enzymes break down food into simpler molecules that our body can use for building materials and energy. Some molecules like water and glucose require no digestion. Most larger molecules like proteins, complex carbohydrates, and fats undergo considerable change in the digestion process. They need to be broken down into smaller molecules so that they can pass through the lining of the intestine and into the blood or lymph systems for transportation throughout the body.

The stomach produces enzymes and other chemicals that cause the breakdown of food. One of the chemicals it produces in large amounts is hydrochloric acid, which activates enzymes needed to digest protein and destroy bacteria. Unfortunately, if too much of this acid is produced, it can result in health problems like ulcers. The stomach contributes very little to the absorption of foods into the blood. Alcohol, however, is one chemical that does pass through the stomach wall, quickly entering the bloodstream. This explains why the effects of intoxication are observed so soon after drinking.

About 90 percent of the digestion of food and absorption of nutrients into the blood take place in the small intestine. The secretions released in the small intestine, in contrast to the stomach, contain bicarbonate which is basic or alkaline. These secretions neutralize the acid from the stomach which prevents damage to sensitive tissues. The wall of the small intestine is very thin with lots of folds that provide for a large inner surface area through which nutrients are absorbed into the bloodstream. In an average adult, the total surface area of the small intestine is over 200 square feet

(18.6 m$^2$). Breakdown products of foods and other substances ingested travel into tiny capillaries that eventually carry the products to the liver.

The liver is a large organ that processes chemicals found in the blood. Blood traveling to the liver from the intestines contains the absorbed chemicals from the digestion of food. The liver converts foods into other chemicals, destroys toxins, manufactures protein, and stores glucose. This makes it the most complex chemical factory of the body. When the liver is not functioning, these necessary chemical processes cease. Severe liver disease may result in tremors, confusion or coma due to toxic chemicals reaching the brain. These toxic chemicals may be normal body chemicals that a properly functioning liver renders harmless by breaking down or combining with other chemicals. When the liver is severely damaged, harmful chemicals are not broken down and can build up in the body.

Some chemicals tend to be stored in the liver. Polar bears provide an interesting example since their livers store large amounts of vitamin A from the fish they eat. There are reports of humans being poisoned from eating polar bear livers (refer to Figure 3-12 for the LD$_{50}$ of vitamin A). The liver also has the ability to destroy certain poisons like nicotine and alcohol. Although the liver is remarkable in its ability to detoxify substances like alcohol, repeated use of a substance over a period of time can cause damage to the hard-working liver cells. When liver cells die they are often replaced by fibrous tissue. This condition is called cirrhosis. Cirrhosis of the liver is often associated with death due to liver failure.

In the last section we learned that the bloodstream carries various waste products as well as the substances necessary for life. The components of our body fluids must be maintained within certain narrow limits. The chemicals that need to be balanced include potassium, sodium, chloride and calcium ions, and blood acids. About 60 percent of our body weight is water, and about half of this is contained in cells. The rest of this water bathes the cells of the body in a weak saline solution. The job of the kidneys is to maintain this critical chemical balance of our body fluids. They also serve as a complicated filtering system that eliminates waste products from the bloodstream.

Filtering occurs as blood circulates through millions of microscopic filter tufts in the kidneys. This filtration separates water, waste products, glucose, and salts from the cells and proteins in the blood. The glucose is almost always totally reabsorbed. The amount of water and salt reabsorbed varies depending on the specific needs of the moment. If you drink a large quantity of liquid you will lose much of this through the kidneys if it is not needed to replace water lost through perspiration or other mechanisms. Waste products are not reabsorbed and leave the body with some water and salt as urine. We must drink sufficient water in order for urine to be made in the amounts necessary to maintain proper chemical balances. When body water loss reaches over 10 percent of body weight, cells will no longer function correctly and death is likely to follow. This can occur in a few days. We excrete many of the body's unwanted substances in urine. This is why urine samples are collected for drug testing.

Kidney malfunction can result from toxic chemicals in the bloodstream. This in turn may cause an accumulation of the body's waste which can result in coma and death. In addition to disease of the kidney caused by chemicals, kidney cancers are known to be associated with exposure to some industrial chemicals. Minimata Disease, the result of the poisoning of Japan's Minimata Bay by industrial pollution, is a tragic example of chemicals altering the critical functions of the kidney. Waste dumped into the bay contained large concentrations of mercury. Bacteria in the water converted the mercury into methyl mercury. Methyl mercury accumulated in the fish eaten by the people of the fishing village. This poison directly affects the ability of the kidney to balance the body's chemicals which then blocks proper nerve transmission. Children of mothers that had eaten the contaminated fish were born with serious physical abnormalities.

## ■ The Nervous System and Sensory Organs

### Concepts

- The nervous system controls all body functions.

- The delicate tissue of the eyes can easily be damaged by chemicals.

- Breathing levels of oxygen below 16% is very dangerous.

Our bodies are remarkably complex, requiring a multitude of functions to take place simultaneously in order to survive. We have to keep breathing, which means our hearts must keep pumping. Our other organs must also keep functioning in the proper manner to keep the body alive and well. In addition to performing these mechanical functions, we think, have emotions, and respond physically to various conditions. We are not even consciously aware of many of these processes. All of these complicated activities that your mind and body carry out are controlled by the nervous system.

The control center for all body functions is called the central nervous system and is made up of the brain and spinal cord. Nerves that leave and return to the central nervous system, called the peripheral nervous system, extend to every part of the body. These nerves are of two types; motor (movement) and sensory nerves. The autonomic nervous system takes care of all necessary body functions that seem to be "in the background," occurring without our conscious awareness.

Nerve functions can be altered or destroyed by some industrial chemicals. Pesticides and metals such as mercury and lead can interfere with the chemical transfer of information necessary for an impulse to be sent through the nervous system. This may result in fewer or no impulses passing through, which can cause paralysis, tremors, loss of reflexes, and loss of feeling. The Mad Hatter in Lewis Carroll's *Alice in Wonderland* is a humorous portrayal of a serious occupational disease of that time. Many hat makers suffered from

**Figure 3-25:** The Mad Hatter from Lewis Carroll's *Alice in Wonderland*.

mental derangement and nervous system disorders brought about by their exposure to mercury which they used to shape felt hats.

The brain is a most delicate and vital part of the body. It must have a constant supply of oxygen. Warnings of insufficient oxygen to the brain should always be heeded! Figure 3-26 shows the effects of various low levels of oxygen.

| Physiological Effect of Oxygen Deficiency | | |
|---|---|---|
| | 21 – 16 | Nothing abnormal |
| | 16 – 12 | Loss of peripheral vision, increased breathing volume, accelerated heartbeat, impaired attention and thinking, impaired coordination. |
| % Oxygen by Volume | 12 – 10 | Very faulty judgment, very poor muscular coordination, muscular exertion causes fatigue that may cause permanent heart damage, intermittent respiration. |
| | 10 – 6 | Nausea, vomiting, inability to perform vigorous movement or loss of all movement, unconsciousness, followed by death |
| | <6 | Spasmodic breathing, convulsive movements, death in minutes. |

**Figure 3-26:** Physiological effects of oxygen deficiency.

The senses are the windows of the nervous system. They provide the information about the world around us that is processed by the nervous system. The eyes are very sensitive tools of perception. The eye itself is surrounded by fat, blood vessels, and motor and sensory nerves. The eyelids provide protection and keep the eye lubricated. Inflammation and infection of the mucous-membrane lining of the eyelids and eyeball can be caused by irritation from chemical pollutants and smoking. Unfortunately, eye injuries from chemicals are common, yet most are easily preventable.

Acids and bases are corrosive and can easily cause serious tissue damage in the eyes. Bases are particularly dangerous because they can penetrate to the interior of the eye very rapidly, usually within seconds or minutes. Lime, which is used in wall plaster, creates a serious problem for construction workers since it can easily get into their eyes. Many do not realize that it is alkaline enough to cause blindness if it is not quickly and thoroughly removed from the eye. Acids burn the eye but do not penetrate as easily as bases. Nonetheless, they can cause permanent damage. Many other chemicals can cause serious damage to the eye. Some cause significant visual loss. Methanol, or wood alcohol, can cause total blindness from damage to the optic nerve.

The skin is the largest organ of the body with a surface area of approximately 19 square feet. Three layers of tissue make up the skin: the epidermis, the dermis, and the subcutaneous layer (Figure 3-27). The dermis, sometimes

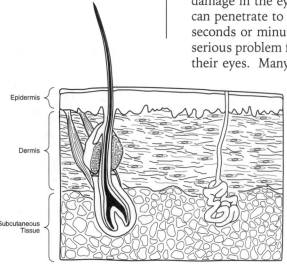

Epidermis

Dermis

Subcutaneous Tissue

**Figure 3-27:** Layers of Skin.

called the "live skin," is just beneath the epidermis and on top of the subcutaneous. This layer contains blood vessels, nerves, nerve receptors, hair follicles, sweat glands, and oil glands.

The skin performs a number of important jobs. It protects the body against invasion by bacteria, injury to more sensitive tissues within the body, rays of the sun, and loss of moisture. The skin is also an organ of perception for the nervous system. It senses pressure, pain, and temperature and regulates body temperature through the blood vessels and by producing moisture from the sweat glands. It provides the first line of defense against industrial hazards of every type. In spite of its many protective properties, the skin can be injured by exposure to many chemicals. Corrosive chemicals can dissolve naturally protective coatings and/or react with the skin. Some chemicals, particularly those that can dissolve fats, like solvents, are absorbed directly into the bloodstream through the skin.

## ■ Checking Your Understanding (3-2)

1. Give one example each of a local and a systemic effect that you have personally experienced or have read about and explain why you categorized them as you did.

2. In medicine, *incubation period* is the time between the disease-causing microbe first entering the body and when the signs or symptoms of disease first appear. Make a table showing where this concept is the same as latency period and where it differs.

3. Explain the difference between an additive effect, synergism, potentiation, and antagonism.

4. Classify the following into the three physical states of matter: acetone vapors, sawdust, fumes from welding, and fine mist from a sprinkling system.

5. Describe the effect a strong sensitizer may have on an employee handling it.

6. Explain what breathing is.

7. List the concentration of the three main components of air in ppm.

8. Describe the function of the alveoli.

9. Describe the role of hemoglobin.

10. Which organ is the "chemical factory" of the body?

11. Which organ eliminates waste products through a complex filtering system?

12. At what concentration of oxygen do we typically see the first effects of oxygen deficiency?

13. Which physical property tells us how easily a substance can be dissolved in another substance?

## Science & Technology

# Chemical/Physical Properties

The physical and chemical properties of a substance tell us how we can expect it to behave. **Physical properties** refer to the appearance of a substance and all of its behaviors except how it will react with other chemicals. **Chemical properties** refer to its behavior during chemical reactions.

Physical properties include descriptive information about the substance such as its color and smell. For instance, chlorine is a greenish gas that has a very noxious and irritating odor. They also include measurable properties such as the temperature at which the liquid state turns into a gas, or how dense it is.

Some physical properties that are useful when assessing the toxic danger of a chemical include:

■ *Melting Point* – the temperature at which a solid changes to the liquid state. Remember: liquids are typically more easily absorbed through skin than solids.

■ *Boiling Point* – the temperature at which a liquid changes to the gas state. Remember: gases are easily and quickly absorbed through inhalation.

■ *Vapor Pressure*–how readily a liquid will go into a vapor by the escape of molecules from its surface. Liquids with high vapor pressures evaporate easily; they are **volatile**. Remember: the more volatile the liquid, the more that is available to inhale.

■ *Solubility* – the ability for a substance (*solute*) to dissolve in another substance (*solvent*). Remember: substances that dissolve the fat in our cells typically pass easily through the skin and are absorbed.

■ *Density* – the mass per unit volume. Think about a 1-inch cube of balsa wood versus a 1-inch cube of steel. Although they are the same volume, the steel has the greater mass and therefore has the greater density. Remember: A gas that is more dense than air could collect in low spots where workers could be asphyxiated from lack of oxygen.

Chemical properties are important because they help us predict what will happen when the substance reacts. A **chemical reaction** is any change that rearranges the atoms or molecules of a chemical, resulting in the formation of one or more new substances.

In order for the forces holding the original chemicals together (they are called chemical bonds) to be broken during a reaction, energy is needed. Energy can be in the form of heat, light, or electricity. For instance, methane, which is the main component in natural gas, is a molecule made up of four hydrogen

atoms bonded to one carbon atom. When energy is applied in the form of a burning match or a spark, the methane and oxygen react with each other. The heat released during the reaction provides the energy necessary to continue the reaction with other molecules.

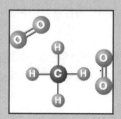

This reaction results in all the bonds between the atoms being broken.

New bonds then form to make two new substances – carbon dioxide and water – and heat is released.

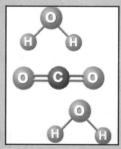

Chemical properties are different for each substance and depend on the type of bonding it has, its structure, the configuration of its electrons, and energy levels. Some of these properties will be explored in future chapters.

# 3-3 Carcinogenicity

## ■ What is Carcinogenicity?

### Concepts

■ A carcinogen is an agent that can cause cancer.

■ Mutations are changes in the genetic information of a cell that can be transmitted to and reproduced in future generations.

■ Teratogens cause damage to the developing fetus in the mother's womb.

**Carcinogenicity** is the tendency for cancer to occur. **Cancer** is the uncontrolled growth and spread of abnormal cells in the body. An agent that can cause cancer is called a **carcinogen**. Cancer is a disease state caused by the unregulated overgrowth of cells which is first indicated by the occurrence of **malignant tumors**. A malignant tumor tends to invade surrounding tissue and spreads from its site of origin to additional distant sites elsewhere in the body. The presence of a tumor is not in itself an indication of cancer. A **benign tumor** is well-defined and does not invade surrounding tissue or spread to additional distant locations in the body.

During the 1970s, the public became more aware of the potential for chemicals to cause cancer. Newspapers reported regularly of chemicals being banned from use because they caused cancer in laboratory animals. Many chemicals were banned based on the accepted belief that as carcinogens they exhibited no threshold for "safe" levels of use or consumption. This belief follows from the "one-hit" model. This model suggests that exposure to a single particle of a carcinogen can initiate the events that result in the formation of a cancer. The list of chemicals in this category includes pesticides, drugs, and food additives.

For most people, cancer is a very frightening disease. The prospect of getting cancer is so fearsome that it has caused some people to "cry wolf" and believe that everything causes cancer if taken in large enough doses. Studies of chemicals have indicated that only a relatively small number of the thousands of chemicals in commercial use cause cancer. This tendency to attribute blame to chemicals may stem from the need on the part of the public to find a plausible explanation for why cancer rates are so high. Chemicals provide a handy scapegoat. Currently, according to the American Cancer Society, a staggering one in three people will develop cancer in some form during their lifetime, yet only an estimated 10 to 15 percent of these are from occupational exposure to chemicals.

Researchers continue to learn more about the causes of cancer and how to prevent it. Over the years, research has identified chemicals that are known to cause cancer in humans. These chemicals have received this dubious distinction because they have been

■ Aflatoxins

■ 4-Aminobiphenyl

■ Analgesic mixtures containing Phenacetin

■ Arsenic and certain Arsenic compounds

■ Asbestos

■ Azathioprine

■ Benzene

■ Benzidine

■ Bis (Chloromethyl) Ether

■ 1,4-Butanediol Dimethylsulfonate (Myleran)

■ Chlorambucil

■ 1-(2-Chloroethyl)-3-(4-Methylcyclohexyl)-1-Nitrosourea (MeCCNU)

■ Chromium and certain Chromium compounds

■ Conjugated Estrogens

■ Cyclophosphamide

■ Diethystilbestrol

■ Erionite

■ Melphalan

■ Methoxsalen with Ultraviolet A Therapy (PUVA)

■ Mustard gas

■ 2-Naphthylamine

■ Thorium Dioxide

■ Vinyl Chloride

**Figure 3-28:** Some known human carcinogens. (Source: U.S. Department of Health & Human Services, National Toxicology Program, *Sixth Annual Report on Carcinogens, 1991*)

shown to cause cancer in the human populations exposed to them. Figure 3-28 lists some of the known human carcinogens. Other chemicals are suspected of causing cancer based on tests conducted on laboratory animals. There are about 30 chemicals that are considered human carcinogens and about 200 that are suspect based on animal studies.

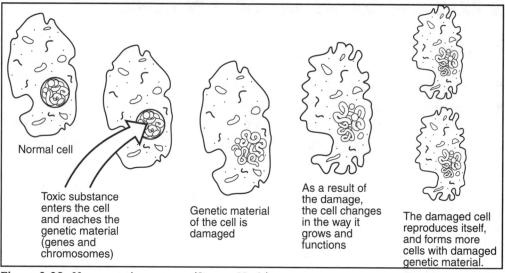

Normal cell

Toxic substance enters the cell and reaches the genetic material (genes and chromosomes)

Genetic material of the cell is damaged

As a result of the damage, the cell changes in the way it grows and functions

The damaged cell reproduces itself, and forms more cells with damaged genetic material.

**Figure 3-29:** How mutations occur.  (Source: Hesis)

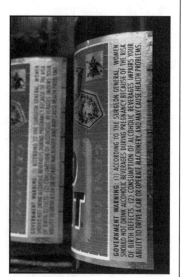

**Figure 3-30:** Beer labels warn of reproductive harm.

Toxic chemicals have other potential effects that are often confused with carcinogenicity. One of these effects is the ability to cause **mutations**. Mutations are changes in the genetic information of a living cell that can be transmitted to offspring and reproduced in future generations (Figure 3-29).  Genetic information is contained in genes found on chromosomes. This information directs proper functioning and reproduction of a cell. Some chemicals can damage genes or chromosomes causing a scrambling of the information they contain. Reproduction of genetically altered cells will cause the formation of more similarly damaged cells with faulty genetic information. This can ultimately lead to a group of cells which do not function correctly.  A chemical that has the ability to damage genetic material in this way is called a **mutagen** and the tendency for genetic mutations to occur is referred to as **mutagenicity**. Most carcinogens are also mutagens; however, not all mutagens cause cancer.

The second effect that is sometimes confused with carcinogenicity is **teratogenicity**. This refers to the tendency for specific interference with development of an unborn child. An agent that causes abnormal development (birth defects) in the fetus is called a **teratogen**. Most chemicals have not undergone tests to determine their reproductive effects in animals. Currently, only a few chemicals are known to cause reproductive harm.

# Box 3.1

## Alice Ottoboni – The One-Hit and Threshold Theory

*Alice Ottoboni, Ph.D., has written numerous books and articles on the subject of toxicology. This Viewpoint consists of excerpts from a paper she presented to regulatory agency personnel.*

The one-hit theory of chemical carcinogenesis, also known as the chemical bullet theory, has become the popular and fashionable explanation for how chemicals initiate or set the stage for the development of cancer. The one-hit theory is accepted unquestioningly by governmental health regulatory agencies and is used as a model for estimating the risk to humans posed by exposure to low levels of chemicals either known or suspected of causing cancer. Thus, this model serves as the basis for governmental decisions on standards of permissible human exposure to such chemicals.

---

**The chemical bullet theory is in contrast to the threshold theory . . .**

---

The one-hit theory is patterned after the theory of radiation-induced carcinogenesis. Simply stated, it postulates that one molecule of a cancer-causing chemical, like one unit of radiation, is capable of hitting a crucial molecule within a cell, thereby initiating a cancer process. Hence, the terms "one-hit" and "chemical bullet." According to this theory, since one molecule can cause a cancer, there is no such thing as a safe dose of a chemical carcinogen. The chemical bullet theory is in contrast to the threshold theory which considers that there is a safe dose for all chemicals, a threshold dose, below which no adverse effects, including cancer, will occur. The threshold theory is based upon the principles which form the foundation of the sciences of pharmacology and toxicology. Those who support the one-hit theory believe that these laws and principles do not apply to the process whereby chemicals cause cancer.

The translation from radiation-induced carcinogenesis to chemical carcinogenesis assume that radiation and chemicals behave in the same manner in their pathways through our bodies. There is little scientific evidence that they do, and much scientific evidence that they do not. There is no dispute that certain radiations can cause cancer, nor that some chemicals can cause cancer. The question is the validity of equating the mechanisms of radiations and of chemicals in the cancer process. The fact that they both produce the same end result does not require that the manner in which they do so be the same. There are innumerable examples, in all phases of our experience, of identical results being achieved by different means. We can arrive in New York City from the West Coast by train, plane, automobile, or boat; we can obtain ethyl alcohol by fermenting sugar or by synthesizing it in a laboratory. The ends are the same, but the means are quite different.

The fact that certain classes of chemicals could cause cancer was not widely recognized until about thirty years ago. In the early days of interest in cancer causation by chemicals, it was natural to look to the wealth of data available on radiation-induced carcinogenesis for clues that might shed light on the mechanism of chemical carcinogenesis.

One striking similarity between chemicals and radiations that strongly reinforced the justification for using radiation as a model for chemical carcinogenesis is the fact that cancers produced by both kinds of agents have relatively long latency periods. The molecular events that occur during the latency period are not understood. Obviously, exposure to a carcinogen can set in motion a process that remains obscure for a period of time until, finally, clinical symptoms become apparent. The length of the latency period for a given carcinogenic agent varies with the dose; the greater the dose, the shorter the time between exposure and clinical disease. When exposure to a carcinogen is repeated, day after day, for many years, the latency period is obscured. In fact, the only way we can know that latency periods exist at all is by controlled studies in which exposure to a carcinogen is followed by a long observation period during which no exposure occurs.

---

**The latency period for a given carcinogenic agent varies with dose.**

---

Another similarity, only recognized within recent years, is that a major fraction of the chemicals known to be carcinogenic in animals or man, like ionizing radiations, are capable of producing mutations. The theory for the carcinogenic action of ionizing radiation is based on their ability to induce mutations. The demonstration that a number of carcinogenic chemicals are also mutagenic for certain strains of microorganisms has been ac-

## Box 3.1

cepted by some as further evidence for the validity of using radiation as a model for chemical carcinogenesis.

The differences between chemicals and ionizing radiations in their interactions with living tissues are more numerous and of greater consequence than the similarities. Ionizing radiations are forces, units of energy, that pierce our bodies like bullets and follow a straight course in their passage through tissues and organs. Their pathways are independent of and unchanged by the structure or function of the matter they penetrate. Ionizing radiations that shoot through us may or may not do damage on the way. Some radiations that are extremely small with very high energy, such as cosmic rays, may not collide with any structure in their flight through a human body; at the submolecular level our tissues are mostly empty space. By chance, some radiations that pass through us may collide with one or more molecules in their paths, losing some of their energy with each collision. Radiations that penetrate and give up all of their energy in the collision process do not exit from the body. They cease to exist, just as the light ceases to exist when the flame is snuffed out.

Chemicals are not forces, but matter. They do not impinge on the surface and travel through our bodies in bullet-like fashion. Chemicals are substances that enter into us only if they can pass the barrier at the site of exposure – the lungs, the skin, or the digestive tract. Once transported across one of these barriers, a molecule of a chemical follows well-defined anatomic, physiologic, and biochemical pathways through the body. In its travels through the body, a molecule meets many barriers, some of which it can pass and some of which it cannot pass. There is absolutely no randomness or chance about where a molecule goes or what other molecules it comes into contact with in its passage through a human organism, as there is with radiations. Thus, the majority of foreign molecules that enter our bodies probably never come into contact with chromosomal molecules, because their pathways do not cross. Collisions of radiations with chromosomal molecules occur regularly by chance.

---

There are few if any universal carcinogens...

---

Chemical carcinogenesis is more complicated and less understood than radiation carcinogenesis, probably because chemicals are so much more varied and complex than radiations in their reactions and interactions with living systems. Some chemicals cause cancer only in one or a few species,

in one sex, in one age group, or in individuals with marginal nutrition. There are few, if any universal chemical carcinogens – chemicals that cause cancers in all groups of individuals regardless of species, sex, age, or diet. Ionizing radiations do not make such distinctions. Most chemical carcinogens are capable of causing cancer only by one or two routes of exposure but not by other routes. Sugar, for example, when injected under the skin of certain species of rodents can cause a malignant tumor at the site of exposure, but fortunately for those that have a sweet tooth, sugar is not carcinogenic in any species by ingestion.

Scientific evidence and practical observations do not support a simple one-hit theory for chemical carcinogenesis. All humans are exposed throughout their lifetimes to countless numbers of carcinogens, both natural and synthetic, yet not all humans develop cancer. To illustrate, benzopyrene is a naturally occurring and relatively potent carcinogen that is omnipresent in our environment. It is a product of combustion of organic matter. It has been determined that there are about 50 micrograms of benzopyrene in a kilogram of charcoal-broiled steak. One kilogram is equal to 2.2 pounds; thus a generous 7-ounce portion of steak weighs one-fifth of a kilogram. One-fifth of a kilogram of charcoal-broiled steak would contain one-fifth of 50 micrograms of benzopyrene, or 10 micrograms. Ten micrograms is a very, very small quantity, as demonstrated by the fact that there are over 28 million micrograms in one ounce. But when one considers how many molecules are contained in one microgram, that seemingly insignificant quantity takes on really formidable proportions. In 10 micrograms there would be $2.4 \times 10^{16}$ molecules, or 24,000,000,000,000,000 molecules. To give this astronomic figure a name, there would be 24 quadrillion molecules of benzopyrene in a 7-ounce portion of charcoal-broiled steak.

We all do not eat charcoal-broiled steaks every day, but considering that benzopyrene is so widespread in our environment, it is most reasonable to assume that we all are exposed to at least billions of benzopyrene molecules every day of our lives. The most common site of cancer caused by ingestion of benzopyrene is the stomach, yet in the United States cancer of the stomach is one of the less common cancers, accounting for only about 10–15 percent of all malignancies.

There are hundreds of examples of carcinogens which, like benzopyrene, we encounter daily. These examples cast serious doubt on the validity of a simple one-hit theory of chemical carcinogenesis; if one molecule is capable of causing cancer, how can "hits" by billions and billions of carcinogenic chemicals fail to cause cancer?

# ■ How is Carcinogenicity Determined?

## *Concepts*

- Epidemiological or animal studies are used to determine carcinogenicity.

- Epidemiological studies involve human populations and can be retrospective or prospective.

- Extrapolation of high animal doses to low human doses can call into question estimations of carcinogenicity based on animal studies.

There are two ways to examine and gather evidence on substances that are potentially cancer-causing. One way is to conduct an epidemiological study or survey and the other is to perform experiments using laboratory animals.

**Epidemiology** is the field of science dealing with the study of disease occurrence in human populations. Epidemiological studies, if done in a scientifically sound manner, are far more reliable than animal studies since the effects of a substance on animals cannot be shown to correspond exactly with its effects on humans. There are two types of epidemiological studies – retrospective, which looks at the past; and prospective, which is ongoing.

A museum's retrospective of the works of a famous artist displays earlier works; similarly, a retrospective study looks at the exposure of a specific group of workers in the past. In a case where the artist is dead, he or she cannot benefit from the heightened appreciation and renown that the showing may bring. In the same manner, a retrospective epidemiological study occurs after the fact and can only benefit those coming after the test group. One concern about retrospective studies is that often it has been difficult, if not impossible, for researchers to obtain detailed medical histories of their subjects that include lifestyle risk contributors such as smoking, alcohol consumption, obesity, and the like. These important pieces of missing information can skew the results.

A prospective study maintains environmental data as well as exposure and medical records on workers as they are exposed. These studies can be used to closely monitor workers at sites with known or suspected carcinogens and other sites where excess cancer risk is not suspect. The final results of these studies, however, are completed far in the future; so again, they may not benefit the study group. Prospective studies are especially difficult in industries where there is a large turnover of employees. This is especially critical since the latency period of many cancers is over 20 years. For this reason, much of our epidemiological information has been acquired in foreign countries that typically don't have high worker turnover rates.

If a worker population has a higher than normal incidence of a specific kind of cancer, and the cancer-causing agent is unknown, an epidemiological study is the only option. Sometimes the type of exposure and combinations of potential cancer-causing agents can-

## High-to-Low Dose Extrapolation

| Lifetime Daily Dose | Lifetime Incidence of Liver Cancer in Rats | Lifetime Probability of Liver Cancer |
|---|---|---|
| 0 mg/kg/day | 0/50 | 0.00 |
| 125 mg/kg/day | 0/50 | 0.00 |
| 255 mg/kg/day | 10/50 | 0.20 |
| 500 mg/kg/day | 25/50 | 0.50 |
| 1000 mg/kg/day | 40/40 | 0.80 |

## Use of Mathematical Models

| Model applied | Lifetime risk at 1.0 mg/kg/day |
|---|---|
| One-hit | 1 in 17,000 |
| Multistage | 1 in 167,000 |
| Multi-hit | 1 in 230,000 |
| Weibull | 1 in 59 million |
| Probit | 1 in 5.2 billion |

**Figure 3-31:** High-to-Low Dose Extrapolation and Use of Mathematical Models. The data in the top part of the table represents the results of an animal study for a hypothetical chemical and the bottom part plugs the data from the study into various mathematical models to present what each model would estimate the lifetime risk to be for rats from a dose of 1.0 mg/kg/day of the chemical.

not be adequately duplicated in the laboratory. This, too, is a situation where epidemiological studies are a must.

The second way to obtain evidence of a substance's ability to cause cancer is from administering it to experimental animals in a laboratory setting. Results that would be considered positive include an increase in the amount of tumors in the animal population or a more rapid onset (decrease of the latency period) of tumors. Animal studies are usually performed on rodents because they are inexpensive and easy to keep in the laboratory. Dogs, apes or monkeys may be used in special situations, but these tests are much more expensive. The life span of most rodents is from two to three years. Since developing cancer sometimes takes nearly the entire life span of the animal, animal testing is very expensive to conduct. Although one animal study may take more than three years to complete, it seems relatively short when compared with the time required for an epidemiological study involving humans.

Several scientific organizations have made recommendations and have endorsed procedures for conducting animal studies. Regulatory agencies, such as the EPA, that use study findings to manage human and environmental risks also have approved procedures. Included in these accepted procedures are requirements for choosing the species and number of animals, how the agent is to be administered, the dose levels, duration of the testing, and how the results will be statistically analyzed. These are very important procedures. Failure to follow them has resulted in several studies being called into question. In one case, trace amounts of contaminants were discovered in the test agent that could have contributed to the growth of tumors. In the case of early asbestos studies, it was found that many of the samples contained trace metals that hindered an enzyme believed to initiate increased asbestos cancers in those who smoke.

Testing on animals is done using "doses and under experimental conditions likely to yield maximal tumor incidence" (Interagency Regulatory Liaison Group, 1979, Washington, DC). This is done by estimating the highest dose that will be tolerated by the experimental animal without causing it to become sick or die from reasons other than cancer. Any agent that is determined to cause cancer in laboratory animals can only be considered suspect as a carcinogen in humans. Applying the results of animal studies to humans requires extrapolation, a process that estimates a result outside of known data. In this case, statistical analysis is used to estimate the cancer risk of low doses for humans in the general population from known data about high dose effects in animals. This is extremely controversial due to the complexities and uncertainties of the assumptions involved. Extrapolation relies on the use of statistical manipulation through mathematical models. Different models give widely different estimates of the bottom line risk (see Figure 3-31 for some examples). Attempting to use mathematics on a question with so many complex variables has been called into question over the years. Unfortunately, government agencies often need to base regulatory decisions on available information and cannot, due to political pressures, wait until actual risk can be determined.

The use of the industrial chemical ethylene dibromide (EDB) as a fumigant for fruit provides an example of the uncertainty of extrapolation. Experiments showed increased gastric tumor production in rats receiving doses of 40 mg/kg per day of EDB. The Carcinogen Assessment Group of the EPA plugged the information into a conservative "one-hit" model and estimated nearly a 100 percent incidence of cancer over their lifetime for workers exposed to less than 1 ppm on average. Workers that had been exposed at this level have been monitored since the early 1980s, and the predicted results have fortunately not come to pass. Although there have been cancers in workers at plants with EDB, nothing near the expected level of cancer incidence has occurred. The reason for this is that the "one-hit" model assumes that it takes only a small amount of a carcinogen and one hit to start the chain reaction of events leading to the development of cancer. Some scientists believe there are a number of flaws in this model, including the fact that it does not take into account the human body's defense mechanisms.

## ■ Checking Your Understanding (3-3)

1. Explain what cancer is.

2. Why are some chemicals classified as suspected carcinogens and others as known human carcinogens?

3. Explain the difference between a malignant and benign tumor.

4. Describe the differences between a mutagen and a teratogen.

5. What do the one-hit and threshold theories refer to?

6. Describe the two ways carcinogenicity can be determined.

7. What is the difference between retrospective and prospective studies?

8. What is an epidemiological study?

9. Describe the role of extrapolation in the analysis of animal studies.

# 3-4 Risk

## ■ What is Risk Management?

### Concepts

■ Risk-Benefit Analysis does not rely strictly on science.

■ Risk is the probability of loss or injury.

■ We tend to perceive voluntary risks as less risky than involuntary risks.

The newspaper headlines read "Trihalomethanes in Drinking Water Linked to Cancer" or "Breathing Southern California Air Shortens Life Expectancy." These headlines are disturbing and provoke a variety of initial reactions from individuals. At one end of the spectrum, there are those who dismiss these "alarms" entirely. People perceive that "doomsday" events have often been predicted and did not occur. Therefore, all such predictions should be viewed as overreaction. This attitude leads people to wait until a predicted disaster materializes before they take any action. People in the opposite camp might issue an immediate call to arms, requiring a new restriction or change in public policy to control the hazard. So, which way is most correct? The direction of current governmental policy indicates that the answer will depend on what we as a society decide is the level of acceptable risk as determined through a process known as **risk management**. Risk management, being more political than scientific, is society's decision on how to minimize or control a risk. This contrasts with **risk assessment**, which uses scientific principles to determine what level of risk actually exists. For example, in recognition of the scientific evidence demonstrating the health hazards associated with breathing secondhand smoke, several states have now chosen to minimize the exposure of its citizens by designating all public buildings as "no smoking" areas.

William W. Lowrance, in his paper on risk for the Committee on Science and Public Policy of the National Academy of Sciences, states:

> We are disturbed by what sometimes appear to be haphazard and irresponsible regulatory actions, and we can't help being suspicious of all the assaults on our freedoms and our pocketbooks made in the name of safety. We hardly know which cries of "Wolf!" to respond to; but we dare not forget that even in the fairy tale, the wolf really did come.

The feeling of anxiety and confusion about health risk issues takes on a more frustrating twist when there is disagreement within the scientific community. One noted and respected scientist says, "Yes, saccharin poses a significant threat to health and should be banned" and another, of equal distinction and qualification, says that "In the quantities consumers would use, the risk is so minimal that the benefits outweigh them." Which scientist is correct?

The question of how much risk we are willing to accept is at the core of most questions involving environmental contaminants. Questions such as "How clean is clean?" at the cleanup of a contaminated site, "What concentration of contaminants can be in drinking water?" for a community's water supply, and "What level of controls of the contaminants emitted from industrial processes should we impose?" all deal with risk. These types of questions, surprisingly to many, are not currently answered strictly on health terms; other benefits are also a factor. For instance, OSHA, in modifying the benzene standard (the maximum average concentration allowed for workplace exposures), admits that worker risk is still higher than would be desirable for health reasons, but that at this time it is not feasible to achieve the recommended levels. The National Institute of Occupational Health and Safety (the research arm of OSHA) has recommended, based on research into benzene's ability to cause leukemia and other diseases of the blood and blood-forming organs, that workplace permissible exposure levels be lowered from 1 ppm to 0.1 ppm. In the Federal Register on Friday, September 11, 1987, OSHA responded:

*A 0.1 ppm permissible exposure limit does not, on the evidence now before OSHA, appear to be technically feasible to achieve. To attempt to achieve it with engineering controls and work practices would appear to require major redesigns in large, capital intensive facilities such as refineries, coke operations, and petrochemical plants and small businesses as well. Many large-scale operations would need to be isolated or automated. Making major modifications for a large percentage of facilities in a number of the affected industries would not appear to be technically feasible at this time.*

The process illustrated in this example is called "risk/benefit analysis" and is being utilized in the decision-making processes of many regulatory agencies. Risk/benefit analysis weighs potential risks against benefits. This is a complex task that calls for more than quality scientific data. Subjective concerns such as politics, lifestyle, freedoms, economics, and progress also figure into the equation. Making this analysis is neither easy nor absolute, with tremendous opportunity for objection and controversy at every turn.

Have you stopped to think how you make risk/benefit decisions each day? Although we constantly do this, we are not usually conscious of the process. Driving a car, for instance, we typically don't think of the risk involved unless it is higher than usual as in a snowstorm or icy road conditions. A man in Michigan who drove across the state in a blinding snowstorm to acquire a collectible piece of furniture admitted it was "crazy" given the risk, but the benefit of having the piece made it worth it to him at the time. If you did not have an interest in antique furniture, you may not have made the same risk/benefit decision. Skydiving is another case in point. Obviously, the thrill and challenge of the experience for some is a benefit. Others do not find the thrill sufficiently enticing to outweigh the risk.

## How the Public Ranks Selected Environmental Risks

High Risk

- Chemical waste disposal
- Water pollution
- Chemical plant accidents
- Outdoor air pollution

Medium Risk

- Oil tanker spills
- Exposure to pollutants on the job
- Eating pesticide-treated food
- Other pesticide risks
- Contaminated drinking water

Low Risk

- Indoor air pollution
- Exposure to chemicals in consumer products
- Genetic engineering (biotechnology)
- Waste from strip mining
- Non-nuclear radiation
- "Greenhouse effect" ($CO_2$ and global warming)

Figure 3-32: How the public prioritizes environmental risks.

These examples also touch on the perception of risk. We tend to perceive voluntary risks as less perilous than those we are forced to take. This explains in part the seeming contradiction of someone speaking out against a certain level of contaminant in their drinking water, while still smoking three packs of cigarettes a day. Even when confronted with data indicating that the greater risk lies with their personal lifestyle habits, their perception of the risk is lower since smoking was their choice.

Given so many variables, how are risk/benefit analyses done? In order to answer this, we must first determine what people consider to be safe. To some people, "safe" means without risk. We realize that zero risk does not and cannot exist. There is some element of risk in everything we do. A simple, usable definition of risk, then, is the probability of loss or injury.

Examining the risk involved in a given decision is difficult enough, but assessing the benefits is usually even more difficult. This is due to the highly charged, individualized, emotional aspects of this process. Since risk/benefit analysis is now more generally accepted (and oftentimes required) in the regulatory processes of state and federal government agencies, it is useful to look at some of the types of benefits that may be examined in the process.

Economics – Saving money is often an uncomfortable benefit to address; however, it plays an increasingly important role in agency decision-making with the realization that the amounts of monetary resources needed to achieve cleanup goals are limited. For example, if it will cost $2.5 million to remove and treat 99 percent of the contamination at a site, but cleaning up that last 1 percent would cost an additional $10 million, it may be decided – after looking at the risk level of the remaining 1 percent – that saving the additional millions is a benefit that outweighs the risk.

- Lesser of the Evils – Sometimes the alternative to the risk being evaluated has other, undesirable effects. An example is that earlier in this century, chlorofluorocarbons (CFC's) were hailed as one of the great chemical discoveries for their ability to replace the highly flammable and toxic materials used as refrigerants at the time. We now are replacing CFC's because of their role in reducing the stratospheric ozone layer.

- Improvement of Health – Many prescription and over-the-counter medicines have risks of side-effects, but the potential to cure or relieve one's illness oftentimes outweighs the potential risk. For example, aspirin relieves pain, but may cause bleeding in the stomach.

- Emotional Satisfaction – Race-car driving, mountain-climbing, and hang-gliding are all examples of activities with high risks that many are not willing to take, but that provide some individuals with strongly desired benefits.

- Provides New Technology – The potential benefit of science promoting progress and solving problems has historically called for risk-taking. For example, the Wright brothers crashed, but their activities lead to modern air travel, which is now the safest method of travel.

■ Saves Lives – A good example of this is vaccines against many life-threatening diseases. There is risk associated in receiving the vaccines, but they have saved many lives. A small number of people receiving vaccines will have reactions to them and die; however, millions will be protected from disease and death.

■ Making Life Easier and More Pleasant – Power tools and equipment have enabled us to spend less time in backbreaking manual labor but have also increased the risks of certain jobs. Carpal Tunnel Syndrome is related to the use of keyboards, but computers replace tedious record keeping and copying activities.

Of the benefit topics discussed, the one that seems to be the most controversial is economics. It is certainly uncomfortable to think we are putting a price on human life, however, there are already procedures that appear to do this. Elyse M. Rogers, in her article "Life in the Balance" describes an example pertaining to highway safety:

> Engineers and scientists know quite a bit about accident prevention and what it would take to make our highways safer than they are. Three major highway modifications would go a long way toward making auto travel safer – eliminating blind curves, building more pedestrian over-crossings, and installing road overpasses at railroad crossings. This may sound reasonable until we learn that the cost of railroad overpasses alone is estimated in the billions of dollars. Because the price is too high, we live with the railroad-crossing risk. This doesn't mean that we approve of highway deaths or dangerous intersections, just that the cost of totally eliminating a particular degree of risk does not seem worth the price. . . . In fact, the Federal Highway Administration ranks defects in highway design by the cost-per-life saved by removing the defect. It then encourages the states to make those improvements that will save the most lives . . .

Other economic concerns involved in these risk decisions are the hidden costs of reducing risk. Some examples include: higher prices for goods, our ability to remain competitive in a global marketplace, and loss of jobs.

The decisions on when and how to reduce the risks of environmental contaminants come from the government, through our elected officials, regulatory agencies, and the courts. The political ramifications of these decisions are summed up by James Martin's comment while a member of the House of Representatives. At issue was the safety of saccharin and the effectiveness of the Delaney Clause, which states that there shall be no substance(s) in food that is known to cause cancer.

> . . . (any politician) would be terrorized and quickly wilted by the simple suggestion that "your constituents deserve better than to be represented by someone who favored a little bit of cancer."

We, as concerned citizens, must understand the methods and consequences of risk/benefit-based decision-making. Ultimately, the kinds of decisions we, as a nation, and a global community make

## How EPA Experts Rank Selected Environmental Risks

Overall High/Medium Risk
■ "Criteria" air pollution from mobile and stationary sources (includes acid precipitation)
■ Stratospheric ozone depletion
■ Pesticide residues in or on foods
■ Runoff and air deposition of pesticides

High Health: Low Ecological and Welfare Risk
■ Hazardous/toxic air pollutants
■ Indoor radon
■ Indoor air pollution other than radon
■ Drinking water as it arrives at the tap
■ Exposure to consumer products
■ Worker exposure to chemicals

Low Health: High Ecological and Welfare Risk
■ Global warming
■ Point and nonpoint sources of surface water pollution
■ Physical alteration of aquatic habitat (including estuaries and wetlands) and mining waste

Overall Medium/Low Risk (Groundwater-Related Problems)
■ Hazardous waste sites – active (RCRA)
■ Hazardous waste sites – inactive (Superfund)
■ Other municipal and industrial waste sites
■ Underground storage tanks

Mixed and/or Medium/Low Risk
■ Contaminated sludge
■ Accidental releases of toxic chemicals
■ Accidental oil spills
■ Biotechnology (environmental releases of genetically altered materials)

Figure 3-33: How the EPA prioritizes environmental risks.

now will influence our future. Making wise use of our resources by prioritizing risks is a regulatory goal that needs more thoughtful input and participation by the public. A paper, "Identifying and Controlling Pulmonary Toxicants," prepared for the Congress of the United States by the Office of Technology Assessment in June 1992 touched on the issue of the difficulty of risk management:

> *The potential for chemicals and materials used in industry, transportation, and households to be simultaneously beneficial and toxic to human life creates a legislative and regulatory dilemma. The challenge of balancing a strong economy, one that delivers products people need and desire, with the health and safety of the populace sometimes seems to be a tremendous burden.*
>
> *Technological advances add to the weight of that burden. Thousands of new, potentially toxic substances enter the market annually. Advanced instruments help scientists measure the presence of new and existing substances in minute quantities. Substances formerly unknown or undetected suddenly become worrisome as technology provides the means to predict human risks from these substances.*

## ■ Risk Assessment

### Concepts

- Safe does not mean "without risk."

- Risk management involves different activities than risk assessment.

- There are four basic steps required to conduct a risk assessment.

The term "safe," in its common usage, means "without risk." In technical terms, however, this common usage is misleading because there is no such thing as zero risk. Classifying substances as either "safe" or "unsafe," although common, is therefore misleading. As we know, all substances, even those which we consume in high amounts every day, can be made to produce a toxic response under some conditions of exposure. In this sense, all substances are toxic and therefore pose some risk. The important question is not simply that of toxicity, but rather that of risk – i.e., what is the probability that the toxic properties of a chemical will cause damage to health under actual or anticipated conditions of human exposure? To answer this question requires far more extensive data and evaluation than the determination of toxicity; it requires conducting a risk assessment.

We have already defined risk as the probability of injury, disease, or death under specific circumstances. It may be expressed in quantitative terms, taking values from zero (certainty that harm will not occur) to one (certainty that it will). In many cases risk can only be described qualitatively, as "high" or "low." All human activities carry some degree of risk. Many risks are known with a relatively high

| Annual Risk of Death From Selected Common Human Activities | | | |
|---|---|---|---|
| Activity | Number of Deaths in Representative Year | Individual Risk/Year | Lifetime Risk 2 |
| Coal mining accident | 180 | $1.30 \times 10^{-3}$ or 1/770 | 1/17 |
| Black lung disease | 1,135 | $8 \times 10^{-3}$ or 1/125 | 1/3 |
| Motor Vehicle | 46,000 | $2.2 \times 10^{-4}$ or 1/4,500 | 1/65 |
| Truck Driving | 400 | $10^{-4}$ or 1/10,000 | 1/222 |
| Falls | 16,339 | $7.7 \times 10^{-5}$ or 1/13,000 | 1/186 |
| Home accidents | 25,000 | $1.2 \times 10^{-5}$ or 1/83,000 | 1/130 |
| 1. Selected from Hutt (1978) "Food, Drug, Cosmetic Law J," 33:558-589. | | | |
| 2. Estimated based on 70-year lifetime and 45-year work exposure. | | | |

**Figure 3-34:** Annual risk of death from select common human activities.

degree of accuracy because we have collected data on their historical occurrence. Figure 3-34 lists the risks of some common activities.

The risks associated with many other activities, including exposure to various chemical substances, cannot be readily assessed and quantified. Although there are considerable historical data on the risks of some types of chemical exposures (e.g., the annual risk of death from intentional overdoses or accidental exposures to drugs, pesticides, and industrial chemicals), such data are generally restricted to those situations in which a single, very high exposure resulted in an immediately observable injury, leaving little doubt about the cause. Assessment of the risks from chemical exposures that do not cause immediately observable injury or disease is far more complex.

As defined by the National Academy of Sciences, risk assessment is the scientific activity of evaluating the toxic properties of a chemical and the conditions of human exposure to it in order to determine:

– The likelihood that exposed humans will be adversely affected, and

– To describe the nature of the effects they may experience.

The concept of environmental risk assessment is used broadly by regulatory agencies to mean the scientific activity in which facts and assumptions are used to estimate the potential for adverse effects on human health or the environment resulting from exposure to pollutants or other toxic agents.

The values and policy preferences of decision makers, risk assessors, and representatives of industry, labor unions, environmental organizations, and public interest groups often differ. Scientists disagree about the nature of scientific evidence. These differences explain some of the past controversies over the regulation of specific chemicals and the development of agency policies.

In 1983, a committee of the National Research Council recommended the development of uniform guidelines for conducting risk assessments. The committee described what they believed to be several advantages in the areas of promoting quality control, consis-

**Figure 3-35:** Which of these activities involves zero risk?

tency, predictability, public understanding, administrative efficiency, and improvements in methods. The plan for conducting risk assessments they developed consists of four basic steps:

1. **Hazard Evaluation.** Involves collecting and evaluating information about the toxic properties of contaminants. As we know, there are two principal sources of information about toxic properties: epidemiological or clinical studies and experimental data. This step involves looking carefully at what we already know about the chemical.

2. **Dose-Response Relationships.** Involves describing the relationships between known exposure levels and the appearance of toxic effects in exposed populations. We know that, in general, for a given duration of exposure, the frequency at which toxic effects appear in an exposed population increases with increasing dose.

3. **Exposure Assessment.** Involves estimating the dosage of contaminants received by exposed populations, identifying the exposed population, and identifying the body sites at which toxic effects are produced. Oftentimes, determining the intake of contaminants is difficult because of the multiple pathways along which exposure occurs as well as the potential for fluctuating concentrations of contaminant. The expected behavior of the agent must also be known so that its movement and persistence in the environment, as well as how readily it breaks down, can be taken into account.

4. **Risk Characterization.** Involves determining a numerical risk factor. This final step is designed to ensure that exposed populations are not at significant risk. Although the calculation of these values for any given contaminant involves many simplified assumptions and approximations, an additional limitation is that these estimates treat contaminants individually and independently of each other. In most instances, however, populations are exposed not to individual contaminants but to complex and possibly time-varying mixtures. How and where contaminants interact with each other to produce toxic effects is complicated and poorly understood. Some evidence suggests that such interactions are significant. The health risks from exposure to combinations of contaminants may differ substantially from health risk from exposure to individual contaminants.

Risk assessments are required by regulatory agencies when contaminants have been released into the environment. The assessment of potential risk to the environment and public health aids in determining acceptable cleanup levels. The EPA has stated, for instance, that risk assessment is their major analytical tool in supporting environmental decision-making. Risk assessments are also used in determining whether certain consumer products such as food additives can be used. The approach is relied upon in the evaluation and regulation of carcinogenic substances. It is important to remember that risk assessment is different from risk management. Risk assessment describes the characteristics of the adverse health effects of human exposure to environmental hazards while risk management involves choosing regulatory options.

### ■ Checking Your Understanding (3-4)

1. Explain the difference between risk management and risk assessment.

2. List three risks in your life and describe how you "manage" them.

3. Discuss the problem with the concept of "zero risk" in terms that a sixth grader could understand.

4. Of the following lifetime risk values, which represents the highest risk – 1/6, 1/99, or 1/765?

5. Describe how the four steps of the risk assessment process would be used to assess the risk to a small community with gasoline-contaminated drinking water.

PRESERVING
THE LEGACY

4

# Basic Ecology

## ■ Chapter Objectives

After completing this chapter, you will be able to:

1. **Define** principal ecological terms.

2. **Describe** typical terrestrial and aquatic food chains.

3. **Explain** the significance of energy flow through successive trophic levels.

4. **List** factors that contribute to loss of biodiversity.

5. **Describe** the cycles that are crucial to sustain life.

6. **List** key global ecological concerns.

7. **Describe** the major benefits of wetlands.

8. **Explain** the purpose and implementation of NEPA.

## ■ Terms and Concepts to Remember

Atmosphere
Bioaccumulation
Bioamplification
Biological Diversity
Biome
Biosphere
By-product
Carbon Cycle
Community
Consumer
Decomposer
Ecology
Ecosystem
Endangered Species
Environmental Assessment
Finding of No Significant
   Impact (FONSI)

Food Chain
Food Web
Habitat
Hydrologic Cycle
Incinerator
Landfills
Niche
Nitrogen Cycle
Organism
Oxygen Cycle
pH
Population
Producer
Runoff
Species
Succession
Trophic Level

## ■ Chapter Introduction

In the last chapter we learned about the types of effects that toxic substances can have on the health of living things. We also explored the challenges involved in both assessing and managing the risks these substances pose. Since there is no such thing as zero risk, society must often make difficult risk-benefit decisions when it comes to risk management.

This chapter looks at health and risk on a broader scale – that of the health of the environment. All living things are affected at every moment by their environment. We know how quickly we are seriously affected if we are deprived of the oxygen normally found in the air around us. This chapter will explore the relationships among all components – air, water, the earth's crust, and living things – and how the activities of humans can impact their fragile balance.

**Figure 4-1:** Living things exist in a precise balance

# 4.1 Relationships

## ■ Life and the Environment

### Concepts

- A community and its environment are an ecosystem.

- Each organism maintains a niche within an ecosystem.

- Living things are involved in a complex web of interdependence.

**Figure 4-2:** Earth, the only planet known to support life.

*The most wonderful mystery of life may well be the means by which it created so much diversity from so little physical matter.*

Edward O. Wilson

The complex web of living things, if examined closely, shows remarkable organization. Living things are called **organisms** and include all types of plant and animal life, ranging from bacteria in the soil to a fern or a giraffe. An interconnectedness and interdependency exist between the smallest of organisms in the biological world and seemingly unrelated large animals and plants. Understanding the basic parts of these natural systems allows us to see how living things relate to and depend upon each other and the other components in their surroundings. We call everything that surrounds or affects an organism its environment. The environment is made up of living and nonliving things. Nonliving components of the environment include light, heat, water, soil, and air. The study of this relationship between living things and their environment is called **ecology,** which is a word derived from Greek words meaning the study of the home.

An *Apollo* astronaut gazing upon the blue planet that is our home remarked how very fragile it looked. We recognize the poignancy of this observation when we consider the relatively thin layer of the planet that sustains life. The part of the earth that supports life is called the **biosphere.** This layer includes water, the rock and soil of the earth's crust, and the **atmosphere.** The atmosphere is the gaseous envelope held close to the earth by gravity. Water in the biosphere is found in the air as water vapor, on the earth's surface in water bodies such as lakes and oceans, and in the pores of earth materials as groundwater. These components of the biosphere provide the unique conditions enabling the earth to support life. Other planets in our solar system have some of the components of a biosphere, but do not have the proper balance of components that provide the narrow range of conditions upon which life depends.

For life to exist, the environment must contain water, at least some of which is in liquid form. The fundamental building block of all organisms is the cell which is composed of 85 to 90 percent water. A variety of chemical elements must be present in order to form the complex molecules found in cells. The most important is carbon, which combines with itself to form long chains of carbon atoms and also combines with other elements to form the various organic molecules typical of living things. Even the size of the planet and its location in the solar system are important. If the earth were smaller, it would not have the gravitational pull needed to hold our atmosphere close to the earth. As we all know, temperature affects the physical state of water. If we were closer to the sun, water would boil, and if further away, it would freeze. Finally, sunlight provides a continuous supply of needed energy. Reflecting on the precise characteristics that create these conditions, we realize how truly miraculous the presence of life is.

Studying the biosphere as a whole poses a challenge; the complexity is enormous. The world of living things, with its complicated, interdependent relationships, is like an intricate tapestry. You can see the pattern from a distance, but when you get close you recognize that the pattern is created by the perfect placement of individual threads. If you tried to duplicate the tapestry you would need to look at a small section to see how the individual threads compose the pattern before attempting to understand the entire tapestry. The more we understand about the parts, the closer we come to understanding the whole. The same is true of the biosphere. We need to break it down into manageable units that lessen its complexity. These smaller, distinct areas of the biosphere are called **ecosystems** (Figure 4-3). An ecosystem is a group of organisms interacting with each other and with their nonliving surroundings. Each ecosystem may have a very different and distinguishing climate, soil, vegetation, and animals; however, all parts contribute to and are interrelated into a whole. Each component relies on the others in such a way to hold the ecosystem in a delicate balance. In thinking about an ecosystem as a complex tapestry, we do not always know exactly how many or which of the threads, if broken, will cause significant portions, or even the whole work to become unraveled.

Some examples of ecosystems include: deserts, temperate forests, tropical rain forests, oceans, mountains, grasslands, rivers, and lakes. The largest terrestrial ecosystems are referred to as **biomes.** They are distinct regional ecosystems identifiable by similar types of soil, climate, plants and animals regardless of their location on the planet. For example, deserts, rain forests, and tundra are all biomes. Their location does not change their category. A desert located in North America, for instance, has much in common with a desert located in Africa.

Ecosystems can be of varying sizes. Even a rotting tree is an example of a very small, specialized ecosystem. A group of living things within an ecosystem is called a **community.** An example is the variety of organisms found in a mountain lake. The lake's organisms may include a species of trout as well as other fish, plants, and animals, including microorganisms. These organisms may perish if the chemistry or temperature of the lake changes. Even if only one organism is immediately affected by the change, other organisms may depend on it for food or it may eat other organisms that will rapidly multiply and consequently deplete their food supplies. Relatively small changes can upset the balance of an entire community and begin a process of change.

**Figure 4-3:** An artificial ecosystem.

Within a community, organisms typically reproduce with their own kind. A group of organisms able to breed freely with one another, but not with members of other groups, under natural conditions to produce offspring is called a **species.** Elephants can only breed with other elephants to produce more elephants; therefore, elephants are a distinct species of organism. Members of the same species sharing a habitat are called a **population.** Populations are inventoried to determine whether certain species are threatened or endangered with extinction.

Each kind of organism requires certain environmental features to provide what it needs to live. These places are called **habitats.** A habitat is where specific plant and animal species are naturally found. Organisms are adapted to particular habitats. The type of vegetation in a habitat depends on the climate, humidity, the amount of sunlight and rainfall that are available, and the types of soil. Since

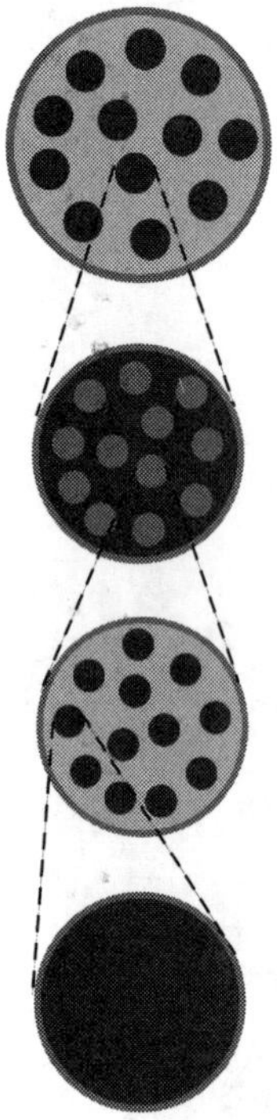

**Figure 4-4:** System of biological organization.

plants provide food for other organisms, their appearance or disappearance has far-reaching effects on the other living things in a habitat.

The organisms in a community influence their surroundings in a variety of ways. The effects an organism has on its surroundings and how the surroundings affect the organism are collectively referred to as its ecological **niche.** An analogy might be that the niche represents an organism's profession. A niche includes the ways an organism interacts with other living and nonliving components of its habitat. For example, what it eats and how its activities and behaviors affect other living things are all part of its niche. Describing the many dimensions of an organism's niche is difficult. To do this, one would have to include how its life is influenced by the nonliving components of its environment such as sunlight, temperature, rainfall, and soil conditions, as well as how it assists, competes with, and otherwise interacts with all other living organisms sharing the habitat.

Changes occur in communities that affect the organisms found within them. If a corn field in the Midwest were left undisturbed for a time, changes would begin to take place. We might immediately notice the appearance of some annual weeds. Animals such as field mice would move in. Within two or three years, grasses and perennial weeds would begin to grow in this habitat. Insect and bird populations would increase as would larger animals like foxes and woodchucks. After several more years, we might see the first shrubs and bushes. After even more time has passed, we might see some small oak and hickory trees which will eventually become large enough to provide shade needed by other species. This gradual change from annual weeds to forest is a slow process. For example, it takes 100 years or more for mature forests to develop on abandoned farmland.

This orderly progression that eventually results in the establishment of a stable community is called ecological **succession.** At the final stage of succession, a community reaches stability and is referred to as a "climax community;" the weed stage, the grass stage, and the shrub stage were just temporary. In the example of the abandoned farmland above, the oak-hickory forest is the climax community. A climax community is not always a forest. Factors such as water and soil determine the kind of climax community. A very dry area with low rainfall cannot be expected to become a forest; it will become a desert or a grassland.

# ■ How do the Different Levels Within the Organization Interact?

## Concepts

■ Each organism is positioned at a trophic level.

■ Organisms are divided into three nourishment groups: producers, consumers, and decomposers.

■ A food web is an interwoven group of food chains.

So far, we have seen how each organism fits into a system of biological organization. Now lets look at how organisms interact with one another. In order to do this, we need to look at links between organisms. The most obvious link between living things in a community is the food they eat. We can see that an aquatic plant, a fish, and a bear are linked because the fish eats the plant and the bear eats the fish. This type of link is called a **food chain** since each of the organisms is food for the next in line.

Organisms can be classified according to the way they are nourished or the way they provide nourishment to other organisms. All living things get their energy directly or indirectly from the sun. Plants use the sun's energy *directly* to make their own food through a process called photosynthesis. This is why they are referred to as **producers,** or *autotrophs* (meaning self-nourishment). By utilizing the sun's energy they manufacture the complex organic molecules needed to nourish themselves and create or produce food for the next classification of organism. Producers such as plants, algae, and some bacteria are at the first level of the food chain.

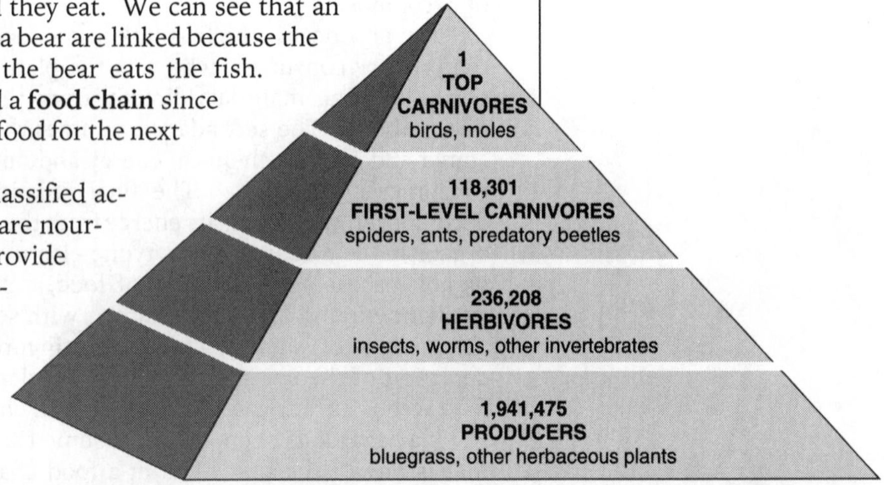

**Figure 4-5:** A biotic pyramid. Producers form the supporting base and second-level carnivores are on top.

**Consumers,** also called *heterotrophs* (meaning different nourishment), are unable to manufacture or produce food for themselves, so they rely on producers to provide their food energy. For example, a deer is a consumer since it eats plants. Consumers can be divided into three groups: primary, secondary, and tertiary consumers. Primary consumers get their food energy directly from the producer by eating it. These plant-eating animals are called herbivores. Animals like rabbits or cows are primary consumers. Secondary consumers are those animals who eat the primary consumers, thus getting their food energy from the producers *indirectly*, or through secondary means. Meat-eating animals are carnivores. Animals like lions and some birds are secondary consumers. Tertiary consumers obtain their food energy through the most indirect means. These animals eat a secondary consumer who ate a primary consumer who, in turn, ate a producer. An example of a tertiary consumer would be a Cooper's hawk that ate a warbler that ate an insect that ate the leaves on a tree. Omnivores are those animals that rely on both plants and animals for their food and therefore do not readily fit into any one consumer category. In reality, many animals which we tend to think of as solely herbivores or carnivores are often omnivores; mice eat insects, squirrels eat eggs, bears and foxes eat roots and fruits.

**Decomposers** are organisms that nourish themselves by breaking down dead organic matter. Organic matter contains carbon.

Their activities not only provide them with energy and nutrients, they provide several vital functions in an ecosystem. They perform important cleanup duties, and their feeding process releases some of life's essential elements to be used again by producers. This enables the cycle to continue which ultimately enables more food to be created for decomposers. Bacteria and fungi are good examples of decomposers.

The producer is the first link in the food chain. The next link is the primary consumer, who eats the plant and obtains energy and tissue building materials directly from the molecules of the producer's body. The secondary consumer then eats the primary consumer and obtains chemical energy and nutrients from the primary consumer's molecules. The final link in the food chain is the decomposer, which gets its energy from the waste products and dead organisms left behind by everyone else in the chain. Because few organisms eat only one kind of food, a simple food chain is not frequently found in nature. Rather, with several food options available to most organisms, a **food web** (Figure 4-6) is a more accurate depiction of the complex nutritional cycles that exist in nature. A food web is an intricate network of food chains.

If a hazardous chemical is consumed and stored by an organism that is one of the first links in a food chain, its toxic effects may become more severe as it is passed to other consumers higher in the chain. **Bioaccumulation** is the storage of chemicals within an individual organism at higher levels than found in the environment. **Bioamplification** occurs as chemicals accumulate in organisms in increasingly higher concentrations at successive levels of the food chain. The processes work together like this: A chemical spilled into a river or lake is ingested and stored by small organisms like plankton; small fish eat the plankton; and larger fish eat the smaller fish. Each organism accumulates a certain amount of the toxin. However, since secondary and tertiary consumers are eating organisms that already have accumulated the toxin, the level of toxin in their bodies rises even more. The effect of the toxin is amplified at higher levels in the food chain. As the process works its way up the food chain, the chemical may become thousands of times more concentrated in the tissues of the large fish than in the plankton. This is the reason why some fish from parts of the Great Lakes are considered unsafe to eat.

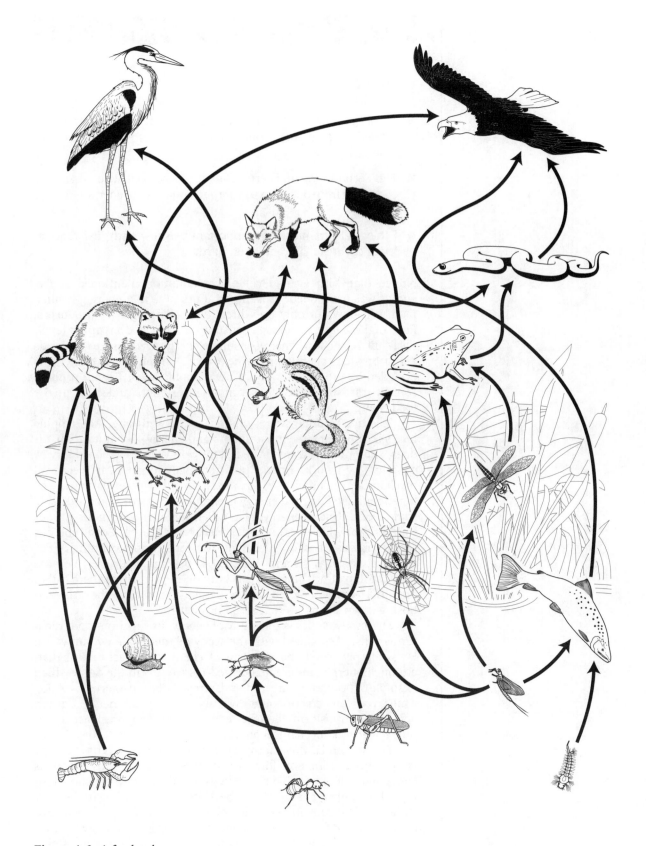

**Figure 4-6:** A food web.

# ■ Why is Energy Flow Important?

## Concepts

■ The First Law of Thermodynamics says that energy cannot be created or destroyed – only transformed from one form to another.

■ The Second Law of Thermodynamics says that when energy is converted from one form to another, some useful energy is always lost.

■ Ecological pyramids show important reasons for the existence of fewer coyotes than rabbits on Earth.

Notice that with every level of consumption mentioned above, energy was mentioned. It is important to note that energy within a food web flows in only one direction, beginning with producers. Each of these levels of energy consumption is called a **trophic level.** Each link in the food chain consumes energy according to its needs and it provides energy for the next level. Trophic level one, T1, is occupied by producers. At the next trophic level, T2, are primary consumers. Next, at T3, are secondary consumers, and so on, up the food chain.

In order to understand why energy flow through the trophic levels is important, we must understand energy – what it is, what it does, and how it is transformed. Energy is defined as the ability or capacity to do work. Energy takes on many forms: heat, electrical, mechanical, or chemical. It can exist as either potential energy or kinetic energy. Potential energy is energy that is stored and available to do work. Kinetic energy, also called the energy of motion, does the work for which the potential energy was stored. So, as you can see, energy can and does change from one form to another. For instance, you may sit and think about washing your car. You are fully capable of washing the car, but you're just sitting there full of potential energy. When you're actually scrubbing the grill and hosing down the hood, that's kinetic energy!

Thermodynamics is the study of energy, its functions and transformations. Associated with the study of energy are two universal laws. Simply stated, the First Law of Thermodynamics tells us that, although energy can be transformed (as we saw in the car-washing example), it cannot be created or destroyed. For instance, if you jog a mile, you don't create a mile's worth of mechanical energy. Rather, you simply transform chemical energy from your breakfast into a mile's worth of mechanical energy.

The Second Law of Thermodynamics says that each time energy is converted from one form to another, some of the energy is converted to less valuable heat energy and released to the surroundings. Remember, energy cannot be destroyed (according to the First Rule), but it is converted into a less useful form of energy heat. For example, when fossil fuels are converted into kinetic energy to move your car, you know that the engine generates a great deal of heat which must be carried away from the engine by the cooling system.

This heat was once usable energy in the form of fossil fuel. However, when it was transformed into kinetic energy, some of it was converted to heat energy – still energy, but without the ability to accomplish much work.

As with energy transformation, energy flow from one trophic level to the next results in a significant loss of useable energy. Let's look, for example, at a food chain starting with an acorn that is eaten by a mouse, which in turn is eaten by an owl. The acorn represents the stored energy of the oak tree. When it is eaten by the mouse, only about 10 to 15 percent is stored in the mouse's body. The rest is used for fueling the activities of life such as breathing, moving the body, and digesting food. In addition, the mouse releases a great deal of heat to its surroundings. When an owl eats a mouse, heat is lost by the owl and the remaining energy is used for its life activities. Thus only 10 to 15 percent is stored for use by the next predator up the food chain.

Ecological pyramids depict this energy loss. Green plants, the producers in a food chain, make up the first, and widest level of the pyramid. Animals that eat only plants make up the second level. Flesh-eating animals are at the highest levels. Each of these successive levels of the pyramid contain significantly less available energy than the previous. Most of the energy in an ecosystem is in the producers. The least amount of energy is at the highest level of consumers. The pyramid shape helps us to see that it takes a large number of producers to supply energy for a smaller number of consumers.

There are three types of ecological pyramids, each graphically depicts specific data:

■ *A pyramid of numbers* indicates numerically how many organisms occupy each trophic level within an ecosystem. At the top of the pyramid is the highest trophic level carnivores. There must be fewer carnivores than herbivores or the entire food chain would lose an essential link (the herbivores) when the carnivores consumed them all.

■ *A pyramid of biomass* shows the total biological mass at each trophic level. This type of pyramid shows the great quantities required at each trophic level to sustain those at the next level of the pyramid.

■ *A pyramid of energy* is closely related to a pyramid of biomass in that, at each trophic level, there is a 90 percent reduction in biological mass, as well as a 90 percent reduction in energy content.

**Figure 4-7:** A pyramid of biomass. Increasing quantities of energy are required at each successively lower trophic level.

■ Checking Your Understanding (4-1)

1. Starting with organism, list all of the levels of biological organization in a meaningful order.

2. Explain the functions of producer, consumers, and decomposers in a food chain.

3. Compare and contrast a food chain and a food web.

4. Explain why it is impossible for there to be as many coyotes as mice on earth.

5. Explain in your own words the First Law of Thermodynamics.

6. Explain in your own words the Second Law of Thermodynamics.

# 4-2 CYCLES

## ■ The Carbon and Oxygen Cycles

### Concepts

■ Carbon is one of the elements essential to all life forms.

■ $CO_2$ is taken in by plants during photosynthesis.

■ Cell respiration, combustion, and erosion return carbon to the abiotic environment.

In the last sections, we have examined the relationships that exist between living organisms within an ecosystem. The cycles that tie living organisms to the physical environment are another essential component of an ecosystem's relationships. These cycles are called *biogeochemical* cycles. You may notice the root *bio* in the word. Broken down, biogeochemical means life-earth-chemical cycles. Essentially, in these processes, matter is cycled through the nonliving environment and back to living organisms.

Carbon is an essential element in living organisms. Over 80 percent of all known compounds contain carbon. Carbohydrates, which act as fuel for living bodies, are made up of carbon compounds. Amino acids, the building blocks of protein, are carbon compounds that also contain nitrogen. Carbon dioxide ($CO_2$) from the atmosphere is one of the components in the process of photosynthesis. The **carbon cycle** is the term used to describe how carbon circulates through the air, plants, animals, and soil. Because it is a cycle, we can break in at any phase to explain it. So, let's start with the photosynthesis process plants use to convert energy from the sun and $CO_2$ into usable chemical compounds.

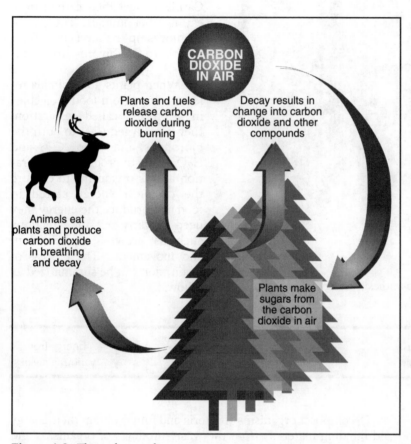

**Figure 4-8:** The carbon cycle.

During photosynthesis, plants take in $CO_2$ from the atmosphere through their leaves and obtain water from the soil through their roots. They combine this $CO_2$ and water ($H_2O$) using light energy from the sun to make the sugar, glucose ($C_6H_{12}O_6$). This sugar molecule is a source of potential energy for the plant and any other organism that consumes it. In one sense, the sugar molecule is storing sunlight energy as chemical energy. Oxygen is released as a **by-product** of photosynthesis. The process of photosynthesis can be summarized as follows:

| Carbon Dioxide from air | + | Water from soil | + | Light energy from sun $\longrightarrow$ | Sugar stored | + | Oxygen released |
|---|---|---|---|---|---|---|---|

The sugar molecule produced during photosynthesis is a basic building block for many other kinds of compounds which the plant uses to maintain itself and grow. As plants grow, they store more

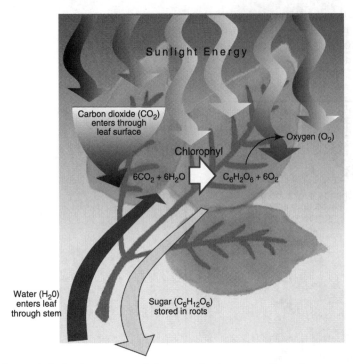

**Figure 4-9:** In the process of photosynthesis, energy from the sun

and more carbon-containing compounds in their structures. Herbivores obtain the carbon their bodies need by eating the carbon-containing compounds of plants. Carnivores get their carbon from eating other animals. Thus, carbon moves up the food chain from plants to herbivores to carnivores.

When plants and animals release energy from food molecules in a process called respiration, they use oxygen ($O_2$) to burn the molecules and expel $CO_2$ and $H_2O$ as waste products. Respiration allows organisms to capture the chemical energy stored in food and produce the energy they need to carry out life functions such as growth, reproduction, and movement. The process of respiration can be summarized as follows:

| Sugar and other food molecules | + | Oxygen from air | ⑧ | Water released | + | Carbon Dioxide released to air | + | Energy from various activities |
|---|---|---|---|---|---|---|---|---|

Decomposer organisms (bacteria and fungi) obtain their energy and building materials from the waste products of organisms and dead organisms by this same process of respiration. Thus, even the carbon in wastes and dead organisms can be recycled.

In some situations biological matter does not decay because decomposer organisms are not present or cannot live under certain conditions. This often occurs when organic matter like trees, mosses or bacteria is buried, preventing access to oxygen. This can occur in swamps and bogs or at the bottoms of lakes or oceans. Since decomposers require oxygen for respiration, the carbon in the tissues of buried organisms is not cycled back to the atmosphere as $CO_2$. If these deposits are buried, heat and pressure can convert the organic matter into coal, oil, or natural gas, which are known as fossil fuels. These fuels are the products of photosynthesis that took place millions of years ago. As these sources are tapped, refined, and utilized through combustion, carbon dioxide is released into the atmosphere and the carbon cycle for these deposits is complete. In like manner, the shells of marine organisms contain carbon. These organisms die in the ocean, and their shells settle to the ocean floor and are covered by other deposits and sediment. After many thousands of years and the accumulation of millions of organisms, these carbon-rich shells fuse together to form a layer of limestone. Over millions of years, this limestone layer becomes exposed to the elements, where it is subjected to erosion and slowly wears away, back into the waters from which it came. Through this process, it releases its ample stores of $CO_2$ into the atmosphere for reuse in the carbon cycle.

The **oxygen cycle** involves the circulation of oxygen through the various environmental compartments. Living things require oxygen, which can be obtained from the air or water for respiration. Oxygen is released into the atmosphere by green plants during photosynthesis. Because of its role in these two processes, the circulation of oxygen is tied closely to the circulation of carbon.

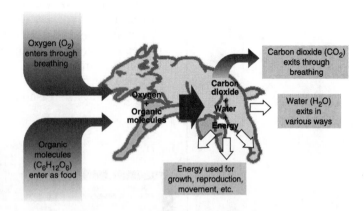

**Figure 4-10:** The process of respiration. Organic molecules react with oxygen, producing energy in usable form – carbon dioxide and water.

# ■ Nitrogen Cycle

## Concepts

- Nitrogen fixation gets its name from the fact that nitrogen is fixed into a usable form.

- Denitrifying bacteria release nitrogen back to the atmosphere.

- Plants take in carbon dioxide and produce oxygen during photosynthesis.

The circulation of nitrogen through plants and animals and back to the atmosphere is called the **nitrogen cycle.** At 78 percent by volume, nitrogen is the most plentiful gas in our air; it is one of the essential elements for life as we know it. Nitrogen is required by plants to manufacture proteins and other nitrogen-containing molecules that are vital for growth and reproduction. Nitrogen also plays a role in the production of nucleic acids, an essential component in DNA and RNA synthesis. Since 78 percent of the atmosphere is composed of nitrogen, it seems unlikely that we will run out any time soon. However, atmospheric nitrogen, in its gaseous state, is not readily usable by living organisms. It must first undergo a process called *nitrogen fixation,* which combines nitrogen with other elements to form compounds called ammonia, nitrates, and nitrites.

Although nitrogen fixation can occur in several ways, it is accomplished most often through biological means. Special nitrogen-fixing bacteria, algae, and some lichens, that live in the soil, convert atmospheric nitrogen into ammonia and similar compounds. Some of these nitrogen-fixing bacteria live in the roots of such plants as peas, beans and clover. Other soil bacteria convert ammonia to nitrite ($NO_2$) and nitrate ($NO_3$). Plant roots can take up ammonia and nitrates which they then use to make protein. Plants provide bacteria with carbohydrates for food, and the bacteria in turn release nitrogen that has been converted into a form that plants can use. When animals eat plants, they obtain their nitrogen in the plant proteins. Nitrogen compounds return to the soil through animal waste and through the decay of dead animals and plants. Nitrogen can be returned to the atmosphere by the action of denitrifying bacteria that have the ability to convert nitrites to nitrogen gas.

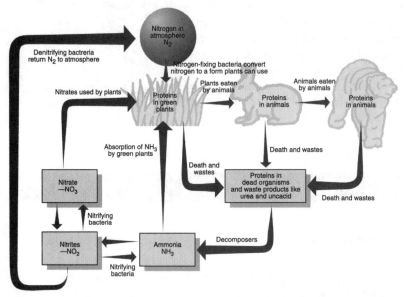

**Figure 4-11:** The nitrogen cycle. Nitrogen-fixing bacteria convert atmospheric nitrogen into a form that plants use to make protein. Proteins pass to other organisms when one organism is eaten by another. The decay of dead organisms and wastes produces ammonia which is reused by plants or converted into nitrogen compounds by bacteria. Denitrifying bacteria convert inorganic nitrogen compounds back into atmospheric nitrogen.

# ■ The Hydrologic Cycle

## Concepts

■ The hydrologic cycle describes water's circulation through the environment.

■ Evaporation, transpiration, runoff, and precipitation describe specific water movements.

The cycle most familiar to us is probably the **hydrologic cycle,** which means water-earth cycle (Figure 4-12). Water in all three forms – solid, liquid, and vapor – is constantly moving through the environment. There are several components to this cycle.

Precipitation involves the movement of water from the earth's atmosphere to the earth's surface in the form of rain, sleet, snow, or hail. Evaporation is the movement of water vapor from the earth's surface (lakes, soil, oceans) back to the atmosphere. When these processes occur, the water is not only moving, but it is also changing forms. Liquid water must become water vapor in order to enter the atmosphere. Notice the word broken down: e-VAPOR-ation.

Water that ends up on land can also be returned to the atmosphere by a process called transpiration. As water seeps through the soil, it becomes available for use by plants. Only about two percent of the water absorbed into plant roots is used in photosynthesis. Nearly all of it travels through the plant to the leaves where it is transpired to the atmosphere to begin the cycle again. The term transpiration literally means "across-breathe."

Water that continues flowing or seeping downward through the soil can become groundwater, or it may become runoff to feed lakes and streams and eventually end up in the ocean. Plants and animals temporarily absorb some of these water molecules, but are constantly exchanging water with their surroundings.

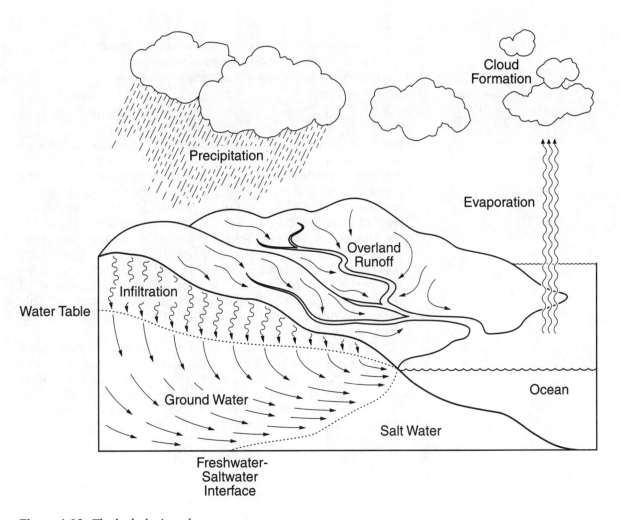

**Figure 4-12:** The hydrologic cycle.

## ■ Checking Your Understanding (4-2)

1. Explain the basic components of the carbon and oxygen cycles.

2. Explain the basic components of the nitrogen cycle.

3. Explain the basic components of the hydrologic cycle.

4. Explain what would happen if the cycles – carbon, oxygen, nitrogen, and hydrologic – did not exist.

5. Explain the difference between evaporation, transpiration, and precipitation.

# 4-3 Ecological Concerns

*Environmental problems do not stop at national boundaries. In the past decade, we and other nations have come to recognize the urgency of international efforts to protect our common environment.*

Jimmy Carter

President Carter made this statement in his Environmental Message to the Congress on May 23, 1977. In this statement, he issued a directive that called for a study of the probable changes in the world's population, natural resources, and environment through the end of the century. The Council on Environmental Quality and the Department of State, working in cooperation with other agencies such as the EPA and the National Science Foundation, were to prepare the study. The resultant three-volume report is called the *Global 2000 Study* and was completed in 1980. The *Global 2000 Study* is significant in that it is the first U.S. Government effort that looks at all three issues from a long-term global perspective that not only recognized their interrelationships but attempted to make connections between them.

This section will provide an overview of major global ecological issues such as acid deposition, ozone depletion, species loss, and global warming. This, however, is not an exhaustive list. Other issues of great importance on a global scale that are not discussed here include population, soil erosion, food and water demands/production, deforestation, and energy use. (Population was surveyed in Chapter 1.)

# ■ Acid Deposition (Acid Rain)

## Concepts

■ Acid rain is a result of reactions to atmospheric pollutants.

■ Rain with a pH of less than 5.6 is acid rain.

■ The eastern part of the U.S. has been hit hardest by acid rain.

Until recently, air pollution was considered a local problem. Now we know that winds can carry air pollutants hundreds of miles from their points of origin. These transported air pollutants can damage aquatic ecosystems, crops, and forests and may pose risks to human health. One of these pollutants is called acid deposition (or more commonly, acid rain).

Acid rain, which results from the reaction of pollutant gases and moisture in the atmosphere, has irreversibly damaged the environment in many parts of the world. It is now recognized as a serious long-term air pollution problem for many industrialized nations. The process of acid deposition begins with emissions of sulfur dioxide (primarily from coal-burning power plants) and nitrogen oxides (primarily from motor vehicles and coal-burning power plants). These pollutants interact with sunlight and water vapor in the upper atmosphere to form acidic compounds (sulfuric acid and nitric acid). These compounds often fall to earth as acid rain or snow, but the compounds can also join dust or other dry airborne particles and fall as dry deposition.

Over 80 percent of sulfur dioxide emissions in the United States originate in the 31 states east of or bordering the Mississippi River. Prevailing winds transport emissions hundreds of miles to the northeast, across state borders and into Canada. Thousands of lakes and tens of thousands of miles of streams in the eastern United

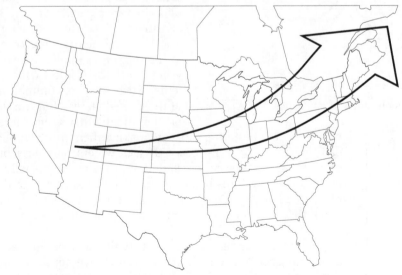

**Figure 4-13:** Accumulation of pollutants. As air moves across the continent from west to east, each major population area adds to the total pollution in the atmosphere.

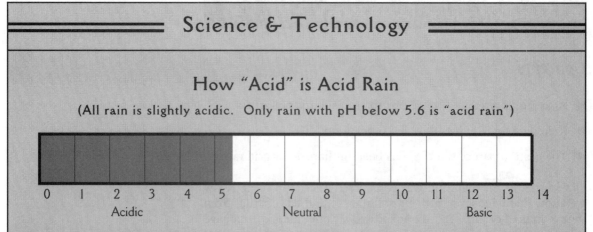

# Science & Technology

## How "Acid" is Acid Rain

### (All rain is slightly acidic. Only rain with pH below 5.6 is "acid rain")

| 0 | 1 | 2 | 3 | 4 | 5 | 6 | 7 | 8 | 9 | 10 | 11 | 12 | 13 | 14 |

Acidic                      Neutral                      Basic

The level of acidity of acid rain is measured by its **pH**. A substance with a pH of 7 is neutral, that is, it is neither acidic nor basic. Values lower than 7 indicate that the substance is acidic. Values higher than pH 7 are bases (or alkaline). The pH scale ranges from 0 to 14 with a pH of 0 representing the most concentrated acid and a pH of 14 representing the most concentrated base. Since the scale is logarithmic, each unit represents a ten-fold change (either increase in acidity going in the direction toward zero or increase in alkalinity in the direction toward 14). For example, a substance with a pH of 4 is ten times more acidic than a substance with a pH of 5 and 100 times more acidic than a substance with a pH of 6. Both acids and bases can be corrosive. How "acid" is acid rain? All rain is slightly acidic. Only rain with a pH below 5.6 is called "acid rain," although rain events with a pH of less than 3 have been recorded.

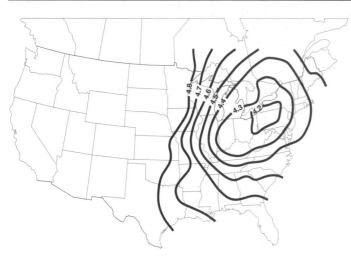

**Figure 4-14:** Areas with significant acid rain deposition.

States and Canada are vulnerable to the effects of acid deposition. Some of these have already been harmed. Acid deposition, along with other stresses such as ozone and natural factors such as drought, may also account for the declining forest productivity observed in parts of the East.

The extent of the damage caused by acid rain depends on the total acidity deposited in a particular area and the sensitivity of the area receiving it. Areas with acid-neutralizing compounds in the soil, for example, can experience years of acid deposition without problems. Such soils are common in much of the United States. But the thin soils of the mountainous Northeast have very little acid-buffering capacity, making them more vulnerable to damage from acid rain. Since surface waters have less buffering capacity than soils, acid precipitation problems are usually noticed in aquatic ecosystems before they are noticed in terrestrial ecosystems. Acid deposition may increase the acidity of surface waters so much that it reduces or eliminates their ability to sustain aquatic life.

Forests and agricultural crops are vulnerable because acid deposition can leach nutrients from the ground, hamper the microorganisms that nourish plants, and release toxic metals that would normally be tied up in the soil at higher pH.

# ■ Ozone Depletion

## Concepts

■ CFCs behave differently in the troposphere and the stratosphere.

■ CFCs have had wide industrial application for many years.

■ One chlorine atom can destroy many ozone molecules.

Increasing concentrations of the synthetic chemicals known as CFCs and halons are breaking down the ozone layer in the upper atmosphere. The ozone layer filters out some of the ultraviolet light from the sun. Destruction of the ozone layer allows more of the sun's ultraviolet rays to penetrate to the earth's surface. Increased ultraviolet radiation can lead to greater incidence of skin cancer and cataracts, as well as decreased crop yields and reductions in the populations of certain fish larvae and plankton vital to aquatic food chains. Increased ultraviolet radiation also reduces the useful life of outdoor paints and plastics.

CFCs are compounds that consist of chlorine, fluorine, and carbon. First introduced in the late 1920s, these gases have been used as coolants for refrigerators and air conditioners, propellants for aerosol sprays, agents for producing plastic foam, and cleansers for electrical parts.

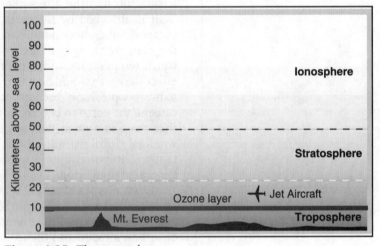

Figure 4-15: The atmosphere.

CFCs do not degrade easily in the first layer of the atmosphere – called the troposphere – which extends about seven miles above the earth's surface and contains the gases that support life. As a result, they rise into the second layer called the stratosphere where they are broken down by ultraviolet light. The stratosphere, which extends roughly 731 miles above the earth's surface, contains the ozone molecules that filter out ultraviolet light. The chlorine atoms from CFCs react with ozone to convert it into two molecules of oxygen (Figure 4-16). Each chlorine atom can destroy many ozone molecules, over 10,000 by some estimates.

Halons are an industrially-produced group of chemicals that contain bromine, which destroys ozone in a manner similar to chlorine. Halons are used primarily in fire extinguishing foam.

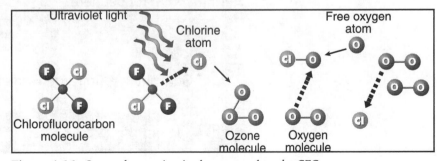

Figure 4-16: Ozone destruction in the stratosphere by CFCs.

## ■ Global Warming

### Concepts

■ Carbon dioxide is the main greenhouse gas.

■ Many human activities produce carbon dioxide.

■ The potential consequences of global warming are staggering.

The "greenhouse effect" is a natural phenomenon caused by carbon dioxide and other gases in the atmosphere. Their effect on the earth is comparable to that of the glass in a greenhouse. Visible light passes through the atmosphere to the earth's surface. When the light is absorbed by the earth, it is converted to heat. Some heat escapes, but carbon dioxide and other gases in the troposphere trap the rest, warming the earth. This allows the earth to maintain a warm temperature and support life. If there were no greenhouse effect the earth would be much colder (Figure 4-17). In recent years, many people have become concerned that human activities are causing the earth to become too warm. It is important to recognize that there is a great deal of disagreement among scientists about whether global warming is actually occurring. However, it is certainly an important political issue which has shaped legislation and generated regulations.

Certain types of air pollutants may be producing long-term and perhaps irreversible changes to the global atmosphere. Industrial growth since the mid-nineteenth century has released large amounts of carbon dioxide into the troposphere. Burning fossil fuels such as coal, oil, and natural gas also releases large amounts of carbon dioxide. The clearing of rain forests by burning the wood contributes

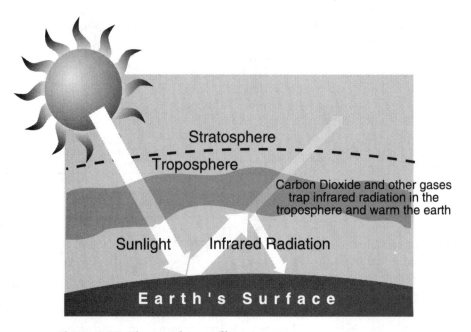

**Figure 4-17:** The greenhouse effect.

carbon dioxide and other greenhouse gases to the atmosphere. Moreover, the clearing of large areas of rain forests means that less carbon dioxide is removed from the air by plants. Recent studies and computer models indicate that by increasing the amount of carbon dioxide in the atmosphere, we may have initiated a warming trend that may raise average global temperatures between 2°F and 8°F by the year 2050.

Global warming may change weather patterns and regional climates. Many important agricultural areas of the United States, for example, could become less productive. Natural ecosystems would also be affected. One consequence of global warming, rising sea levels of one foot in the next 30 to 40 years, and two to seven feet by the year 2100, would inundate 50 to 80 percent of the U.S. coastal wetlands, erode all recreational beaches, and increase salinity of estuaries and groundwater.

# ■ Species Loss (Including Habitat Destruction)

## Concepts

- Diversity varies within ecosystem, species, and genetic levels.

- The earth's loss of biological diversity is increasing.

- Loss of diversity is happening most rapidly in the tropics.

- Solutions are complex, and international in scope.

- Nearly half of all prescription medications are derived from wild plant species.

The reduction of the earth's biological diversity has emerged as a public policy issue in the last several years. Growing awareness of this planetary problem has prompted increased study of the subject and led to calls to increase public and private initiatives to address the problem.

**Biological diversity** refers to the variety of and variability among living organisms and the ecological complexes in which they occur. Diversity can be viewed as the number of different items and their relative frequency within a set. Biological diversity occurs at many levels ranging from complete ecosystems to the chemical structures that are the basis of heredity.

The examples below describe diversity within ecosystem, species, and genetic levels:

- Ecosystem diversity: A landscape interspersed with croplands, grasslands, and woodlands has more diversity than a landscape of wheat fields or lawns.

- Species diversity: A rangeland with 100 species of annual and perennial grasses and shrubs has more diversity than the same rangeland after heavy grazing has eliminated or greatly reduced the frequency of the perennial grass species. Concerns over the loss of biological diversity have been defined almost exclusively in terms of species extinction.

■ Genetic diversity: Economically useful crops are developed from wild plants by selecting valuable inheritable characteristics. Thus, many wild ancestor plants contain genes not found in today's crop plants. An environment that includes both the domestic varieties of a crop (such as corn) and the crop's wild ancestors has more genetic diversity than an environment in which the wild ancestors have been eliminated to make way for domestic crops.

The earth's biological diversity is being reduced at a rate that is

likely to increase over the next several decades. This loss of diversity measured at the ecosystem, species, and genetic levels is occurring in most regions of the world. It is most pronounced, however, in particular areas, most notably in the tropics. The conversion of natural ecosystems to human-modified landscapes is the principal cause for reduction in biological diversity. Alterations in natural ecosystems provide considerable benefit when the land's ability to sustain development is preserved, but many scientists and policy makers feel that the resultant rapid and unintended reductions in biological diversity are undermining society's capacity to respond to future opportunities and needs. It is important to recognize that many industrialized nations have already modified their ecosystems and caused extinction of some of their native organisms. Since much development is taking place in nations in the tropics, current attention is focused on the impact of human activities in the tropical rain forests. Most scientists and conservationists working in this area believe that the problems there have reached crisis proportions. Still, many people remain skeptical and maintain that this heightened level of concern is based on exaggerated or insufficient data.

The abundance and complexity of ecosystems, species, and genetic types have defied complete inventory and thus the direct assessment of changes. As a result, an accurate determination of the rate of loss is not currently possible. Determining the number of species that exist, for example, is a major obstacle in assessing the rate of species extinction. The most recent estimates of total species approaches 30 million. Edward O. Wilson, one of the leading authorities on the subject, has calculated that we may already be losing a staggering 17,500 species a year.

**Figure 4-18:** All species of life can be endangered. White Rhinoceros (top); *Trillium reliquum* (middle); Blunt-Nosed Leopard Lizard (above).

Reduced diversity may have serious consequences for civilization. It may eliminate the possibility of using untapped resources for agricultural, industrial, and medicinal development. Diverse genetic resources have accounted for about 50 percent of productivity increases and annual contributions of about $1 billion to U.S. agriculture. For instance, two species of wild green tomatoes discovered in an isolated area of the Peruvian highlands in the early 1960s have contributed genes for increased fruit pigmentation and soluble-solids content of the tomato. These genetic improvements are currently worth nearly $5 million per year to the tomato processing industry. Future gains depend on genetic diversity.

The loss of plant species could mean the loss of billions of dollars in potential plant-derived pharmaceutical products. Tropical rain forests, which harbor an extraordinary diversity of species, and desert ecosystems, which harbor genetically diverse vegetation, are of particular concern. The loss of potential medicines has human consequences that go beyond potential economic benefits. For example, alkaloids from the rosy periwinkle flower (*Catharantus roseus*), a tropical plant, are used in the successful treatment of several forms of cancer, including Hodgkin's disease and childhood leukemia. Another threatening aspect of diversity loss is the disruption of environmental functions that depend on the complex interactions of ecosystems and the species that support them. Humans value diversity for reasons other than its utility. Aesthetic and ethical motivations to avoid the irreversible loss of unique life forms have played an important part in promoting initiatives to maintain diversity.

Forces that contribute to the worldwide loss of diversity are varied and complex. Historically, concern for diversity loss focused on commercial exploitation of threatened or **endangered species**. Increasingly, however, attention has been focused on indirect, non-selective threats that are more fundamental and sweeping in scope, such as air and water pollution. Most losses of diversity are unintentional consequences of human activity resulting in habitat destruction (e.g., clearing of rain forests).

# ▪ Why are Wetlands and Rain Forests Important to Us?

## *Concepts*

- Wetlands and rain forests are among the most beneficial biomes on Earth and are being destroyed at staggering rates.

- Wetlands control flood waters.

- Species diversity in the rain forests is unmatched by any other biome on Earth.

Wetlands, everglades, swamplands, marshes. There was a time when these words would conjure up visions of humid, murky land that was not beneficial or useful. Their image as a fertile breeding ground for disease-carrying pests like mosquitoes added the element of threat to human health. Consequently, over 55 percent wetlands in the United States have been converted into industrial, commercial, residential, and farming meccas. However, within the last two decades, a growing awareness of the many environmental benefits and services provided by wetlands has led political and environmental leaders to take a stand against the continued destruction of the wetlands. The major benefits of sustaining wetlands are:

1. Species diversity: Collectively the various kinds of wetlands (swamps, bogs, freshwater marshes, and saltwater marshes) display a great diversity of plants and animals. Because wetlands have been so heavily impacted by human activity, it is not surprising that nearly half of the nation's species of endangered animals and over one-quarter of the nation's endangered plants are wetlands species. In addition to its full-time inhabitants, the wetlands also provide a seasonal sanctuary for many species of migratory birds.

2. Flood control: When heavy rains threaten extreme flood conditions, wetlands provide a natural flood control system, as well as a constant source of fresh water for adjacent rivers. When rivers become filled to capacity, the wetlands catch the overflow and, as the water level in the rivers begins to decrease, the water being stored in the wetlands slowly drains into the rivers, thus maintaining an adequate year-round supply of fresh water.

3. Filtration: As the first destination for much water **runoff**, including water polluted by industrial and farming activity, wetlands collect and trap pollutants. Some substances that could play a negative role, if allowed to flow into the rivers, are actually beneficial to the plants in the wetlands in moderation.

Of all of the biomes on Earth, the tropical rain forests are unmatched in biodiversity. Although the rain forests cover only about seven percent of the earth's surface, more than 50 percent of the world's plants and animals live here. Species we have not studied yet, or even know to exist, may someday be found to be a cure for cancer or they may enhance our ability to produce crops that are more resistant to diseases and pests.

This tropical diversity has also given us such foods as bananas, chocolate, papayas, and a newly-discovered high protein soybean that could feed millions of people. In addition, cosmetics such as deodorant, mouthwash, hand lotion and lipsticks, and industrial products like rubber, gums, resins, and dyes are derived directly from wild tropical species. Potential new products being evaluated for development from the rain forests include biodegradable insect repellent, low-calorie protein sweetener, leaf protein supplement and gasohol – a renewable fuel source.

The rain forests play a vital role in the world's weather patterns which in turn is crucial to ecosystems. They absorb rainfall and release moisture into the atmosphere, thus maintaining the hydrologic cycle. The forests are thought to absorb a large percentage of the worlds $CO_2$ emissions, thus maintaining the necessary temperature range for our continued survival. They create essential oxygen as a by-product of photosynthesis and control soil erosion and landslides, as well as moderate the effects of floods and droughts.

## ■ Checking Your Understanding (4-3)

1. Explain what pH has to do with acid rain.

2. Where does acid rain come from?

3. Why is acid rain a bigger problem in the eastern U.S.?

4. What are CFCs?

5. Describe the greenhouse effect.

6. What has caused carbon dioxide levels to increase over the last decades?

7. List three potential effects of global warming.

8. Explain how the earth's climate might be negatively affected if the rain forests were completely destroyed.

9. Other than aesthetic reasons, explain why we, in developed countries, should be concerned if species diversity is minimized in the rain forests and wetlands.

10. Explain the difference between ecosystem diversity, species diversity, and genetic diversity.

11. List two environmental benefits of wetlands.

# 4-4 Environmental Protection

## ■ History and Background of NEPA

### Concepts

■ The NEPA provides for a national policy for environmental protection.

■ A main component of the NEPA is the requirement for an EIS for government actions.

■ FONSIs are findings of no significant impact.

The previous sections of this chapter demonstrate the importance of a national policy that states our intent to achieve productive and enjoyable harmony between the activities of humans and the environment. This is the goal of the National Environmental Policy Act (NEPA), which was signed into law by President Nixon on New Year's Day in 1970. In ushering in the new law the President stated: "The 1970s absolutely must be the years when America pays its debt

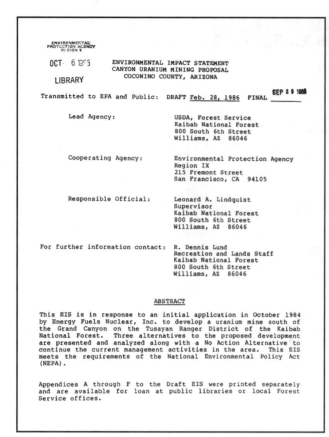

ENVIRONMENTAL IMPACT STATEMENT
CANYON URANIUM MINING PROPOSAL
COCONINO COUNTY, ARIZONA

SEP 2 9 1986

Transmitted to EPA and Public: DRAFT Feb. 28, 1986 FINAL _____

Lead Agency:  USDA, Forest Service
Kaibab National Forest
800 South 6th Street
Williams, AZ 86046

Cooperating Agency:  Environmental Protection Agency
Region IX
215 Fremont Street
San Francisco, CA 94105

Responsible Official:  Leonard A. Lindquist
Supervisor
Kaibab National Forest
800 South 6th Street
Williams, AZ 86046

For further information contact:  R. Dennis Lund
Recreation and Lands Staff
Kaibab National Forest
800 South 6th Street
Williams, AZ 86046

ABSTRACT

This EIS is in response to an initial application in October 1984 by Energy Fuels Nuclear, Inc. to develop a uranium mine south of the Grand Canyon on the Tusayan Ranger District of the Kaibab National Forest. Three alternatives to the proposed development are presented and analyzed along with a No Action Alternative to continue the current management activities in the area. This EIS meets the requirements of the National Environmental Policy Act (NEPA).

Appendices A through F to the Draft EIS were printed separately and are available for loan at public libraries or local Forest Service offices.

**Figure 4-19:** Table of Contents from an EIS document.

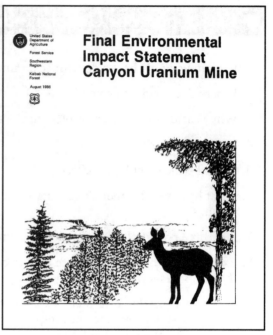

United States
Department of
Agriculture

Forest Service

Southwestern
Region

Kaibab National
Forest

August 1986

**Final Environmental Impact Statement Canyon Uranium Mine**

**Figure 4-20:** Cover of an Environmental Impact

to the past by reclaiming the purity of its air, its waters and our living environment. It is literally now or never."

The NEPA is a statute found in 42 U.S. Code Annotated 4321-4347. Guidelines for its implementation by the EPA can be found in 40 CFR Parts 1500–1508. A peculiar aspect of this law is that no enforcement agency is assigned to enforce its provisions. It is essentially enforced by lawsuits from outside the government. The NEPA did, however, establish a new office in the Executive Branch called the Council on Environmental Quality (CEQ). Although not an enforcement agency, the responsibilities of this office include: studying the condition of the nation's environment; developing new environmental programs and policies; coordinating the myriad federal environmental efforts; assuring that all federal activities take environmental considerations into account; and assisting the President in assessing and controlling environmental problems. Another peculiarity of the NEPA is that its application is limited to government actions. If something doesn't require government approval, it is outside the realm of the NEPA.

The cornerstone of the NEPA is a requirement that the environmental impact of governmental actions be assessed. All government agencies must complete an **Environmental Assessment** prior to beginning a new project or issuing a recommendation. If the proposed action is found to present no significant impact to the environment, a report called a **Finding of No Significant Impact (FONSI)** must be filed. This finding must be based on an initial study that was reviewed by the public and other agencies. The lead agency involved must approve the finding and provide public notice of this determination. If it is determined that a particular activity will have a significant environmental impact, a public document called an Environmental Impact Statement (EIS) must be prepared. The EIS must assess in detail the potential environmental impact of a proposed action. The law states that each federal agency must prepare an EIS *in advance* of each major action, recommendation, or report on legislation that may significantly affect the quality of the human environment.

Examples of federal actions that would require an EIS include new highway construction, harbor dredging or filling, nuclear power plant construction, munitions disposal, siting of a hazardous waste disposal site facility, municipal **landfill** siting, large-scale aerial pesticide spraying, river channeling, and the list goes on and on. Private industry may also be required to complete the EIS process. This would be true if government funding or regulations are involved. Industry routinely does environmental assessments and environmental impact statements.

The primary purpose of the EIS is to reveal the environmental consequences of a proposed action. A draft statement is initially prepared that is reviewed by appropriate federal, state, and local environmental agencies as well as the public. After comments are received from all the interested parties, the EIS is prepared in final form. The final form must address all comments and objections received on the draft version and must state how issues of concern have been resolved. The draft and final EISs are filed with the CEQ and are available to the public. The goal of this disclosure process is to alert agencies, the public, the President, and Congress to environmental risks associated with projects in which the Federal government has decision-making powers.

NEPA dictates whether an action requires an EIS based on its environmental impact. The following guidelines were developed in order to determine whether an action qualifies as *major and environmentally significant*:

– Actions whose impact is significant and highly controversial on environmental grounds;

– Actions that are precedents for much larger actions which may have considerable environmental impact;

– Actions that are decisions in principle about major future courses of action;

– Actions that are major because of the involvement of several Federal agencies, even though a particular agency's individual action is not major; and

– Actions whose impact includes environmentally beneficial as well as environmentally detrimental effects.

Actions is used very broadly to include:

– Agency recommendations on their own proposals for legislation.

– Agency reports on legislation initiated elsewhere but concerning subject matter for which the agency has primary responsibility.

– Projects and continuing activities which may be:

    a. undertaken directly by an agency;

    b. supported in whole or in part through federal contracts, grants, subsidies, loans, or other forms of funding assistance; or

    c.  part of a Federal lease, permit, license, certificate or other entitlement for use.
- Decisions of policy, regulation, and procedure-making.

    Draft EISs must be circulated for comment at least 90 days before the proposed action they pertain to is scheduled to take place. The final statement must be made public at least 30 days before the proposed action. Any deviations from these requirements must be reviewed and approved by the CEQ. It is the responsibility of each agency to obtain the views of other agencies and interested parties at the earliest feasible time in the development of program and project proposals.

    NEPA directs each agency to establish procedures for dealing with environmental impact statements. Agencies must also make sure their standard operating procedures are in line with national environmental goals. Each federal agency must consult with and obtain the comments of any other agency (including state and local as well as federal) which has jurisdiction by law or special expertise with respect to any environmental impact involved in an impact statement. Guidelines for the development and review of EISs state that the following areas must be reviewed by appropriate agencies:

- air quality and air pollution control
- weather modification
- environmental aspects of electric energy generation and transmission
- natural gas energy development, generation, and transmission
- toxic materials
- pesticides and herbicides
- transportation and handling of hazardous materials
- coastal areas: wetlands, estuaries, waterfowl refuges and beaches
- historic and archeological sites
- flood plains and watersheds
- mineral land reclamation
- parks, forests, and outdoor recreational areas
- soil and plant life, sedimentation, erosion and hydrologic conditions
- noise control and abatement
- chemical contamination of food products
- food additives and food sanitation
- microbiological contamination
- radiation and radiological health
- sanitation and waste systems
- shellfish sanitation
- transportation and air quality
- transportation and water quality
- congestion in urban areas, housing and building displacement
- environmental effects with special impact on low-income neighborhoods

- rodent control
- urban planning
- water quality and water pollution control
- marine pollution
- river and canal regulation and stream channelization
- wildlife

## ■ The Roles of Agencies and Interested Parties

### Concepts

■ EPA is a major reviewer of EISs.

■ The input and disclosure processes built into NEPA give the public an opportunity to participate in the decision-making process.

■ Many states have statutes that parallel NEPA.

The Council on Environmental Quality assists other agencies in determining whether NEPA is applicable to a particular action. It also develops agency procedures and resolves any conflicts that may arise as a result of NEPA provisions. Its role, therefore, is that of a coordinator. It does not actually comment on EISs, nor does it approve or disapprove of comments made by other parties. As long as the CEQ has an EIS, it can fulfill its role as internal advisor to the Executive Branch, as designated in NEPA.

In contrast, the EPA reviews EISs that deal with its responsibilities for environmental protection, including but not limited to air and water pollution, drinking water supplies, solid waste, pesticides, radiation, and noise. In addition to reviewing federal statements, the EPA frequently reviews statements filed by states and other jurisdictions. The EPA's obligation to review proposed federally supported actions extends beyond that of other agencies because of its role as the principal federal regulator of environmental control matters.

For instance, the Clean Air Act states that the EPA must review and comment in writing on the environmental impact of any legislation, action or regulation proposed by any federal agency if it affects matters related to the EPA's jurisdiction. If EPA determines that any proposed activity is unsatisfactory from the standpoint of public health or welfare or environmental quality, that determination must be published and the matter referred to the CEQ.

Periodically, the EPA lists EISs it has reviewed in the *Federal Register* and provides a summary of the nature of its comments. The EPA can indicate that it endorses the proposed action, requests more information to determine environmental damage, or objects to the action on environmental grounds. The EPA has no authority to stop a project sponsored by another federal agency. It can only act in an advisory capacity to other agencies, the CEQ, and the President.

The public disclosure and input process built into NEPA give the public an opportunity to participate in federal decisions that may have an effect on the environment. Procedures to reach the public include: announcements in the *Federal Register*, press releases at various stages in the EIS development and review process, publications placed in federal repository libraries, and agency bulletins/regulatory subscription services.

Interested members of the public at large may submit comments to agencies on any impact statement issued by that agency. If an individual believes a draft statement is inadequate, he or she may offer written comments on the draft. If the handling of any comment(s) in the final statement is inadequate, the commenter may notify the agency involved and inform the CEQ. In addition to these mechanisms for public participation, many agencies also hold public hearings.

Federal courts have added substance to the law's requirement for EISs. Suits charging violation of NEPA have dealt essentially with two significant questions: (1) whether a federal agency should be required to write impact statements; and (2) to what extent courts should review the content of statements already written. A number of early court cases resulted in the U.S. Supreme Court greatly extending the range of considerations the courts were allowed to review. Court decisions have temporarily and permanently halted projects due to decisions on lawsuits pertaining to environmental considerations.

About half of all environmental impact statements have been written for road building actions undertaken by the Federal Highway Administration. The statements have resulted in significant planning changes including increased landscaping, the creation of hiking and bicycle trails along roadways, and the integration of mass transit routes into highway corridors. The second largest number of statements have been prepared for watershed protection and flood control projects. A portion of these projects were developed by the U.S. Army Corps of Engineers. NEPA led the Corps to establish citizen advisory committees and increase its environmental staffing by adding resource planners, landscape architects, biologists, and foresters.

One of the first projects to be canceled based on its EIS was a proposed 1,760-foot pier for ocean research that was to extend into the Atlantic Ocean from Maryland's Assateague Island National Seashore. The pier was slated to be built in the early 1970s. Through the EIS process it was disclosed that the area was one of the few remaining natural barriers along the nation's eastern coastline. Because of numerous opposing comments on the statement, the Army Corps of Engineers canceled all construction plans.

Another early project canceled based on EIS comments was a proposal by the Department of Health, Education, and Welfare for an **incinerator** to handle wastes from the Bethesda Naval Hospital, Walter Reed Army Medical Center, and the National Institutes of Health Bethesda Campus. After the draft impact statement showed several preferable alternative means of disposal, the plans for the incinerator were scrapped.

Since the passage of NEPA, we also recognize the need to analyze the environmental impact of actions taken by other groups in addition to federal agencies. Many states have passed statutes requiring environmental impact statements on state actions similar to the statements NEPA requires of federal agencies.

**Figure 4-21:** Planned projects like this incinerator at the New York City Sanitation Department can be halted based on the contents of an EIS.

## ■ The Contents of an Environmental Impact Statement

### Concepts

■ The law states the minimum content of an EIS.

■ Final EISs must discuss concerns that were identified during draft stages.

The law states that each environmental impact statement must include:

- A detailed description of the proposed action including information and technical data adequate to permit a careful assessment of environmental impact.

- Discussion of the probable impact on the environment, including any impact on ecological systems and any direct or indirect consequences that may result from the action.

- Any adverse environmental effects that cannot be avoided.
- Alternatives to the proposed action that might avoid some or all of the adverse environmental effects, including analysis of costs and environmental impacts of these alternatives.
- An assessment of the cumulative, long-term effects of the proposed action including its relationship to short-term use of the environment versus the environment's long-term productivity.
- Any irreversible or irretrievable commitment of resources that might result from the action or which would curtail beneficial use of the environment.

A final impact statement must include a discussion of problems and objections raised by other federal, state, and local agencies, private organizations and individuals during the draft statements review process.

An EIS varies in length according to the complexity of the proposed action under review. Each statement ordinarily is introduced by a summary sheet suggesting the nature of its contents. This summary sheet will name the agency responsible for the action. It will give a brief description of the action, indicating states and counties particularly affected. The environmental impact and adverse environmental effects will be summarized, and all the alternatives that were considered will be listed. In the draft stage, all the federal, state, and local agencies from which comments have been requested will be listed. In the final statement, those agencies from which written comments have been received will be listed. Finally, dates that the draft and final statements are filed with the CEQ and made available to the public will be noted.

## ■ Checking Your Understanding (4-4)

1. Explain how NEPA is enforced.

2. Define FONSI.

3. List three pieces of information you would expect to find in an EIS.

4. List a potential action in your community that would require an EIS.

5. Describe the role of the public in the EIS process.

# 5

# Examining the Environmental Compartments

## ■ Chapter Objectives

After completing this chapter, you will be able to:

1. **List** the major sources of air, water, and land pollution.

2. **Describe** basic geologic and hydrologic terminology associated with an aquifer.

3. **Describe and give examples** of the types of environmental and health damage caused by air, water, and soil pollution.

4. **Examine** the impacts of human activities on air, water, and soil quality.

5. **List** factors that influence land-use decision-making.

6. **Relate** how the environmental compartments are interconnected.

7. **Describe** primary processes that control the fate of pollutants in the environment.

## ■ Terms and Concepts to Remember

Air Pollution
Airborne Particulates
Ambient Air Quality
Anthropogenic sources
Aquifer
Aquitard
Avogadro's Law
Bioassay
Biogenic sources
Biomonitoring
Boyle's Law
Charles' Law
Confined Aquifer
Confining Bed
Criteria Pollutants
Diffusion
Dispersion
Dredging
Effluent
Gay-Lussac's Law
Geogenic Sources
Gradient
Hydrocarbon
Igneous Rocks
Impoundment
Infiltration
Injection Well
Kinetic Molecular Theory
Lagoon
Leachate
Metamorphic Rocks

Mitigation
Mobile Sources
National Ambient Air Quality
  Standards (NAAQS)
Nitrogen Oxides (NOx)
Nonattainment
Nonpoint Sources (NPS)
Particulate Matter (PM-10)
Pathogens
Permeability
Photochemical Reactions
Plume
Point Sources
Porosity
Primary Pollutant
Saturated Zone
Secondary Pollutant
Sedimentary Rocks
Sludge
Smog
Soil
Soil Profile
Stationary Sources
Temperature Inversion
Troposphere
Unsaturated Zone
Vadose Zone
Volatile Organic Compounds
  (VOCs)
Wastewater
Well

## ■ Chapter Introduction

In the last chapter we learned about the delicate and fragile balance between livings things and their environment. We also learned that humans are part of the environment and their activities impact it dramatically. We now realize that environmental concerns are global. Pollutants travel across the geographic boundaries that nations establish and cause damage in a number of ways.

This chapter examines the air, water, and land compartments of the environment to explore how pollution in one affects the others. The sources of pollution, pollution effects, and movement of pollutants for each compartment will be examined since ultimately, pollutants in one compartment will invariably find passage into others.

# 5-1 Air Pollution

## ■ Introduction and Overview

### Concepts

- Human activities are major sources of air pollution.

- Environment, health, and welfare are harmed by air pollution.

- Smog in cities is now generally caused by photochemical reactions.

For anyone living in a crowded city, it's hard to envision a world without smog. The word **smog** is a relatively recent term describing a type of pollution associated with urban settings and smokestack industries. The word itself, a blend of the words *smoke* and *fog*, was first used in the early 1900s by a French physician in an address to a group of public health professionals in London. In his presentation,

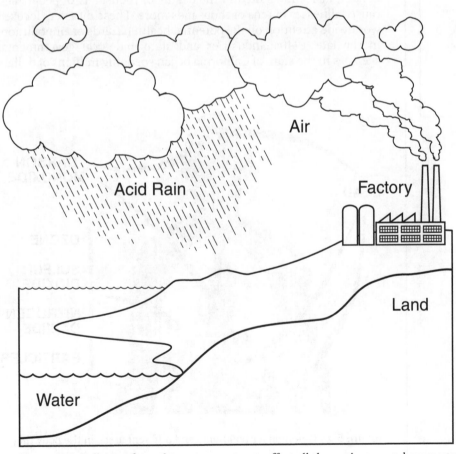

**Figure 5-1** Pollution from the same source can affect all the environmental compartments

**Health research has established that air pollution:**

- aggravates cardiovascular and respiratory illnesses;

- adds stress to the cardiovascular system, forcing the heart and lungs to work harder in order to provide oxygen;

- speeds up the natural aging process of the lungs, accelerating the loss of lung capacity;

- damages cells in the airways of the respiratory system;

- damages the lungs even after symptoms of minor irritation disappear;

- contributes to the development of disease including bronchitis, emphysema, and possibly cancer.

Dr. H. A. des Voeux stated that it required no science to see that there was something produced in great cities which was not found in the country, and that was smoky fog, or what was known as smog. A newspaper reporter actually thanked Dr. des Voeux for benefiting the public by providing a new word for London's smoky fog. Later in this section we will learn how the meaning of this term has changed in modern usage.

**Air pollution** refers to the accumulation of substances in the atmosphere that can cause harmful health effects to living things or can negatively affect the public welfare. Negative effects of public welfare include the economic impact of damage to crops or property, such as buildings or works of arts. Air pollution is the result of human activity as well as naturally-occurring phenomenon. Transportation, power and heat generation, industrial processes, and the burning of solid wastes are the major sources of pollution due to human activities. Volcanic eruptions and naturally occurring fires such as those caused by lightning storms are natural causes of air pollution.

Initially, the most obvious effect of air pollution was that it reduced visibility; however, the most important reason for reducing levels of air pollution is the health problems that it causes. The first air pollution laws in the United States were passed by cities mainly to control smoke emissions. Chicago and Cincinnati enacted regulations in the 1880s, and New York and Pittsburgh followed suit in the next decade. Killer fogs in Donora, Pennsylvania in 1948 and in London in 1952 resulted in the deaths of thousands of people and caused illnesses in tens of thousands more. These disasters focused worldwide attention on the potential health hazards of air pollution. In the late 1940s, industries and state and local governmental agencies in the state of California began spending millions of dollars

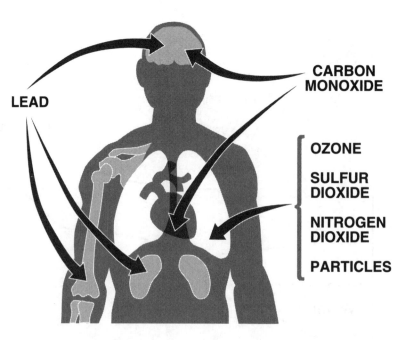

**Figure 5-2:** Common air pollutants and different areas in the body they

to study the causes and effects of smog. In 1952, Oregon became the first state to pass comprehensive statewide legislation and establish a state air pollution control agency.

The EPA has identified air pollution as one of the greatest risks to human health and the environment in our country. It also represents a significant risk on a global scale. The long list of health problems brought on or aggravated by air pollution includes: lung diseases, such as chronic bronchitis and pulmonary emphysema; cancer, particularly lung cancer; neural disorders including brain damage; bronchial asthma and the common cold, which are most persistent in places with highly polluted air; and eye irritation. Environmental problems range from damage to crops and vegetation to acid rain, which we learned in the previous chapter has increased the acidity of some lakes making them unlivable for fish and other aquatic life, to global warming and destruction of the stratospheric ozone layer, which can potentially have devastating effects on our planet. Air pollution is a part of everyday life for millions of people the world over who live in highly industrialized and populated urban areas. They may be regularly exposed to air pollution levels that can cause nausea, headaches, dizziness, and shortness of breath, even among healthy adults.

While air pollution affects everyone to one degree or another, some people are extremely susceptible to severe health damage. Those at high risk include young children whose respiratory systems are still developing, those who suffer from diseases of the heart or respiratory systems, and healthy adults who exercise vigorously. In combination, these groups represent a significant portion of the population.

Today, when we speak of smog we generally no longer mean the visible smokestack emissions of soot and smoke that the term once referred to exclusively. As the use of coal has gradually been replaced by petroleum products, a new kind of smog has become predominant in many cities. Now, the term is also applied to air pollution caused by the **photochemical reactions** of certain common pollutants which form ozone. The prefix *photo* means light; therefore, these reactions require energy in the form of sunlight in order to occur.

The proliferation of ozone is one of the most widespread environmental problems and one of the most difficult to manage. Just about every major urban area of the country exceeds the health-protective limits for ozone established by EPA. Ozone is a colorless gas with a pungent odor, and the chief component of urban smog. Chemically, it is a form of oxygen with three oxygen atoms ($O_3$) instead of the two normally found in oxygen. This makes it very reactive, which means it combines with practically every material it comes into contact with. This reactivity explains the health effects it causes, since it has the ability to break down biological tissues and cells as it reacts with them.

Ozone levels high enough to cause health problems for people are also high enough to damage crops and vegetation. Ozone can also damage building materials and cultural treasures such as priceless and irreplaceable works of art and architecture.

Ozone is produced in the atmosphere when sunlight triggers certain chemical reactions. The precursors, or chemicals, that are initially needed for this reaction to take place include **volatile** organic compounds (VOCs) and nitrogen oxides (NOx). VOCs are released into the air when petroleum products are combusted, handled, or processed. Nitrogen oxides are also produced by combustion.

Ozone levels are typically highest during daytime hours in the summer months when heavy morning traffic releases large amounts of VOCs and nitrogen oxides in the presence of sunlight. The sunlight provides the energy to fuel the reactions between these compounds and naturally occurring atmospheric gases, resulting in the production of ozone. The fact that automobile use continues to grow and has become such an essential part of our society makes this pollutant a difficult one to control. For example, in the four years between 1980 and 1984, Americans increased their driving by almost two billion vehicle miles!

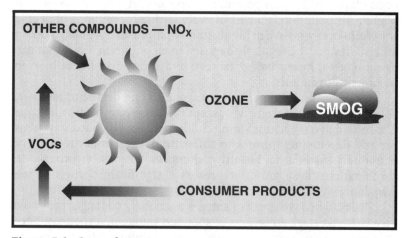

**Figure 5-3:** Ozone formation.

In some urban areas, ozone levels are high enough to trigger health advisories or smog alerts during the summer months. When these levels are reached, even healthy adults and children are advised to avoid or reschedule sustained strenuous outdoor exercise such as soccer and long-distance running. Individuals with heart or lung problems are further advised to reduce their activity and exposure.

Recent research, discussed in more detail in the Air Pollutants and Human Health section of this chapter, provides compelling evidence to dispute the belief that the respiratory system fully restores itself from exposures to ambient air pollution. Related investigations are underway or planned which will provide further information regarding the risk to human health that may be associated with long-term exposure to air pollution. Recent findings indicate that some pollutants, especially ozone, are more harmful in lower concentrations than previously thought. This research has determined that high pollution levels can cause immediate health problems, and that chronic exposure to lower concentrations may be the basis for lifelong, permanent health damage.

California provides one of the best examples of the challenges of controlling air pollution. The combination of population growth, climate, and geography in California make its atmosphere Mother Nature's perfect smog chamber. Emissions in that state very quickly

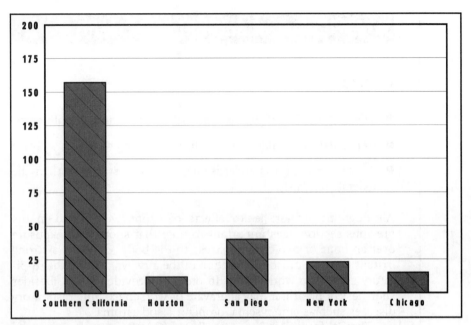

**Figure 5-4:** Number of days per year when ozone levels exceeded 0.12 ppm. (Source: ARB)

transform into major air quality problems; greater problems, in fact, than the same amount of emissions would make in some other part of the nation.

It is because of this natural phenomenon that California has the twin distinctions of having the most severe air quality problem in the nation and the country's most advanced air pollution controls. The EPA lists more than 100 cities with ozone levels above acceptable health standards. Nine California metropolitan areas are routinely listed within the top 20 spots. Many of the listed cities outside California, however, only violate air quality standards occasionally (Figure 5-4). None have chronic air quality problems as severe as those in California.

In fact, when the real extent of health hazard violations is measured, and the number of people exposed to excessive pollution counted, the result is a different perspective on the severity of the nation's air quality problems. Three-fourths of the nation's health problems from excessive ozone or smog are concentrated in the greater Southern California area alone. All told, California's air quality problems outrank those from the remaining 49 states combined.

There has been dramatic progress toward cleaner air in California, the result of an air pollution control program that is reputed to be the strictest in the world. California's unique emissions standards for cars, trucks, buses, motorcycles, and other motor vehicles have dramatically cut emissions, as have tight controls on non-automotive pollution sources. The battle is not won, however, since continued growth in population and overall vehicle use offset much of the gain.

As we have seen in the case of ozone, air quality is affected dramatically by human activities. Nowhere is the link between resource abuse and degradation more obvious than in a discussion of air pollution. The next section looks at specific air pollutants in order to more fully explore their effects on air quality.

## ■ Air Pollutants and Human Health

### Concepts

- The EPA has health-based standards for common air pollutants.

- Very small particulates are implicated in lung damage.

- States' air quality standards may be more stringent than the federal standards.

Although the major health effects of ozone were listed in the previous section, it is important to note that short-term exposure over an hour or two can add stress to the body. Ozone is a strong irritant that can cause constriction of the airways, forcing the respiratory system to work harder in order to provide oxygen. Besides shortness of breath, it may aggravate or worsen existing respiratory diseases such as emphysema, bronchitis and asthma.

The California Air Resources Board (CARB) recently performed research that has provided preliminary evidence that some degree of permanent lung damage may occur in young adults, aged 14 to 25, who have been lifelong residents of Southern California. This pilot study examined the lungs of young accident and homicide victims. The frequency of certain types of lung damage observed in this young population raises concerns regarding the health effects of long-term exposure to air pollution, including high levels of ozone, found in Southern California and other metropolitan areas.

The broadest finding among these subjects was some degree of chronic inflammation of the bronchial tubes. Nearly all of the lungs examined had some form of chronic bronchitis and 76 percent showed some degree of inflammation. In addition, about one-third of the subjects had some degree of chronic interstitial pneumonia, a form of the disease found deep within lung tissue. Chronic exposure to ozone can damage deep portions of the lung, even after symptoms such as coughing or a sore throat disappear. Ozone can damage the alveoli and cause the loss of lung capacity.

This information led CARB to develop standards which require health advisories when one-hour ozone concentrations reach 0.15 ppm. These smog alerts provide warnings so that residents can take precautions to protect their health from excessive ozone levels, usually by avoiding strenuous exercise or outdoor exposures. The federal standard for allowable ozone in the ambient air is 0.12 ppm, averaged over one hour.

In addition to ozone, common pollutants that were among the first to be regulated throughout the country include: carbon monoxide, airborne particulates, sulfur dioxide, lead, nitrogen oxides, asbestos, beryllium, mercury, vinyl chloride, arsenic, radionuclides, benzene, and coke oven emissions. Their major health effects are listed in Figure 5-5.

Carbon monoxide (CO) was introduced in Chapter 3 as a serious health hazard due to its special ability to bind with hemoglobin and cause chemical asphyxiation. The result is reduced oxygen reaching the heart, brain, and other tissues. This can be critical for people with heart disease, chronic lung disease or anemia, as well as unborn children. Even healthy people who are exposed to excessive CO can experience headaches, fatigue, slow reflexes and dizziness. Health damage caused by CO is of greater concern at high elevations where the air is less dense, aggravating the consequences of a reduced oxygen supply.

CO is an invisible, odorless gas that is a product of incomplete fuel combustion, primarily from motor vehicle exhaust. The highest concentrations are found in areas with congested or high volumes of traffic and during the winter months. The EPA's air quality standards are 35 ppm averaged over one hour and 9 ppm averaged over eight hours. The standards are designed to prevent chest pain in moderately exercising people who have heart problems, but other types of health damage can result from higher concentrations. In recent years, carbon monoxide levels have declined about 40 percent, the result of stricter tailpipe emission limits for cars, trucks, and buses.

Sulfur dioxide ($SO_2$) is produced primarily by the combustion of coal, fuel oil, and diesel fuel. EPA's standard is 0.14 ppm averaged over 24 hours. Sulfur dioxide causes a constriction of the airways and poses a particular health hazard for asthmatics. The air quality standard was set to protect people from breathing difficulties during and after short periods of exercise. Children exposed to sulfur dioxide experience increased respiratory tract infections and healthy people may experience sore throats, coughing and breathing difficulties when exposed to high sulfur dioxide concentrations.

Nitrogen dioxide ($NO_2$) is frequently a by-product of combustion and is emitted from sources such as motor vehicles, industrial boilers and heaters. It is an irritating gas that may increase the susceptibility to infection and may constrict the airways of asthmatics. It is one of the pollutants from the group known as nitrogen oxides (NOx), which are a major component of urban smog and are responsible for its reddish-brown haze. In winter months, however, when the photochemistry that forms ozone is lowest, nitrogen dioxide concentrations remain high. The EPA air quality standard for nitrogen dioxide is 0.053 ppm, averaged over one year.

**Particulate matter**, such as diesel soot and products resulting from wood burning, can be emitted directly into the air. It can also be produced through photochemical reactions among polluting gases, primarily sulfur oxides and nitrogen oxides, resulting in corrosive sulfate or nitrate ions. Although all particles

| Health Effects of Common Air Pollutants | |
|---|---|
| Criterial Pollutants | Health Concerns |
| Ozone | Respiratory tract problems such as difficult breathing and reduced lung function. Asthma, eye irritation, nasal congestion, reduced resistance to infection, and possibly premature aging of lung tissue. |
| Particulate Matter | Eye and throat irritation, bronchitis, lung damage, and impaired visibility. |
| Carbon Monoxide | Ability of blood to carry oxygen impaired. Cardiovascular, nervous, and pulmonary systems affected. |
| Sulfur Dioxide | Respiratory tract problems; permanent harm to lung tissue. |
| Lead | Retardation and brain damage, especially in children. |
| Nitrogen Dioxide | Respiratory illness and lung damage |
| Hazardous Air Pollutants | |
| Asbestos | A variety of lung diseases, particularly lung cancer. |
| Beryllium | Primary lung disease, although also affects liver, spleen, kidneys, and lymph glands. |
| Mercury | Several areas of the brain as well as the kidneys and bowels affected. |
| Vinyl Chloride | Lung and liver cancer |
| Arsenic | Causes cancer |
| Radionuclides | Causes cancer |
| Benzene | Leukemia |
| Coke Oven Emissions | Respiratory cancer |

Figure 5-5: Health effects of common air pollutants.

can pose a potential health problem, the greatest concern is for microscopic, invisible particles which are the greatest health threat. These particles are less than 10 microns (a micron is a micrometer, or 1/1,000,000th of a meter) in diameter, about one-fifth the size of a human hair, and are known as **PM-10**. The EPA's standards for these small particles are 50 ug/m$^3$ averaged over a year and 150 ug/m$^3$ averaged over 24 hours.

Concerns about these particles are based on their ability to bypass the body's natural filtering system, posing a threat to the respiratory tract. Short-term exposures can lead to increased bronchial disease. In addition, some of the directly emitted particulates, such as diesel soot and wood smoke, can be carriers for other toxic compounds including benzene and dioxins, which increase potential cancer risks.

Even though all particles 10 microns or less in diameter present a health problem, they affect different parts of the respiratory tract depending on their size. Particles from 2.5 to 10 microns in diameter tend to collect in the upper portion of the respiratory system, affecting the bronchial tubes, nose and throat. Those particles 2.5 microns and smaller in diameter can infiltrate deeper portions of the lung and remain there longer, increasing the risks of long-term disease.

Airborne lead is a pollutant that has been dramatically reduced in recent years as a direct result of the reduction of lead in gasoline that was required by EPA regulations. Currently, at least 70 percent of the lead in the air is due to industrial sources.

Lead particles small enough to be inhaled into the lungs are readily absorbed into the blood and circulated throughout the body. The most crucial area damaged by lead is the brain. At relatively low levels, lead exposure can result in a permanent decrease in the IQ of children. At higher levels, anemia can occur in both adults and children. The EPA air quality standard for lead is set at 1.5 ug/m$^3$ averaged over a calendar quarter.

The primary purpose of the EPA's air quality control program is to reduce outdoor pollution. This is referred to as **ambient air quality**. The Clean Air Act identifies **National Ambient Air Quality Standards (NAAQS)** for carbon monoxide, lead, nitrogen oxides, ozone, particulate matter, and sulfur dioxide, which are known as **criteria pollutants**. When an area does not meet these national standards, they are said to be in **nonattainment** and the federal government can impose sanctions to gain compliance. The next chapter will cover these and other provisions of the Clean Air Act. It is important to note that California and some other states have imposed even stricter air quality standards than the federal standards. As an example, Figure 5-6 compares federal and California standards for several pollutants.

| Ambient Air Quality Standards | | | | | | |
|---|---|---|---|---|---|---|
| Pollutant | Averaging Time | California Standards | | National Standards | | |
| | | Concentration | Method | Primary | Secondary | Method |
| Ozone | 1 Hour | 0.09 ppm (180 ug/m$^3$) | Ultraviolet Photometry | 0.12 ppm (235 ug/m$^3$) | Same as Primary Standard | Ethylene Chemiluminescence |
| Carbon Monoxide | 1 Hour | 9.0 ppm (10 mg/m$^3$) | Non-dispersive infrared spectroscopy (NDIR) | 9 ppm (10 mg/m$^3$) | | Non-dispersive infrared spectroscopy (NDIR) |
| | 8 Hour | 20 ppm (23 mg/m$^3$) | | 0.053 ppm (100 ug/m$^3$) | | |
| Nitrogen Dioxide | Annual Average | | Gas Phase Chemiluminescence | 0.053 ppm (100 ug/m$^3$) | Same as Primary Standard | Gas Phase Chemiluminescence |
| | 1 Hour | 0.25 ppm (470 ug/m$^3$) | | | | |
| Sulphur Dioxide | Annual Average | | Ultraviolet Flourescence | 80 ug/m$^3$ (0.03 ppm) | | Pararosoaniline |
| | 24 Hour | 0.05 ppm (131 ug/m3) | | 365 ug/m (0.14 ppm) | | |
| | 3 Hour | | | | 1300 ug/m3 (0.5 ppm) | |
| | 1 Hour | 0.25 ppm (655 ug/m$^3$) | | | | |
| Suspended Particulate Matter (PM$_{10}$) | Annual Geometric Mean | | Size Selective Inlet High Volume Sampler and Gravimetric Analysis | | | Inertial Separation and Gravimetric Analysis |
| | 24 Hour | | | 150 ug/m$^3$ | Same as Primary Standard | |
| | Annual Arithmetic Mean | | | 50 ug/m$^3$ | | |
| Sulfates | 24 Hour | 25 ug/m$^3$ | Turbidimetric Barium Sulfate | | | |
| Lead | 30 Day Average | 1.5 ug/m$^3$ | Atomic Absorption | | | Atomic Absorption |
| | Calendar Quarter | | | 1.5 ug/m$^3$ | Same as Primary Std. | |
| Hydrogen Sulfide | 1 Hour | 0.03 ppm (42 ug/m$^3$) | Cadmium Hydroxide STRactan | | | |
| Vinyl Chloride (chloroethene) | 24 Hour | 0.010 ppm (26 ug/m$^3$) | Tedlar Bag Collection, Gas Chromatography | | | |
| Visibility Reducing Particles | 8 Hour (10 am to 6 pm, PST) | In sufficient amount to produce an extinction coefficient of 0.23 per kilometer due to particles when the relative humidity is less than 70 percent. Measure in accordance with ARB Method V. | | | | |
| Applicable Only in the Lake Tahoe Air Basin | | | | | | |
| Carbon Monoxide | 8 Hour | 6 ppm (7 mg/m$^3$) | NDIR | | | |
| Visibility Reducing Particles | 8 Hour (10 am to 6 pm, PST) | In sufficient amount to produce an extinction coefficient of 0.07 per kilometer due to particles when the relative humidity is less than 70 percent. Measure in accordance with ARB Method V. | | | | |

**Figure 5-6:** Ambient air quality standards.

# ■ Smog and Crops

## Concepts

■ Smog can significantly impact the economy because of agricultural damage.

■ Ozone is the most serious threat to vegetation.

■ National forests, native plants, and range grasses are also affected by air pollution.

| Air pollution affects plants by: |
| --- |
| ■ injuring leaves, stems, and roots; |
| ■ reducing yield, cutting fruit size and weight; |
| ■ cutting market value by spotting leaves and fruit; |
| ■ causing plant death. |

The welfare of the people of a state not only hinges on the healthfulness of their environment but also on the condition of the economy. Air pollution can negatively impact a state's economy. For example, California, which has the world's richest and most productive farmland (agribusiness provides about one-half of the nation's fruits and vegetables), has been severely impacted economically. The state's farmers produce an average of $17 billion in crops each year and one job out of every six to California's economy. In addition, related industries such as packing, canning, textiles, and machinery make the total value of California agriculture more than $70 billion a year. California's number one industry, however, may be losing more than $300 million each year to air pollution. Smog damage is a major reason why crops such as spinach, celery, lettuce, tomatoes, string beans, and cucumbers, as well as many ornamental plants, are no longer grown commercially around metropolitan Los Angeles. For this reason, much research was done in California on the effects of air pollution on plants. This section will highlight some of the key findings of this research.

Smog does not just damage crops; it also damages forests, range and pasture lands that produce another $700 million in revenue for California each year. These natural ecosystems account for approximately 85 percent of California's land area and provide recreation and watershed land as well as support for the timber and livestock industries. Extensive injury to trees was discovered in the mountains east of Los Angeles during the 1960s and tree injury in Southern California parks and in the Sierra Nevada mountains was first reported during the 1970s.

During the 1940s, Southern California researchers were puzzled by what was called the X disease that damaged trees, but whose source could not be traced. By the early 1950s, however, a clear link had been established between Los Angeles smog and the mysterious plant disease. Studies confirmed that pollutants emitted from Los Angeles factories and freeways were blown to farming areas downwind. By the mid-1960s, crop damage from smog was apparent in all of the state's most important agricultural regions.

California's geography and climate coupled with population growth and heavy dependence on cars is at the root of its air pollution problems. The mountains surrounding valleys form basins that trap and hold air pollution, allowing the accumulation of high concentrations of pollutants. As pollution spreads throughout these valleys, a phenomenon referred to as a **temperature inversion** puts a lid on those valley, preventing the air pollution from escaping. The trapped air pollution can then bake under the state's sunny skies and can be converted into ozone. Nearly all of the state's major crop-producing areas are located in valleys or basins. Furthermore, when trapped air pollution does climb the valley walls, it is carried to the mountain regions, where it can damage trees and grasses above the valleys.

Ozone is the most serious threat to vegetation. It attacks leaves, causing them to yellow, develop dead spots and drop early (Figure 5-7). Low level ozone exposure over long time periods can reduce a plant's growth and fruit yield and increase its susceptibility to disease and insect attack. It also interferes with photosynthesis. When leaves yellow and develop dead areas, the surface area available for photosynthesis is reduced. California reduced its ozone standard in 1987 from 0.10 ppm to the current 0.09 ppm to better protect the state's crops and natural vegetation from the effects of continued ozone exposure. Standards that protect public health are also expected to lower vegetation damage.

CARB research shows that sulfur dioxide ($SO_2$) may also damage plants. Short-term, high concentration sulfur dioxide exposure can reduce root and stem weight, as well as cut protein and carbohydrate content and ultimately result in plant death. Other less common pollutants also affect vegetation. Fluorides, ammonia and ethylene, by-products of industrial processes, as well as boron and hydrogen sulfide, emitted from geothermal operations, can injure leaves and reduce plant growth.

One of the best examples of a crop significantly affected by air pollution is cotton, which is the state's biggest single agricultural product, worth $1 billion per year. Through the use of computer models, scientists estimate that the average cotton yield loss from ozone during 1988 was about 16 percent, with the highest reductions estimated at about 44 percent in the southern San Joaquin Valley.

Other crops, such as grapes, potatoes, beans, and lettuce have also suffered air pollution damage. Statewide, the average loss estimate for all grapes was approximately 25 percent during 1988. One study demonstrated losses of more than 40 percent in total potato number and yield in smoggy air. The leaves of beans develop dead yellow spots in smoggy air and the plants die sooner than those grown in clean air. Even more important, the effects of ozone and sulfur dioxide reduce the weight, number of seeds and pods and yield of beans. Lettuce exposed to polluted air produces smaller, lighter heads and develops dead areas which lessen their market value. These effects to crops occur below the current California air quality standard for ozone of 0.09 ppm for one hour.

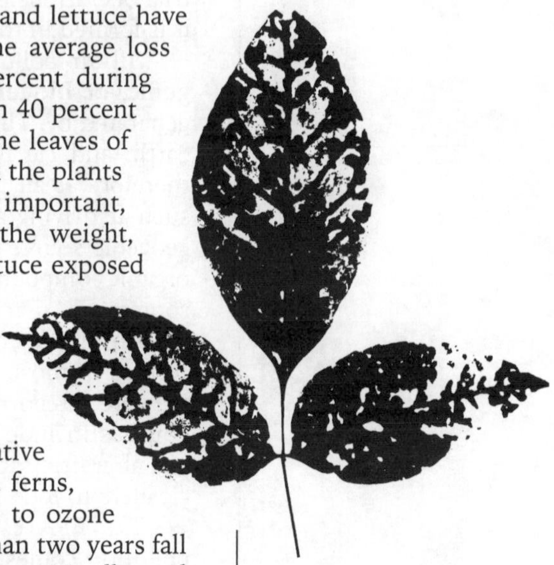

Air pollution is known to harm all major native plant groups, including flowering plants, conifers, ferns, mosses, lichens and fungi. Pine needles exposed to ozone develop yellow, blotchy marks and needles older than two years fall off, giving branches a scraggly, whiskbroom appearance. Needles and debris from trees killed by smog not only increase the risk of forest fire, but reduce seed germination and the chances of seedling survival. The most devastating effect to native plants is a reduced ability

**Figure 5-7:** Discolored leaves are a result of air pollution that damages crops and reduces yield.

to cope with drought, disease, and insects. Air pollution may put these plants at a reproductive disadvantage causing them to produce fewer seeds. These conditions can lead to changes in succession resulting in a totally different plant community occupying a site. This in turn can affect an entire ecosystem due to alteration of food chains.

# ■ Sources and Fate of Air Pollutants

## Concepts

■ There are two basic categories of air pollutant – primary and secondary.

■ Air pollution can originate from anthropogenic, geogenic, and biogenic sources.

■ Dispersion of pollutants is impeded by temperature inversions.

■ There are set relationships between the temperature, volume, and pressure of a gas.

Air pollutants fall into two categories: primary pollutants and secondary pollutants. **Primary pollutants** differ from **secondary pollutants** in that they enter the atmosphere directly as opposed to being formed from another substance that was released into the atmosphere. Oxides of carbon and nitrogen, and hydrocarbons are examples of primary pollutants. A major source of these are motor vehicle emissions (fossil fuel combustion). In addition, large quantities of sulfur oxides and particulates are directly emitted by industrial operations. An example of a secondary pollutant is ozone since it is formed in the atmosphere by photochemical reactions.

All air pollutants originate from either **anthropogenic, geogenic**, or **biogenic sources**. The suffix *genic* means produced or generated by. The prefix *anthropo* refers to human beings, *geo* to the earth, and *bio* to living organisms. Anthropogenic air pollution, therefore, is air pollution that arises from the activities of humans, such as driving automobiles. A volcano would be an example of a geogenic source of pollution, and pine trees, which release volatile organic compounds such as turpentine, are an example of a biogenic source.

Anthropogenic sources of air pollution are by far the most serious in most areas of the world. For most regulatory purposes, they are categorized as either mobile or stationary sources. **Mobile sources** include passenger cars, trucks, buses, motorcycles, boats, and aircraft. **Stationary sources** can range from oil refineries to dry cleaners to food processing plants to gas stations.

The EPA estimates that more than half of the nation's air pollution comes from mobile sources. Exhaust from such sources contains carbon monoxide, VOCs, NOx, particulates, and lead. Stationary sources generate air pollutants mainly by burning fuel for energy and as by-products of industrial processes. Electric utilities, factories, and residential and commercial buildings that burn coal, oil, natural gas, wood, and other fuels, are sources of pollutants such as sulfur dioxide, NOx, carbon monoxide, VOCs, and lead. Toxic air pollutants also come from a variety of industrial and manufacturing processes.

Some operations, such as burning and grinding, also generate particulate matter. In Oregon, for instance, controlled burning by the agricultural and forestry industries is a common practice. This type of burning is the largest source of PM-10 emissions in the state, generating seven times more than industrial sources. Burning, in fact, generates nearly 100,000 tons of PM-10 in Oregon each year. This has been an ongoing controversy in the state and serves as another example of the conflict between environmental and economic concerns.

Toxic pollutants, such as asbestos, benzene, mercury, vinyl chloride, and radionuclides, are one of the most serious, emerging air pollution problems. Despite their low concentrations, toxic chemicals emitted into the air by human activities may have serious short-term and long-term effects on human health and the environment. Many sources, both mobile and stationary, emit toxic chemicals into the atmosphere: industrial and manufacturing processes, solvent use, waste water treatment plants, hazardous waste handling and disposal sites, municipal landfills, incinerators, and motor vehicles. For example, smelters (used to melt or fuse metals), manufacturing processes, and stationary fuel combustion sources emit toxic metals such as cadmium, lead, arsenic, chromium, mercury, and beryllium. Organic toxic materials such as vinyl chloride and benzene are released by a variety of sources including plastics and chemical manufacturing plants and gas stations. Highly toxic chlorinated dioxins are emitted by some chemical processes and the high-temperature burning of plastics in incinerators.

One of the most notable examples of the problems caused by toxic air emissions is found in an 85-mile industrial corridor that stretches from Baton Rouge to New Orleans, Louisiana. Approximately one-fifth of the nation's petrochemicals are produced in this area which has regrettably come to be known as Cancer Alley. Health statistics show that the population in this area has an unusually high rate of several types of cancers. Although direct cause-and-effect has not been proven, many people believe that the high cancer rate is due to air pollution, particularly by vinyl chloride, one of the most prominent chemicals produced in the corridor. Vinyl chloride is a colorless gas used to manufacture polyvinyl chloride, a major industrial plastic. It has been shown to cause liver cancer, and there is evidence linking it to lung cancer, nervous disorders, and other illnesses.

As mentioned earlier, the EPA's main regulatory focus is ambient air quality. Most people, however, spend 90 percent of their time indoors, where they are also exposed to chemicals and pollutants. Indoor pollutant levels are frequently higher than outdoors, particularly where buildings are tightly constructed to save energy. The EPA estimates that as many as 30 percent of new and remodeled buildings may have indoor air quality problems. These problems typically manifest themselves in complaints of eye, nose, and throat irritations, fatigue, headaches, nausea, irritability, or forgetfulness. Long-term health effects range from impairment of the nervous system to cancer. The term used to describe this type of problem is "sick building syndrome."

Harmful indoor air pollutants include: airborne pathogens (disease-causing organisms) such as viruses, bacteria, and fungi; as well as radioactive gases such as radon; inorganic compounds like mercury and lead; and an array of organic compounds such as formaldehyde, chloroform, and perchlorethylene. These pollutants may come from sources such as tobacco smoke, building materials, furnishings, space heaters, gas ranges, wood preservatives, consumer goods such as air fresheners, and solvents in cleaning agents.

The EPA and state air quality regulatory agencies have increased research efforts to study indoor pollution levels, prompted by concern that exposures are significant and can affect how people react

# Science & Technology

## The Behavior of Gases

**The Kinetic Molecular Theory** states that the particles (atoms, ions and molecules) which make up all matter are in constant, chaotic motion. According to the theory, the particles in solids are packed very closely together so their movement is limited to no more than a shaking or vibrating motion. In liquids, they are still closely packed together, but the attraction between the particles is less than in solids, resulting in groups of them being able to slip past each other. This allows them to take the shape of their container, but still have a fixed volume. The particles in gases have little or no attraction for each other and will move far apart. This enables one gas to move in a random and chaotic path through another. When one gas moves through another, it is called **diffusion**, and explains why the odor of air freshener sprayed in one room is soon smelled throughout the whole house.

When heated, solids, liquids, and gases tend to undergo expansion. Because the particles in both solids and liquids are close together, only a very small amount of expansion occurs. Liquids expand about ten times more than solids. Gases, however, can expand hundreds of times more than liquids.

Unlike solids and liquids, gases do not have a fixed volume. They are said to take the shape and size of their container. To determine the amount of gas present, you must know its pressure and temperature as well as the size of its container. For example, oxygen gas is sold in a heavy walled cylinder where its volume does not change as the gas is released. The difference between a full and empty cylinder is measured by the pressure change

(2,000 psi = full and 15.7 psi = empty). The empty cylinder is still full of gas, but now there are just fewer molecules exerting the pressure. The pressure a gas exerts on the container can also be changed by changing the temperature. As you heat a gas, the energy is converted into the speed of the particles. If the size of the container does not change, then they will hit the wall of the container more often resulting in higher pressure. This is why heating a substance such as hair spray in a thin walled container can be very dangerous. As the molecules move faster and push harder against the sides of the container they may eventually cause the can to burst with an explosive force.

There are several gas laws that help us understand and predict the behavior of gases:

■ **Boyle's Law** states that at constant temperature, the volume of a gas is inversely proportional to the pressure, which is another way of saying the more we squeeze a gas, the smaller it gets.

■ **Charles' Law** defines the relationship between the temperature and the volume of a gas. It states that at constant pressure, the volume of a gas varies proportionally to its temperature. In other words, when you heat a gas it will expand.

■ **Avogadro's Law** states that if we have different gases all at the same temperature, pressure and volume, they will contain the same number of particles. This is true even though one gas may be composed of atoms (He) and the other of molecules ($CH_4$).

to outdoor air quality. In addition, even when levels of indoor pollutants are low, long-term exposure to them can cause a significant health risk.

Indoor pollution can be generated by everyday activities including cooking or the use of common household products such as cleaning agents, paints, and hair spray (Figure 5-8). In addition, common building materials and home furnishings can be a source of toxic vapors. Recent development of miniature monitoring equipment will help in measuring the total amount of many chemical compounds that people are exposed to in a typical day. This will help scientists estimate the total amount of pollution that people breathe, and to develop more effective approaches to reducing total exposures.

There have been many examples in previous sections of the mobility of pollutants in the ambient air. Trees and other plant life, for instance, are oftentimes harmed by smog from urban areas that are far away. In fact, of the environmental compartments, air can move contaminants the fastest, on the order of several hundreds of miles per day. Understanding basic atmospheric processes can help us to understand why pollutants sometimes concentrate and accumulate in certain areas before moving on. Atmospheric science is very complex. Trying to predict the fate of air contaminants requires heavy reliance on the use of modeling to take into account the multitude of factors that influence their transport.

After release into the atmosphere, pollutants are diluted by the air and their ultimate concentration in any given location is affected by air motion and turbulence. Larger particles tend to settle out quickly. Where they settle depends on wind speed and motion. **Dispersion** is the term used to describe the distribution of the pollutant into the atmosphere. Horizontal dispersion is caused by winds. The extent of the dispersion is dependent on the velocity of the wind and how turbulent or agitated and disturbed the wind is. A very smooth wind, for instance, may carry pollutants downwind in a narrow ribbon-like pattern which would not result in additional dilution, and dangerous concentrations could move intact for considerable distances.

Turbulence results in vertical dispersion which is very beneficial because it causes pollutants to be carried upward where wind velocities are generally higher than near the ground. This results in faster and more effective dilution of the pollutants. If vertical dispersion does not occur, troublesome pollutant concentrations are likely to occur near the ground. The conditions that impede vertical dispersion include gentle, or nonexistent winds, tempera-

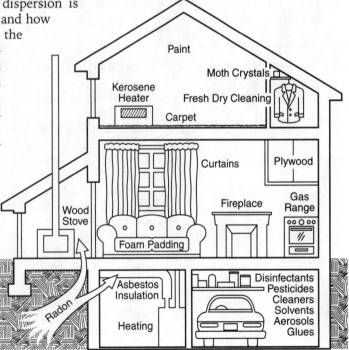

**Figure 5-8:** Air pollution in the home.  (Source: USEPA)

ture inversions, and heavy layers such as smoke that interfere with normal heating of cold air near the ground.

Temperature inversions have been mentioned several times. They are the critical factor in the severe smog conditions in areas like Southern California and Denver, Colorado. Basically, the atmosphere is divided into layers depending on temperature. The troposphere is the layer closest to the earth and is about seven miles in thickness. It is in this layer where weather occurs. Air temperatures in the troposphere typically decrease about 5.4° Fahrenheit for every 1,000 foot rise in altitude, unless there is an inversion.

An inversion denotes that this normal cooling with altitude is reversed. The temperature actually increases with height and the warm air forms a blanket, trapping the cool air beneath it. Under this condition, pollutants will not rise or disperse to any great extent (Figure 5-9). It's easy to see how persistent inversions lasting several days can be very dangerous. This is, in fact, what happened in the historical episodes of killer air pollution cited early in the chapter.

Los Angeles is surrounded on three sides by mountain ranges and on the fourth side, to the west, by the ocean. Breezes blow off the ocean into the basin causing inversions to occur about 320 days of the year. The heights of inversions are variable and early attempts at cures involved elevating smokestacks. However, we now know the most effective measure is to reduce air pollution emissions in the first place.

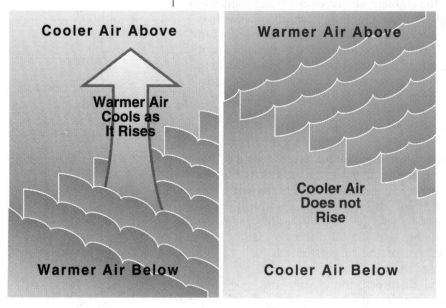

**Figure 5-9:** Normal air movement compared with temperature inversions.

Figure 5-10 shows a view of downtown Los Angeles in the early 1960s. The photograph shows the area engulfed in smog with the estimated base of the inversion about 1500 feet above ground level.

Inversions are not a new phenomenon. Probably the first recorded description of one was in a London resident's diary entry in January, 1684:

*The frost still continuing more and more severe . . . London, by reason of the excessive coldnesse of the aire, hindering the ascent of smoke, was so filled with the fuliginous steame of the Sea-Coale, that hardly could one crosse the streete, and this filling the lungs with its grosse particles exceedingly obstructed the breast, so as one could scarce breathe.*

**Figure 5-10:** View of downtown Los Angeles during rush hour, October 8, 1980.

## ■ Checking Your Understanding (5-1)

1. Explain why California presents the nation's highest potential health threat from smog.

2. List four potential health impacts of air pollution.

3. Define *anthropogenic, biogenic,* and *geogenic.*

4. Explain what a temperature inversion is and how it affects air quality.

5. Contrast primary and secondary pollutants.

6. Describe how ozone is formed.

7. List three ways air pollution can affect public welfare.

# 5-2 Water Pollution

## ■ Introduction and Overview

### Concepts

- Significant percentages of surface and groundwaters are contaminated.

- There are many pollution sources within the four categories; municipal, industrial, nonpoint, and dredge and fill.

- NPS pollution is a serious challenge to water quality and to policy-makers.

- Aquifers are underground sources of fresh water.

On Earth Day in 1994, Carol Browner, Administrator of the USEPA, took a field trip with a group of elementary school students. She conducted some simple science experiments with them on a boat on the Anacostia River in Washington, D.C. They caught some fish and took samples of the river water to take a look at the overall health of the ecosystem. Neither Browner nor the students liked what they saw. One fish they caught had a tumor, and garbage abounded both along the banks and on the water. Unfortunately, a report on the state of the nation's surface waters that Browner released the same day did not dispute these impressions that America's waters are in trouble.

The report, which was based on data collected in 1991–92, indicated that approximately 40 percent of the waters studied by the states were not suitable for beneficial uses such as swimming or fishing. In fact, over 10 percent, or 17,000 of the nation's usable waterways were found to be significantly polluted. The fact that this study also found routine dumping of toxic wastes into the waterways by 250 city wastewater treatment facilities and 627 industrial operations could have something to do with that number.

Drinking water did not fare any better. Serious contamination of drinking water supplies was also found throughout the country. For instance, it was discovered that more than 10 percent of all large and medium-sized public water systems in the United States provide drinking water that contains lead levels exceeding health-based limits. These high lead levels are especially dangerous to small children, whose developing nervous systems can be severely harmed. Untreated water obtained from both ground and surface sources may contain contaminants that in the parts per million or parts per billion ranges can pose acute and chronic threats to human health.

Granted, we are not now seeing the gross contamination of the past, when industries built factories next to a river so they would have a plentiful supply of water and a ready-made disposal outlet. These activities have been curbed through regulation and enforcement of the Clean Water Act. We no longer have rivers, like the Cuyahoga River in Ohio, bursting into flames from flammable materials indiscriminately dumped into it. Our new problems stem from insidious, less blatant pollution sources.

Reading any newspaper in the country will reveal water quality problems. Many coastal towns have had to close beaches to recreational use due to shoreline pollution. In Puget Sound, wastewater has contaminated oyster populations making them unfit to eat, and harbor seals have been found to have high levels of PCBs in their systems. Wetlands, which we have already learned are crucial wildlife habitats, are being destroyed at a rate of between 350,000 to 500,000 acres per year.

In addition to uneasiness over water quality, there is also concern over the adequacy of supplies. You may look at a globe and see that three-quarters of the earth's surface is covered with water. However, almost all of that water is oceanic salt water and not potable, or drinkable. Even most of the earth's fresh water is not readily available to us; it is in the form of polar ice. Less than one percent of the world's fresh water is in a usable form.

Of the existing fresh water sources, many are either not accessible or not fit for any further use. Of the water that is usable, much of it is available only in limited quantities. These limited, and therefore, precious water resources have become increasingly threatened by the effects of human activities such as the application of pesticides and fertilizers to the land, uncontrolled hazardous waste disposal, leaking underground storage tanks, and use of septic tanks and drainage wells.

The average American uses more than 180 gallons of fresh, clean water a day, while many rural villagers in Third World nations spend up to six hours a day obtaining their water from distant, polluted streams. Nearly 10 million people die each year as a result of intestinal diseases caused by unsafe water. While the magnitude of water use in most western countries gives the appearance of an unlimited supply, it is only an illusion. In the 30 year period between 1950 and 1980, the U.S. population grew by 50 percent. During the same time period, water consumption increased by 150 percent. The Water Resources Council projects that of the 106 water supply regions in the U.S., water supplies to 17 of them will be seriously inadequate in the next decades.

The job of cleaning up and protecting the nation's water resources is made complex by the variety of sources of pollution that affect them (Figure 5-11). In general, water quality problems are caused by one or more of four major categories of pollution: municipal, industrial, nonpoint, and dredge and fill activities.

| | Common Pollutant Categories | | | | | | | |
|---|---|---|---|---|---|---|---|---|
| | BOD | Bacteria | Nutrients | Ammonia | Turbidity | TDS | Acids | Toxics |
| **Point Sources** | | | | | | | | |
| Municipal Sewage Treatment Plants | X | X | X | X | | | | X |
| Industrial Facilities | X | | | | | | | |
| Combined Sewer Overflows | X | X | X | X | X | X | | X |
| **Nonpoint Sources** | | | | | | | | |
| Agricultural Runoff | X | X | X | | X | X | | X |
| Urban Runoff | X | X | X | | X | X | | X |
| Construction Runoff | | | X | | X | | | x |
| Mining Runoff | | | | | X | X | X | X |
| Septic Systems | X | X | X | | | | | X |
| Landfill/Spills | X | | | | | | | X |
| Silviculture Runoff | X | | X | | X | | | X |

**Figure 5-11:** Pollutants and their sources. (Source: modified from the 1986-305(b) National Report Abbreviations: BOD = Biological Oxygen Demand; TDS = Total Dissolved Solids)

Municipal wastewater is primarily made up of household liquid wastes from toilets, sinks and showers. This waste, which may be contaminated by organic materials, nutrients, bacteria, and viruses, goes to city wastewater treatment plants. Toxic substances used in the home, including household cleansers, paint, and pesticides also make their way into the wastewater collection systems. In many areas, vast numbers of industrial facilities are also hooked into the municipal wastewater system and may discharge toxic metals and organic chemicals into the systems. Storm water from downspouts, streets, and parking lots enters street collection systems and may carry contaminants with it.

Industrial processes, such as the manufacturing of steel or chemicals, produce billions of gallons of wastewater daily. Undeniably, industrial and mining operations contribute the greatest quantities of certain pollutants to the nation's waters. Pesticides and fertilizers, steel, paper, plastic and petroleum, all enhance our standard of living but not without a price.

**Nonpoint sources (NPS)** of water pollution are widely dispersed sources as opposed to **point sources**, which are attributable to a single point source, such as a discharge pipe from a factory. An example of a nonpoint source of pollution is rainwater carrying topsoil and chemical contaminants into a stream. Some of the major sources of nonpoint pollution include water runoff from farming, urban areas, forestry, and construction activities. Runoff may carry a variety of toxic substances, nutrients, as well as bacteria and viruses with it. Nonpoint sources now comprise the largest source of water pollution, contributing an estimated 65 to 76 percent of the contamination in quality-impaired rivers and lakes.

**Dredging** is the process of making a waterway wider or deeper. The dredging action stirs up bottom sediments and any pollutants, such as PCBs and heavy metals, that are bound to the sediments. This actually gives pollutants a second chance to do their damage. When the dredged materials are piled on the shore they can seriously harm sensitive wetland areas.

Water, in its various forms circulating through land and air, inextricably links the environmental compartments. Under natural circumstances, the earth is able to circulate water through the compartments in proportions that maintain equilibrium, or balance. However, as human activities encroach upon natural processes such as the hydrologic cycle described above, nature's cycles are forced to accommodate humans, not the other way around. For instance, in the U.S., more than 80 percent of total water use is devoted to cropland irrigation. Much of this water is pumped from **aquifers** (underground fresh water sources), and in many instances, transported from areas of adequate supply to arid and semiarid locations. Normally, aquifers are renewable water sources. However, in many areas of the Great Plains and the Southwest, water from these aquifers is being withdrawn more rapidly than it can be replenished. By overdrawing water, a condition called **subsidence**, or irreversible settling of the land, is occurring. Once an aquifer settles and collapses, it can never again be used. In coastal regions, overdrawing of an aquifer can pull salt water into the aquifer. The aquifer won't collapse, but the introduction of salt water generally renders the aquifer unfit for use. This effectively eliminates a valuable water resource.

In the hydrologic cycle, water continuously evaporates from the oceans into the atmosphere. Water vapor moves through the atmosphere and eventually returns to the ocean through one or more routes. This water vapor can carry air pollutants and deposit them on land or in water bodies far away from their original source, through precipitation. For instance, one study showed that pesticides used only in Texas and the Southwest had been transported in the atmosphere and had contaminated the Great Lakes. Precipitation also becomes the runoff that flows back to the oceans in the form of streams and rivers and infiltrates the ground. This water can then carry or push contaminants down through the soil into the water table.

## ■ Groundwater

### *Concepts*

■ Significant percentages of Americans get their drinking water from underground sources.

■ The saturated zone is where water fills all the voids in rocks and soil.

■ Confined aquifers are beneath a confining bed or aquitard.

■ Porosity indicates how much water can be stored in pore spaces of rocks and soil.

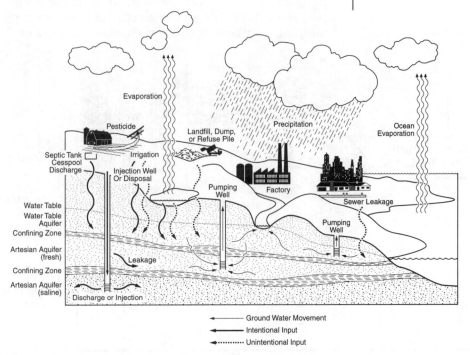

**Figure 5-12:** Water movement and sources of groundwater contamination.

## Major Sources of Groundwater Contamination Reported by States

| Source | No. of States Reporting Source | No. of States Reporting as Primary Source |
|---|---|---|
| Septic Tanks | 46 | 9 |
| Underground Storage Tanks | 43 | 13 |
| Agricultural Activities | 41 | 6 |
| On-site Landfills | 34 | 5 |
| Surface Impoundments | 33 | 2 |
| Municipal Landfills | 32 | 1 |
| Abandoned Waste Sites | 29 | 3 |
| Oil and Gas Brine Pits | 22 | 2 |
| Saltwater Intrusion | 19 | 4 |
| Other Landfills | 18 | 0 |
| Road Salting | 16 | 1 |
| Land Application of Sludge | 12 | 0 |
| Regulated Waste Sites | 12 | 1 |
| Mining Activities | 11 | 1 |
| Underground Injection Wells | 9 | 0 |
| Construction Activities | 2 | 0 |

**Figure 5-14:** Major sources of groundwater contamination reported by states. (Source: USEPA)

Half of all Americans and 95 percent of rural Americans use groundwater for drinking water. It is also used for about half of the nation's agricultural irrigation needs and about one-quarter of the nation's industrial needs. Obviously, all these uses put a high demand on withdrawals from groundwater sources. In fact, in the 30 years from 1950 to 1980 groundwater withdrawals increased by 150 percent. Early in this century we believed that groundwater was safe from contamination. We know better now.

In the mid-1980s the EPA reported over 8,000 drinking water wells throughout the nation were no longer fit to be used due to contamination. The difficulty in monitoring the complex groundwater situation nationwide is due to the vast numbers of potential sources of contamination (Figure 5-13). There are over 30,000 hazardous waste sites that are in need of cleanup; millions of septic tanks since one-fourth of U.S. homes use them; over 180,000 surface impoundments such as pits, ponds, and lagoons; nearly 20,000 municipal landfills; 5 to 6 million underground storage tanks, a large number of which may be leaking or have over-spillage; thousands of underground injection wells; and millions of tons of pesticides and fertilizers spread on the ground.

This vast network of sources contain untold numbers of chemicals that could find their way to groundwater which in turn could contaminate drinking water wells. In the mid-1980s, nearly one-tenth of the public water supplies tested in an EPA study exceeded drinking water standards for inorganic substances, typically fluoride and nitrates. It was also

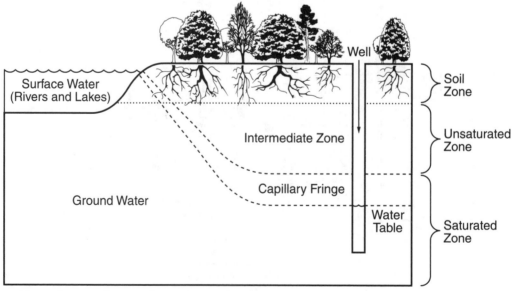

**Figure 5-13:** Underground water. (Source: USGS)

found that at least 13 chemicals that are carcinogens were detected in some groundwater drinking sources.

Obviously, the main concern with groundwater contamination is its adverse impact on public health. However, we are now more cognizant of its impacts on the environment as well. Fish, vegetation, and wildlife can also be negatively affected. For instance, it is estimated that 15 percent of endangered species rely upon groundwater for maintaining their habitat.

Since most drinking water in this country is obtained from groundwater sources, it is important to understand the terminology used to describe various zones and characteristics of the different types of groundwater. A common misconception is that groundwater is obtained from underground lakes. Actually, wells tap into water that fills the voids between earth materials.

Most of the rocks near the earth's surface are composed of both solids and voids. The solid parts are more obvious than the voids, but without the voids there would be no water to supply wells and springs. Water-bearing rocks consist either of unconsolidated (soil-like) deposits or solid rocks. All water beneath the land surface is referred to as underground water or subsurface water. This underground water occurs in two different zones: 1) the zone which is found immediately below the land surface in most areas and contains both water and air is referred to as the **unsaturated zone**, or **vadose zone**; 2) the unsaturated zone is nearly always underlain by a zone in which all interconnected openings are full of water which is referred to as the **saturated zone**.

Water in the saturated zone is the only underground water that is available to supply wells and springs and is the only water to which the name groundwater is correctly applied. Recharging of the saturated zone occurs by percolation of water from the land surface through the unsaturated zone.

All rocks that underlie the earth's surface can be classified either as aquifers or as **confining beds**. An aquifer is a permeable geologic unit with the ability to store, transmit, and yield water in usable quantities to a well or spring (Figure 5-15). A confining bed is a rock

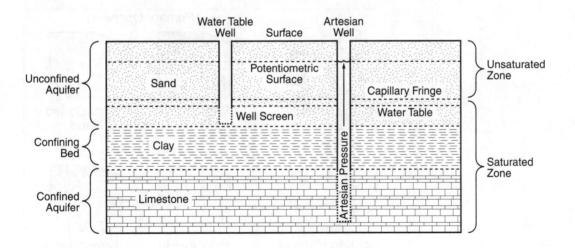

**Figure 5-15:** Aquifers and confining beds.  Source:  USGS

unit that restricts the movement of groundwater either in or out of adjacent aquifers.

Groundwater occurs in aquifers under two different conditions. When water only partly fills an aquifer, the upper surface of the saturated zone is free to rise and fall. The water in such aquifers is said to be unconfined, and the aquifers are referred to as unconfined aquifers.

When water completely fills an aquifer that is overlain by a confining bed, the water in the aquifer is said to be confined. Such aquifers are referred to as **confined aquifers** or as artesian aquifers.

Wells open to unconfined aquifers are referred to as water-table wells. Wells drilled into confined aquifers are referred to as artesian wells. The water level in artesian wells stands at some height above the top of the aquifer but not necessarily above the land surface. If the water level in an artesian well stands above the land surface, the well is a flowing artesian well.

A perched groundwater aquifer is also considered unconfined. A perched groundwater aquifer occurs when the precipitation moving downward through the unsaturated zone is intercepted by a layer of relatively low permeability, which means it is not easily penetrated, such as a layer of clay. The water accumulates on the top of this restricting layer and forms a perched water table. Perched groundwater depends on infiltration at the surface and can disappear during long dry seasons. Other terms for perched groundwater are: perched aquifer and perched water table. An **aquitard** is a confining layer having low permeability. It is essentially impermeable, meaning water cannot pass through it.

An aquifer performs two important functions: storage and transport. The pores, fractures, and other openings in the water-bearing strata serve both as storage spaces and as a network for transport.

**Porosity** of the earth material is the percentage of the total volume of the material that is occupied by pores or other openings (Figure 5-16). Porosity is an index of how much total groundwater can be stored in the saturated material; however, it does not indicate how much water the porous material will yield. Permeability indicates how much water the formation will yield.

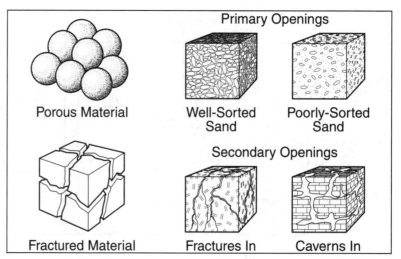

Figure 5-16: Porosity of water-containing earth materials.

The direction of the slope of the water table is important because it indicates the direction of groundwater movement and hence, the direction in which carried contaminants are moving. This is referred to as the groundwater **gradient**.

# ■ Surface Water

## Concepts

■ Industrial and municipal effluent is a serious toxic threat to surface waters.

■ NPS pollution presents a new and complex challenge to water quality.

■ Biomonitoring is useful to determine environmental impact of pollutants.

Pollutants in surface water bodies can harm aquatic life, threaten human health, or result in the loss of recreational or aesthetic potential (Figure 5-17). Contaminants come from industries or treatment plants discharging wastewater into streams or from non-point sources of pollution. Some pollutants cause obvious decline in the quality of water bodies. For example, nitrates and phosphates from fertilizers and cleansers can cause severe changes in lake chemistry.

Although some problems are readily visible, some, such as those caused by toxic contaminants in the ppb or ppm ranges, present more of a challenge because they are not so obvious. Toxic pollutants can be discharged by industry and through the **effluent** (wastewater that is the result of a process) of municipal wastewater treatment plants. In fact, a study in 1986 found that approximately 37 percent of the toxic compounds finding their way to surface waters did so by passing through treatment plants.

If wastewater discharged from treatment plants is not sufficiently treated, it

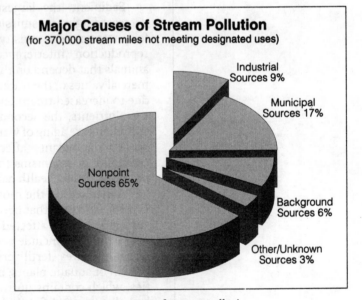

**Figure 5-17:** Major causes of stream pollution.

can easily affect the quality of both surface and ground water, as it does in many parts of the country. The nutrients in sewage accelerate the growth of algae and other aquatic plants. When the excessive plant life dies, the decay process uses up an inordinate amount of the dissolved oxygen in the water leaving less for the fish, who will consequently die. Also, if treatment is not adequate, the discharged water may contain bacteria and chemicals that are in themselves detrimental to aquatic life.

The residue left from wastewater treatment is called **sludge**. In this country, the volume of this sludge continues to grow, doubling every dozen years or so. If a sludge is clean, or free of toxic substances, it can actually be beneficially used as a soil conditioner. But if it has pollutants or **pathogens**, which are disease-causing organisms, it presents an entirely different problem. In fact, disposal of contaminated sludges is a challenging problem due to the relatively limited options and expense of proper disposal.

Industrial wastewater discharged directly into waterways or, as previously mentioned, indirectly through wastewater treatment plants, presents one of the most significant sources of pollution of surface waters. Pretreatment is necessary for industrial discharges into municipal treatment plants for several reasons. Flammable wastes can cause fires or explosions, corrosive wastes can corrode pipes and leak into the environment, and toxic wastes may interfere with the operation of the plant and pass through into surface and ground waters. Pretreatment is designed to reduce or eliminate these hazards, thereby controlling health and environmental risks.

Control of nonpoint-source pollution (NPS) is a relatively new challenge being undertaken in the protection of water quality. Unlike pollution from point sources, it is very diffuse in both its origin and how it enters surface and ground waters. It is the result of a variety of human activities taking place over a wide geographic area. Part of the difficulty in controlling it stems from the manner in which it finds its way into water in sudden surges, often in large quantities associated with rainfall and snowmelt.

Sediment, the largest contributor to NPS problems, causes decreased light transmission through water. Since light is necessary for photosynthesis, this type of pollution results in decreased plant reproduction, interference with feeding and mating patterns of animals that depend on the plants, decreased recreational and commercial values of the water body, and increased drinking water costs due to increased treatment costs.

Nutrients, the second most common NPS pollutant, promote the premature aging of water bodies. Pesticides hinder photosynthesis in aquatic plants, affect aquatic reproduction, increase the susceptibility of organisms to disease, accumulate in fish tissues, and present a human health hazard through fish and water consumption.

Agriculture is the most common culprit in NPS pollution. The USEPA estimates that between 50 to 70 percent of quality-impaired surface waters are affected by NPS pollution from agricultural activities. Pollutants include sediments from eroded cropland and overgrazed pastures; fertilizers or nutrients, which promote excessive growth of aquatic plants; animal waste from confined animal facilities, which contains nutrients and bacteria that can cause shellfish bed closures and fish kills; and pesticides, which can be toxic to aquatic life and humans.

Urban and mining runoff is the second most common source of NPS pollution. USEPA estimates that pollutants carried by runoff from streets and roadways, parking lots, and commercial sites affect between 5 to 15 percent of surface waters. Pollutants from urban runoff include salts, oily residues, and a variety of toxic materials, including gasoline and asbestos. Abandoned mines can affect surface waters through acid drainage and pollution from mill tailings. Other resource-extraction operations such as abandoned or improperly sealed oil and gas wells also contribute to water pollution.

Commercial timber cutting and other forestry operations cause eroded soil and debris to find its way into streams and lakes. These sensitive habitats can be permanently altered, affecting a wide range of plant and animal species. Land development projects, including highway construction, produce sediment and toxic materials that can ultimately find their way to water bodies.

One of the techniques used to accurately take the biological pulse of a surface water ecosystem is called **biomonitoring**. Biomonitoring essentially uses actual plant and animal species as reference points for environmental quality. It serves two major needs: it allows us to screen waters for signs of degradation with fewer resources than would be needed to conduct chemical testing, and it can confirm the results of chemical monitoring data. Biomonitoring can indicate when chemicals have reached toxic levels even if the identity of the chemical is not known. It also enables us to assess the cumulative effects on aquatic life of multiple sources of pollution. Biomonitoring may be a controlled laboratory experiment, where test organisms are exposed to water containing a specific chemical or a complex mixture of chemicals. This type of experiment is known as a **bioassay**. It is used to evaluate the relative potency of a chemical or mixture of chemicals by evaluating its effects on living organisms under controlled conditions. Biomonitoring can also take place in the field.

## ■ Drinking Water

### Concepts

■ Waterborne pathogens still present a concern for safe drinking water.

■ Special contaminants of concern for drinking water are lead, radionuclides, and disinfection by-products.

Historically, some of the most severe public health effects from contaminated drinking water were diseases such as cholera and typhoid. For the most part, these types of dangers have been eliminated in the United States. They still present a severe problem in many parts of the world. A recent episode in this country, however, serves as a reminder that we cannot let down our guard.

In 1993, Milwaukee suffered an outbreak of cryptosporidiosis, a waterborne disease, which cost the city tremendously in human suffering, lost wages and productivity. In all, nearly 400,000 people were stricken with the disease. Some lost their lives because of it. It is this nation's worst drinking water disaster to date.

Upstream feedlots and wastewater treatment plant discharges along with an alleged illegal discharge from a slaughterhouse are thought to be responsible. This, coupled with an outdated and poorly designed water treatment plant, made typical safeguards that would prevent an episode like this ineffective.

Besides this case, other problems with contamination by disease-causing microbes have been documented. The Centers for Disease Control reported 112 disease outbreaks linked to contaminated water from 1981 through 1983. Parasites, such as Giardia lamblia, were the cause in 43 of the cases, while bacteria and viral pathogens were the causes in 22 cases. Microbial contamination continues to be a national concern because contaminated drinking water systems can rapidly spread disease.

In addition to these types of serious pathogenic hazards, other contaminants in our drinking water, namely chemicals, present yet another cause for concern. The USEPA has identified several special contaminants of concern in drinking water. They are: lead, radionuclides, and disinfection by-products.

The primary source of lead in drinking water is corrosion of plumbing materials, such as lead service lines and lead solders, in water distribution systems and in houses. Virtually all public water systems serve households with plumbing containing lead solders of varying ages, and most faucets are made of alloys that can contribute some lead to drinking water.

Radionuclides emit radiation as they decay. The most significant radionuclides in drinking water are radium, uranium, and radon, all of which occur in nature. Ingestion of uranium and radium in drinking water may cause cancer of the bone and kidney. While radium and uranium enter the body by ingestion, radon, which is a gas, is usually inhaled after being released into the air during showers, baths, and other activities such as washing clothes or dishes. The main health risk of concern due to inhalation of radon is lung cancer. Radionuclides are found primarily in those drinking water systems that use groundwater. Naturally occurring radionuclides are seldom found in surface waters.

Disinfection by-products are a fairly new concern. They are produced during water treatment by the chemical reactions between disinfectants and naturally occurring or synthetic organic materials present in untreated water. Trihalomethanes are disinfection by-products that are linked to cancer. Obviously, disinfectants provide needed protection against pathogenic organisms which make this a tricky problem for the EPA. They are currently looking at ways to minimize the risks from by-products.

## ■ Checking Your Understanding (5-2)

1. List three sources of pollution affecting water resources.

2. Contrast nonpoint sources and point sources of pollution.

3. Define aquifer.

4. Explain the differences between water in the unsaturated zone and water in the saturated zone.

5. Describe the difference between porosity and permeability, using a sponge as an example.

# 5-3 Land Pollution

## ■ Introduction and Overview

### Concepts

- Leachate can move through the soil to contaminate groundwater.

- Contamination of the land affects all the other compartments.

- Land use decisions are critical in the prevention and control of pollution.

Historically, land has been used as the dumping ground for wastes, including those removed from the air and water. Early environmental protection efforts focused on cleaning up air and water pollution, which were easier to connect directly to damage to public health. It was not until the 1970s that there was much public concern about pollution of the land, which impacts all the other compartments. We now recognize that contamination of the land threatens not only future uses of the land itself, but also the quality of the surrounding air, surface water, and groundwater. Pollutants on the surface of the land or in the soil frequently move to surrounding air and water, particularly groundwater. Sometimes this contamination is the result of a direct application, say of pesticides or fertilizers, onto the land; it can also result from improper storage, handling, or disposal of toxic substances.

More than six billion tons of agricultural, commercial, industrial, and domestic waste are produced in the United States each year. Much of this waste presents few health or environmental problems. Half the total, for example, is agricultural waste, primarily crop residues, most of which is plowed back into the land. Other waste, particularly that from industrial sources, can imperil both public health and the environment. Leaks from underground storage tanks and chemical spills also contribute to contamination of the land and groundwater.

If not properly disposed, even common household wastes can cause environmental problems ranging from foul-smelling smoke from burning trash to breeding grounds for rats, flies, and mosquitoes. Even something as seemingly harmless as disposing of one quart of used motor oil improperly can cause hundreds of thousands of gallons of water to be contaminated at levels deemed undrinkable. This is because the land and atmospheric water in the form of rain provide linkages to groundwater.

Even at properly run disposal sites, land contamination can contribute to air and water pollution because small quantities of toxic substances may be dumped with other household wastes. Rain water seeping through these buried wastes may form **leachate**. Leachate is the liquid that results when water moves through any non-water media and collects contaminants. Common examples would include water as it trickles through either wastes or soils where agricultural pesticides or fertilizers have been applied. This leachate can then percolate down through the soil and may result in contamination of the groundwater.

Other organic wastes such as garbage and paper products decompose and can form explosive methane gas which, because it is lighter than air, tends to rise through the soil and into the atmosphere. Instances of houses near municipal landfills collecting explosive levels of methane gas in crawl spaces or basements have been documented.

The main sources of land pollution are municipal waste disposal, illegal hazardous waste disposal, abandoned hazardous waste sites, and underground tanks. There are estimated to be over two million underground tanks in this country that contain substances ranging from common gasoline or diesel fuel to not so common, but highly toxic substances. A Canadian study of underground tanks found that older, unprotected bare steel tanks had a life-expectancy of 15 years on average. This is because of the relatively rapid corrosion of tanks underground, unless they are afforded proper corrosion protection. Even a small leak can cause large amounts of chemicals to be released in subsurface locations that are not readily visible. In addition to the potential poisoning of drinking water supplies, there is a very real threat of fire or explosion from tanks containing flammable substances such as gasoline.

Industrial wastes may present particularly troublesome problems. Most people don't realize that industrial processes used to make common everyday products generate huge amounts of hazardous waste. Figure 5-18 lists some examples. Many components of these wastes, such as dioxins, present serious health or environmental threats by themselves; others are hazardous only in combination with other substances. Potential health effects range from headaches, nausea, and rashes, to acid burns, serious impairment of kidney and liver functions, cancer, and genetic damage.

| The products we use | The potential hazardous waste they generate |
|---|---|
| Plastics | Organic chlorine compounds, organic solvents |
| Pesticides | Organic chlorine compounds, organic phosphate compounds |
| Medicines | Organic solvents and residues, heavy metals (mercury and zinc, for example) |
| Paints | Heavy metals, pigments, solvents, organic residues |
| Oil, gasoline, and other petroleum products | Oil, phenols and other organic compounds, heavy metals, ammonia, salt acids, caustics |
| Metals | Heavy metals, fluorides, cyanides, acid and alkaline cleaners, solvents, pigments, abrasives, plating salts, oils, phenols |
| Leather | Heavy metals, organic solvents |
| Textiles | Heavy metals, dyes, organic chlorine compounds, solvents |

**Figure 5-18:** Wastes generated from manufacture of common products. (Source: USEPA)

# ■ Soil

## Concepts

- There are four major components to soil.

- A soil profile indicates the depth and thickness of soil layers.

- The shape contaminants take in a compartment is called a plume.

It is necessary to understand basic principles related to the nature of soils in order to understand the movement of contaminants in this medium. We have been using the word "land" in a general sense to describe the ground of the earth. We use the term **soil** to refer to the thin layer of Earth's crust that we live on and that has been affected by weathering and decomposition of organisms. Soils have developed under different climatic conditions and vegetation covers, and from different parent rock material. Soils vary in their color, thickness, in the number of layers they have, and in the amount of clay, salts, and organic matter they contain. These factors affect where and how fast pollutants will move through the soil. Some prediction of how deep contaminants will move in the soil can be made from knowledge of characteristics such as clay content.

Soil is critical to life. Plants get their nutrients from it and, through the food chain, all animals rely on it and could not survive without it. Soil is formed from rock that over long periods of time is broken down by weathering. Three types of rock comprise the crust of the earth. **Igneous rocks** are solidified molten material. **Sedimentary rocks** are the result of the weathering of preexisting rocks. **Metamorphic rocks** were originally igneous or sedimentary rocks that were modified by temperature, pressure, and chemically-active fluids. These rocks make up what we call the ground or land.

The formation of soil is a continuous process in which living organisms and climate play a role. Freezing and thawing, for instance, result in expansion and contraction of rock materials causing them to break into smaller and smaller pieces over time. Living organisms produce acidic by-products that can cut into the surface of rocks causing tiny cracks. There are many complex chemical, physical, and biological activities constantly going on to form soil. Soil is more than just small bits of rock. Soil scientists distinguish soil from other geologic materials by its four components: mineral particles (rocks and clays), organic matter, water, and air. Darker soils, for instance, indicate high organic matter content.

Decaying plants and animals provide the organic matter in soil. Bacteria and fungi are constantly at work decomposing this material into a rich, dark material referred to as humus. Earthworms, termites and ants also help to break down humus. The process of decomposing organic matter releases essential minerals and nutrients into the soil where they can be utilized by plants.

A major characteristic of soils is texture, which refers to the sizes of the particles making up the soil. Particles are classified as various grades of gravel, sand, silt and clay, in decreasing order. The U.S. Department of Agriculture has set up standard definitions of soil texture classes in which the proportions of sand, silt and clay are given in percentages. Soil texture is important because it largely determines how much water the soil can retain and how easily other substances can move through it. This is why, when conducting an investigation into the extent of soil contamination, a **soil profile** is developed. This profile indicates the depth and thickness of various soil layers.

Other soil properties, such as structure and pore size distribution, are also major factors in how quickly water and other liquids move through the soil. Organic matter enhances the water-holding capacity and infiltration of the soil. On the other hand, formation of

## Box 5.1

## Ted James on Land Use Decisions

*Ted James is the Planning Director for the Kern County Department of Planning and Development Services. He has been involved in land development planning issues in the Central Valley of California since 1977. As Planning Director, he is responsible for administering the General Plan, zoning ordinances, land division ordinances of state and local laws, and administering the California Environmental Quality Act, which deals with compliance with environmental requirements of land use projects.*

**Q:** Where, in your list of priorities in making a land use decision, do environmental issues come in?

**A:** They come early in the process and, because of the structure of state and federal environmental laws, are of paramount importance. They are the basis for determining the conditions which must be satisfied for project approval. When an applicant comes in with a development project, he also has to submit an application to process a CEQA document, which is the California Environmental Quality Act process. The document is circulated and we function as a clearing house to receive comments from different agencies at the federal, state, and local levels related to that project. The comments we get back are put into the form of **mitigation** measures which become required conditions that must be satisfied in order to get the project approved. Many of the comments we get are environmental issues and are built into the project as it is presented for approval by the decision makers.

**Q:** When does the *Environmental Impact Report* enter this process?

**A:** An EIR comes in as a development project is being considered. It is prepared before the project can go out for public hearing and before the decision maker can take any action on the project. It is provided in conjunction with the application. The EIR provides a great deal of beneficial information to the acting decision makers regarding the environmental impacts that may be encountered if the applicant should receive approval for the project.

**Q:** Over the period of time that you have been making these decisions, have you seen changes in priorities?

**A:** Oh, very much so. It's an evolving process. As we move forward with more and more growth, we see a change in terms of issues that become sensitive to the community or to the decision makers. Groundwater issues are an example of a recent change that has come about. Early on, groundwater wasn't an issue. But as more growth projects occur, the agricultural water community has more concern about groundwater draw down. We're having to retool our land use policies to make sure there is no draw down from new developments. It's always an evolving situation where, as a result of laws or concerns or some federal, state, or local action, we have to respond in new ways to address some of these issues.

**Q:** Early on, what were the environmental concerns that you faced?

**A:** Well, I think one that's been of importance has been endangered species. Over the last ten years we're spending more and more time and energy trying to address issues of gaining local compliance with state and federal endangered species laws. We're trying to come up with comprehensive programs to adopt locally that will help applicants through the process of gaining compliance with these laws.

**Q:** It sounds like what you're saying is that of the four environmental compartments, your current primary environmental concern is biota.

**A:** No, I think that air is probably the primary environmental concern, with water and biota as a close second. I say that air is the primary concern because the issues of upcoming compliance with the Clean Air Act are very significant and will have a very dramatic impact on growth as we move forward. I think more and more we'll see regulations that address land use concerns in what's called "indirect source issues." That's basically when you have land uses that, because of their nature, generate or attract traffic. We're moving in a direction where you have to be sensitive as a land use planner so that you're doing whatever can be done to reduce traffic. Some large generators of traffic such as shopping centers, large industrial employers, and large residential projects, all either create or generate transportation concerns. Applicants are saying, "You can't make me bear all the financial burden because the people already in the community are taking advantage of my development as well." The applicants are very frustrated because they already have a variety of other mitigation requirements they have to address like water, school, sewer, and endangered species. When you add air quality, they really get frustrated. Then you've got the existing community saying, "Your development is creating

## Box 5.1

a burden to the transportation system. Why should we increase the community's air problems at all?" The local decision-making bodies are in the middle trying to balance the two concerns.

**Q:** It sounds like your concerns over traffic and transportation are clear manifestations of the major concern over air quality.

**A:** Yes. Air quality and traffic are major issues and are closely related. We're going to continue to see a lot of life-style changes as a result of federal and state environmental regulations. By gradually intermixing commercial, industrial and residential sites proximate to each other, we're trying to make land use decisions that will create a situation where people don't drive all the way across town for shopping or to work. The only problem with that is you can never predict how the human animal will react. Designing so that people can be close to their shopping and employment is one thing. Predicting that people's personal preferences will fall in line with those designs is another. That's not necessarily the human instinct.

**Q:** As population increases, which natural resources are of greatest concern with regard to their depletion?

**A:** There are concerns about preserving farmland and our mineral resources. Those types of finite commodities will always be a concern. We are mandated by state and federal laws to provide for habitat preservation, which in turn provides for open spaces. That is a mutually supportive situation. There are other laws that conflict with each other. For instance, the Endangered Species Act is very much a concern to the farming community and urban developers. These people continually plow their land so that endangered species will not reinhabit the area to be farmed or built upon. But, by constantly plowing their land, they are creating PM-10 problems and that's a problem from the standpoint of the Clean Air Act. So, in the environmental arena there are conflicting laws that work at cross-purposes because they were developed independently of each other.

**Q:** What is the biggest challenge that you face in planning?

**A:** Right now, my biggest challenge is the funding dilemma in terms of local government being able to provide the services that it has provided in the past. Everything rolls downhill in terms of requirements and mandates. Some are redundant and maybe were designed to address a situation in one part of California but are applied to the entire state.

**Q:** With that in mind, do you see any environmentally-oriented mandates that you think are unnecessary?

**A:** Yes. A good example of that would be the redundancy in endangered species regulations. There is a federal Endangered Species Act and a state Endangered Species Act. When an applicant gets comments back on his project, he gets one set from the U.S. Fish and Wildlife Service and another from the State Department of Fish and Game. Both agencies are spending moneys addressing the same issues and sometimes the agencies conflict. I think the Federal Government should delegate the responsibility to the state for implementation because the state is closer to the people who are affected by it. Many air laws or rules are not sensitive to local issues.

**Q:** How do special interest groups, i.e. environmental, and the public affect the decisions you make?

**A:** These groups have a very strong influence in the decisions that we come up with. Whenever we make a land use decision, we are using our adopted land use plans as one part of the input. As a second form of input we evaluate comments from agencies and the public, and the final input is the public hearing process. All three of these come together to determine what we as a planning staff decide as a recommendation.

**Q:** Where do you see land use decisions going in the future with regard to environmental issues?

**A:** Growth issues will become more important as we go on. As we grow, resources become more limited. Minimizing air pollution will become more of a concern. The groundwater that we have, either surface water entitlement or what's in the ground, is a finite resource. Transportation resources will be a major concern. Habitat preservation for endangered species will continue to require complex decisions that will balance needs.

**Q:** So, it sounds like when it comes to land use planning it's appropriate that we use the term environmental "compartments" since it appears that these issues are viewed in a very compartmentalized fashion. There doesn't seem to be much concern for how a particular regulation will affect the other compartments.

**A:** Yes, that's true. Local government is often challenged with making the various laws fit together. These are some of the challenges we are trying to overcome.

less permeable layers, such as clay, substantially reduce the movement of water.

When a liquid contaminant is released onto the surface of the ground, several things start to happen. If it is a volatile substance, some of it will escape into the atmosphere. Liquid portions will move into the soil pores and continue to move downward under the force of gravity until it attenuates, or reduces in concentration due to its storage within or adherence to the soil particles. In other words, if we had a 500 gallon underground tank containing gasoline, a leak would coat soil particles beneath the tank and move through the pore spaces of the soil until there was no free liquid left to move downward. Depending on the type of soil, this leak could move only a few feet or hundreds of feet. One relatively small underground tank leak migrated over 100 feet in a formation that was predominantly cobbles. Had the soil regime been mostly clay, the owners that had to pay for the cleanup would have been much happier. This final shape of the contaminated area is called a **plume**. The term is used to describe contaminant shapes for air, water, and land.

## ■ Checking Your Understanding (5-3)

1. Define leachate.

2. Describe two factors that affect the rate at which a plume of gasoline will move through the soil.

3. Discuss land use decisions in your community that have or will have an effect on the environment.

# 6

# Air

## ■ Chapter Objectives

After completing this chapter, you will be able to:

1. **Summarize** the basic provisions of the Clean Air Act.

2. **Describe** the role of federal, state, and local agencies in implementing the provisions of the Clean Air Act.

3. **Explain** the "Bubble" policy for emissions trading.

4. **Compare** regulation of point versus non-point sources.

5. **List** examples of air emission control technologies.

6. **Discuss** strategies that can be used to effectively manage and control air pollution.

## ■ Terms and Concepts to Remember

Air Pollution Control District
Air Quality Control Region
Alternative Fuels
Ash
Best Available Control
  Technology (BACT)
Calibration
Clean Fuels
Continuous Emission
  Monitoring Systems (CEMS)
Control Techniques
Emission
Emission Allowances
Emission Standard
Emissions Trading
Federal Reference Methods
Gravimetric
Hazardous Air Pollutants
  (HAPs)
Inspection and Maintenance
  Program (I/M Program)

Market-based Approach
Maximum Achievable Control
  Technology (MACT)
Monitor
National Emissions Standards
  for Hazardous Air Pollutants
  (NESHAP)
New Source
New Source Performance
  Standards (NSPS)
Nonattainment Pollutants
Offset
Oxygenated Fuel
Permit Fees
Primary Standard
Reformulated Gasoline
Scrubber
Secondary Standard
State Implementation Plan
  (SIP)
Vapor Recovery Nozzles

## ■ Chapter Introduction

In the last chapter, we learned that pollution in one environmental compartment can damage public health and cause harm in other environmental compartments. We learned that air pollutants can cause sickness; both long-term, like cancer, birth defects, and brain and nerve damage; and short-term, like burning eyes and noses, and irritation to breathing passages. Air pollution's effects on the environment include acid deposition, loss of stratospheric ozone, and potentially, global warming. It can also cause property damage, eating away at buildings, monuments, and statues. The haze of some air pollution also causes reduction in visibility, which can interfere with communications and navigational equipment as well as reduce the aesthetic appeal of our environment. Air pollutants are emitted from stationary and mobile sources and can travel great distances to cause damage far from their source. The primary source of many important pollutants, such as sulfur dioxide, nitrogen oxides, and particulate matter is the burning of common fuels such as coal, oil, gas, and wood. Particulate matter also comes from dust at construction sites, grain operations, and farming.

This chapter will provide information on current approaches to improving air quality. It will cover the regulatory framework of the Clean Air Act and management and control tactics for reduction of various air pollutants.

# 6-1 The Clean Air Act

**Figure 6-1:** George Bush signed the Clean Air Act into effect in 1990.

## ■ History and Overview

### Concepts

- The Clean Air Act (CAA) regulates NAAQSs and NESHAPs.

- States are required to develop SIPs to show how they will implement the CAA.

- The Amendments of 1990 are the latest major changes in the CAA.

The original federal Clean Air Act predates the formation of the EPA by 15 years. Its enactment in 1955 primarily authorized technical and financial assistance to the states in order for them to look at air pollution issues. It has been amended several times in the 1960s, in 1970, 1977 and most recently, in 1990. The amendments in the 1960s granted greater federal authority, including *control* over vehicle **emissions** and a limited ability to regulate interstate pollution. An emission is a release of pollutants into the air from a source such as an automobile, industrial boiler, farm, or power plant. The Clean Air Act is found in *U.S.C. Sections 7401-7642*. Regulations for its implementation are found in *CFR 40, Part 50*.

Amendments in 1970 required the federal government to set national standards for ambient air quality which regulated levels of pollutants. This directive resulted in the development of the National Ambient Air Quality Standards, or NAAQS, which we learned about in the previous chapter. These standards define the principal types of pollution found throughout the country, such as carbon monoxide, lead, nitrogen oxides, and sulfur oxides, and the levels of each that should not be exceeded. The 1970 amendments also set separate standards for new cars and new stationary sources of pollution.

A new category of pollutants was added by the 1970 amendments. The EPA was directed to set national standards for **hazardous air pollutants (HAPs)** or *air toxics*, defined as those not covered by national ambient air quality standards but which "may reasonably be anticipated to result in increased mortality or an increase in serious irreversible, or incapacitating reversible, illness." It quickly became apparent that hazardous air pollutants had a significant detrimental impact on public health in areas where they were emitted. They were determined to cause health effects such as cancer, birth defects, nervous system problems and death. The standards that regulate air toxics are called **National Emissions Standards for Hazardous Air Pollutants, or NESHAPs**. The earliest substances covered under these standards included arsenic, asbestos, benzene, beryllium, mercury, radionuclides, radon-222, and vinyl chloride.

The 1970 amendments also required states to draw up **state implementation plans (SIPs)** for achieving the ambient air standards in each "**air quality control region**" in the state. This is still an important part of the law as we know it today. A state's plan must include an inventory of specific pollution sources, with estimates of how much of each type of pollutant is emitted annually. Information on the numbers of motor vehicles and their level of use must also be included. New stationary sources of pollution, such as factories or power plants, must meet a number of conditions which include using the best available pollution control methods *before* they start operating. Under the amendments of 1977, new facilities can be built in regions that do not meet national air standards if existing sources of pollution agree to reduce their emissions more than enough to compensate for the new ones.

The original Clean Air Act was less than 50 pages long. In 1990, the Clean Air Act underwent a major revision known as the 1990 amendments. These amendments made significant changes in the Act causing it to impact the way business is conducted in this country and even some of the choices we have as individuals and consumers on a much larger scale. The 1990 amendments are close to 800 pages in length. These amendments address new areas of air pollution that were not covered in previous versions. Some examples include acid rain emissions and phasing out the production of chemicals contributing to depletion of the stratospheric ozone. A new concept of major significance incorporated into the emissions limits established by the law is a system of *tradable emissions credits*. An example of this is a facility that reduces emissions below the required standard or accomplishes reductions before the required compliance date. The facility would earn emissions credits that could be applied to future emissions or sold to another facility.

The 1990 Clean Air Act (referred to as the "Act" in the remainder of this chapter) is currently the basic law of the land for air pollution control. William K. Reilly, who was EPA Administrator when the 1990 Act was passed, stated that once fully implemented it would: "... remove 56 billion pounds per year of pollution from the air (224 pounds per year for every man, woman, and child in the United States); reduce by 50% emissions causing acid rain; reduce by 75% emissions and resulting risks from toxic air pollutants; result in cleaner cars, fuels, factories, and power plants; assure that all areas of the country meet ambient air quality standards; and reduce damage to lakes, streams, parks, and forests."

In developing an implementation strategy for the Act, the EPA was directed to concentrate on policies that ensure a healthy environment while supporting national economic and energy policies. This could be achieved by adding market incentives and other innovative approaches to traditional command and control tactics. Also, the EPA's rule making was to include the greatest possible consultation and consensus by the public, state and local agencies, and other interested parties.

The states do much of the work to carry out the provisions of the Act. The law recognizes that it makes sense for states to take the lead, because pollution control problems often require special understanding of local industries, geography, housing patterns, climate, and other site-specific factors. The bottom line, however, is that the Act provides a mechanism for the EPA to set limits on how much of a pollutant can be in the air anywhere in the United States. In this respect, it is a minimum standard designed to ensure that everyone has the same basic protections. Individual states may, of course, have stricter standards, but never less stringent ones.

Air pollution does not observe political boundaries such as state lines. In fact, it often travels from its source in one state to another state. In many areas, two or more states share the same air; that is, they are in the same air basin defined by geography and wind patterns. In our mobile society it is not uncommon for persons to live in one state and work or shop in another. Their vehicles cause air pollution that can spread throughout several states. Some pollutants, such as the combustion products that cause acid rain, may travel over several states before they do their damage. The Clean Air Act provides for interstate commissions on air quality. These commissions have the responsibility for developing regional strategies for cleaning up air pollution.

Air pollution also moves across national borders. The Act addresses pollution originating in Mexico and Canada that drifts into the United States as well as pollution originating in this country that crosses into our neighbors' air space. The Act does this through provisions for cooperative efforts to reduce pollution that originates in one country yet affects another.

Although the full Act is very complex, an outline of the major topic areas will help the student understand the comprehensive nature of its amendments. There are nine Titles in the Act: *Title I* covers ambient air quality including SIPs and nonattainment areas for the criteria pollutants; *Title II* covers motor vehicles including fuels programs and emissions control devices; *Title III* covers air toxics including sources and listings of specific hazardous air pollutants; *Title IV* covers acid rain chemical reductions and planning programs; *Title V* covers permit programs; *Title VI* covers stratospheric ozone protection strategies and listing of ozone-destroying chemicals; *Title VII* covers enforcement provisions of the Act; *Titles VIII – IX* cover miscellaneous topics such as greenhouse gases, international issues, and disadvantaged business concerns. Key areas will be further discussed in the remainder of this section.

## ■ Protection of Ambient Air Quality

### *Concepts*

- The EPA reviews and approves SIPs.

- Primary NAAQSs address protection of public health, and secondary NAAQSs address public welfare.

- Areas not meeting primary NAAQSs are in nonattainment.

- Use and trading of offsets are market approaches to achieve better air quality.

A portion of *Title I* of the Act addresses the requirements and approval process for SIPs. Since 1970, states have had to develop SIPs that explain how they will achieve the goals of the Act, for each of its amended versions. The SIP itself is a collection of the regulations the state will put into place to clean up polluted areas. Throughout the process of developing these regulations, the state must involve the public by holding hearings and providing opportunities to comment on the measures being taken. The EPA takes an active role in reviewing and approving these SIPs. If a state's SIP is unacceptable, the EPA will take over enforcement of the Act in that state. These decisions are published regularly in the *Federal Register*. The federal government also provides assistance to the states, through the EPA, in conducting research studies, developing engineering designs, and providing money to support clean air programs.

The majority of *Title I* addresses the criteria air pollutants. Remember, these pollutants are sulfur dioxide ($SO_2$), particulate matter (PM-10), carbon monoxide (CO), ozone ($O_3$), lead (Pb), and nitrogen oxides (NOx). They have been designated as such by the EPA because of their widespread effects on public health and welfare and because of their numerous sources. For each criteria pollutant, the EPA establishes a National Primary and Secondary Ambient Air Quality Standard (NAAQS). The word *ambient* means surrounding, so these standards have to do with the air around us as opposed to the air in a confined space or coming out of a smokestack. The **primary standard** specifies maximum acceptable ambient concentrations for the protection of *public health*. The **secondary standards** specify maximum ambient concentrations acceptable for the protection of the *public welfare*. "Public welfare" includes concerns such as crop damage, animal health, and materials deterioration. Any area that does not meet the Act's primary standard for any criteria pollutant is in *nonattainment* and must be classified as such. Even though the EPA has had limits for these pollutants since the 1970 version of the law, many urban areas are currently classified as nonattainment for at least one of the pollutants. The EPA estimates that about 90 million Americans, approximately 35 percent of the U.S. population, live in nonattainment areas.

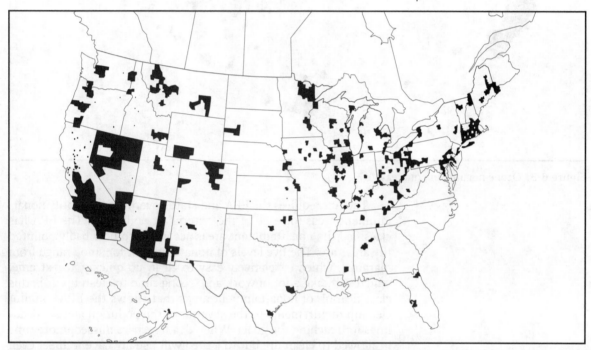

**Figure 6-2:** Areas in nonattainment with the EPA standards for particulate matter – 1985. (Source: *Maps Depicting Non- Attainment Areas Pursuant to Seciton 107 of the Clean Air Act – 1985*, USEPA)

One of the primary targets for pollution control is urban smog, which we learned in the last chapter is primarily made up of ground-level ozone. Remember that ground-level ozone is produced by photochemical reactions that occur when pollutants from many sources, including smokestacks, cars, paints, and solvents, are released into the atmosphere. When you drive your car or use paint thinner to clean your brushes after dabbling in your fine arts project, smog-forming pollutants are released. The wind often blows these pollutants some distance. It also takes time for the chemical reactions to occur that form the various components of smog. This explains why smog is often more severe several miles from the source of the pollutants that form it. Weather and geography play an important role in the harmful effects of smog. Temperature inversions cause smog to stay in place for days at a time. Smog can travel across county and state lines, so if a metropolitan area covers more than one state, their air pollution control agencies must work cooperatively to solve air pollution problems. An example of this situation is the New York metropolitan area which includes parts of New Jersey and Connecticut.

**Figure 6-3:** Ozone nonattainment areas.

The Act requires the EPA and state governors to identify nonattainment areas for each of the criteria air pollutants. The EPA has classified the nonattainment areas according to how badly polluted the areas are. The five levels of nonattainment for smog range from marginal, which is relatively easy to clean up quickly, to extreme, which will take a lot of work and a long time to clean up. It is this classification of nonattainment areas that allows the EPA to tailor cleanup requirements to the severity of the pollution and set deadlines for reaching the goals. When deadlines are missed, more time is allowed to clean up, but that area will usually have to meet even

| Classification of Areas | | | |
|---|---|---|---|
| | Class | Level – PPM | Attainment Date |
| Ozone | Marginal | .121 to .138 | 3 years |
| | Moderate | .138 to .160 | 6 years |
| | Serious | .160 to .180 | 9 years |
| | Severe 1 | .180 to .190 | 15 years |
| | Severe 2 | .190 to .280 | 17 years |
| | Extreme | .280 and above | 20 years |
| Carbon Monoxide | Moderate | 9.1 to 16.4 | 12/31/1995 |
| | Serious | 16.5 and up | 12/31/2000 |
| | For ozone and CO: adjustment possible based on 5% Rule; EPA may grant two one-year extensions of attainment date | | |
| PM-10 | Moderate | | 12/31/1994 |
| | Serious | | 12/31/2001 |
| | Possible extension of attainment date up to five years for serious area | | |

**Figure 6-4:** Classification of areas.

| Carbon Monoxide Nonattainment Areas | |
| --- | --- |
| State | City/Area |
| Alaska | Anchorage area |
| | Fairbanks-North Star |
| Arizona | Phoenix |
| California | Chico |
| | Fresno |
| | Lake Tahoe-South Shore |
| | Los Angeles South Coast Air Basin* |
| | Modesto |
| | Sacramento |
| | San Diego |
| | San Francisco-Oakland-San Jose |
| | Stockton |
| Colorado | Colorado Springs |
| | Denver-Boulder |
| | Fort Collins |
| | Longmont |
| Connecticut | Hartford-New Britain-Middletown |
| District of Columbia | Washington, D.C. |
| | D.C. Metropolitan area-Maryland, Virginia |
| Massachusetts | Boston |
| Maryland | Baltimore |
| | Washington, D.C. Metropolitan area-Maryland, DC, Virginia |
| Minnesota | Duluth |
| | Minneapolis-St. Paul |
| Montana | Missoula |
| North Carolina | Raleigh-Durham |
| | Winston-Salem |
| New Jersey | New York Metropolitan area-Connecticut, New York, New Jersey |
| | Camden-Philadelphia, Pennsylvania |
| New Mexico | Albuquerque |
| Nevada | Las Vegas |
| | Reno |
| New York | New York Metropolitan area-New York, Long Island, New Jersey, Connecticut |
| | Syracuse |
| Ohio | Cleveland |
| Oregon | Grant's Pass |
| | Klamath Falls |
| | Medford |
| | Portland-Vancouver, Washington |
| Pennsylvania | Philadelphia-Camden, New Jersey |
| Tennessee | Memphis |
| Texas | El Paso |
| Utah | Ogden |
| | Provo-Orem |
| Virginia | Washington, D.C. Metropolitan area-Virginia, D.C., Maryland |
| Washington | Vancouver-Portland, Oregon |
| | Seattle-Tacoma |
| | Spokane |

* Los Angeles - South Coast Air Basin is the only carbon monoxide nonattainment area classified as "serious"; all others are classified as "moderate." All areas listed as "moderate" must meet the CO standard by December, 1995. The "serious" area must meet the CO standard by December, 2000.

**Figure 6-5:** Carbon monoxide nonattainment areas.

stricter cleanup requirements. States take these factors into account when implementing their SIPs. They use a system of permits to make sure pollution sources in the state do their part to help the state meet cleanup goals.

The Act takes a comprehensive approach to reducing criteria pollutants. It covers a wider variety of sources than ever before and incorporates a number of cleanup methods and technologies. Many of the smog cleanup requirements target motor vehicles. Also, as pollution problems worsen, pollution controls are required for smaller and smaller sources. For example, in the case of PM-10, reducing particulates requires pollution controls on power plants and restrictions on smaller sources like wood stoves, agricultural burning, and dust from fields and roads.

| Particulate (PM-10) Nonattainment Areas | |
|---|---|
| **MODERATE** | |
| Areas must meet PM-10 standards by December 31, 1994 | |
| Alaska | Anchorage, Juneau |
| Arizona | Santa Cruz, Pima, Maricopa and Gila counties: Yuma, Paul's Spur, Nogales |
| California | Inyo, San Bernardino, Kern, Mono, Stanislaus, Madera, Riverside (eastern part), San Bernardino (part) counties |
| Colorado | Archuleta, Adams, Denver, Boulder, San Miguel, Prowers, Pitkin, Freemont counties |
| Idaho | Ada (Boise), Shoshone, Bannock, Power (Pocatello) Bonner counties |
| Illinois | Cook (Chicago), LaSalle, Oglesby, Madison counties |
| Indiana | Lake (Gary, Hammond, East Chicago) county |
| Maine | Arostock (Presque Isle) county |
| Michigan | Wayen (Detroit) county |
| Minnesota | Ramsey, Olmstead counties |
| Montana | Flathead, Lincoln, Lake Missoula, Libbey, Rosebud, Silver Bow, Butte counties |
| New Mexico | Dona Ana county |
| Nevada | Washoe (Reno) county |
| Ohio | Cuyahoga (Cleveland), Jefferson (Steubenville) counties |
| Oregon | Jackson (Ashland-Medford), Josephine (Grant's Pass), Klamath Falls, Lane (Eugene), Union (LaGrande) counties |
| Pennsylvania | Allegheny (includes Clairton) county, Liberty-Lincoln-Portview-Glassport boroughs-Clairton |
| Texas | El Paso county |
| Utah | Salt Lake (Salt Lake City), Utah county |
| Washington | King (Seattle), Pierce (Tacoma), Spokane, Yakima, Thurston (Olympia, Tumwater), Walla Walla counties |
| West Virginia | Brooke (near West Virginia border at Steubenville) county |
| Wyoming | Sheridan county |
| **SERIOUS** | |
| Area must meet PM-10 standards by December 31, 2001 | |
| California | San Joaquin Valley, Owens Valley, South Coast Air Basin, Coachella Valley |
| Nevada | Las Vegas |

**Figure 6-6:** Particulate (PM-10) nonattainment areas.

*Title I* of the Act also addresses the situation where a company needs to make modifications or expand in such a way that will result in the increased emission of a criteria pollutant. In this case, an **offset**, or reduction of the criteria pollutant by an amount somewhat greater than the planned release, must be obtained somewhere else. That "somewhere else" can be at the same plant or at another plant owned by the same company or even another company in the nonattainment area. In this way, permit requirements are met and the nonattainment area keeps moving toward attainment. For example, the company might be able to install a tighter air pollution control on a stack somewhere else at the plant that provides more reduction than the total new emissions. Since total pollution will continue to go down, trading offsets among companies is allowed and is considered one of the market approaches to achieving better air quality.

## ■ Cleaning Up Air Pollution From Mobile Sources

### Concepts

■ Motor vehicles are responsible for over half of the smog-forming VOCs and nitrogen oxides.

■ People in the United States are using automobiles more than ever before.

■ In smoggy areas, reformulated and oxygenated fuels must be used.

■ Inspection and maintenance programs as well as vapor recovery nozzles are two methods used to reduce air pollution from cars.

*Title II* of the Act covers motor vehicles, which represent a major source of air pollution. Since the late 1970s, significant gains have been made in reducing certain pollutants, notably carbon monoxide and lead. Even more staggering gains have been made since the 1960s. In fact, the EPA estimates that currently, cars produce 60 to 80 percent less pollution than did cars in the 1960s. Leaded gasoline has been phased out, which has resulted in dramatic declines in levels of lead in the air. Unfortunately, even with these dramatic gains, we are far from the level of improvement from mobile sources necessary to significantly improve the air quality. In fact, motor vehicles are still responsible for almost half of the smog-forming VOCs and nitrogen oxides. Motor vehicles release more than 50 percent of the hazardous air pollutants and they release up to 90 percent of the carbon monoxide found in urban air.

The reasons for the seeming discrepancy (cars are cleaner but air is still dirty) is that more people are driving more cars more miles than ever before. In 1970, Americans traveled 1 trillion miles in motor vehicles, and by the year 2000 we are expected to drive 4 trillion miles each year! The move to suburbia in the 1950s meant that many people began to live farther away from their work than they did before. Most areas do not have adequate public transportation from suburban areas to the workplace and even if they do it is sometimes difficult to entice people away from the sense of freedom and control that driving to work offers. Also, most people still drive to work alone, even when carpool lanes and company van pools are available. Buses and trucks, which produce significant pollution, have not been required to clean up their engines and exhaust systems as much as cars. Amazingly, removing the lead from gasoline has actually caused the fuel to be more polluting in other ways. The formulation used to make up for the octane loss when lead was removed made gasoline more likely to release smog-forming VOCs into the air.

Even though cars have had pollution control devices since 1970, they were initially only required to function for 50,000 miles. Since most cars are driven for at least twice that many miles, the control devices were inadequate. The 1990 amendments took a more comprehensive approach to reducing pollution by providing for cleaner fuels as well as cleaner running vehicles. Auto inspection requirements were built into the law to insure that vehicles are kept well-maintained. The Act also includes policy changes relating to transportation that can help air pollution.

The phase-out of lead from gasoline was completed prior to January 1, 1996. The process of refining diesel fuel is changing in order to reduce the amount of sulfur in the fuel, which contributes to acid rain and smog. Gasoline refiners must produce **reformulated gasoline** to be sold in the smoggiest areas. It must contain less VOCs such as benzene, which is also a hazardous air pollutant. Benzene causes cancer and aplastic anemia, a potentially fatal blood disease. Other polluted areas can request that the EPA include them in the reformulated gasoline marketing program. In some areas, additional carbon monoxide is caused by people starting their cars in winter. In these areas, refiners will have to sell **oxygenated fuel,** gasoline blended with one or more of its components containing oxygen to make the fuel burn more completely, thereby reducing the carbon monoxide emissions.

All gasolines will have to contain detergents which will prevent build-up of engine deposits. This will help keep engines working smoothly and burning fuel cleanly. Low VOC, oxygenated fuel and detergent gasolines are already sold in several parts of the country. The Act also encourages development and sale of **alternative fuels** such as alcohols, liquefied petroleum gas (LPG) and natural gas. Lastly, gasoline stations in smoggy areas must use **vapor recovery nozzles** on gas pumps. These nozzles dramatically cut down the amount of vapor released to the atmosphere while refueling a vehicle.

New cars must have some type of onboard warning system to let the driver know that the pollution control devices are working properly. Pollution control devices must be designed for 100,000 miles of usage rather than the original 50,000 miles. Engines of some cars must be designed to operate on alternative **clean fuels**, including alcohol, that release less pollution from the tailpipe. Manufacturers are encouraged to build electric cars, which are low-pollution vehicles. Since California has the worst smog problems in the country, a pilot project has been initiated there whereby at least

500,000 clean fuel cars will be available for sale each year in the state. Other states can require that cars meeting the California standards be sold in their states. Companies and government agencies that have fleets of cars in smoggy areas will be required to buy new, cleaner cars starting in the late 1990s.

Even if auto manufacturers build cleaner cars that use cleaner burning fuels, if they are not adequately maintained to keep emissions low, much of the benefit is lost. That is why the law adds a third part to the mobile source compliance program, vehicle **inspection and maintenance programs (I/M Programs)**. The 1990 Act includes very specific requirements for inspection and maintenance programs since they are considered one of the most effective means of reducing air pollution from cars.

Before the 1990 Act went into effect, 70 cities in this country and several states already had auto emission inspection programs. The amendments require these programs in even more areas; 40 metropolitan areas, including many in the northeastern part of the country, have to start these programs. Some of the areas that already had programs are now required to enhance their emission inspection machines and procedures. These enhanced programs are designed to give better measurements of the pollution a car releases when it is actually being driven, rather than just parked at the inspection station.

Starting in 1994, diesel truck engines were required to be built to reduce particulates by 90 percent. Buses must reduce their particulate releases even more than trucks. This means that companies and governmental agencies that own buses or trucks will need to buy new, cleaner models. Small trucks (pickups) have requirements similar to cars.

The EPA is studying the need to regulate locomotives, construction equipment, outboard motors, and lawn mowers. The EPA must issue regulations if studies show that controls on these sources would help cut pollution.

The metropolitan areas that are the smoggiest must incorporate transportation policies that discourage auto use and encourage efficient commuting. This is usually done through a combination of van pooling requirements and high occupancy vehicle lanes. States can add surcharges to parking fees in order to raise revenues to implement these policies.

## ■ Solving the Air Toxics Problem

### *Concepts*

- Sources of air toxics emissions are categorized as major or area.

- Sources are required to use MACT to reduce air toxics emissions.

- Offsets are also allowed for hazardous air pollutants.

- Incentives are provided to industry to help speed reductions of air toxics.

*Title III* of the Act covers air toxics, or hazardous air pollutants. These chemicals are released from a variety of sources. For example, gasoline contains toxic chemicals like benzene. You can often see the vaporization, or evaporation of the vapors, from gasoline as you fill your car gas tank. Those wavy lines in the air at the pump nozzle as well as the characteristic odor are telltale signs that gasoline vapors are in the air. When motor vehicles burn gasoline, hazardous air pollutants also come out of the tailpipe. These air toxics are combustion products, or chemicals that are produced when a substance is burned. Cleaner fuels and engines are promoted in the law to reduce the large amounts of hazardous air pollutants coming from motor vehicles.

Air toxics are also released from small stationary sources like dry cleaners and auto paint shops, and large stationary sources like chemical factories. The Act deals more strictly with large sources but there are standards imposed on small ones as well.

In order to reduce air toxics, Congress felt that more study had to be conducted on their identity, sources, and amounts. The EPA received the authority under the Act to first develop a list of air toxics and then to regulate them. Up until 1990, the EPA regulated only seven toxic chemicals in the air. The 1990 Act included a list of 189 air toxics Congress selected to be regulated based on their ability to cause health and environmental damage. The Act also allows the EPA to add to this list as necessary.

The first step in regulating hazardous air pollutants was to identify categories of sources that would typically emit one or more of the 189 chemicals on the initial list. Each category was identified as a *major* (large) emitter or an *area* (small) emitter. Each small source contributes to the pollution load of the area. Typical major sources include coal-burning power plants and chemical plants; area sources include gasoline stations, dry cleaners, and body shops.

Once sources were categorized, the EPA could then issue regulations, which in some cases detailed exactly how to reduce pollutant releases. Congress made clear that in doing this, companies should, wherever possible, have the flexibility to choose how they will meet the requirements. Sources are required to use the **maximum achievable control technology,** or **MACT,** to reduce pollutant releases. MACT is a control technology which can be used effectively on a specific piece of equipment to achieve maximum control of emissions. It obviously provides a high level of pollution control.

The EPA was charged with developing regulations for major sources first, and then issuing regulations for smaller sources. Priorities were set based on health and environmental hazards, production volume, and whether requirements are achievable.

This portion of the CAA also allows a company to use offsets to deal with situations where it may be necessary to increase emissions from a particular source. As long as the total hazardous air pollutant releases do not go up, offsets provide flexibility to look at the entire operation rather than each particular source independently. If offsets are not available, they may choose to install pollution controls to keep pollutants at the required level.

If a company reduces its releases of a hazardous air pollutant by about 90 percent before the EPA regulates the chemical, the company will get extra time to clean up the remaining ten percent. This is called the early reduction program and is designed to produce rapid reduction in the levels of several key hazardous air pollutants.

The Act does not neglect small, neighborhood polluters like paint shops and dry cleaners as it once did. The EPA is required to study these sources after the larger sources are regulated and determine whether they require further regulation.

A Chemical Safety Board was established by the Act that investigates instances where accidental releases from industrial plants occur. A release in 1986 of methyl isocyanate in Bhopal, India that killed thousands prompted the drafters of the Act to require that businesses with the potential to release very dangerous air toxics develop plans to prevent their accidental release. The Board's function is very similar to the National Transportation Safety Board, which investigates plane and train crashes.

## ■ Controlling Acid Rain

### Concepts

- Power plants burning coal with high sulfur content are considered a major cause of acid rain.

- By the year 2000, annual sulfur dioxide emissions are to be 40 percent of those in 1980.

- EPA's program of Emissions Allowances will restrict the amount of sulfur dioxide emissions from power plants.

*Title IV* of the Act includes an innovative program to reduce pollutants that react to form acid deposition, which can be acid rain, acid snow, acid fog, or acid dust. Since acid rain is by far the most common and troublesome, we will use that term to refer to all forms of acid deposition.

The type of acid rain receiving the most attention in this country is caused mainly by pollutants from big coal-burning power plants in the Midwest and Northeast. Coal from the Midwest and Appalachian area of the country has a much higher sulfur content than western coal. When burned, the sulfur in the coal becomes sulfur dioxide ($SO_2$). In addition to sulfur dioxide, these power plants emit nitrogen oxides. We have already learned that these are the two chemicals responsible for combining with other gases in the atmosphere to produce sulfuric and nitric acids.

These oxides of nitrogen and sulfur, released from Midwestern power plants, are carried by winds toward the eastern seaboard and Canada. When winds blow the acid-producing chemicals into areas where there is wet weather, the acids become part of the rain, snow, or fog. Chapter 4 described some of the severe damage caused by the acidification of lakes and streams. Both plant and animal life is affected. The pollutants that cause acid rain can also make the air hazy or foggy, affecting the enjoyable views popular with vacationers and potentially affecting communications and navigation.

Health, environmental, and property damage can also occur when sulfur dioxide gas pollutes areas close to its source. Sulfur dioxide pollution has been found in areas where paper and wood pulp are processed and in areas close to some power plants. The 1990 Act's sulfur dioxide reduction program strengthens the health-based limits that were in place and protects the public and environment from both nearby and distant sources of pollution.

The 1990 Act takes a new approach to controlling the acid rain problem. It sets up a market-based program to lower sulfur dioxide pollution levels. Beginning in the year 2000, annual releases of sulfur dioxide are to be about 40 percent lower than 1980 levels. The expectation is that reducing sulfur dioxide emissions will cause a major reduction in acid rain.

Phase I of the acid rain reduction program went into effect in 1995. Coal-burning boilers at 110 power plants in 21 Midwest, Appalachian, Southern, and Northeastern states had to reduce releases of sulfur dioxide. Phase II begins in the year 2000, requiring even further reduction of sulfur dioxide from the big coal-burning sources as well as covering other smaller polluters. Total sulfur dioxide releases for power plants will be permanently limited to the level set for the year 2000.

These reductions in sulfur dioxide emissions will be obtained through a program of **emission allowances**. The EPA will issue these allowances to power plants covered by the acid rain program. Each allowance permits one ton of sulfur dioxide to be released from the smokestack. In order to obtain reductions in sulfur dioxide pollution, allowances are set below the current level of releases. Plants are only allowed to release as much sulfur dioxide as they have allowances. If a plant needs more allowances it can get more by buying them from another plant that has reduced its emissions below its number of allowances or through brokers involved in the allowances market. Allowances can be traded and sold nationwide. There are stiff penalties for plants that release more pollution than they have allowances.

The acid rain program also gives bonus allowances to power plants for installing clean coal technology designed to reduce sulfur dioxide releases or for using renewable energy sources such as solar or wind and for encouraging reduced energy use by customers. All power plants covered under the acid rain program must have continuous emissions monitoring systems. These systems keep track of the amount of sulfur and nitrogen oxides the plant is releasing. Details on how the plant will comply with its emission limits will be listed on its permit, which is filed with the state and with the EPA.

Reducing nitrogen oxide emissions requires more than targeting power plants. The goal of Title IV is to reduce NOx emissions by approximately two million tons. This requires stringent new NOx emissions rates for various types of utility boilers and cars. New cars must be designed to reduce these emissions.

# ■ The Permit Program

## Concepts

- The Permit Program affects certain sources.

- Permits must include information on pollutants and how they are being monitored and reduced.

- Permit fees pay for administering the Act.

The 1990 Clean Air Act established a permit program for certain sources of air pollution. A source can be anything that releases pollutants into the air, for instance, motor vehicles or power plants. We learned in the last chapter that sources that stay in one place are referred to as stationary sources, and those that move around are called mobile sources. Mobile sources are divided into two groups: road vehicles, which include cars, trucks, and buses and non-road vehicles, which include trains, planes and lawn mowers. A permit is a document that resembles a license and is required by the Act for major sources of air pollution, such as power plants, chemical factories, and in some cases, smaller polluters.

Although the permit program is considered one of the major breakthroughs in the 1990 amendments to the Clean Air Act, it is certainly not a new idea. Before these amendments went into effect, 35 states already had permit programs. The Act requires that states issue permits, unless they have been determined to be inadequate in their responsibilities to do so, in which case the EPA will step in and do the job.

Permits must include information on the specific pollutants that are being released, including how much of each is being released. They must also detail the methods that the owner or operator of the source is taking to reduce the amount of pollution. Lastly, they must include plans to **monitor**, or measure, the emissions. Certain large polluters must perform enhanced monitoring to provide an accurate picture of their pollutant emissions. Enhanced monitoring programs may involve keeping records of materials used by the source, periodic inspections, and installation of **continuous emission monitoring systems (CEMS)**. Continuous emission monitoring systems will measure, on a continuous basis, how much pollution is being released into the air. The Act requires states to monitor community air in polluted areas to check on whether the areas are being cleaned up according to the schedules set up in the law.

Since many businesses are covered by more than one part of the law, this permit system is expected to reduce paperwork. A permit will make it possible to get information about all the air pollution produced by a source from one document (see Figure 6-7). The permit system identifies a business' obligation for cleaning up all types of air pollution to which they contribute. An electric power plant, for example, is affected by many parts of the law. It could be covered by the acid rain, hazardous air pollutant, nonattainment, or smog sections of the Act. Their permit would have all the detailed information required by all these separate sections of the Act in one place. Permits are required both for the operation of plants, called operating permits, and for the construction of new plants.

Under the Freedom of Information Act, all of these permits, as well as permit applications, are available to the public for review. A great deal of information is available on these permits, including how much pollution is being released and monitoring data. The Act even requires the EPA to set up clearinghouses to collect and give out technical information. These clearinghouses are charged to serve the public as well as states and other air pollution control agencies.

Businesses required to obtain air quality permits must pay **permit fees**. Permit fee money helps to pay for the administration and enforcement necessary for the Act to work. The Act gives important new enforcement powers to the EPA. As we learned in Chapter 2, enforcement refers to the legal methods available to regulatory agencies to make polluters comply with the law. It used to be very difficult for the EPA to penalize a company for violating a provision of the Clean Air Act. Even a minor violation required going to court. This was very costly in time and money. The 1990 amendments provided a means for the EPA to fine violators. The EPA can issue a citation, which is much like a traffic ticket. The purpose of this new authority is to speed up compliance with the law and reduce court time and cost. Other enforcement tools include court-imposed fines and jail terms. The amendments increased the penalties for violating the law.

The Act also sets deadlines for the EPA, states, local governments, and businesses to come into compliance. The EPA feels that these deadlines are realistic and can be met, but some states and local authorities feel they are ambitious and do not take into account local factors such as climate and population growth.

Public participation in the rule-making process is built into the 1990 Act. Congress felt it very important to give the public as many opportunities as possible to help determine how the law will be carried out. Citizens can take part in hearings on state and local plans for cleaning up air pollution as well as comment on state and EPA reviews of permit applications. Individuals and special interest groups can sue the government or a source's owner or operator to get action if it is felt that the EPA or the state has not adequately enforced the Act. Individuals can request action by the local, state, or federal enforcement power against violators.

The 1990 Act has incorporated new, flexible programs to clean up air pollution with the goal being to do it as efficiently and inexpensively as possible. These new programs are called **market-based approaches**. For instance, the program that is designed to reduce acid rain gives businesses some choices in how they can reach their pollution reduction goals including trading, buying, and selling pollution allowances. The Act has built-in economic incentives for cleaning up pollution. For instance, a gasoline refiner can get credits if able to produce cleaner gasoline than is required at a given time. They can later use these credits if their gasoline doesn't meet future clean-up requirements.

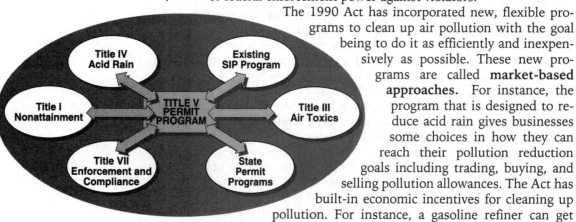

**Figure 6-7:** Title V – Operating permits.

# ■ Controlling Damage to the Stratospheric Ozone

## Concepts

- The 1990 Clean Air Act sets a schedule for phasing out production and use of ozone-destroying chemicals.

- Products with ozone-destroying chemicals must be properly labeled and recycled.

- Equipment, appliances, wood burning stoves/fireplaces and consumer products are affected.

In Chapter 4 we learned about the damage to the stratospheric ozone layer by CFCs. Ozone in the upper atmosphere serves to filter out harmful ultraviolet rays of the sun. Exposure to these harmful rays has been linked to skin cancer and cataracts and also damages agriculture. Scientists for many years have noted thin spots in the ozone layer that have been referred to as "holes." *Title VI* of the Act contains provisions that are designed initially to slow down and ultimately solve this problem.

It was in the mid 1970s that scientists first suggested a connection between CFCs and damage to stratospheric ozone. CFCs were in wide usage as aerosol propellants in consumer products such as hair sprays and deodorants and had a number of uses in industry, too. Because of concerns about their harmful effects to stratospheric ozone, the U.S. government banned their use as propellants in aerosol cans in 1978.

Scientists continue to measure the amount of ozone in the stratosphere and have been alarmed by the persistent thinning. Between 1978 and 1991, there was a 4 to 5 percent loss of ozone in the stratosphere over the United States. Thinning has also been noted in northern Europe. Ninety-three nations have agreed to cooperate in phasing out production and use of CFCs and other ozone-destroying chemicals since it has become clear that the thinning is occurring more rapidly than anticipated. Unfortunately, this will take longer than we would like. The CFCs already in the stratosphere coupled with their continued, albeit lessened release, means that damage will continue to occur for at least an estimated 20 years.

The 1990 Act sets a schedule for phasing out production and use of ozone-destroying chemicals. The worst culprits will be phased out first and the schedule can be accelerated if scientific indicators warrant this move. Congress listed CFCs, halons, HCFCs (hydrochlorofluorocarbons) as ozone-destroying chemicals in the 1990 Act. The EPA can add other chemicals that destroy ozone to the list.

The EPA issues allowances to control the manufacture of chemicals that are being phased out. Companies can sell unused allowances to other companies still making the chemicals, or they can use the allowances, with certain limits, to make a different, less ozone-destroying chemical on the phase-out list.

In the interim, before the phase-out is complete, the law takes other steps to reduce the release of ozone-destroying chemicals. The law requires recycling and labeling of products with these substances. The law also encourages "ozone-friendly" substitutes for ozone-destroying chemicals.

Car air conditioners are the largest single source of ozone-destroying chemicals. Starting in 1993, all car air conditioner servicing had to use equipment that recycles CFCs and prevents their escape into the atmosphere. Larger automotive service shops had to start using this equipment in 1992. CFCs in small cans used in servicing air conditioning units will only be sold to specially trained and certified repair persons.

As the group of ozone-destroying chemicals is phased-out, numerous appliances that use them will have to be changed. One common example is refrigerators which will have to be redesigned

---

## Box 6.1

## Daniel J. Nickey

*Daniel J. Nickey is a Waste Reduction Specialist for the Iowa Waste Reduction Center. This center is run by the University of Northern Iowa to assist small businesses in reducing their waste. Daniel specializes in helping businesses with air quality permits and compliance. (For contact purposes: University of Northern Iowa, 75 Biology Research Complex, Cedar Falls, IA 50614-0185, (319)273-2079, Fax (319)273-2926*

**Q:  Daniel, tell us more about your job duties.**

A:  The majority of my time is spent helping small business comply with state and federal air regulations. This involves primarily conducting air audits and providing follow-up technical assistance. The center assists small business in obtaining air permits and reducing air emissions. Most of our clients don't have the technical or financial ability to properly complete air permit applications. We also educate small businesses on new air regulations that they will have to comply with and technologies that will reduce their air emissions.

**Q:  What got you interested in this type of work?**

A:  I have always enjoyed being in the outdoors and wanted to do something that would contribute to its betterment. I also wanted a career where I would have the opportunity to interact with the business community. In college and high school, science courses were the classes that interested me the most. These courses introduced me to the idea of a career in the environmental profession.

**Q:  What prepared you for your work?**

A:  In college I took a wide variety of science and engineering courses which covered all aspects of the environment. During the summers in college I worked and did internships in the environmental field. This experience prepared me the most for my career. It also provided me with a direction on the exact area of the broad environmental field I wanted to work in. My internship also gave me valuable contacts, which helped me obtain a job after graduation.

**Q:  What do you find most satisfying about your work?**

A:  We work with businesses that might not otherwise be able to obtain environmental assistance. Since we are non-regulatory and free, businesses are very honest about their environmental problems. Our clients want assistance and are glad we are there. This is a very rewarding working atmosphere. I enjoy getting businesses in regulatory compliance and reducing the amount of pollutants they emit.

**Q:  What do you see in the future for your profession?**

A:  In the future, more businesses will have their own environmental staff in-house. With the amount of requirements they must meet, it will be more economical to employ a full-time environmental staff person. Small businesses, however, will still require free non-regulatory assistance. More programs like ours will have to be formed to address this need. As time goes on, I believe that more environmental professionals will be required to assist all businesses with the ever-growing amount of environmental regulations.

or configured to use alternative refrigerants. In the meantime, servicing of refrigerators will involve methods to prevent the escape of these substances. Another example of a chemical that will be phased out is methyl chloroform, also known as 1,1,1,-trichloroethane, which is a commonly used solvent found in everyday products such as brake cleaners and spot removers. Its use must be phased out by 1996, which will lead to many changes in operations and consumer products.

The law also addresses the question of substitute products. The EPA must approve the products as safe before production and sales can take place. There has been much concern about the health and environmental safety of these alternative products.

Even consumer products are regulated under the 1990 Act. Many consumer products including hair sprays, paints, foam plastic products, and carburetor additives release smog-forming VOCs and ozone-destroying chemicals. Starting in May of 1993, the chemicals most harmful to the ozone layer required a warning notice on the label stating "WARNING: contains or manufactured with (name of chemical), a substance which harms public health and the environment by destroying ozone in the upper atmosphere." Products containing less destructive ozone-destroying chemicals will require this type of labeling by 2015. The EPA has the authority to implement other regulations of consumer products including labeling, repackaging, formulation changes, fees, or any other procedures that will result in reduction of VOCs and ozone-destroying chemicals starting with the worst polluters.

The 1990 Act reaches even closer to home with its identification of wood stoves and fireplace inserts as major contributors of particulates and hazardous air pollutants, including some cancer-causing chemicals. Gone are the Norman Rockwell-type images of a warm, cozy family sitting next to a roaring fire as a positive picture of America. Wood smoke is much dirtier than smoke from oil or gas-fired furnaces. Most people are not aware of the price they pay in additional air pollutants when they look at the relatively cheap costs to operate a wood burning stove. In fact, in some areas of the country, local governments have had to restrict the use of wood stoves and fireplaces under certain weather and pollution conditions.

Controlling wood smoke pollution has involved redesigning the burning system in wood stoves. Newer wood stoves put out much less pollution than older models. Under the Act, the EPA has issued guidelines for reducing pollution from home wood-burning. At this point they are guidelines, not requirements.

## ■ Checking Your Understanding (6-1)

1. Contrast criteria pollutants and hazardous air pollutants.

2. Explain what *offsets* are.

3. Describe the various air pollutants that result from use of automobiles.

4. List three ways air pollution from automobiles can be reduced.

5. Explain how market-based approaches provide industry with incentives to reduce emissions.

6. Describe how emissions allowances will be used to reduce acid rain.

7. List three requirements the 1990 CAA imposes on use or manufacture of ozone-destroying chemicals.

8. Define *monitor*.

# 6-2 Emission Control Technologies

## ■ Sampling and Analysis Overview

### Concepts

■ Emissions can be sampled in the ambient air and at their source.

■ NESHAPs and New Source Performance Standards require sampling at the source.

■ Federal Reference Methods describe sampling and analysis of criteria pollutants in ambient air.

In the overview of the provisions of the 1990 Clean Air Act, the importance of monitoring air pollutants in order to improve air quality was emphasized. Monitoring is central to any air pollution control program because it helps determine where the air pollutants are, what they are, and how concentrated they are. This is essential to the development and enforcement of regulations as well as to the evaluation of the effectiveness of air pollution control programs. The monitoring process, as described in the Act, refers to sampling and reporting the results.

There are two basic areas where actual sampling can take place, in the ambient air and at the source. Pollutants in the ambient air are generally in diluted form. Source emissions from a known source, either stationary or mobile, may be highly concentrated, and as they move away from the source they become diluted to their ambient concentrations.

Ambient monitoring data helps us to see trends in ambient air quality over time. This, of course, is useful in assessing the progress made toward attainment of air quality standards, which are based on ambient air. The data provides baseline air quality information before new sources begin operating and help in developing and evaluating computer models depicting movement and fate of air pollutants in the air basin. These data can also be used to identify potential *episodes*, or times of unusually high pollutant concentration, so that emergency control programs can be activated. Scientific research, including finding correlations between air pollutant concentrations and effects on human health, always requires good monitoring data.

Source emissions data provide more specific, useful information on equipment and compliance. Such data are necessary in order to evaluate compliance with source emissions regulations and to determine whether pollution control equipment is efficient and effective.

Although all of the criteria pollutants are monitored in the ambient air, some are monitored as source emissions as well. Hazardous air pollutants are monitored at their source under the provisions of the NESHAPs standards. Other harmful pollutants are required to be monitored in source emissions as specified by the EPA in its *Standards of Performance for New Stationary Sources*, commonly known as **New Source Performance Standards (NSPS).** Examples of such regulated pollutants are acid mists, hydrogen sulfide, total reduced sulfur, hydrogen fluoride, and VOCs.

The monitoring process consists of two parts – sampling and reporting data, also called data management. The EPA has a variety of sampling and analysis procedures depending on the pollutant, its physical state at the time of sampling, and its expected concentration. Determining the best method is important because the ultimate goal is to determine the actual concentration of the pollutant in the sample. This concentration is expressed as mass per unit volume, usually micrograms per cubic meter, or ug/m$^3$.

The EPA has identified **Federal Reference Methods** for the sampling and analysis of criteria pollutants in ambient air. The Federal Reference Methods provide manual and/or automated methods for collecting data on ambient air. Manual methods require two steps; the sample is first collected, then analyzed. Automated systems are continuous. The sample is collected and analyzed at the same time by an apparatus that links the sampling equipment to the analyzer. The EPA's reference methods change from time to time and updated lists are published in the *Federal Register*.

The EPA has also identified Federal Reference Methods for source emission measurement. These methods are used for the sampling and analysis of sulfur dioxide, carbon monoxide, particulate matter, nitrogen oxides, and a number of other pollutants. These methods are all manual and are used in conjunction with the New Source Performance Standards, which provide specific information about variations in sampling procedures for each type of source. The Federal Reference Methods specify detailed sampling and analysis procedures that in turn provide necessary data about pollutant concentrations. The important next question is what to do with these data.

First, a determination is made as to whether the data gathered is useful and valid. This is accomplished by reviewing the quality control and quality assurance of everything from sample collection to data gathering. The key aspect of quality control is **calibration,** which helps to ensure that processes and equipment accurately sample and analyze pollutants. Calibration is the checking and adjusting of an instrument to ensure it is giving accurate readings by comparing its readings with a known standard. It is so important that the Federal Reference Methods give very specific instructions about it. These methods provide specific procedures for sampling, analysis, calibration, and calculation that *must* be followed for any monitoring activity related to compliance with the Clean Air Act. These reference methods are listed as appendices to *CFR 40, Part 50*.

Quality assurance is very involved – the EPA has a three-volume *Quality Assurance Handbook* to explain it. Quality assurance, which has been called the "quality control on the quality control," involves an audit process that compares data from air samples with data from standard samples, and compares several different analyses of the same air sample. Audits are designed to verify the accuracy of calibration and in turn the accuracy of the monitoring activity.

This monitoring data is provided to the EPA by the States in quarterly and annual reports depending on the data and the surveillance program for which they are to be used. Once data is verified, it is entered into EPA computer databases. One example of data reporting is the annual air quality data summary that must be submitted to the EPA by each state. Data reported to the EPA may be used to determine whether the National Ambient Air Quality Standards are being met or maintained. State and local agencies may also use such data to develop and revise the State Implementation Plans or to demonstrate the effectiveness of control strategies. In addition, the EPA currently requires that agencies provide a daily air quality index report in all urban areas with a population over 500,000.

## ■ A Look at Some Reference Methods

### Concepts

■ A gravimetric approach is a manual method of sampling particulates.

■ Automated methods of sampling and analysis involve the use of specialized equipment and instruments such as the atomic spectrometer.

■ NSPSs differ from NAAQS in that they regulate sources.

■ Samples can be isokinetic, grab, integrated, or continuous.

This section will provide some basic information about a number of methods used to sample criteria air pollutants. The information is designed to provide a general idea of what these methods typically involve without getting bogged down in the high level of detail found in the standards. For this reason, and because these are subject to change, any specific information on these methods that you may need requires a review of the previously mentioned appendices in *40 CFR, Part 50*.

Each description addresses whether a method is manual or automated, what the principle of measurement is, and the general procedure used to conduct the measurement. Specific standards and types of standards are not discussed since there are so many. It is beyond the scope of this introductory text to go into the calculations that determine concentration, calibration procedures, monitoring frequency, and sampling duration as specified in the reference methods. It is important to know, however, that this level of detail and specification does appear in the standards. The next few methods are characteristic of those used in ambient air for criteria pollutants.

A manual method that can be used to measure particulate matter involves a **gravimetric** approach, which is a term referring to measurement by weight. In this method, called a *high volume method*, the mass of suspended particulate matter is determined by weighing the particles collected on a filter. A filter covers an opening through which air must pass as it is drawn in by a high volume blower. The filter is weighed before and after use of the apparatus and the weight gain represents the suspended particulate matter collected (Figure 6-8).

A manual method for measuring lead also involves use of the high volume sampler, only in this case, the lead must be *extracted* from the other particulate matter on the filter. This is done with acid, and the lead content in the resulting solution is determined using a device called an atomic absorption spectrometer, or AA (Figure 6-9).

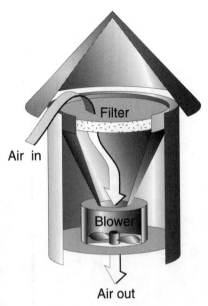

**Figure 6-8:** A high-volume sampler.

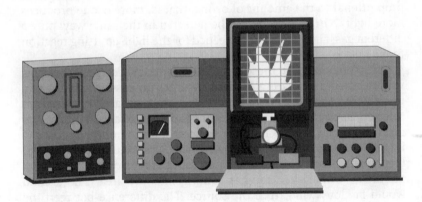

**Figure 6-9:** An atomic absorption spectrometer.

An automated method for measuring carbon monoxide involves an infrared spectrometric principle. Carbon monoxide absorbs infrared radiation at characteristic wavelengths. This absorption is measured with an instrument called an infrared spectrometer. The spectrometer has two matched cells, one of which is used for reference and the other for the sample. The sealed reference cell contains a nonabsorbing gas. An infrared light source continually transmits infrared light waves through both cells. The sample cell has an inlet and outlet to allow air to flow continuously through the cell. If there is no carbon monoxide, the signals observed at the detector are zero. When carbon monoxide is present, it absorbs the infrared radiation and the resulting change is detected electronically (Figure 6-10).

An automated system for monitoring ozone involves an approach called gas phase chemiluminescence, which refers to chemical reactions that produce light. Sample air is drawn through a tube

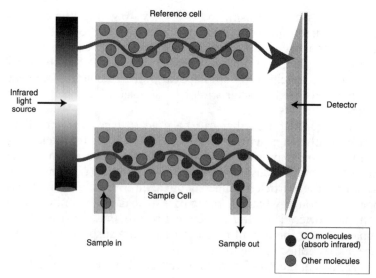

**Figure 6-10:** An infrared spectrometer.

and mixed with a high concentration of ethylene. Ozone reacts chemically with ethylene to emit light which is detected by a photomultiplier tube. If ozone is present, the photomultiplier tube detects the light and sends a signal to a recorder. The amount of light is proportional to the amount of ozone; that is, more ozone produces more light. Nitrogen oxide can be detected in the same way; only a different gas, such as ozone, is needed for the light-emitting reaction to take place.

It is obvious that these types of sample collection techniques would have to be carefully sited in order to assure validity and usefulness of the data obtained. *Where* and *how* sampling sites are selected is not specified in the Federal Reference Methods. For instance, if the goal of monitoring is to determine the air quality impact of an isolated source, two areas would need to be measured. One would need to be upwind from the source in order to measure pollutants from other sources coming into the area. The second would be downwind from the source. The difference between the pollutants measured at each should reveal the effects on air quality of the source in question.

Characteristics of the site such as proximity to trees, buildings, and roadways are also important. These local conditions can easily affect the accuracy of the sampling. For example, if the objective of a monitoring station is to measure public exposure to carbon monoxide on urban streets, the monitor must be close to street level, not subject to accidental destruction or vandalism, and far enough from walls and vegetation so that it will take a representative air sample at that location.

The EPA has published a number of guidelines for siting monitors. The number of sites and the pollutants to be monitored at each are determined by individual states in order to monitor criteria pollutants according to their needs for reporting to the EPA. It is generally expected that reduction of source emissions will result in the reduction of ambient pollution levels. For this reason, the EPA has established *New Source Performance Standards* for emissions and sampling methods at a variety of identified emissions sources.

A **new source** is defined as one where construction begins *after* the EPA publishes standards in the *Federal Register* for its source category. A source constructed prior to the published standards is normally subject only to state regulations, unless the source is modified so as to increase their emission levels.

The list of New Source Performance Standards covers a variety of sources. Some examples are: fossil fuel-fired steam generators; incinerators; nitric acid plants; waste water treatment plants; lead smelters; petroleum refineries; and asphalt/concrete plants, among others. Each standard contains five parts that specify: applicability of the standard; definitions; standards for pollutants; emission monitoring; and test methods and procedures. From this outline, it is probably apparent that these standards are quite a bit more complex than the ambient air quality standards.

The National Ambient Air Quality Standards specify the maximum concentration of criteria pollutants allowable to protect health and welfare under a uniform standard for the entire country. Under federal law, the same maximum allowable concentration for ozone applies in Maine and in Texas. The New Source Performance Standards are different in that each applies to a specific source, the type of pollutants that should be monitored, the concentration allowed for each pollutant, the kinds of monitoring that are required, and which reference method for source testing must be used.

The major difference between NAAQSs and NSPSs is that the latter emphasizes the source of emissions. For instance, sulfur dioxide and acid mists have been identified as some of the pollutants of concern for the source category "sulfuric acid plants." For zinc smelters, a partial list of pollutants includes particulate matter, sulfur dioxide, and visible emissions. Even if the same pollutant is identified for different categories, such as sulfur dioxide in the examples noted, the concentrations allowed for each source may not be the same. Under NSPSs, the allowable concentration of a pollutant may vary with the source. The NSPSs also cover a much greater variety of pollutants than do the ambient standards. They also specify the sampling procedures and reference methods for each pollutant.

A number of specialized techniques are used in sampling sources. These techniques differ in time, continuity, and number of locations utilized per sample. Isokinetic sampling refers to the fact that the velocity of the gas at the sampling probe nozzle is the same as the velocity of the gas stream in the stack. A grab technique measures the volume of a sample by taking it at only one location during one short, continuous period of time. Integrated sampling takes samples from different locations over an extended period of time that is not necessarily continuous. Continuous sampling refers to measuring a sample continuously using a source emission monitor.

The following example provides some basic information on representative reference methods for pollutants from point sources. Figure 6-11 is a diagram of the equipment used for EPA Standard Method 6, a manual integrated sampling technique coupled with a calorimetric measurement, used to determine sulfur dioxide emissions from a stationary source. By means of a probe (1–3) into the stack, an integrated gas stream sample is extracted at a rate of one liter per minute for twenty minutes. The filtered (4–5) gases collected then pass through four midget impingers or bubblers (8–11). The $SO_2$ present in the stack gas is separated from the other compounds such as $SO_3$ and acid mist as the sample passes through the first (8) midget bubblers which contains isopropyl alcohol. The next two bubblers (9–10) contain a solution of hydrogen peroxide which collects the $SO_2$. Since water vapor is usually present in stack gases and its presence interferes with valid data collection, it must be removed. The last impinger tube (11) is dry. The ice bath (7) surrounding the impingers helps condense the water vapor and prevents hot stack gases from vaporizing the collection solutions. The final silica gel tube (13) further dries the sample, eliminating the remaining water vapor. The sample containing the removed $SO_2$ (9–10) is then titrated from orange to pink and the mass of $SO_2$ calculated. Titration is a method of determining the concentration of a substance in solution by adding a standard reagent of known concentration to it in carefully measured amounts until a reaction is completed, as shown in this case by a color change, and then calculating the unknown concentration. The amount of sampled gas is measured by the dry gas meter.

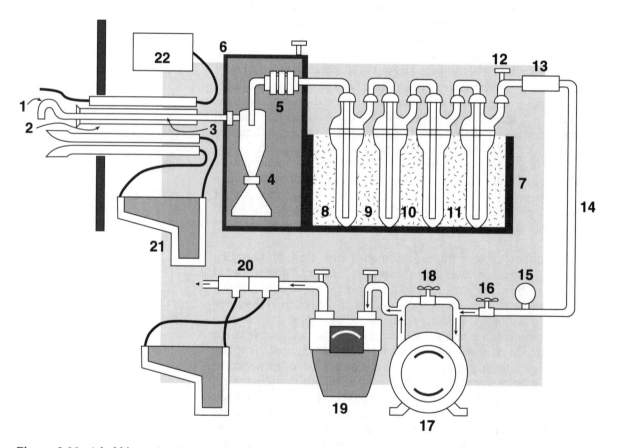

**Figure 6-11:** A bubbler or impinger mechanism.

| | |
|---|---|
| 1–3: Sampling probe and line | 12–13: Thermometer and silica gel drying tube |
| 4–6: Cyclone and stack filter | 14–18: Vacuum pump, lines, and valves |
| 7: Impinger case containing ice | 19–20: Gas meter and manometer |
| 8–11: Impingers or bubblers | 21–22: Manometer and stack thermometer |

# ■ Air Pollution Control Techniques

## Concepts

- Control techniques are equipment, processes or action to reduce air pollution.

- "Bubbles" can be used in emissions trading to look at a facility's sources as a whole.

- The four general methods to control gaseous emissions are absorption, adsorption, condensation, and combustion.

- Adsorption systems can be regenerative or non-regenerative.

Air pollutants are either gaseous or particulate, which makes a difference in how they can be controlled. Some common gaseous pollutants include sulfur oxides, acid gases, volatile organic compounds, nitrogen oxides, and carbon monoxide. Common particulate pollutants include cement dust, smoke, acid mists, metal fumes, and fly **ash**. Particulate emissions, such as smoke, are typically easier to see with the naked eye than gaseous emissions.

Air pollution **control techniques** are the equipment and processes or actions used to reduce air pollution. The range of control technologies and measures employed to reduce pollution varies tremendously. In general, control technologies and measures that do the best job of reducing pollution are required in areas with the worst pollution. For example, in a serious nonattainment area for particulates, a criteria pollutant, a **Best Available Control Technology (BACT)** will be required. A similarly high level of pollution reduction will be required for sources releasing hazardous air pollutants by utilizing Maximum Achievable Control Technologies (MACT).

The latter part of this section will highlight control devices; however, there are several processes or actions that can be used to reduce source emissions that do not require special devices. One technique is process change. For instance, if a fossil fuel-fired industry were to change to solar energy, it would significantly reduce the amount of air pollutants it generates. Changing the type of fuel can also be effective in reducing emissions. If a plant burning high sulfur fuel changes to a low sulfur fuel, specific pollutants will be reduced considerably. Alternative fuels are particularly desirable when they possess the ability to increase energy efficiency while at the same time reducing polluting emissions. Using reformulated gasoline, which is a specially-refined product that yields lower levels of smog-forming VOCs and hazardous air pollutants when burned, is also a way to reduce emissions. In the smoggiest areas, the 1990 Act requires the sale of reformulated gasoline and encourages the development and sale of alternative fuels.

Good operating practices, including housekeeping and maintenance, can also yield emissions savings. For example, auto Inspection and Maintenance Programs are required in some polluted areas. These annual or biennial inspections are to determine if the car and its pollution emissions control systems are being maintained in proper working order. Vehicles that do not pass inspection must be repaired. As of 1992, 111 urban areas in 35 states already had I/M programs. Under the 1990 Act, some especially polluted areas will have to implement enhanced inspection and maintenance programs, using special equipment that can check for things such as how much pollution a car produces during actual driving conditions.

As mentioned before, implementing restrictive programs on individuals can also reduce pollution. These programs include restrictions on operation of fireplaces and wood burning stoves in areas where they make major contributions to pollution.

Even the EPA's **Emissions Trading Policy**, published in the *Federal Register* on December 4, 1988, is designed to result in a net reduction of emissions. The policy sets forth the procedures that the EPA will use to evaluate emissions trades under the Clean Air Act. Emissions trading is considered to be a controversial aspect of this nation's strategy for cleaner air. Questions and concerns continually arise as to whether the trading system is functioning as Congress expected.

Under the policy, based on specific formulas, industries are granted emission allowances. The allowance is a limited authorization to emit a ton of $SO_2$. These allowances can be traded or "banked" for future use or sale. In Title IV, one of the key areas in which emission trading is specified, reduction in emissions that cause acid rain is addressed. Utility companies generally must either reduce emissions or acquire allowances from another utility to make up the shortage. With some exceptions, new power plants that began operating after the passage of the 1990 Clean Air Act must obtain their allowances from those already holding them. The allowance system provides incentives for conservation, the use of renewable energy sources and early use of control technologies.

Anyone can trade allowances – bankers, environmental groups, and private citizens. There are no geographical restrictions, so trades can be conducted nationwide. Anyone can participate in these auctions as either a buyer or a seller. Private parties selling allowances are allowed to set a minimum sale price. The Act allowed the EPA to delegate the administration of these auctions, and the EPA chose the Chicago Board of Trade.

The monitoring of emissions is a very important element in the emissions allowance system. Regular reporting to the EPA is required, and power plants have only 30 days at the end of the year to obtain the allowances they need to cover their emissions for the previous year. The bottom line, however, is that a power plant may not emit $SO_2$ levels that violate the national or state standards, even if they have ample allowances.

As can be imagined, the tracking and recording of this vast allowance system is complex. It is conducted by an automated system managed by the EPA. This system keeps track of the deduction of allowances, as well as the issuance, transfer, and tracking of allowances. The system operates much like a bank.

Title IV also has a provision for penalizing a company that does not have enough allowances to cover its emissions. The automatic fine is $2,000 per ton of excess $SO_2$. This amount was set because it is several times more than the estimated average cost of reducing

$SO_2$ emissions. In addition, the penalty extends into the next year since the facility will lose one allowance for each excess ton of $SO_2$ emitted. After the passage of the 1990 Amendments, the EPA "geared up" and held two allowance auctions to initiate the trading. Those were held in 1993 and 1994 and continue to be held yearly. Title IV further states that the agency is to retain 2.8 percent of the allowances to be auctioned for direct sale.

The idea for the $SO_2$ allowance trading program came from earlier trading programs for air emissions and lead. In the late 1970s emissions were commonly traded under the "bubble policy." Bubbles were a group of sources, within a facility, in which emissions were looked at as a whole, as if they were enclosed in a single glass bubble. For example, under the bubble policy, several different point sources of emissions, such as a number of stacks within the facility, could be managed for compliance purposes as if they were *one source*. Remember that the offset provision allows the siting of new polluting sources or increased pollution by existing sources in areas that are in nonattainment for the NAAQS. Under the bubble policy, owners and operators of those sources could more easily offset increased pollution sources in the area, usually at a greater than one-to-one ratio. They could, then, "bank" reductions for later use or sale.

An example of an EPA-approved bubble plan is Dow Chemical's Plaquemine, Louisiana facility. At this facility, Dow Chemical was able to reduce the emissions from the process unit from nearly 600 to 455 tons per year. They were then allowed to use this reduction to offset the VOCs released from four storage tanks at the same facility. The EPA official that was in charge of overseeing this project stated that the bubble option has not been used much since the passage of the 1990 Act, because it contains an economic incentives program that is inadequate to help most companies comply.

There are four general categories of control equipment for gaseous emissions: adsorption, absorption, condensation, and combustion. Adsorption is the removal of a gaseous pollutant from a gas stream by allowing the pollutant molecules to become attached to a solid surface like activated carbon. Absorption is the dissolving of a gaseous pollutant in a liquid solvent. Condensation is the process by which a gas or vapor is changed into a liquid by increasing the pressure or cooling. Combustion is the combining of a combustible material with oxygen, usually resulting in heat and light.

Adsorption systems can be nonregenerative or regenerative. Regenerative systems can be re-used. Figure 6-12 is a simple diagram of a regenerative system. Non-regenerative systems usually present a waste disposal problem due to the increased volume of waste generated by both the pollutant and the adsorbent, which have to be disposed of together.

Absorption equipment maximizes the opportunity for gas-liquid mixing. Some examples of absorption equipment include spray towers, spray chambers, venturi **scrubbers**, and packed columns. Figure 6-13 is a simplified diagram of a packed column.

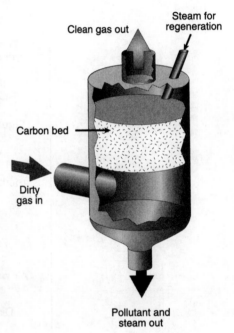

**Figure 6-12:** A regenerative system.

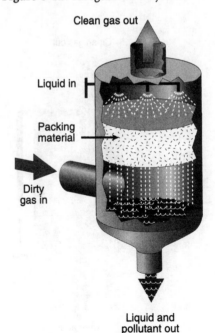

**Figure 6-13:** A packed column.

Condensation occurs when pressure is increased or heat is extracted from a system. The most economical of the two is temperature reduction. Condensers are usually used in conjunction with other devices such as afterburners, absorbers, and adsorption units. Figure 6-14 is a simplified diagram of a surface condenser that can use water or air as a cooling medium.

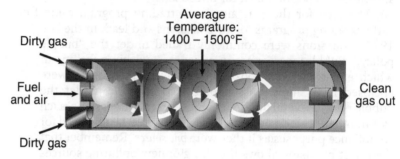

**Figure 6-15:** A thermal incinerator.

Combustion equipment includes flares, thermal incinerators, and catalytic incinerators. Figures 6-15 and 6-16 depict a thermal incinerator and a catalytic incinerator, respectively.

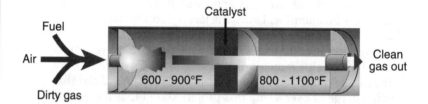

**Figure 6-16:** A catalytic incinerator.

Many questions must be answered about a pollutant to choose the appropriate control technique. If considering combustion, the pollutant must burn and the products of combustion must be taken into account. For absorption, it is necessary to know whether the pollutant is soluble in water or another solvent. Condensation requires a pollutant that can be easily condensed. Economic factors must be looked at as well. The control cost, including equipment and utilities, must be looked at in relation to the price of the product.

Particulate matter can be solid or liquid. In order to design control techniques for particles their size, the gas flow rate, temperature, moisture content, and any significant characteristics such as whether the particles are abrasive, acidic, or explosive must be known.

Different devices have different efficiencies in removing particulates. The efficiency of a device is equal to the weight of the particles entering the control device minus the weight of the particles leaving it, divided by the total weight entering

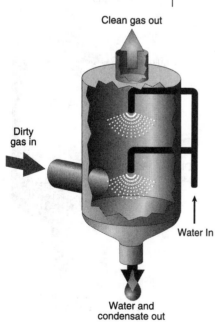

**Figure 6-14:** A surface or contact.

$$\text{Efficiency} = \frac{\text{weight entering - weight leaving}}{\text{weight entering}}$$

the control device. Some common particulate control devices include settling chambers, cyclones, wet scrubbers, electrostatic precipitators, and fabric filters also known as baghouse filters. Particulate control systems often require two or more control devices to obtain the highest possible efficiency. For example, a cyclone (Figure 6-17) might be used to collect large particles followed by an electrostatic precipitator to collect the remaining small particles.

Although researchers continue to improve methods of collecting gaseous and particulate pollutants after they are generated, new emphasis is being given to designing processes that *produce* lower emissions. Carbon dioxide is one pollutant about which attitudes have changed considerably. Until relatively recently, $CO_2$ was considered an inevitable and benign product of combustion since most fuels contain carbon and produce $CO_2$ when they burn. Now, with the focus on $CO_2$ as a major player in the global warming controversy, researchers are looking at ways to capture and dispose of it. This is a difficult problem given the very high concentrations of $CO_2$ in typical gas stacks.

Another common control device is a vapor recovery nozzle. These are special gasoline pump nozzles that reduce the release of gasoline vapor into the air when people put gasoline into their cars. There are several types of vapor recovery nozzles, so nozzles may look different at different gas stations. The 1990 Clean Air Act requires installation of vapor recovery nozzles at gas stations in smoggy areas.

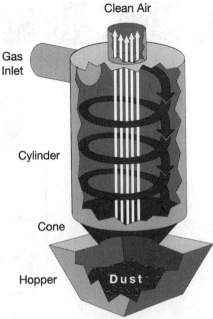

**Figure 6-17:** A cyclone collector.

## ■ Checking Your Understanding (6-2)

1. Explain the difference between New Source Performance Standards and NAAQSs.

2. Define *calibration*.

3. Describe what the Federal Reference Methods are used for.

4. Explain what a *gravimetric* approach is.

5. Describe the difference between isokinetic, grab, integrated, and continuous monitoring.

6. Explain the difference between manual and automatic methods of sampling.

7. List and describe four control techniques for air pollution control.

8. Describe how *bubbles* are used to manage air pollution.

9. Explain the difference between *adsorption* and *absorption*.

10. Explain the difference between *condensation* and *combustion*.

11. Contrast non-regenerative and regenerative absorption systems.

12. Describe what a vapor recovery nozzle is used for.

# 6-3 California, A Case Study

## ■ Overview of California's Regulatory Framework

### Concepts

■ California's air quality management program is the strictest in the nation.

■ The state is divided into Air Pollution Control Districts that work with the ARB to implement air pollution controls.

■ There are strict fines and penalties for violating state standards.

■ Although vast improvements in air quality have been made in the last 30 years, Southern California is still in nonattainment for several pollutants.

California's program to control emissions is the most stringent in the nation due to the severe air pollution in the state. The state's air quality standards, which reflect its definition of healthful air, differ from that of the EPA and the rest of the country. In an effort to better protect the health of the state's most sensitive persons, California's goal is to prevent violations of the standards. The California Clean Air Act of 1988 is the nation's most comprehensive state air quality legislation.

The California Clean Air Act requires that the state's air pollution control program meet higher health standards than the EPA's version. In addition, to ensure constant progress toward healthier air quality, the law requires that emissions be reduced an average of 5 percent per year until those standards are achieved. This is a strong challenge in light of the constant growth predicted for most urban areas. In the South Coast Air Basin, which comprises a large portion of Southern California, population is expected to grow by 25 percent and vehicle traffic by 68 percent between now and the year 2010.

California's Health and Safety Code incorporates implementation of the Federal Clean Air Act, the California Clean Air Act, as well as additional provisions relating to the control of toxic air contaminants, acid deposition, and both vehicular and non-vehicular sources of emissions. California's Legislature mandated both state-wide and local efforts to investigate and regulate all sources and types of air pollution. The State Air Resources Board (ARB) was assigned the responsibility for ensuring effective pollution control by local and regional **air pollution control districts**. Under the law, the ARB and local air pollution control districts share responsibility for improving air quality by pursuing new and better control strategies for the pollution sources in their respective control districts. Governmental agencies that make land-use decisions and transportation agencies have important roles as well. They assist the districts

**Figure 6-18:** The State of California.

in drafting the transportation and growth management portions of their plan for attainment of air quality goals.

When implemented, the California Clean Air Act required each area of the state that had not met the state air quality standards for ozone, carbon monoxide, nitrogen dioxide and sulfur dioxide to submit a plan to the ARB by June 30, 1991. The areas must further endeavor to achieve and maintain state ambient air quality standards for these gasses by the earliest possible date. Areas expecting to meet the state standards by 1994 were classified as "moderate." If they did not expect to meet the standards until between 1994 and 1997, they were classified as "serious," and those areas which are not expected to attain the standards until after 1997 were classified as "severe." The plans to achieve these goals were to include a wide range of control measures, including transportation performance standards, such as car-pooling.

The plans were also to meet progressively more stringent requirements depending upon the extent of air pollution in the district. The ARB was directed to review each plan to determine if it was designed to achieve district-wide emission reductions of 5 percent or more per year for each of the pollutants that had not met the air quality standards. These pollutants are called **nonattainment pollutants**.

Local air pollution control districts have the primary responsibility for implementing their plans. Local districts are composed of either single county districts or two or more county districts that have merged into a unified or regional district. Each district must adopt rules and regulations to implement the plan and each has the power to establish a permit system for any "article, machine, equipment, or other contrivance which may cause the issuance of air contamination." If a district does establish a permit system, the system must ensure that permitted articles will not prevent the attainment or maintenance of air quality standards. A district also has the power to grant variances from its rules and regulations under specified circumstances.

To control non-vehicular air pollution, the California Legislature mandated comprehensive state and local programs and a general prohibition on annoying or harmful air pollution. California law directs the ARB to divide the state into air basins of similar meteorological and geographical characteristics. Each basin then establishes its ambient air quality standards, considering human health, aesthetic value, interference with visibility, and economic effects of pollution. Violation of any of the statute's provisions concerning non-vehicular air pollution of any order of rule issued under them is a misdemeanor punishable by up to a $1000 fine and/or six months in the county jail. Civil penalties up to $25,000 may also be imposed.

The 5 percent annual emission reduction goal required by the California Clean Air Act is to be the combined result of the automotive and fuels standards set by the ARB, non-automotive controls adopted by air pollution control districts, and the transportation and growth management decisions of cities, counties, and regional planning agencies.

The ARB has nearly exclusive authority to regulate motor vehicle pollution. Local and regional air pollution control districts have primary responsibility for regulating most other sources of air pollution. A notable exception is in the case of toxic air contaminants, where the ARB identifies the specific substance and establishes its control measures. The air control districts are then required to adopt these or more stringent measures.

The ARB estimates that the state's efforts will provide for an average of 85 percent of the carbon monoxide reductions needed to meet the air quality standard. In addition, 56 percent of the future hydrocarbon reductions and 72 percent of the nitrogen oxide cutbacks needed to meet the state's ozone standard will also be the product of the state's programs. By region, however, the state's share of future emission reductions will vary, depending on the makeup of local pollution sources. In areas such as Sacramento and San Diego, for instance, where automobiles are a large part of the pollution problem, state programs will provide only about 80 percent of the cutbacks needed to attain the ozone standard. By contrast, in a region such as Kern County, in the Central Valley, where much of the air pollution is generated by oil development and other industrial facilities, enforcing state regulations will provide only about 23 percent of future emission reductions. The remaining pollution control will have to come from local efforts.

The most dramatic progress in cleaning up California's air has been made in the metropolitan Los Angeles area, home to half of the state's population and half of its overall air quality problems. Despite dramatic improvements, Los Angeles remains the site of the nation's worst air quality. During the 1980s emissions of smog-forming hydrocarbon and nitrogen oxide were cut by 27 percent (767 tons per day) and carbon monoxide by 40 percent (2400 tons per day). By the year 2000, those emissions are expected to be half of the current levels.

In Los Angeles, the highest hourly ozone concentrations are three times greater than the state's health standard (0.09 ppm/1 hr), but only half of the levels were measured in the later 1960s, when levels of 0.50 ppm and higher were recorded. Overall, the annual amount of ozone above health standards to which the average person is exposed was cut in half during the late 1970s and 1980s. Unfortunately, even with the successful reductions, the South Coast Air Basin still violates EPA air quality standards for photochemical smog (ozone) approximately 196 days a year. This is compared to 20 to 40 days per year for the rest of California, 50 days per year in Houston and only 15 days per year in New York City. California's stricter health standard for ozone, however, is exceeded 211 days in a typical year. While the South Coast Air Basin is 20 years away from meeting health standards for ozone, levels are expected to be about 50 percent lower by the year 2000 as stricter **emission standards** are adopted and cleaner cars are phased into production. The number of health alerts, currently about 60 per year, is also expected to be dramatically reduced.

Violations of the carbon monoxide health standard were reduced 60 percent during the 1980s, occurring about 40 days a year. Tighter emission standards for motor vehicles are expected to continue that trend, and the ARB predicts that there may be no violations of that health standard by the year 2000.

## ■ Improving Air Quality in California

### Concepts

- California has conducted studies on the effects of acid rain, radon, and photo chemical pollution to better understand how to improve air quality.

- Understanding air pollution requires many detailed scientific measurements and relies on the use of computer models.

- In many polluted urban areas, cleaner vehicles are not enough to achieve air quality goals and transportation systems must be improved as well.

California has conducted many studies in an effort to understand how to most efficiently control air pollution in the state. Concerned that acid rain was causing increased water pollution, declining agricultural and forest land productivity, degradation of natural ecosystems, damage to public health, and acceleration of metal wood, paint, and masonry decay, the Legislature enacted the Kapiloff Acid Deposition Act in 1982. The Kapiloff Act did not give any agency the power to take action to reduce acid deposition; it was only a research measure. The Kapiloff Act required the ARB to develop a comprehensive research and monitoring program as the foundation for future regulation.

Even though acid rain was once thought to be exclusively an East Coast environmental problem, ARB documented high acidity levels in California in the late 1970s. Research through the early 1980s measured acidity in Sierra Nevada lakes which were thought to be particularly vulnerable to acid damage. Under the Kapiloff Act, an $18 million effort over five years studied the potential effect of acidity on soil, stream sediment, aquatic life and vegetation in the Sierra Nevada watershed.

Those studies uncovered acidity levels as high as concentrations measured elsewhere in the country that had caused short-term damage to some aquatic life and vegetation. The research concluded, however, that the acidity did not cause long-term, permanent damage, suggesting that Sierra Nevada lakes had the ability to recuperate from doses of acidity.

In 1988, the ARB began its second phase of acidity research to refine its knowledge of acid rain. The five-year, $15 million program, building on earlier results in the Sierra Nevadas, is studying the potential health consequences of that acidity. Earlier research found that acidity can cause up to $500 million in damage to man-made structures each year in Southern California, especially plastic, steel, glass and paint. Up to $3.75 million is dedicated to further study of this damage.

The Kapiloff Acid Deposition Act was replaced by the Atmospheric Acidity Protection Act of 1988. It requires the ARB to determine the extent to which public health is adversely affected by atmospheric acidity, document the long-term trends, and develop and adopt standards to protect public health and sensitive ecosystems. The California Clean Air Act specifies the minimum protection which the ARB must provide.

ARB research also conducted the first statewide survey of indoor radon levels, prompted by concern over its potential health threat in other regions of the country. The yearlong survey of 385 homes found levels of radon, a colorless and odorless gas radiated by some rock formations, to be far below levels that trigger the concern of health officials. In most cases, the levels were half of the national average and up to 1,000 times lower than the extreme levels measured in the northeast U.S. that had sparked EPA testing. ARB testing is continuing in a few, isolated areas where granite rock formations produce higher than average radon levels.

The basic factors that create California's photochemical pollution, such as the chemical reaction among emissions that creates the pollution that people breathe and the effect of climate and geography on that process, are well known. But beyond this simplistic understanding, the state's air quality problems have gone through some very complex changes. As emissions are reduced, through tighter emission standards and better quality fuels, the relative amounts of different pollutants change, which can in turn affect photochemistry. Changes in traffic patterns or industrial development can affect the pattern of emissions throughout an air basin.

Throughout the 1980's, the ARB research program conducted extensive and sophisticated studies to better understand these changes, in preparation for fine-tuning emission control programs in the future. The projects represent up to $25 million of research, jointly funded in a unique partnership among state, local and federal governments, with matching funds from private industrial and research groups as well. Research teams from around the world participated in these studies, hoping to perfect monitoring technology or analytical methods in an effort to better understand air pollution in other countries as well.

The research was conducted in many major metropolitan areas. In each instance, every aspect of the area was studied intently, giving air pollution researchers millions of pieces of detailed information that, when analyzed in computer simulations, will give researchers more detailed clues about what makes each region's air quality problems unique. In each region, researchers used highly advanced monitors and other equipment that, in some cases, was perfected specifically for the studies. The changing ratio of pollutants was explored and information on weather patterns and their effect on photochemistry was updated. Because of the sensitivity of the monitoring instruments, researchers were even able to measure minute concentrations of some pollutants for which information had been scarce. The studies are expected to continue to provide detailed information that will be useful for many years as we learn more and more about the intricacies of air pollution and its control.

Methods for reducing future emissions will be extensions of current programs by continually tightening emissions limits on traditional pollution sources. How much they will be able to be tightened will depend on the development of new technology and improved industrial practices. Nonetheless, some methods of reducing future emissions will focus on more untraditional pollution sources. California's air quality program has moved beyond simple limits on tail pipes and smokestack emissions. One of the results will be greater emphasis on the relationship between an area's transportation system and air quality.

Even with stricter emission standards for cars, trucks, buses and other types of motor vehicles, polluted urban areas like Southern California will be enhanced by changes in transportation systems. While individual vehicles are made to run cleaner, decisions by local cities, counties, and planning agencies can change the shape of communities and how people move within them. Moving more people in fewer vehicles and reducing traffic throughout the day to avoid congestion improves mobility. Those efforts also pay dividends in better air quality.

Even though California cars certified to ARB standards are the world's cleanest, they do not pollute the same amount for each mile they are driven. Cars pollute much more when the engine and emission control system are cold and still reaching normal operating temperature. Also, the stop-and-go driving in congested rush hour traffic jams in urban areas is much more polluting than driving at a constant speed.

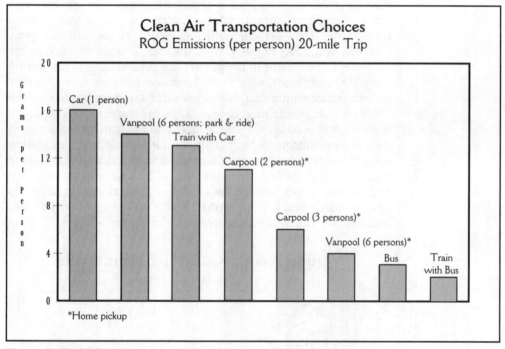

**Figure 6-19:** ROG emissions per person. (Source: ARB)

Currently in California, cars in rush hour traffic carry an average of 1.1 people. Increasing the average to 1.5 people could significantly reduce congestion and pollution. Almost half of the smog-forming hydrocarbon emitted during a typical 20 mile commute is produced in two portions of the trip; the first five miles, while the engine is reaching its normal operating temperature (cold start) and while the engine is cooling down after the car is parked (hot soak). (See Figure 6-19.) Consolidating errands or trips to avoid these driving cycles can significantly reduce emissions.

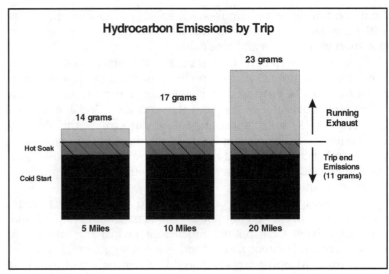

**Figure 6-20:** Hydrocarbon emissions by trip. (Source: ARB)

According to ARB estimates, the total amount of miles being driven in the state is increasing at about 5 percent per year, twice as fast as population gains. In addition, congestion is increasing three times faster than vehicle traffic, at about 15 percent per year. Congestion is a fact of life in all of the state's urban areas. Roughly 10 percent of all traffic in the greater Los Angeles area is run during congested conditions, which could increase to 50 percent if current growth trends continue. Congestion can increase emissions by the constant changing of driving speeds. A car traveling at 5 mph, for instance, will produce three times more smog-forming hydrocarbon than one cruising at 55 mph (Figure 6-20). In addition, congestion can add to emissions by increasing the amount of time that an engine runs. A car that takes 30 minutes to drive 10 miles will produce two and one-half times more pollution than a car making the same trip in only 11 minutes. In other words, 19 minutes of delay will increase hydrocarbon emissions 250 percent! (See Figure 6-21.)

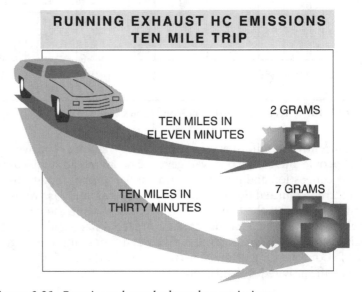

**Figure 6-21:** Running exhaust hydrocarbon emissions.

## Box 6.2

# 21 Ways You Can Reduce Air Pollution

*We have seen many ways smog affects both our health and welfare. Since we all help create it, we should all be knowledgeable about how we can help get rid of it. The following list of 21 ways you can clean the air at home, on the road, and in your community is printed here courtesy of the San Joaquin Valley Unified Air Pollution Control District.*

<u>In the Home</u>

1. **Paint with water-based paints.** On an average day, drying paint releases a considerable amount of smog-forming compounds to the air. Oil-based paints contain 3 to 5 times more toxic solvents than water-based latex paints. Close lids tightly – an open gallon of paint can emit up to three-and-a-half pounds of volatile organic compounds (VOCs).

2. **Paint with brushes or rollers.** Spray paints from cans are double trouble. The gas that propels the spray and any paint that dries before impact or misses its target create additional pollution. When painting, try to use only what you need.

3. **Buy products in their non-aerosol forms.** The propellants used in aerosol cans of hair spray, deodorants, and household products contribute to smog. Alternative, non-aerosol products are readily available and safe.

4. **Select products with less packaging.** Minimizing the paper you have to toss away clears the air. Packaging consumes energy when it's made, generates harmful VOC's when it's printed, and produces carbon dioxide when it's burned. $CO_2$ is one of the "greenhouse gases," and, if that weren't enough, packages represent half of the nation's cost of waste disposal.

5. **Plant trees and low-maintenance landscaping.** Trees add oxygen to the atmosphere, break down some pollutants and reduce dust. A total of 300 trees can counterbalance the amount of pollution one person produces in a lifetime. Water-conserving native plants save three valuable resources – water, energy, and the air – because they require less care. If you push that lawn mower by hand or use an electric-powered model, you'll cut down on more emissions. Try an old-fashioned rake instead of a leaf blower, it's good for you and the environment.

6. **Fire up your barbecue in a different way.** The typical lighter fluid used for backyard barbecues also creates pollution. Experiment with alternate ways of igniting charcoal. Even better, convert that barbecue to natural gas which is cleaner burning, easier to use, and leaves no chemical taste on the food.

7. **Conserve energy.** A year's worth of electricity in the average home sends 4.5 tons of carbon dioxide and other pollutants into the air, adding to global warming and day-to-day smog. Turn electrical appliances off during peak times between noon and 5 p.m. Set the temperature of your water heater to 130 degrees. This will conserve energy and save you roughly $25 a year.

8. **Insulate your home.** The less energy used for heating and cooling, the less pollution from electric power plants and the burning of oil and natural gas. If you can't add attic or wall insulation, you can still caulk and weather-strip doors and windows. And close off unused rooms.

9. **Go solar for home and water heating.** Water and space heating account for more than 50 percent of household energy use. Installing solar energy sources reduces the need to burn fossil fuels. Even without solar, turn off the thermostat at night. And use cold water instead of hot whenever possible.

10. **Reduce physical activity during heavy smog periods.** Everyone's health is at risk from smog, especially in peak episodes. Children, the elderly, pregnant women, and athletes are particularly vulnerable. Watch for the daily air quality forecast on the local news and cut down on outdoor exposure when smog is at its worst. If high smog levels are predicted, try to carpool and postpone smog-producing activities such as painting.

## Box 6.2

### On the Road

11. **Rideshare**. You can cut down on the pollution coming from motor vehicles by sharing a ride. Carpooling or vanpooling also relieve congested roadways.

12. **Leave the driving to them**. Use public transportation. The bus reduces car repairs and commuting costs, along with stress.

13. **Ride a bicycle**. Or walk. Cycling and walking are great for short trips. Human power produces no emissions and exercise has benefits of its own.

14. **Combine errands into one trip**. Instead of hopping in the car whenever you need something, set aside time to plan your errands. Cluster as many as possible.

15. **Make sure your car's air conditioner is working properly**. Leaky air conditioners in cars are the single biggest source of CFCs. Check your auto air conditioner to make sure it's leakproof. And, if it needs to be recharged, have its CFCs recycled.

16. **Prevent gasoline spillage**. Never top off your tank. Topping off overfills your tank with gasoline and the air with contaminants. Remember, gasoline is another source of smog since it contains pollutants such as benzene.

17. **Keep your car engine well-tuned**. Dirty carburetors and injectors, clogged air filters, dirty engine oil and worn out plugs not only waste gasoline and lower engine performance, they cause increased emissions of all pollutants.

18. **Drive on radial tires**. Radial tires produce less air-damaging particulate matter per mile than bias-ply tires do. Properly inflated tires can save up to 10% of your fuel consumption.

19. **Cooperate with the state's smog check program.** By obtaining a smog check on your car and making the necessary repairs, you'll be helping to reduce emissions. Remember, in some states removing a catalytic converter or other smog control equipment is illegal.

### In the Community

20. **Recycle**. Recycled materials give the air a break because they don't need to be burned or buried. Discard used motor oil at a recycling center – if dumped onto the ground it can harm water supplies and release polluting vapors.

21. **Write your legislators and local officials**. Elected officials need to know you support clean air programs.

## ■ Checking Your Understanding (6-3)

1. Explain why California's air quality has not achieved federal standards even though reductions have been made.

2. Describe the difference in responsibility between federal, state, and local regulating agencies in California.

3. Explain the benefits of gathering more information about air pollutants.

4. List five ways you can help reduce air pollution.

# 7

# Water and Wastewater

## ■ Chapter Objectives

1. **Describe** the major provisions of the Safe Drinking Water Act; the Clean Water Act; and the Marine Protection, Research and Sanctuaries Act.

2. **Describe** the basic components of water supply systems.

3. **Describe** the basic unit processes of a water treatment system.

4. **Describe** the types of treatment technologies that are most commonly used to meet water supply maximum contaminant levels.

5. **Describe** the basic principles involved in wastewater collection systems.

6. **Describe** the basic unit processes of a wastewater treatment system.

7. **Describe** the types of treatment technologies that are most commonly used to meet NPDES limits under the Clean Water Act (e.g. physical, chemical, biological).

## ■ Terms and Concepts to Remember

Activated Sludge
Adsorption
Aeration
Aerobic Bacteria
Anaerobic Bacteria
Biochemical Oxygen Demand
  (BOD)
Biological Treatment
Chemical Treatment
Chlorine Demand
Coagulants
Coagulation
Community Water System
Cross Connection
Disinfection
Dissociate
Effluent Limitations
Electrodialysis
Electronegativity
Estuaries
Facultative Bacteria
Filtration
Flocculation
Hydrogen Bonding
Hygroscopic
Influent
Ionic Bond
Land Application
Maximum Contaminant Levels
  (MCLs)
National Pollutant Discharge
  Elimination System (NPDES)

Near Coastal Water Initiative
Nonbiodegradable
Nonpolar Covalently Bonded
Oxidation
Oxidizer
Physical Treatment
Polar Covalent Bond
Polar Covalent Molecule
Pretreatment
Primary Drinking Water
  Standards
Primary Treatment
Publicly Owned Treatment
  Works (POTW)
Receiving Waters
Recharge
Reference Dose (RfD)
Reverse Osmosis (RO)
Screening
Secondary Drinking Water
  Standards
Secondary Treatment
Sedimentation
Solvated
Surface Tension
Tertiary Treatment
Total Suspended Solids (TSS)
Turbidity
Water Softening
Watershed
Wellhead Protection

## ■ Chapter Introduction

In Chapter 5 we learned that significant percentages of surface and groundwaters are contaminated and that this contamination stems from many different pollution sources. Of particular concern for safe drinking water are waterborne pathogens and chemical contaminants. Surface water sources are increasingly threatened by a new and complex challenge to water quality – non-point source (NPS) pollution.

The Safe Drinking Water Act requires that drinking water be treated to prevent diseases that are caused by microbes as well as certain chemical contaminants. The Clean Water Act requires protection of surface waters by treatment of wastewaters. The Marine Protection, Research and Sanctuaries Act protects marine ecosystems by regulating ocean dumping.

This chapter provides information on the specific requirements of these three laws. It also covers the treatment methods for both water and wastewater that are currently used to satisfy legal requirements.

# 7-1 Water Quality Laws

## ■ Regulatory Framework

### Concepts

■ Water bodies and groundwater have been contaminated, and coastal and marine habitats have been lost due to human activities.

■ The three basic areas of concern for protection of water quality are drinking water, critical water habitats, and surface waters.

> *We are a highly compartmentalized agency, organized to control and clean up pollution, medium-by-medium, chemical-by-chemical – not to prevent it. As a result, we often have simply been cycling problems through our system, seldom really solving them. We took toxics from smokestacks, turned them into sludge, dumped the sludge somewhere on the landscape, and then watched the inevitable runoff and leachate contaminate our water. Indeed, nowhere has this frustrating cycle been more apparent than in our efforts to deal with water quality."*

> William K. Reilly, referring to the EPA, and written while he was EPA Administrator under President Bush

In Chapter 5 we learned that the nation's water bodies have sustained damage from the poisonous by-products of our modern age. This has resulted in damage to ecosystems, damage to drinking water sources, and the loss of beneficial uses of water such as swimming or fishing. The potential sources of contamination that are to blame result from both commercial and household activities. These activities have resulted in direct contamination from discharges to water and indirect contamination from runoff and atmospheric deposits. Drinking water supplies have also been affected, with a significant percentage exhibiting serious contamination. Hundreds of chemicals have been found in groundwater, and the list of potential sources for this contamination is long. Waste and contaminants have been injected into the ground; pesticides have leached through the soil; landfills, surface impoundments, and accidental spills have contributed their leached materials; and nonpoint runoff and sewage have found their way into the groundwater. Increasingly high groundwater withdrawals, which have more than doubled since 1950 to more

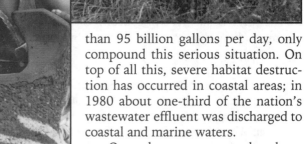

**Figure 7-1:** Three scenes of water quality protection. Preventing degradation and destruction of critical water habitats (above); reducing pollution of surface waters (top right); protecting drinking water quality (right).

than 95 billion gallons per day, only compound this serious situation. On top of all this, severe habitat destruction has occurred in coastal areas; in 1980 about one-third of the nation's wastewater effluent was discharged to coastal and marine waters.

Over the years, water has been regarded as an infinite resource and a natural repository for any and all contaminants. Water bodies have been used as septic tanks and toxic sinks. Coastal waters bear the biggest brunt of the pollution generated by twentieth century society since over one half of the U.S. population lives on the 10 percent of land falling within 50 miles of coastline. The EPA was criticized in 1990 by the Science Advisory Panel for focusing mainly on risks to human health and not balancing this focus with concern for ecological degradation. In response to these concerns, existing water quality protection laws were amended and new laws were enacted. This section will highlight the main laws relating to protection of water quality.

There are three basic areas targeted by water laws: reducing the pollution of surface waters, protecting drinking water quality, and preventing degradation and destruction of critical water habitats (Figure 7-1). We have three major laws to deal with these themes: The Clean Water Act; Safe Drinking Water Act; and Marine Protection, Research, and Sanctuaries Act.

The goal of the Clean Water Act is to "restore the chemical, physical, and biological integrity of the nation's waters." In order to do this, the act requires reduction of pollutants entering all surface waters, including lakes, rivers, oceans, and wetlands. Newer aspects of the act include stricter requirements for wastewater treatment plants, control of nonpoint source pollution, and tighter controls on toxic pollutants.

The Safe Drinking Water Act protects drinking water resources. The act requires adherence to established drinking water standards and protection of underground sources of drinking water. As part of these goals, the act requires **wellhead protection** programs. Wellhead protection areas are defined as the surface and subsurface areas surrounding a water well or wellfield supplying a public water system that may be contaminated through human activity. A wellhead protection program is designed to protect the wellhead from contamination. Figure 7-2 shows the avenues pollutants can use to enter wells.

The goal of the Marine Protection, Research, and Sanctuaries Act is to protect the marine environment from the harmful effects of ocean dumping. This act requires permits to ensure that ocean dumping of wastes does not cause degradation of the marine environment.

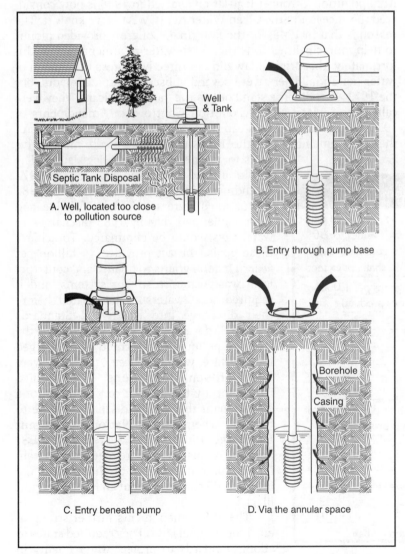

**Figure 7-2:** Ways pollutants can enter a well.

## ■ History and Overview of the Clean Water Act

### Concepts

■ The CWA requires the EPA to set effluent limitations.

■ All dischargers of wastewaters to surface waters are required to obtain NPDES permits.

■ NPDES permits require regular monitoring and reporting.

Federal water pollution legislation dates back to the turn of the century with the Rivers and Harbors Act of 1899. This was a very limited and weak law when it came to protecting the nation's waters from pollution. Stronger legislation enacted in 1948 is now considered the ancestor of the Clean Water Act (CWA) as we know it. The reason is that in 1948, for the first time, Congress provided money to help municipalities construct wastewater treatment plants. Unfortunately, this original law did not directly address the control of water pollution. There were several subsequent amendments, but the 1972 amendments stand out as the most significant. They were called the Federal Water Pollution Control Act Amendments of 1972. These amendments can be considered complete rewrites of the law. Remember that the 1970s were some of the most active years for new environmental legislation. The ambitious and optimistic goals of the 1972 amendments included mandates that by 1983 the nation's waters would be fishable and swimmable, and that by 1985 discharges to waterways would be eliminated. To achieve these goals, the act provided $5 billion per year in federal grants to finance and construct local wastewater treatment systems, and it required wastewaters to be treated before their discharge into waterways. Standards were to be developed by the EPA to define the levels of pollutants that could be discharged into surface waters. These standards were called **effluent limitations.** These limits, which specify the highest concentration of a contaminant that can be discharged, were to be used when issuing **National Pollutant Discharge Elimination System (NPDES)** permits. All dischargers are required to obtain NPDES permits prior to discharging, as the act made discharge without an NPDES permit illegal.

The 1987 amendments further strengthened the original laws. They required states to adopt standards to protect water quality in every stream or surface water body within

| Effluent Limitations | | |
|---|---|---|
| Effluent characteristic | Maximum for any 1 day | Averages of daily values for 30 consecutive days shall not exceed |
| | Metric units (kilograms per 1,000 kg of anhydrous product) | |
| BOD5* | 1.80 | 0.60 |
| COD | 4.50 | 1.50 |
| TSS | 1.20 | .40 |
| Oil and grease | 0.30 | .10 |
| pH | 6.0 to 9.0 | 6.0 to 9.0 |
| | English units (pounds per 1,000 lb of anhydrous product | |
| BOD5* | 1.80 | 0.60 |
| COD | 4.50 | 1.50 |
| TSS | 1.20 | .40 |
| Oil and grease | 0.30 | .10 |
| pH | 6.0 to 9.0 | 6.0 to 9.0 |
| *BOD5 = five-day biochemical oxygen demand | | |

**Figure 7-3:** Example effluent limitations. (Source: 40 CFR, §417.12)

their borders. These standards designated beneficial uses such as fishing or swimming and listed specific requirements to protect these uses. Currently, the requirements include pollutant-specific levels allowed in the water for each identified use. Standards can vary with the particular water body and area of the country. In some instances, controls are needed to reduce pollutant levels. These water quality standards must be reviewed every three years, and revised as needed.

Under the CWA, the EPA sets effluent limitations that are consistent throughout the country. As stated earlier, effluent limitations are restrictions on the amount of specific pollutants that a facility can discharge into a stream, river, or harbor. These are *pollutant-specific* discharge limitations for industrial categories and wastewater treatment plants. Best available technologies and economic feasibility are taken into consideration when developing these limitations. The EPA and the states use these guidelines to establish NPDES permit limitations. The effluent limitations set by the EPA are the minimum standards; states may be stricter in some cases. All industrial and municipal facilities that discharge wastewater directly into surface waters must have a NPDES permit. The NPDES permit requires monitoring and reporting discharge levels and must be renewed at least every five years. Monitoring and reporting requirements contain specific instructions for sampling the effluent in order to determine whether limitations are being met. Instructions include sampling frequency and the type of monitoring required. The permittee may be required to monitor the effluent on a daily, weekly, or monthly basis. The monitoring results are regularly reported to the EPA and state authorities.

In 1989, there were nearly 50,000 industrial and 16,000 municipal facilities that had NPDES permits. State or federal EPA inspectors monitor dischargers to determine if they are in compliance with their permit limitations and conditions. Appropriate enforcement actions, including criminal actions, are taken as needed for noncompliance.

The amended Clean Water Act also has provisions to protect wetlands. To implement these provisions, the EPA joined forces with the U.S. Army Corps of Engineers to regulate the discharge of dredged or fill material into surface waters. EPA's main responsibilities under this program are to develop the extensive environmental criteria necessary to evaluate permit applications and approve or disapprove of permits based on their merits.

The CWA allows for states and the EPA to select and protect **estuaries** of national significance that are threatened by pollution, development and overuse. An estuary is a coastal body of water that is partly enclosed. Because it opens to

---

### pH Effluent Limitations under Continuous Monitoring

(a)  Where a permittee continuously measures the pH of wastewater pursuant to a requirement or option in a National Pollutant Discharge Elimination System (NPDES) permit issued pursuant to section 402 of the Act, the permittee shall maintain the pH of such wastewater within the range set forth in the applicable effluent limitations guidelines, except excursions from the range are permitted subject to the following limitations:

(1)  The total time during which the pH values are outside the required range of pH values shall not exceed 7 hours and 26 minutes in any calendar month; and

(2)  No individual excursion from the range of pH values shall exceed 60 minutes.

**Figure 7-4:** Examples of NPDES monitoring requirements. (Source: 40 CFR § 401.17)

the ocean, fresh and salt water can mix. Estuaries are extremely sensitive and ecologically important habitats, providing sanctuaries for many types of waterfowl. They are also breeding and feeding grounds for many animals and plants. Since they tend to be excellent harbors, most large U.S. ports are located in them. The human impact on estuaries from activities such as landfilling, wastewater outflows, and industrial effluent discharge can destroy the sensitive ecological balance of these areas. To protect selected estuaries, the EPA and the involved state(s) form a management committee consisting of numerous work groups to assess the problems, identify solutions, and develop and oversee implementation of appropriate corrections.

Over 50 percent of the wetlands in the U.S. (not counting Alaska and Hawaii) have been converted to other uses during this country's existence. By the 1950s, the rate of loss had reached staggering proportions – about 458,000 acres annually. This rate, although not increasing, did not decline appreciably through 1990. Protection of critical aquatic habitats is a relatively new concern, and the EPA's actions in this area have increased greatly in the last few years. For example, the EPA has on several occasions used its veto power on permits that had been approved by the Corps of Engineers. The EPA has also made a concerted effort to limit the amount of minimally treated wastewater effluent discharged to the ocean in coastal regions. Areas like New York and New Jersey had dumped sewage sludge in near-shore sites. The EPA requires this dumping to occur in more environmentally sound, deep-water sites.

The CWA authorizes the EPA to provide financial assistance to states to support programs such as the construction of municipal wastewater treatment plants; water quality monitoring, permitting, and enforcement; and implementation of nonpoint source controls. These funds may also support development and implementation of state groundwater protection strategies. In addition, the EPA provides grants to states for the creation of the State Water Pollution Control Revolving Funds.

## ■ History and Overview of the Safe Drinking Water Act

### Concepts

- Community and non-community public water systems are covered by the SDWA.

- The SDWA requires establishment and enforcement of MCLs.

- MCLs for carcinogens and non-carcinogens are to be set as close to MCLGs as possible.

The Safe Drinking Water Act of 1974 (SDWA), which was amended in 1986, requires the U.S. Environmental Protection Agency to identify and regulate substances in drinking water that may have an adverse effect on public health. The SDWA applies to all public water systems, whether privately or publicly owned. It was designed to provide the same basic level of health protection to residents of small communities and large cities as long as the water system provides piped water for human consumption to at least 25 individuals or delivers water through a system having at least 15 service connections. Public water systems fall in either the category of "community water systems" or "non-community water systems." A community water system services, at minimum, 25 year-round residents. Non-community water systems, such as campgrounds, hotels, and restaurants, serve non-residents. Since this type of system is not the principal source of water for the people they serve, some of the SDWA's requirements imposed on community systems are relaxed for non-community systems.

Each state is expected to administer and enforce the SDWA regulations for all public water systems. Public water systems must provide water treatment, ensure proper drinking water quality through monitoring, and provide public notification of contamination problems. The 1986 amendments to the SDWA significantly expanded and strengthened its protection of drinking water. Currently, the SDWA requires six basic activities:

■ Establishment and enforcement of **Maximum Contaminant Levels (MCLs)** – These are the maximum levels of certain contaminants that are allowed in drinking water from public systems. Under the 1986 amendments, the EPA has set numerical standards or treatment techniques for an expanded number of contaminants.

■ Monitoring – EPA requires monitoring of all regulated and certain unregulated contaminants, depending on the number of people served by the system, the source of the water supply, and the contaminants likely to be found.

■ **Filtration** – EPA has criteria for determining which systems are obligated to filter water from surface water sources.

■ Use of lead materials – The use of solder or flux containing more than 0.2 percent lead, or pipes and pipe fittings containing more than 8 percent lead is prohibited in public water supply systems. Public notification is required where there is lead in construction materials of the public water supply system, or where water is sufficiently corrosive to cause leaching of lead from the distribution system or lines.

■ Wellhead protection – The 1986 amendments require all states to develop Wellhead Protection Programs. These programs are designed to protect public water supplies from sources of contamination.

The EPA developed regulations to meet the requirements of the SDWA, which are called the National Drinking Water Regulations. These regulations are found in *40 CFR* and are subdivided into **primary drinking water standards**, affecting public health; and **secondary drinking water standards**, affecting aesthetic qualities relating to the public acceptance of drinking water.

A **contaminant** is defined as "any physical, chemical, biological, or radiological substance or matter in water."

**MCL** is legally defined as the "maximum permissible level of a contaminant in water which is delivered to the free flowing outlet of the ultimate user of a public water system."

In developing and updating the regulations, the most relevant criteria the EPA uses for selection of regulated contaminants are the

| Primary MCLs | |
|---|---|
| Contaminant | Milligrams per liter |
| Arsenic | 0.050 |
| Barium | 1.000 |
| Cadmium | 0.010 |
| Chromium | 0.050 |
| Lead | 0.050 |
| Mercury | 0.002 |
| Nitrate (as N) | 10.000 |
| Selenium | 0.010 |

| Secondary MCLs for public water systems | |
|---|---|
| Contaminant | Level |
| Aluminum | 0.05 to 0.2 mg/l |
| Chloride | 250 mg/l |
| Color | 15 color units |
| Copper | 1.0 mg/l |
| Corrosivity | Non-corrosive |
| Fluoride | 2.0 mg/l |
| Foaming agents | 0.5 mg/l |
| Iron | 0.3 mg/l |
| Manganese | 0.05 mg/l |
| Odor | 3 threshold odor number |
| pH | 6.5 to 8.5 |
| Silver | 0.1 mg/l |
| Sulfate | 250 mg/l |
| Total dissolved solids (TDS) | 500 mg/l |
| Zinc | 5 mg/l |

**Figure 7-5:** Example of primary and secondary MCLs. (Source: 40 CFR)

potential health risks and how readily they may occur in drinking water. This entails evaluation of factors such as their occurrence in the environment, health effects of human exposure, analytical methods of detection, chemical transformations of the contaminant in the drinking water, and calculations of population risks and costs of treatment.

For each of the substances or contaminants that the EPA identifies, there are two methods for developing regulatory measures. The EPA must either establish a Maximum Contaminant Level (MCL) or, if it is not economically or technically feasible to monitor the contaminant level in the drinking water, specify a treatment technique that will effectively remove the contaminant from the water supply or reduce its concentration. The MCLs currently cover a number of volatile organic compounds, organic chemicals, inorganic chemicals, radionuclides, as well as microbes and **turbidity** (cloudiness or muddiness). The MCLs are based on an assumed human consumption of two liters of water per day.

The SDWA Amendments require that the EPA develop both MCLs, which are enforceable standards, and Maximum Contaminant Level Goals (MCLGs), which are not enforceable health goals. For each contaminant, the MCLG development occurs first; and then the final MCL is set as close to the MCLG as is feasible, taking into consideration analytical methods, treatment technology, economic impact (costs), and regulatory impact.

For noncarcinogens, MCLGs are arrived at in a three-step process. The first step is calculating the **Reference Dose (RfD)** for each

|  | Final MCLG* (mg/l) | Final MCL (mg/l) |
| --- | --- | --- |
| Trichloroethylene | zero | 0.005 |
| Carbon tetrachloride | zero | 0.005 |
| Vinyl chloride | zero | 0.002 |
| 1,2-Dichloroethane | zero | 0.005 |
| Benzene | zero | 0.005 |
| p-Dichlorobenzene | 0.075 | 0.075 |
| 1, 1-Dichloroethylene | 0.007 | 0.007 |
| 1, 1, 1-Trichloroethane | 0.200 | 0.200 |

*Final MCLGs were published November 13, 1985. The MCLG and MCL for p-dichlorobenzene were reproposed at zero and 0.005 mg/l on April 17, 1987; comment was requested on levels of 0.075 mg/l and 0.075 mg/l respectively.

**Figure 7-6:** Comparison of MCLGs and MCLs for VOCs. (Source: USEPA Fact Sheet, February 1989)

specific contaminant. The RfD is an estimate of the amount of a chemical that a person can be exposed to on a daily basis that is not anticipated to cause adverse systemic health effects over the person's lifetime. The RfD is usually given in milligrams of chemical per kilogram of body weight per day (mg/kg/day).

The RfD is obtained from a No Observable Adverse Effect Level (NOAEL was covered in Chapter 3), which is a number calculated by using data from animal or human studies. The NOAEL is divided by an *uncertainty factor* to obtain the RfD. This uncertainty factor is designed to take into account the differences between species, limited or incomplete data, the significance of the adverse effect, the length of the exposure, and factors that involve how the chemical is distributed and metabolized in the body. This uncertainty factor can range from 1 to 10,000.

A different assessment system is used for chemicals that are potential carcinogens. Government agencies generally assume there are no thresholds for carcinogens. This means that any exposure is assumed to represent some level of risk (see Chapter 3 for information on risk predictive models). The House of Representatives endorsed this approach in a report that accompanied the SDWA of 1974, indicating that the MCLGs for carcinogens should be set at zero. Today, if toxicological evidence leads to the classification of the contaminant as a human or probable human carcinogen, the MCLG is set at zero.

# ■ The Marine Protection, Research, and Sanctuaries Act (Title I)

## Concepts

■ Ocean dumping can have serious impacts on marine ecosystems.

■ The MPRSA regulates ocean dumping by restricting materials that can be dumped as well as locations and times for dumping.

■ Many governmental agencies are involved in enforcement of the MPRSA.

Ocean dumping is a major source of marine pollution. Over the years many materials were commonly dumped into the ocean. These included dredged material, sewage sludge, and industrial wastes. The sediments dredged from harbors in industrialized areas today are often highly contaminated with heavy metals, toxic organic chemicals like PCBs and petroleum hydrocarbons. When dumped into the ocean, marine organisms absorb these contaminants. Concern over these serious impacts of ocean dumping lead to the passage of the Marine Protection, Research and Sanctuaries Act (MPRSA) in 1972.

Under this act, the EPA regulates dumping materials that could adversely affect human health or the marine environment into the oceans. Ocean dumping requires a permit which may limit the site and times that dumping occurs. Many materials, including radiological, chemical, or biological warfare agents or high-level radioactive wastes, are prohibited from being dumped.

This act brings together the responsibilities of many governmental agencies. The EPA and the U.S. Army Corps of Engineers share permitting authority under this act. The Corps is responsible for the permitting of dredged material, and the EPA is responsible for permitting all other types of materials. The Coast Guard monitors the activities and the EPA is responsible for assessing penalties for violations. The EPA is also responsible for designating sites and times for ocean dumping activities. The Department of State is responsible for international enforcement, and the National Oceanic and Atmospheric Administration and the Department of Commerce are responsible for research to learn more about the effects of ocean dumping.

Since 1972, ocean dumping of dredged materials and industrial wastes has lessened significantly. Approximately 5 million tons of industrial wastes were dumped in the ocean in 1973. In 1986, 0.3 million tons were dumped. One type of waste typically dumped into the ocean, however, has not lessened; the dumping of sewage sludge has increased from 5 million tons in 1973 to 7.9 million tons in 1986. This increase is the result of the number of wastewater treatment plants nationwide. Over time, dumping of this sludge has been moved further out to sea. In New York, for instance, sludge was dumped only about 12 miles offshore until 1987 when the EPA required the dumping be moved about 100 miles offshore. Despite progress in several areas of ocean dumping, a new waste disposal problem has surfaced – the persistence of plastics disposed of at sea. Much of this plastic has been **nonbiodegradable**, which means it does not break down easily in the environment. Plastic and other nonbiodegradable debris in the ocean has caused injury and death to fish, marine mammals, and birds.

The Office of Marine and Estuarine Protection was established in 1984 as the part of the EPA responsible for administering all of the EPA's ocean and coastal programs. The two major laws that provide the framework under which it operates are the Marine Protection, Research, and Sanctuaries Act and the Clean Water Act.

An addition to the scope of the Marine Protection, Research, and Sanctuaries Act is the **Near Coastal Water Initiative**. This initiative was developed in 1985 to provide for management of specific problems in waters near coastlines that are not dealt with in other programs. Near coastal waters include inland waters from the coast to the farthest point inland where the influence of the tides on water level is detected. This category of water includes bays, estuaries, and coastal wetlands, and the coastal ocean at the distance where it is no longer affected by land and water uses in the coastal drainage basin.

At this point, three ongoing projects are looking at innovative and cost-effective techniques for cleaning up and protecting these waters. The EPA is conducting research on the impact of contamination on near coastal waters and studying appropriate regulatory systems to prevent and monitor contamination. Ultimately, these types of programs will take highly coordinated efforts by federal, state, and local governments in order to succeed.

## ■ Checking Your Understanding (7-1)

1. List two examples of restrictions that would reduce pollution of surface waters.

2. Describe the basic differences between the goals of the Clean Water Act and the Safe Drinking Water Act.

3. List two examples of restrictions that would protect drinking water quality.

4. Define "maximum contaminant levels."

5. List two examples of restrictions that prevent degradation and destruction of critical water habitats.

6. Describe the difference between primary and secondary drinking water standards.

# 7-2 Water Treatment

## ■ What is a Water Supply?

### Concepts

■ Water supplies are made up of water sources, storage facilities, conveyance systems, treatment facilities, and transmission and final distribution facilities.

■ A **watershed** is the land area that feeds water into lakes, rivers, and streams.

■ Most water supply water ultimately ends up as wastewater.

A water supply provides water to homes, agriculture, and business for a variety of uses. Just think of the many ways water piped into your community is used. It is used for fire fighting, street cleaning, watering lawns and gardens, industrial processes, and to carry wastes to treatment facilities. Three vital factors determine whether a water supply is adequate – the water's quality, quantity, and the location of the water supply in relation to the points where it is used. Water supply systems are typically made up of:

– One or more supply sources, such as groundwater and rivers

– Storage facility(s), such as a reservoir

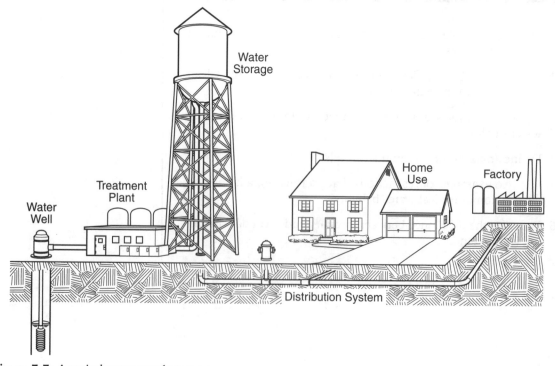

**Figure 7-7:** A typical water supply system.

# Science & Technology

## Water, Water Everywhere, but . . .

From a chemist's point of view, water is not a typical substance. It is true that it covers nearly 75 percent of the earth's surface and is the only common substance to exist in all three (solid, liquid, and gaseous) states of matter under earthly conditions. It is also necessary for all forms of life as we know it, and makes up about 65 percent of our body's weight.

To understand more about this abundant, but unique substance requires a closer look. Relatively speaking, water is a small molecule composed of one oxygen and two hydrogen atoms ($H_2O$). Each of the hydrogen atoms share a pair of electrons with the oxygen atom, however, the oxygen atom with its greater electronegativity tends to hold the shared electrons

Water molecule

closer to its nucleus, resulting in **polar covalent bonds.** What this means is that each of the shared electron pairs are drawn away from the hydrogen nuclei, resulting in partial positive ($\delta+$) charges on each of the hydrogen nuclei. It also means that an excess of electrons now surrounds the oxygen nuclei resulting in it becoming partially negatively ($\delta-$) charged. As a result of this shifting of the electron pairs, the water molecule develops a North and South pole, which makes it behave like a tiny magnet and is referred to as a **polar covalent molecule.** Polar molecules, therefore, feel the presence of other polar covalent molecules, and they tend to align themselves in such a way that their partially negative end will remain close to the partially positive end of a neighboring molecule. This weak force between the polar covalent molecules is called **hydrogen bonding** and is responsible for most of water's unique properties.

For purposes of our discussion here we will consider only four of water's unique properties. First is water's unusually high melting and boiling points which allows it to be mostly in the liquid form at average earthly temperatures. When compared to other pure substances of similar size (formula weights), it would be "normal" for water to have boiled into a gas at room temperature. However, due to hydrogen bonding, approximately ten water molecules $(H_2O)_{10}$ tend to "clump" together giving it an effective formula weight ten times greater. Having a melting point of 0°C and

a boiling point of 100°C is very normal for substances in that (180 g/mole) size range.

Second is water's characteristics upon freezing (turning into a solid). When most pure substances are cooled they tend to become more dense until they become a solid. Water follows that pattern until it reaches 4°C, and then it starts to expand until it freezes at 0°C. The explanation again seems to involve hydrogen bonding. As the molecules of water move closer together, each molecule attempts to have an oppositely charged portion of a neighboring molecule nearby. This results in the water molecules having to move slightly farther apart, resulting in the formation of hexagonal rings. The rings each have some empty space in the center, therefore, the water molecules in ice are less densely packed than in the surrounding liquid. Being less dense, ice floats in its own liquid rather than sinking to the bottom which happens to all other substances. This same reason also explains the "slipperiness" of ice. Normally when pressure is applied to a substance it tends to cause its particles to move closer together, which favors it becoming a liquid or solid. However, putting pressure on ice causes the molecules to have to break out of the hexagonal ice crystal and move closer

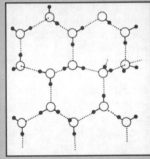

Ice crystals

together which means it has to return to the liquid form. It is the liquid water that forms between your foot and the ice below that provides the lubrication that reduces the friction and makes it slippery walking.

From a water molecule's point-of-view, it is not good to be a part of this surface. Mathematically we now understand that by forming a sphere, a substance can *maximize* its volume while *minimizing* its surface area. This is why drops of water attempt to form perfect spheres; however, under the pull of gravity we normally observe them slightly elongate, or in a "tear drop" shape.

# Science & Technology

Water's third unique characteristic is surface tension. Surface tension is the reason you can fill a glass above its top without the water running off, or why water tends to form droplets. Again it is the hydrogen bonding between polar covalent molecules that explains these properties. In the example of the glass of water, the molecules in the lower layers are holding onto the molecules above them by weak hydrogen bonding. They "care" if their neighbor falls over the edge, because they are involved with them through hydrogen bonding . The network of molecules on the top layer of the water all have molecules below that are pulling on them, but they have no one above pulling in that direction. As a result the top layer of water molecules form a unique "elastic sheet." They are closely packed by the attractions to their side and downward. When jumping into the swimming pool, if you fail to complete your dive you will encounter this more dense packed layer of water molecules with a stinging "belly-flop."

The last unique characteristic of water is its ability to act as a solvent for other polar covalent or ionic substances. Ionic compounds are held together with strong **ionic bonds**. The ions on the surfaces and corners, however, are not completely surrounded by oppositely charged ions. It is, therefore, attractive to them to leave the crystal structure (**dissociate**) and move out into the water, where they can become totally **solvated.** Typical solvated ions may have up to six

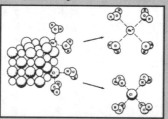

Water dissociating and solvating NaCl

water molecules surround them like fans around a "rock star." Once the ions on the surface are gone, then the next layer becomes the outer layer and the process happens again; soon the crystal has been completely taken apart resulting in a solution composed of oppositely charged ions, each surrounded by its admiring water molecules. Water molecules also finds other polar covalent molecules quite attractive. Many organic substances (e.g., alcohols, amines, and carboxylic acids) that have polar covalent bonds also have some water soluble. It is only the **nonpolar covalently bonded** compounds (e.g., hydrocarbons making up gasoline, diesel fuel, etc.) that water molecules tend to ignore, because they lack any electrical attraction to the polar water molecules.

Try demonstrating the electrical attraction of water molecules yourself by holding a comb that has been run through your hair several times near, but do not allow it to touch, a very small stream of running water from the faucet. Did you find that it is attracted toward the comb? Running the comb through your hair makes it negatively charged by removing electrons from your hair. Each of the water molecules "feels" this excess of electrons and quickly rotates so that its $\delta+$ hydrogens can be closer. It is this electrical attraction between the water molecule and the charge on the comb that causes the water to bend toward the comb.

- A conveyance system, such as an aqueduct to transport water from storage to the treatment facility
- A treatment facility for improving the water quality
- Transmission lines to transport water from the treatment facility to holding tanks or other storage areas prior to being piped to individual consumers
- Final distribution facilities to convey water to the residential or business user.

Water systems can obtain their raw water supply from an underground water source, as discussed in Chapter 5, or from surface waters such as rivers or lakes. As water is taken from a surface water source and processed for human consumption, rainwater runoff replenishes the supply. The land area that drains into a river, river system, or other body of water, is called a watershed.

As part of the hydrologic cycle, we learned that rainwater from the watershed replaces surface water and recharges groundwater.

Source waters naturally contain many inorganic and organic substances, depending on their exposure to various contaminants while cycling through the environmental compartments. Water treatment plants are designed to improve the source water's quality before it is used by the consumer. Treatment plants generally remove disease-causing microorganisms, trace organic compounds, suspended solids, minerals causing hardness, and substances causing disagreeable color, taste, and odor. The sections that follow in this chapter discuss water treatment methods in more detail.

In cities, water is distributed through complex systems of pipes that generally follow the patterns of the streets. These patterns must allow for typical elevation changes throughout the area as well as the service pressures that are needed. The main water pipes are usually 6 to 8 inches in diameter to provide adequate volumes for fire fighting and business/residential uses. Piping from the main supply line to a residence may be as small as 2.5 cm or 1 inch.

Only a relatively small percentage of the water supplied to residences, businesses, and public facilities is actually "used up" by evaporation and processing. It is estimated that the total consumptive use in the United States is about 22 percent of the volume withdrawn. This means that the majority of water at the point of use ends up as wastewater, which is then put back on the land or into water bodies after it is treated. Some areas are starting to use aggressive water reuse systems to lessen dependence on raw water supplies.

## ■ Why Must Drinking Water Be Treated?

### Concepts

- Water treatment is designed to reduce health risks and improve aesthetic qualities of water.

- Cross connections can cause serious contamination of drinking water.

- Physical treatment methods do not result in the formation of new substances, but chemical and biological treatment methods do.

Treatment of water intended for drinking is one of the oldest forms of public health protection. For thousands of years people have treated water to remove particles of solid matter, reduce health risks, and improve aesthetic qualities such as appearance, odor, color, and taste. The earliest recorded water treatment procedures are found in Sanskrit writings from 2000 B.C. The advice in these writings is that, "impure water should be purified by being boiled over a fire, or being heated in the sun, or by dipping a heated iron into it, or it may be purified by filtration through sand and coarse gravel." This is not vastly different from what we do today using chemical and physical techniques that will be described later in the chapter. In

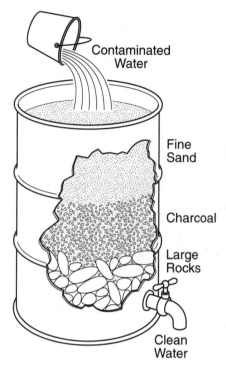

**Figure 7-8:** A primitive water treatment mechanism.

Chapter 5 we learned that the average American uses more than 180 gallons per day (and it is actually closer to 250 gallons per day in populated areas). This shows the sheer magnitude of our water treatment needs.

In the first half of the nineteenth century, scientists began to recognize that specific diseases, such as cholera, could be transmitted by water. Cholera was responsible for nearly 15,000 deaths in London in 1849. The important connection between water and disease transmission lead to discovery of treatment methods to eliminate pathogenic microorganisms in drinking water. Use of these treatment methods has dramatically reduced the incidence of waterborne diseases such as typhoid, cholera, and hepatitis in the United States. For example, in the first decade of the 1900s, approximately 36 out of every 100,000 people died annually from typhoid fever; today there are almost no cases of waterborne typhoid fever in the United States.

The government's requirement for proper treatment has served to greatly improve the quality and safety of drinking water in the United States; however, there are still over 89,000 cases each year of waterborne diseases caused by bacteria, viruses, protozoa (microscopic one-celled animals), helminths (parasitic worms), and fungi. These organisms can contaminate water through surface runoff, which often contains animal wastes; failures in septic or sewer systems; and wastewater treatment plant effluents that have not been disinfected. Microbiological contamination occurs most often in surface water, but it can also occur in groundwater, usually due to improperly protected wells. The first part of this chapter mentioned regulations for safeguarding wells from surface contamination using wellhead protection programs. Contamination can also occur after treated water leaves a treatment plant, through a **cross-connection**, which is a connection between safe drinking water and a source of contamination. An example of a cross-connection is depicted in Figure 7-9, where due to the lack of an anti-siphon valve, the pesticide contaminated water from a lawn applicator can be drawn back into the drinking water system when a nearby fire hydrant is used. Contamination after treatment can also occur if the water comes into contact with microorganisms that are growing in the piping system that distributes water to the final user.

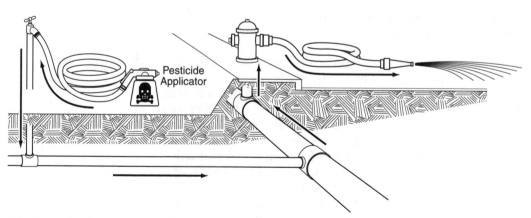

**Figure 7-9:** Example of a cross-connection.

| Waterborne Disease | Causative Organism | Source of Organism in Water | Symptom |
|---|---|---|---|
| Gastroenteritis | Rotavirus | Human feces | Acute diarrhea or vomiting |
| | *Salmonella* (bacterium) | Animal or human feces | Acute diarrhea or vomiting |
| | Enteropathogenic *E. Coli* | Human feces | Acute diarrhea or vomiting |
| Typhoid | *Salmonella typhosa* (bacterium) | Human feces | Inflamed intestine, enlarged spleen, high temperature; sometimes fatal |
| Dysentery | *Shigella* (bacterium) | Human feces | Diarrhea – rarely fatal |
| Cholera | *Vibrio comma* (bacterium) | Human feces | Vomiting, severe diarrhea, rapid dehydration, mineral loss; high mortality |
| Infectious hepatitis | Hepatitis A (virus) | Human feces, shellfish grown in polluted waters | Yellowed skin, enlarged liver, abdominal pain; low mortality; lasts up to 4 months |
| Amoebic dysentery | *EntaY̌oeba histolytica* (protozoan) | Animal or human feces | Mild diarrhea, chronic dysentery |
| Giardiasis | *Giardia lamblia* (protozoan) | Animal or human feces | Diarrhea, cramps, nausea, and general weakness; not fatal; lasts 1 week to 30 weeks |
| Cryptosporidiosis | *Cryptosporidium* (protozoan) | Animal or human feces | Diarrhea, stomach pain; lasts an average of 5 days |

**Figure 7-10:** Waterborne diseases. (Source: Adapted from American Water Works Association, *Introduction to Water Treatment: Principles and Practices of Water Supply Operations*, Denver, Colorado, 1984)

Figure 7-10 lists some of the most common diseases caused by microorganisms found in water supplies. Many of these microorganisms are common in water supplies in developing nations. Several countries in South America, for instance, have reported severe cholera outbreaks in the 1990s. In this country, the protozoan *Giardia lamblia* is the most commonly identified organism associated with waterborne disease. This is a fairly common problem with campers and hikers who drink mountain waters. The organism, which cannot be seen, causes a disease called giardiasis. The disease usually involves diarrhea, nausea, and dehydration that can be severe and can in some cases last for months – not a pleasant experience. Over 20,000 water-related cases of this disease have been reported in the last 20 years, with probably many more cases going unreported. Another protozoan disease, cryptosporidiosis, is caused by *Cryptosporidium*, a cyst-forming organism similar to *Giardia*. Other waterborne diseases common in this country include viral hepatitis, and gastroenteritis.

Chemical contaminants, both natural and from human activities, may also find their way into water supplies. Contamination problems in groundwater, which we learned represents well over half of the water used for drinking, are frequently chemical in nature. Common sources of chemical contamination include minerals dissolved from the rocks that form the earth's crust; pesticides and herbicides used in agriculture; leaking underground storage tanks; industrial effluents; seepage from septic tanks, wastewater treatment plants and landfills; and any other improper disposal of chemicals in or on the ground. In some water systems the water may become slightly corrosive and can leach away materials in the piping

| Process/Step | Purpose |
|---|---|
| **Preliminary Treatment Processes*** | |
| Screening | Removes large debris (leaves, sticks, fish) that can foul or damage plant equipment |
| Chemical pretreatment | Conditions the water for removal of algae and other aquatic nuisances |
| Presedimentation | Removes gravel, sand, silt, and other gritty materials |
| Microstraining | Removes algae, aquatic plants, and small debris |
| **Main Treatment Process** | |
| Chemical feed and rapid mix | Adds chemicals (coagulants, pH adjusters, etc.) to water |
| Coagulation/flocculation | Converts nonsettleable or settleable particles |
| Sedimentation | Removes settleable particles |
| Softening | Removes hardness-causing chemicals from water |
| Filtration | Removes particles of solid matter which can include biological contamination and turbidity |
| Disinfection | Kills disease-causing organisms |
| Adsorption using granular activated carbon (GAC) | Removes radon and many organic chemicals such as pesticides, solvents, and trihalomethanes |
| Aeration | Removes volatile organic chemicals (VOCs), radon, $H_2S$, and other dissolved gases; oxidizes iron and manganese |
| Corrosion control | Prevents scaling and corrosion |
| Reverse osmosis, electrodialysis | Removes nearly all inorganic contaminants |
| Ion exchange | Removes some inorganic contaminants, including hardness-causing chemicals |
| Activated alumina | Removes some inorganic contaminants |
| Oxidation filtration | Removes some inorganic contaminants (e.g., iron, manganese, radium) |
| *Generally used for treating surface water supplies | |

**Figure 7-11:** Basic water treatment processes. (Source: Adapted from American Water Works Assoc., *Introduction to Water Treatment*, V. 2, 1984)

system, possibly introducing lead and other materials into the drinking water. The water treatment process itself might also introduce trihalomethanes, which are chemicals formed as a by-product of the reaction of chlorine with organic materials and other chemical contaminants. These chemicals have been shown to be potent carcinogens. It is important to note that chlorination of drinking water has been the most effective measure ever taken to eliminate waterborne disease transmission – once again, the double-edged sword.

Although safeguarding health is the most important reason for treating drinking water, it is not the only reason. Consumer acceptance is also important. A system might treat water to improve its color, odor, or taste even if it is safe to drink. For example, some systems remove iron and manganese, which can stain laundry and plumbing fixtures. Many communities add fluoride to drinking water to improve dental health. The Safe Drinking Water Act (SDWA) acknowledges both of these reasons.

The treatment needs of a water system differ depending on the quality of the source water. Groundwater sources usually require less treatment than surface water sources. Common surface water contaminants include turbidity, microbiological contaminants, and low levels of a large number of organic chemicals. Groundwater contaminants include naturally occurring inorganic contaminants such as arsenic, fluoride, radium, and radon, and may include a number of specific organic chemicals such as trichloroethylene, which was used extensively as a solvent in dry cleaning operations. Bacteria and viruses can also contaminate relatively shallow groundwater from wastewater and surface runoff.

There are other economic and regulatory considerations in developing a water treatment system. To determine actual treatment needs, the source water's characteristics must be known. These characteristics include pH, temperature, alkalinity, and calcium and magnesium content. Operating costs and waste management issues are important considerations. Most treatment processes concentrate contaminants into a residual stream or sludge that requires proper management. Removal of radon using activated charcoal is a good example. The charcoal to which the radon has adsorbed is a low-level radioactive waste (see Chapter 10 on nuclear waste), which has very strict disposal and management requirements. This may significantly increase disposal costs for the treatment system.

There are a number of basic unit processes used to treat water so that it does not present the threat of waterborne disease(s) to the consumer. These processes typically fit into one of three categories: chemical, physical, or biological. A **physical treatment** process does not produce a new substance. These processes typically concentrate the waste by evaporating water or filtering solids to remove them from a stream. The most common physical treatment methods are: **screening, adsorption, aeration, flocculation, sedimentation,** and **filtration.** A **chemical treatment** process results in the formation of a new substance or substances. The most common chemical treatments include **coagulation, disinfection, water softening,** and **oxidation.** The next section will define and describe the meaning of these terms. **Biological treatment** processes use living organisms to bring about chemical changes. Biological treatment can be viewed as a subset of chemical treatment.

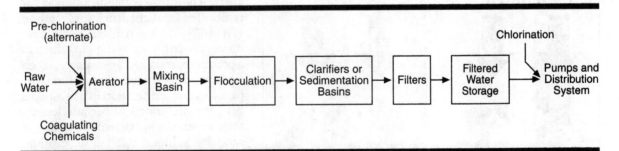

**Figure 7-12:** Typical drinking water treatment steps.

These treatment processes take place at a water treatment plant. The primary goal of a water treatment plant is to furnish water safe for human consumption. A second basic objective is the production of water that is appealing to the consumer. If primary drinking water standards are exceeded, either additional treatment or an alternative water supply source is required to protect the health of those persons using the water. Secondary regulations are based on aesthetic considerations and are not enforceable by the EPA, but may be enforced by the states. Violation of these aesthetic standards should be avoided, if possible, to discourage the consumer from turning to other potentially unsafe water.

The primary drinking water standards regulate levels of inorganic and organic chemicals, turbidity, coliform bacteria and radionuclides. It is important that the applicable standards are met at the customer's tap, except the turbidity standard which must be met at the point of entry into the distribution system. The next section will describe the actual treatment processes used to achieve both primary and secondary drinking water standards.

## ■ Physical and Chemical Processes of Water Treatment

### Concepts

■ Screening and filtration prevent movement of particles through a material.

■ Filtration is a physical process.

■ Chemical pretreatment often improves the water treatment process.

### Screening and Filtration

Screening and filtration are techniques used to capture a wide range of suspended debris from water. Large materials are typically removed by means of a coarse bar screen, with bars 2.5 to 5 cm (1 to 2 inches) apart. The most common method for removal of smaller particles is filtration (Figure 7-13). Mechanical filtration removes the particles by trapping them in a porous material. As an example, when water passes through a bed of sand, the particles are either left in the spaces between the grains or become loosely attached to the surface of the grains of sand by a process called adsorption. Sand filtration will remove particles ranging in diameter from 0.001 to 50 microns and larger; much smaller than the empty spaces between the grains. This phenomenon is due principally to the large surface area of sand. The par-

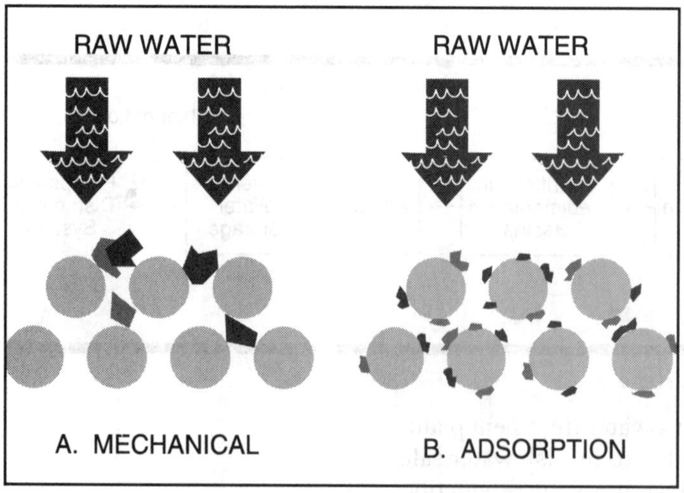

Figure 7-13: Physical and chemical filtration mechanisms.

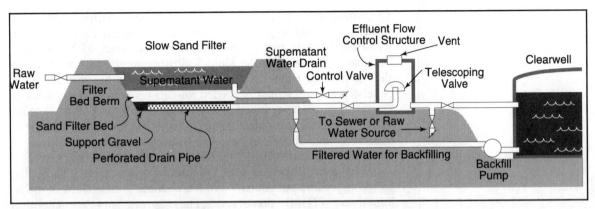

**Figure 7-14:** A typical slow sand filter.

ticles are attracted to the surface of the grains and are held there by relatively weak adsorption forces. This is the reason filtration helps control biological contamination and turbidity. As mentioned earlier, turbidity is a measure of the cloudiness of water caused by the presence of suspended matter, which shelters harmful microorganisms and reduces the effectiveness of disinfecting compounds. In some filtering systems a skin forms on the filter medium that contains microorganisms that trap and break down algae, bacteria, and other organic matter.

Filtration is a process that occurs naturally as surface waters move through the porous layers of soil to recharge groundwater. This natural filtration removes most suspended matter and microorganisms and is the reason many wells produce water that doesn't need any further treatment.

Filtration technologies commonly used in water treatment systems include slow or rapid sand filtration, diatomaceous earth filtration, and package filtration systems. The slow and rapid systems refer to the rate of flow per unit of surface area. Filters are also classified by the type of granular material used in them. Sand, anthracite coal, coal-sand, multi-layered, mixed bed, or diatomaceous earth are examples of different filtering media. Filtration systems may also be classified by the direction the water flows through the medium. Some examples are: downflow, upflow, fine-to-coarse, or coarse-to-fine. Lastly, filters are commonly distinguished by whether they are gravity or pressure filters. Gravity filters rely only on the force of gravity to move the water down through the grains and typically use upflow for washing the filter media to remove the collected foreign material. Gravity filters are free surface filters that are much more commonly used for municipal applications. Pressure filters are completely enclosed in a shell so that most of the water pressure in the lines leading to the filter is not lost but is used to push the water through the filter.

Slow sand filtration was first used in this country in 1872, making it the oldest type of municipal water treatment. This type of system consists of a layer of fine sand supported by a layer of graded gravel. The process is purposefully slow, with a filtration rate of about 0.022 cubic meters per hour per square centimeter (0.90 gal/hr/in$^2$) of filter area. The removal action involves a biological process as well as the physical and chemical processes already mentioned. A sticky coating of biological matter forms on the sand surface. The microorganisms that form the coating biologically degrade the organic matter that comes into contact with this film. Because of the reliance on biological processes, water treated in a slow sand filter cannot have chlorine in it since this would kill the beneficial microorganisms needed for this process. The turbidity of the water must be reduced before using a fine sand filter because suspended particulate matter can quickly clog the filter. A properly operating sand filter can achieve 91 to 99.99 percent removal of viruses and a greater than 99.9 percent removal of *Giardia* cysts.

Slow sand filtration does not remove organic chemicals, dissolved inorganic substances like heavy metals, fine clays or trihalomethane (THM) precursors. These precursors are chemical compounds formed when organic substances dissolve in water that can ultimately form THMs when reacted with chlorine.

The slow sand filter must be regularly cleaned. It is easy to know when to clean the filter because water flows through at a much reduced rate. The length of time between cleanings can range from several weeks to a year, depending on the raw water quality. Cleaning is performed by scraping off the top layer of the filter bed. Once this is scraped, the filtered water quality is bad for the few days it usually takes to produce the functioning biological filter.

Diatomaceous earth (DE) filtration, which is widely used for filtering swimming pool waters, has also been used successfully to remove turbidity and *Giardia* cysts from drinking water. This type of filtration works best in small water systems where the turbidity levels and bacterial counts of the raw water are relatively low. DE filters use a thin layer (3.2 to 6.4 mm, which is $\frac{1}{8}$ to $\frac{1}{4}$ inch) of diatomaceous earth, which is coated on a porous filter element as the filter material. DE filters can effectively remove *Giardia* cysts, algae, and asbestos; and if there is a sufficiently fine grade of DE, can remove bacteria. Pretreatment, consisting of coagulation, will also allow these filters to remove viruses.

Membrane filtration, also known as ultrafiltration, makes use of hollow fiber membranes to treat water (ultrafiltration membranes exclude particles larger than 0.2 microns). These systems can, therefore, remove bacteria, *Giardia*, and some viruses. They are best suited for polishing water with low turbidity that has already been treated by other methods.

Cartridge filters use a ceramic or polypropylene filter element packed into pressurized housing. They use a strictly physical method of filtration, by straining the water through the pores in the medium. Cartridge filters require raw water with low turbidity, and their effectiveness in removing viruses is unknown.

Chemicals are often added to the raw water before filtration. These chemicals improve the treatment process. Examples of pretreatment chemicals include: pH adjusters and **coagulants**. Coagulants are chemicals that cause small particles to stick together to make larger particles that are easier to remove. They do this by reducing the charge on the suspended particles that tend to keep them apart. Alum is an example of a chemical that is frequently used as a coagulant.

Flocculation is the "clumping" of particles as the result of coagulation. For this reason, the two terms are often used interchangeably, but they really are distinct. The actual joining together of the small particles into larger, settleable, and filterable particles is called flocculation. It is a process that causes suspended particles to collide, aggregate and form heavier particles called floc. For this process to be efficient, the water must be agitated gently for a time in order to allow the particles to come into contact with one another.

After flocculation, sedimentation may be used to separate the liquid from the solids. One way to accomplish this is by slowing the velocity of the water, giving the solids a chance to settle. The settled particles, called a sediment, combine to form a sludge that can be removed.

# ■ Disinfection with Chlorine

## Concepts

■ Water treatment requires primary and secondary disinfection.

■ Chlorination can produce potentially harmful by-products.

■ Three forms of chlorine are typically used for disinfection.

Disinfection is a chemical process that kills pathogenic organisms. Primary disinfection refers to the initial killing of *Giardia* cysts, bacteria, and viruses. Secondary disinfection refers to the maintenance of a disinfectant residual which prevents the regrowth of microorganisms in the water distribution system.

For the past several decades, chlorine, dispensed as either a solid $(Ca(ClO)_2)$, liquid (NaClO) or gas $(Cl_2)$, has been the disinfectant of choice in the United States. This is because it is effective and inexpensive and can provide a disinfectant residual in the distribution system. However, under certain circumstances, chlorination may produce potentially harmful by-products such as the trihalomethanes mentioned earlier. Small systems can successfully use ozone and ultraviolet radiation as primary disinfectants, but chlorine or an appropriate substitute must also be used as a secondary disinfectant. This is because ozone and ultraviolet radiation cannot provide a residual effect that will continue to kill microorganisms after the initial treatment.

Large water treatment systems usually use chlorine gas, supplied in liquid form, in high strength, high-pressure steel cylinders. The liquid immediately vaporizes when released from these pressurized containers forming a yellow-green, toxic gas. Chlorine gas irritates the eyes, nasal membranes, and respiratory tract. It is lethal at concentrations as low as 0.1 percent air by volume; therefore, systems using chlorine gas must have several major pieces of safety equipment:

– Chlorine gas detectors to provide early warning of leaks
– Self-contained breathing apparatus for the operator
– A power ventilation system for rooms in which chlorine is housed
– Emergency repair kits

Small systems most commonly use sodium hypochlorite or calcium hypochlorite, because they are simpler to use and have less extensive safety requirements than gaseous chlorine. Sodium hypochlorite provides 5 to 15 percent available chlorine and is the easiest of the three to handle. It is very corrosive, however, and should be handled and stored with care and kept away from equipment that can be damaged by corrosion. Sodium hypochlorite solution is more costly per pound of available chlorine and does not provide the level of protection of chlorine gas.

Calcium hypochlorite is a white solid in granular, powdered, or tablet form containing 65 percent available chlorine. When packaged, calcium hypochlorite is very stable, however, it is **hygroscopic**, which means it readily absorbs moisture. It reacts slowly with moisture in the air to form chlorine gas. It is a corrosive material with a strong odor, and requires proper handling. It must be kept away from organic materials such as wood, cloth, and petroleum products. Reactions between it and organic materials can generate enough heat to cause a fire or explosion.

Complex chemical reactions occur when chlorine is added to water, but these reactions are not always obvious. For example, a chlorine taste or odor in finished water is sometimes the result of too little rather than too much chlorine. It is important that anyone responsible for operating a treatment system understand basic chlorination chemistry and the factors affecting chlorination efficiency. These topics are beyond the scope of this introductory text.

When chlorine is fed into water, the various substances it can react with create a **chlorine demand.** Chlorine demand is a measure of the amount of chlorine that will combine with impurities and therefore will not be available to act as a disinfectant. Impurities that increase chlorine demand may include such substances as organic materials, sulfides, ferrous iron, and nitrites.

$$TCR = TCA - TCD$$

total chlorine residual = total chlorine available − total chlorine demand

Chlorine can also combine with ammonia or other nitrogen compounds to form chlorine compounds that have some disinfectant properties. These compounds are called combined *available chlorine residual.* "Available" means available to act as a disinfectant. The uncombined chlorine that remains in the water after combined residual is formed is called free available chlorine residual. Free chlorine is a much more effective disinfectant than combined chlorine.

Free chlorine is not available for disinfection unless the chlorine demand of the raw water is satisfied. When chlorine dosage exceeds the point at which chlorine demand is satisfied, additional chlorine will result in a free available chlorine residual. The chlorine dosage needed to produce a free residual varies with the quality of the water source.

Whenever chlorine is used for disinfection, the chlorine residual should be monitored at least daily. Samples should be taken at various locations throughout the water distribution system, including the farthest points of the system. Most small systems use a quick and simple colorimetric test, which changes color depending on the amount of chlorine residual in the water.

Five factors are important to successful chlorination: concentration of free chlorine, contact time, temperature, pH, and turbidity levels. The effectiveness of chlorination is directly related to the contact time with and concentration of free available chlorine. At lower chlorine concentrations, contact times must be increased. Maintaining a lower pH will also increase the effectiveness of disinfection. The higher the temperature, the faster the disinfection rate. Lastly, chlorine (or any disinfectant) is effective only if it comes into *contact* with the organisms to be killed. High turbidity levels can prevent good contact and protect the organisms. For this reason as well as aesthetics, turbidity should be reduced where necessary through the coagulation and sedimentation methods previously discussed.

## ■ Other Disinfectants and Disinfectant By-Products

### Concepts

■ Ozone and UV Radiation are other potential water supply disinfectants.

■ Secondary disinfectants like chlorine are still needed when alternative disinfectants do not result in disinfectant residuals.

■ Suspended solids, turbidity, and organic matter interfere with the action of disinfectants.

Ozone ($O_3$) is widely used as a primary drinking water disinfectant in other parts of the world, but is relatively new to the United States. It requires shorter contact time than chlorine to kill pathogens, making it a more powerful disinfectant. Ozone is a toxic gas formed when air-containing oxygen flows between two electrodes. It is unstable and must be generated on site. In addition, it has a low solubility in water, so efficient contact with the water is essential.

A secondary disinfectant, usually chlorine, is required because ozone does not maintain an adequate residual in water. The cost of installing, operating, and maintaining ozonation systems is relatively high. As with chlorine, the ozone demand of the water must be satisfied before the ozone is available for disinfection. The available ozone must be routinely measured.

Ultraviolet radiation (UV) effectively kills bacteria and viruses. As with ozone, a secondary disinfectant must be used in addition to ultraviolet radiation to prevent regrowth of microorganisms in the water distribution system. UV radiation can be attractive as a primary disinfectant for a small system because it is readily available, produces no known toxic residuals, required contact times are short, and the equipment is easy to operate and maintain.

UV radiation does not inactivate *Giardia* cysts so it cannot be used to treat water containing those organisms. It is recommended only for groundwater for which there is no risk of *Giardia* cyst contamination. UV radiation is also unsuitable for water with high levels of suspended solids, turbidity, color, or soluble organic matter. These materials can react with or absorb the UV radiation, reducing the disinfection performance. Substances in the raw water exert a UV demand similar to chlorine demand. The UV demand of the water affects the exposure time and intensity of the radiation needed for proper disinfection.

## Disinfection By-Products and Strategies for Their Control

Earlier in this section it was noted that adding a disinfectant to water might result in the production of harmful by-products. Chlorine, for example, can mix with the organic compounds in water to form trihalomethanes (THMs). One THM, chloroform ($CHCl_3$), is a suspected carcinogen. Other common trihalomethanes are similar to chloroform and may cause cancer.

Several strategies for minimizing harmful chlorination by-products can be used by small systems:

■ Reducing the concentration of organic materials *before* adding chlorine. Common water clarification techniques, such as coagulation, sedimentation, and filtration, can effectively remove many organic materials. Activated carbon, which is described in the next section, might be needed to remove organic materials at higher concentrations or for those not removed by other techniques.

■ Reevaluating the amount of chlorine used – the same degree of disinfection might be possible with lower chlorine dosages.

■ Changing the point in treatment where chlorine is added. If chlorine is presently added before treatment (chemical feed, coagulation, sedimentation, and filtration), it can instead be added after filtration, or just before filtration and after chemical treatment.

■ Using alternative disinfection methods.

A system with a high concentration of chlorination by-products in the treated water might consider alternative disinfection methods; however, as we learned in the previous section, ozonation and ultraviolet radiation, the alternative methods most practical for small systems, cannot be used as disinfectants by themselves. Both require a secondary disinfectant (usually chlorine) to maintain a residual in the distribution system.

Ozonation might also result in the formation of some harmful by-products. Ozone can produce toxic by-products from a few synthetic organic compounds, such as the pesticide heptachlor. If ozone is added to water containing bromide ions, it can form brominated organic compounds such as bromine-containing trihalomethanes.

Also, studies have shown that the addition of ozone followed by chlorine or chloramines can result in higher levels of certain by-products than when these disinfectants are used alone. For these reasons, it is important to know what compounds are in the raw water before choosing ozone as a disinfectant. Researchers continue to study ozonation by-products and their potential health effects.

Ultraviolet radiation is suspected to produce some by-products from organic compounds, but actual by-products have not yet been identified.

| Disinfectant | Advantages | Disadvantages | Application Point |
|---|---|---|---|
| Chlorine | ■ Effective for viruses, bacteria and *Giardia* cysts<br><br>■ Can be used as either a primary or secondary disinfectant<br><br>■ Chlorine residual can be easily monitored<br>■ Available as a gas, liquid, or solid<br>■ Minimal O&M requirements, especially for liquid and solid forms | ■ May result in potentially harmful by-products (THMs)<br>■ Significant safety concerns, especially for gas systems<br><br>■ May result in precipitation of iron and manganese | ■ Variety of application points<br><br>■ To minimize THM formation, generally added to the end of the treatment process |
| Ozone | ■ Effective against viruses, bacteria, and *Giardia* cysts<br>■ Enhances removal of biodegradable organics in slow sand filter | ■ Must be generated on-site<br><br>■ Does not produce a stable, long-lasting residual<br><br>■ May result in harmful by-products<br>■ Low solubility in water<br>■ Complex O&M requirements<br>■ Exhaust gas must be treated to remove ozone<br>■ Difficult to measure residual | ■ Prior to rapid mixing step<br><br>■ Should provide adequate time for biodegradation of oxidation products prior to chlorination |
| Ultraviolet Light | ■ Effective against viruses and bacteria<br><br>■ Minimal O&M requirements | ■ Not effective against *Giardia* cysts<br><br>■ Limited to groundwater systems not directly influenced by surface water supply | ■ Downstream of sedimentation or filtration process |

**Figure 7-15:** Comparison of three disinfectants for small-community drinking water systems. (Source: EPA)

# ■ Treatment of Organic and Inorganic Contaminants

## Concepts

■ Water treatment must address organic and inorganic contaminants as well as pathogens.

■ Volatile organic compounds easily become gases.

■ There are several treatment techniques to address inorganic, organic, and pathogenic contaminants.

Many synthetic organic chemicals (SOCs), which are man-made compounds that contain carbon, have been detected in water supplies in the United States. Some of these, such as the solvent trichloroethylene, are volatile organic compounds (VOCs). VOCs easily become gases and can be inhaled in showers, baths, or while washing dishes. They can also be absorbed through the skin.

Water supplies become contaminated by organic compounds from sources such as improperly disposed wastes, leaking gasoline storage tanks, pesticide use, and industrial effluents. Technologies that can be used effectively to remove these contaminants include activated carbon and aeration. These technologies are discussed in the following section.

The inorganic contaminants in water supplies consist mainly of substances occurring naturally in the ground, such as arsenic, barium, fluoride, sulfate, radon, radium, and selenium. Industrial sources can contribute metallic substances to surface waters. Nitrate, an inorganic ion frequently found in groundwater supplies, is found predominantly in agricultural areas due to the application of fertilizers and feedlot runoff. High levels of total dissolved solids (TDS) might, in some instances, require removal to produce a potable supply.

Inorganic chemicals might also be present in drinking water due to corrosion. Corrosion is the deterioration or destruction of components of water distribution and plumbing systems by chemical or physical action, resulting in the release of metal and nonmetal substances into the water. The metals of greatest health concern are lead and cadmium; zinc, copper, and iron are also by-products of corrosion. Asbestos can be released by corrosion of asbestos-cement pipe. Corrosion reduces the useful life of the water distribution and plumbing systems. It can also provide a surface for microorganism growth, which may result in disagreeable tastes, odors, and slimes.

Treatment alternatives for inorganic contaminants include removal techniques and corrosion controls. Removal technologies, coagulation/filtration, reverse osmosis, ion exchange, and activated alumina, are used to treat source water that is contaminated with metals or radioactive substances (such as radium). Aeration effectively strips radon gas from source waters. Corrosion controls reduce the presence of corrosion by-products such as lead at the point of use (such as the consumer's tap). Treatment technologies for organic contaminants consist of adsorptive methods and aeration. All of these treatment technologies are discussed in the following sections.

## ■ Aeration and Oxidation

### Concepts

■ Aeration refers to processes where water and air are in contact with one another to move volatile substances into or out of the water.

■ Aeration is particularly effective in reducing carbon dioxide, hydrogen sulfide, and in removing odors and in removing iron and manganese.

■ Oxidation refers to a chemical reaction where electrons are lost.

■ Aeration and chemical oxidizers initiate oxidation reactions.

In water treatment, the term aeration refers to any process where water and air are brought into contact with each other in order to move volatile substances into or out of the water. These volatile substances include oxygen, carbon dioxide, nitrogen, hydrogen sulfide, methane and unidentified organic compounds responsible for tastes and odor. Aeration is also commonly called air stripping and is not used at all water treatment plants.

The water source is an important factor in deciding whether aeration is necessary. Surface waters usually have low concentrations of carbon dioxide, no hydrogen sulfide and fairly high concentrations of dissolved oxygen. Consequently, aeration is not required for the removal or addition of these gases. Many surface waters contain traces of volatile organic substances that cause disagreeable tastes and odors. Unfortunately, conventional aeration systems are not particularly effective for removing the compounds responsible for taste and odor problems since these compounds are of low volatility.

Groundwaters may contain excessive concentrations of carbon dioxide, methane, hydrogen sulfide, iron, or manganese. Aeration is recommended as a treatment process for these gases and as an aid in removal of iron and manganese through precipitation.

Water treatment utilizes oxidation for various purposes. Oxidation is a reaction in

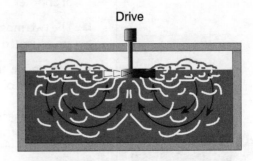

**Mechanical Surface Aerator**

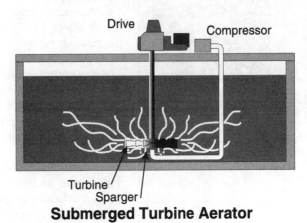

**Submerged Turbine Aerator**

**Figure 7-16:** Mechanical aeration.

which a substance loses electrons, thus increasing its charge. The substance that oxidizes another substance is called an **oxidizer** or oxidizing agent. A number of oxidizers can be used to remove or destroy undesirable tastes and odors, to aid in the removal of iron and manganese, and to help improve clarification and color removal in source water.

The most common and effective chemical oxidizers used in water treatment are chlorine, chlorine dioxide, ozone and potassium permanganate. Chlorine and potassium permanganate far outnumber the others in total usage. Ozone and chlorine dioxide require on-site generation and are relatively expensive.

Using atmospheric oxygen through aeration to oxidize the organic substances responsible for undesirable tastes and odors is usually too slow to be of value. However, if dissolved gases such as hydrogen sulfide are the cause of taste and odor problems, aeration will effectively remove them through oxidation and stripping.

## ■ Adsorptive Methods of Treatment

### Concepts

■ Adsorption results in one substance attaching itself to the surface of another.

■ The two most common adsorptive media used in water treatment are activated carbon and activated alumina.

■ Adsorptive methods are generally most effective for taste and odor control and removal of organic pollutants.

Adsorption is the attraction and accumulation of one substance on the surface of another. Two important adsorptive media used in water treatment are activated carbon and activated alumina. The most important applications of adsorption in water treatment are the removal of arsenic and organic pollutants. Some drinking water sources, particularly from surface water sources, may contain significant concentrations of organic materials. These dissolved organics may affect the taste, odor, or color of the water and result in consumer complaints.

Adsorption of organic impurities using activated carbon has been a common practice in the water treatment field for many years. Activated carbon is especially effective as an adsorbing agent because of its large surface area. Each activated carbon particle contains a huge number of pores and crevices into which organic molecules enter and are adsorbed onto the activated carbon surface (Figure 7-17).

Activated carbon has a particularly strong attraction for organic molecules so it is well suited for the removal of hydrocarbons and those organics responsible for the taste, odor, and color of water. Two forms of activated carbon are used in water treatment: powdered and granular. Powdered activated carbon is often used for taste and odor control. Its effectiveness depends on the source of the

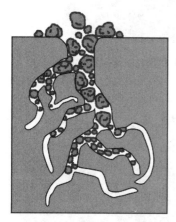

**Figure 7-17:** Internal carbon structure.

undesirable tastes and odors. This type of activated carbon is a finely ground, insoluble black powder. It can be added at any point in the treatment process ahead of the filters.

Activated carbon media must periodically be replaced with new or regenerated activated carbon. Replacement cycles can vary from one to three years for taste and odor removal, down to three to six weeks for removal of organics. Regeneration involves:

– Removing the spent carbon as a slurry

– Dewatering the slurry

– Feeding the carbon into a special furnace where regeneration occurs

– Returning it to use.

Sites that use activated carbon systems can choose to conduct on-site regeneration, purchase new activated carbon, or ship the spent carbon to a regeneration center. On-site regeneration is expensive due to the costs of the necessary equipment. It may also result in air pollution, so in some areas restrictions or requirements for additional controls have been added.

Activated alumina is a highly porous and granular form of aluminum oxide. This material is available from several aluminum manufacturers in various mesh sizes and degrees of purity. Alumina is used in the water treatment industry for removing arsenic and excess fluoride ions. The treatment consists of percolating water

---

## Readily and Poorly Adsorbed Organics

**Readily Adsorbed Organics**

■ Aromatic solvents (benzene, toluene, nitrobenzene)

■ Chlorinated aromatics (PCBs, chlorobenzenes, chloronaphthalene)

■ Phenol and chlorophenols

■ Polynuclear aromatics (acenaphthene, benzopyrenes)

■ Pesticides and herbicides (DDT, aldrin, chlordane, heptachlor)

■ Chlorinated aliphatics (carbon tetrachloride, chloroalkyl ethers)

■ High molecular weight hydrocarbons (dyes, gasoline, amines, humics)

**Poorly Adsorbed Organics**

■ Alcohols

■ Low molecular weight ketones, acids, and aldehydes

■ Sugars and starches

■ Very high molecular weight or colloidal organics

■ Low molecular weight aliphatics

**Figure 7-18:** Readily and poorly adsorbed organics. (Source: Drinking Water and Center for Environmental Research Information, *Technologies for Upgrading Existing or Designing New Water Treatment Facilities,* March 1990)

through a column of the alumina media. Actual removal of the arsenic and fluoride ions is accomplished by a combination of adsorption and ion exchange.

Use of the activated alumina process for the removal of arsenic and fluoride ions from water requires periodic regeneration of the alumina media. When the alumina becomes saturated with arsenic and fluoride ions, it is regenerated by passing a caustic soda (NaOH) solution through the media. Excess caustic soda is neutralized by rinsing the activated alumina with an acid. Since arsenic and fluoride ions are toxic, disposal of these wash waters must be done in accordance with applicable laws.

## ■ Ion Exchange

### Concepts

- Ion exchange moves ions between a solid medium and a solution.

- Demineralization is an ion exchange process that removes the dissolved solids content of a water supply.

Ion exchange is the reversible exchange of ions between a solid medium (resin) and a solution. Ion exchange can be used to remove just cations, called softening. Hardness in water is principally caused by the presence of calcium and magnesium ions; however, iron, barium, aluminum, strontium, and other metal ions also contribute to the total hardness.

Ion exchange resins used for softening water generally have exchange sites containing sodium ions. Hydrogen ion replacement resins are also available, but normally they are not used for softening drinking water supplies. Sodium cation resins replace the ions that cause hard water, resulting in water with a reduced hardness, but an increased sodium ion content. This increase in sodium ions is a consideration for persons on sodium-restricted diets.

The softened water produced can then be blended with unsoftened water to obtain water with the desired degree of hardness. This is usually more practical economically for a small water treatment softening application than for an entire water supply.

Demineralization is an ion exchange process that removes the dissolved solids of a water supply. Dissolved solids contain both cations and anions and therefore require two types of ion exchange resins. Cation exchange resins used for demineralization purposes have hydrogen exchange sites and are divided into strong acid and weak acid classes. The anion exchange resins commonly used contain hydroxide ions and are divided into strong and weak base classes.

The demineralization process should be considered for removal of: arsenic, barium, cadmium, chromium, fluoride, lead, mercury, nitrate, selenium, silver, copper, iron, manganese, sulfate, and zinc. Some general advantages to using ion exchange to remove these contaminants are the low capital investment required and the mechanical simplicity of the process. The major disadvantages are high chemical requirements to regenerate the resins and to dispose of chemical wastes from the regeneration process. These factors make ion exchange more suitable for small systems than for large ones.

Dissolved organics, strong oxidizing agents, and suspended solids are harmful to ion exchange systems. Organics, which may be irreversibly absorbed into the resin, and chlorine, can be removed by carbon adsorption. Suspended solids can inhibit passage of water through the ion exchange unit and prevent appropriate contact between the water and exchange resin.

# ■ Membrane Processes

## Concepts

- Electrodialysis and reverse osmosis are the most common membrane processes.

- Membrane processes are good for demineralization of water.

Two treatment processes involving membranes are commonly used to remove salts (demineralize) from water: **electrodialysis** and **reverse osmosis**.

In normal osmosis, two solutions containing different concentrations of minerals are separated by a semipermeable membrane. Water tends to migrate through the membrane from the side of the more dilute solution to the side of the more concentrated solution. This phenomenon, called osmosis, continues until the build-up of hydrostatic pressure on the more concentrated solution is sufficient to stop the net flow.

In reverse osmosis (RO), the flow of water through the semipermeable membrane is reversed by applying external pressure to offset the hydrostatic pressure (Figure 7-19). This results in a concentration of minerals on one side of the membrane and pure water on the other side.

Electrodialysis (ED) is the demineralization of water using the principles of osmosis, but under the influence of a direct-current electric field. Minerals **dissociate** into cations and anions when they enter water. A positively-charged electrode, called an anode (+), attracts these anions (–) and a negatively-charged electrode, called a cathode (–), attracts the cations (+).

Two types of selectively permeable membranes are used in ED. One can be permeated by cations but not anions. The other can be permeated by anions but not cations. These membranes are arranged in a stack, with cation-permeable membranes alternating with anion-

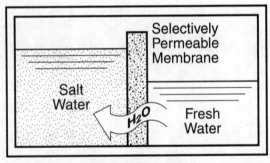

Normal Osmosis:
Water moves from side of lower concentration to side of higher concentration.

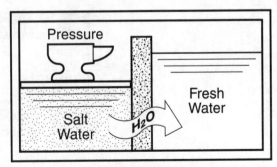

Reverse Osmosis:
Normal osmotic process is reversed due to the application of pressure.

**Figure 7-19:** Process of Reverse Osmosis.

permeable membranes. Water is fed into the spaces between the membranes and a direct-current electric field is applied to the stack, causing the ions to migrate toward the oppositely charged electrodes. This results in the concentration of ions in alternate spaces between membranes, and the water in the other spaces to become depleted in ions, or demineralized. Water is then drawn off from between the membranes in two separate streams, one containing most of the ions and the other relatively free of ions.

The essential element in the reverse osmosis method of demineralization is the semipermeable membrane. Several types and configurations of selectively permeable membranes are currently available. The performance of EO and RO units is highly dependent on a number of water quality parameters. Suspended solids, dissolved organics, hydrogen sulfide, iron, manganese, and strong oxidizing agents (chlorine, ozone, and permanganate) are harmful to membranes. Cellulose membranes, for example, are very susceptible to bacterial attack but are somewhat tolerant of chlorine. Polyamide (nylon) membranes are not as susceptible to biological attack, but are very sensitive to chlorine. In order to remove these undesirable constituents, the water entering the ED unit must often be pretreated. The efficiency of the membranes may also be reduced by scaling deposits, which are caused or contributed to by hardness, barium, strontium, iron, and manganese. Common scale prevention measures include:

- pH adjustment to between 5.0 and 6.5 to prevent hydroxide and carbonate scale formation

- Iron and manganese ion reduction by pretreatment

- Use of a polyphosphate compound to inhibit calcium sulfate $CaSO_4$ scaling

Overall, when confronted with waters with high total dissolved solids concentrations, membrane processes can be considered. Both ED and RO are effective in reducing total dissolved solids concentrations; however, appropriate pretreatment is a major factor in successful operation. ED and RO, like ion exchange, can be used to remove a variety of cations and ions. In addition, RO can be used for reduction of bacteria, radionuclides, and color.

## ■ Stabilization

### *Concepts*

■ Water introduced into distribution systems should not be scale-forming or corrosive.

■ pH adjustment and addition of polyphosphates are the most common stabilization techniques for drinking water.

Drinking water leaving a treatment plant and entering the distribution system should be stable. This means it should not be scale-forming (depositing calcium carbonate) or corrosive (dissolving calcium carbonate) under the temperatures experienced in the distribution system. Two common ways of stabilizing water are pH adjustment and the addition of polyphosphates or silicates. The pH of water is considered to be optimum when raising its pH results in scale formation and lowering it makes it corrosive.

Stabilization of water is most often associated with an upward adjustment of pH to control corrosion; however, there must be sufficient calcium ions present in the water for calcium carbonate to form. If there is a calcium ion deficiency, lime (CaO) can be added to raise the pH. In water where there are sufficient calcium ions, sodium hydroxide or soda ash is added to raise the pH without increasing the hardness.

## ■ Checking Your Understanding (7-2)

1. List two diseases that can result from contaminated drinking water.

2. Describe a situation other than the one depicted in Figure 7-9 that could result in a cross-connection between contaminated water and drinking water.

3. Explain into which of the treatment categories, physical or chemical, each of the following treatment methods fall and why: coagulation, aeration, screening, oxidation.

4. List two reasons why ozonation and ultraviolet light are not as commonly used for disinfection of water supplies as chlorination.

# 7-3 Wastewater

## ■ Wastewater Collection and Disposal

### Concepts

■ BOD measures the oxygen requirement of wastewaters.

■ Anaerobic, aerobic, and facultative bacteria play an important role in wastewater treatment.

■ There are three levels of wastewater treatment: primary, secondary, and tertiary.

■ It is important for industry to pretreat waste before discharging it to collection systems.

We learned in the previous section that the majority of water provided to a community by a public water supply company is discharged to some form of wastewater collection and disposal system. Waste that is not properly treated presents a great health danger. Wastes may emanate foul odors and draw flies and rodents, causing unappealing nuisance conditions. Pathogenic organisms in the waste can contaminate shellfish, swimming areas, and drinking water supplies. When microorganisms consume these wastes, they remove dissolved oxygen The amount of oxygen required for the microorganisms to consume all of the waste is known as **Biochemical Oxygen Demand (BOD).** In his book, *Water and Waste-Water Technology*, Mark Hammer formally defines BOD as the "quantity of oxygen used by a mixed population of microorganisms in the aerobic oxidation of the organic matter in a sample of wastewater." In reality, it is used to measure the "loading" of waste into treatment plants and the effectiveness of the treatment.

Microorganisms + Food + $O_2$ ® More Microorganisms + $CO_2$ + $H_2O$

It takes a large amount of water to move a relatively small amount of waste to systems that will treat it. Treatment in rural areas and in smaller communities may consist of combinations of septic tanks and leaching fields or pits (Figure 7-20). These systems are designed to use the soil as a filtering media for the wastewater. In larger urban communities and in smaller communities where the soil is not appropriate for septic systems, piping to a wastewater treatment facility is installed. Proper engineering on the piping system is critical because the pipes usually have quite a large diameter, on the order of 4 to 8 feet, and rely on gravity to move the

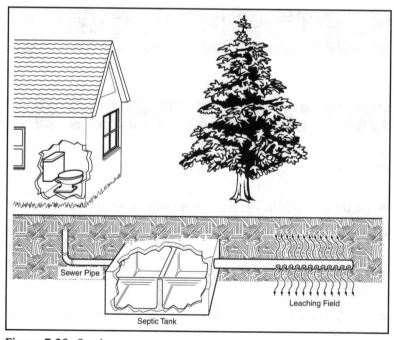

Sewer Pipe

Septic Tank

Leaching Field

**Figure 7-20:** Septic system.

waste toward the treatment plant.

The wastewater treatment plant, also known as a **publicly owned treatment works (POTW),** may use a variety of methods to treat the waste. Like water treatment, the most common systems employ a combination of physical, chemical, and biological methods. There are three general levels of treatment at municipal wastewater treatment plants: primary, secondary, and tertiary. **Primary treatment** is the first step, which involves screening and sedimentation to remove materials that float or settle. **Secondary treatment** uses growing numbers of microorganisms to digest organic matter and reduce the amount of organic waste.

The most common biological processes used in secondary wastewater treatment include: aerated lagoons, activated sludge trickling filters, and rotating biological contactors (RBCs). Bacteria are categorized by their requirements for free oxygen. **Aerobic bacteria** require free oxygen to carry on their metabolism. **Anaerobic bacteria** cannot tolerate free oxygen, but use their food source in metabolism as both the oxidizing and reducing agent. **Facultative bacteria** can function in either way.

In the modern wastewater treatment plant, the bacteria that break down organic matter occur naturally in human waste, but are not disease-causing in themselves. They continue the break-down process until the waste is changed into a material they can no longer use as food. At this point the waste is considered stable. It is important to note that biological treatment by itself does not guarantee that there will be no disease-causing microorganisms in the waste. This is why it must be disinfected. The water that leaves this process is chlorinated to destroy disease-causing microorganisms before release to the ground or a water body.

To compare the different treatment methods, it is useful to look at how much they reduce BOD and suspended solids. Hammer estimates that sedimentation reduces BOD about 35 percent and suspended solids 50 percent. Biological treatment reduces both suspended solids and BOD on the order of 85 percent. Overall, the goal of wastewater treatment is to reduce BOD, suspended solids, total nitrogen, phosphorus and ammonia. Note that toxic chemicals are not listed. They pass through municipal wastewater treatment plants that are not equipped to treat them. In 1986 it was estimated

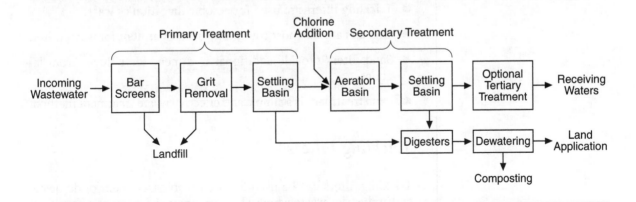

**Figure 7-21:** Primary, secondary, and tertiary treatment.

that nearly 40 percent of the toxic industrial compounds entering surface waters did so by passing through treatment plants.

Treatment methods result in the accumulation of sludge, which is the remaining residue. Sludge management and disposal is a growing problem due to the increases in its generation; and amounts are expected to double again by the year 2000. Sludges vary tremendously in quality. Some sludges are relatively free of toxic substances and can be used as soil conditioners. Other sludges may contain organic, inorganic, or toxic pollutants and pathogens. These types of sludges are difficult and costly to dispose of.

Municipal wastewater treatment plants are required to obtain and abide by the provisions of their NPDES permits. Since municipal wastewater treatment plants are not equipped to remove many types of pollutants, industries must first remove them. This practice of removing pollutants from their wastewaters is called **pretreatment**.

Wastewater discharged by industry often contains toxic substances that are not common to other sources. For instance, it can contain lead from the manufacture of batteries, or cyanide from electroplating shops. Not only are treatment plants not designed to treat these wastes, but they can also interfere with their operation on the substances they were designed to treat. They can also increase the costs and environmental risks of sludge treatment and disposal.

The types of waste problems that can be controlled through pretreatment include: corrosion, fires and explosions, interference with plant treatment system, employee exposures, pass-through of toxic pollutants into surface waters, and sludge disposal problems. Treatment techniques to reduce these concerns in industrial waste streams will be covered in Chapter 11. **Tertiary treatment** is not always required. It consists of advanced cleaning and removes nutrients such as phosphorus and nitrogen and most BOD.

## ■ Secondary Wastewater Treatment Processes

### Concepts

- Trickling filters are used for aerobic digestion of waste.

- Lagoons are the most commonly used treatment for wastewater.

- Sand filters provide additional treatment for effluent from lagoons.

- Land treatment is a simple and effective waste treatment method.

### Trickling Filters

Trickling filters can be used for the aerobic treatment of domestic and industrial wastewater. They are about 85 percent effective in reducing BOD and **Total Suspended Solids (TSS)**, although discharge standards may require even more effective reduction. Trickling filters require a minimum of technical supervision and have lower operating costs than their closest counterpart – activated sludge systems. A trickling filter system typically includes the following components in the order listed: screens, grit removal tank, primary clarifier, trickling filter, secondary clarifier, disinfection system, and sludge treatment and disposal components.

After primary sedimentation, wastewater is piped to a bed of rock or plastic media and allowed to trickle down over it. As it does so, a layer of bacterial slime forms on the media. It is the biological action of this slime layer that removes organic matter from the wastewater. The slime layer continually grows in thickness until portions slough off with the wastewater passing over it. The wastewater and solids are transported to a secondary settling tank where the two are separated. A portion of the treated wastewater is usually recycled back to the trickling filter because recirculating it often improves the quality of the final effluent.

The sludge produced by the trickling filter process comes from the primary clarifiers and the solids, which include biomass, that are continually sloughed off of the filter media and collected in the final clarifiers. This sludge must be digested either aerobically or anaerobically before it is disposed. Usually, it is placed in a landfill or applied to the land for disposal. If solids are not adequately removed from the trickling filter process, poor effluent quality will result.

Because these filters rely on living organisms, cold weather significantly reduces the treatment efficiency. Due to the strict effluent limitations imposed nationwide, trickling filters alone are generally not able to do an acceptable job without installation of additional treatment processes.

## Activated Sludge

The **activated sludge** system is a biological treatment process where additional $O_2$ is used to activate aerobic bacteria. This activated sludge is added to the settled waste solids that are to be treated. The mixture is thoroughly agitated using compressed air from the bottom of the tank. This agitation maximizes absorption of oxygen from the atmosphere.

The microorganisms in the activated sludge oxidize soluble organics and gather particulate solids into larger particles in the presence of dissolved molecular oxygen. The mixture of microorganisms, solid particles, and wastewater (referred to as mixed liquor) is aerated in a basin. The aeration step is followed by sedimentation to separate biological solids from the treated wastewater. A major portion of these biological solids are removed by sedimentation and recycled to the aeration basins to be recombined with the incoming wastewater. The excess biological solids (i.e., waste sludge) must be disposed of.

A variation of the activated sludge process that is applicable in higher flow situations is the oxidation ditch. This system continuously recirculates wastewater in closed loop channels where aeration occurs (Figure 7-22). The wastewater is normally circulated from 18 to 24 hours. Mechanical aerators are typically used for mixing,

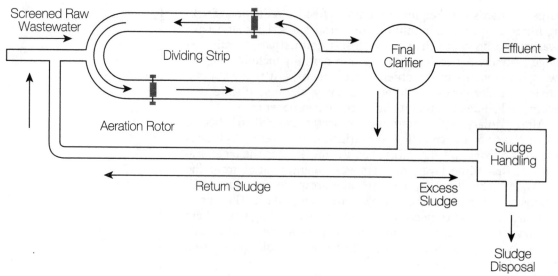

**Figure 7-22:** Extended aeration activated sludge system.

aeration, and circulation of the activated sludge. These aerators must provide not only the oxygen necessary for oxidation of the waste but must also keep everything moving so that the solids remain suspended.

Yet another variation of an activated sludge treatment process is the Sequencing Batch Reactor (SBR). In this system, aeration, clarification, and sludge treatment processes are carried out in sequence in the same tank. These tanks cycle through five operations: filling, aeration, settling (clarification), drawing off the liquid above the settled solids (decanting), and idle time, called "sludge wasting."

## Lagoons

Lagoons of some sort are the most commonly employed treatment technology for wastewater. Most cannot treat wastes to the level required by law without the use of additional processes. All lagoon systems operate under the same principle. They utilize the natural abilities of bacteria and algae to reduce the organic content in wastewater. In the daytime, photosynthesis by the algae results in the release of oxygen necessary for respiration by the bacteria. Variations of lagoon treatment may incorporate additional aeration systems. Lagoon systems typically fall into one of two categories: stabilization ponds or aerated lagoons. Stabilization ponds are typically shallow (no more than about six feet deep) to keep their operation aerobic. They take upwards of 30 days to break down the waste. Aerated lagoons can handle higher loads of organic wastes. This is because of the supplemental oxygen from the enhanced aeration. They require less land area and shorter holding times for the wastes to be broken down.

Facultative lagoons, or ponds, are the most frequently used form of municipal wastewater treatment in the U.S. (Figure 7-23) They fall into the category of stabilization ponds. There are currently over

5,000 of these systems in operation. These lagoons are typically 1.2 to 1.8 meters in depth (4 to 6 feet) and involve no mechanical mixing or aeration. The water near the surface is aerobic due to the oxygen in the atmosphere and the respiration of algae. The layer at the bottom is anaerobic and contains the settled solids. The in-between zone is called the facultative zone. The algae near the surface take up carbon dioxide and release oxygen. This carbon dioxide removal raises the pH to above 10, which helps volatilize the ammonia from the lagoon. The oxygen produced by the algae along with that from the atmosphere is used by the aerobic bacteria to consume the organic material in the upper layer of water. In the bottom layer, anaerobic bacteria ferment the organic matter. Cold temperatures interfere with both types of decomposition so in many parts of the country, effluent cannot be discharged from facultative lagoons during the winter months.

It typically takes 20 to 180 days for the effluent BOD to be reduced to acceptable levels in facultative lagoons. Lagoons require a fair-sized portion of land and can be a single large pond or be divided into two or more cells. If land is not limited, lagoons are a good choice for treating wastewater since they are reliable and easy to operate.

Aerated lagoons are large compartmentalized basins or individual basins that are mixed and aerated using either fixed or floating surface aerators. Good removal of soluble organic matter can be achieved with the proper mix of retention time and aeration. A biomass removal step must follow the aerated lagoon process before effluent can be discharged to the **receiving waters**. This is often accomplished in a large still pond or in a section of the aerated lagoon isolated by baffles or dikes. Sometimes, these lagoons are used as a pretreatment device. If so, the biomass is carried with the liquid to subsequent unit processes. The main differences between this process and the activated sludge system are that the microorganisms in the lagoon are grown in the dispersed state rather than as a flocculant mass, and the biomass is not recycled from the sedimentation step to the aeration step.

The performance of aerated lagoons in removing biodegradable organic compounds depends on several factors, including retention time, temperature, and the nature of waste. Aerated lagoons generally provide a high degree of BOD reduction. In general, problems with aerated lagoons are excessive algae growth, offensive odors if

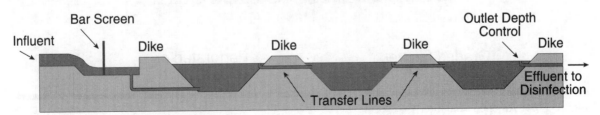

**Figure 7-23:** Wastewater treatment lagoon.

sulfates are present and dissolved oxygen is depressed, and seasonal variations in effluent quality. Aerated lagoons can handle considerable variations in the amounts of organics presented if sized properly, and are less vulnerable to process upsets than most biological wastewater treatment methods.

## Filtration

Intermittent sand filters are most commonly used as an additional treatment for the effluent from a lagoon or septic tank system. These systems are a type of fixed film biological treatment. They can either be single-pass, where the wastewater moves through the sand filter only once, or they can be recirculating, where it passes through the filter more than once. In either case the filter system usually consists of a bed of sand about 3 feet thick that is above a gravel layer. Slotted pipe is installed under the gravel to provide drainage. The floor of the filter, which is usually earthen, is sloped to provide for drainage. The total sand bed area is usually divided into two or more smaller filters. Wastewater is applied to each in alternating cycles that allow the sand filter to drain completely. Complete draining is required for aerobic conditions to be maintained.

A recirculating intermittent sand filter uses a recirculation tank where most of the discharge from the sand filter is diverted into and mixed with other wastewater to be reapplied to the filter. Using this type of system results in the wastewater being diluted so that it is not as likely to clog the filter. Cold weather is likely to adversely

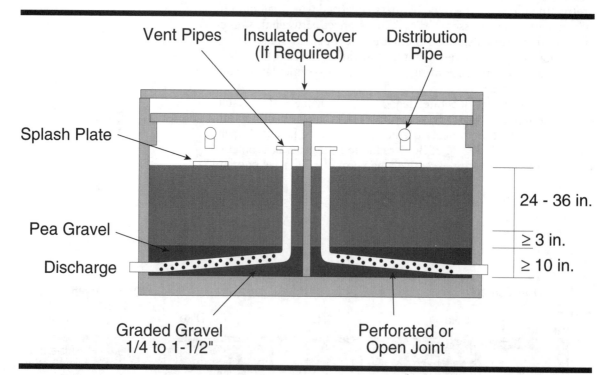

**Figure 7-24:** An open sand filter.

affect these filters. Some sludge is generated whenever the surface of the filter is cleaned.

## Land Treatment Methods

There are a number of **land application** systems that can be used as an effluent disposal system or as further treatment of effluents. Land treatment systems have unique benefits: nutrient recovery, cash crop production, groundwater recharge, and water conservation if used for irrigation of landscaped areas. These types of systems are the most desirable in areas where surface water discharge requirements are strict and land is relatively inexpensive. Soil type and texture must, however, be of a type suitable for land treatment.

Land application is a straightforward, easy method of wastewater treatment. Pretreated wastewater is applied to the land by infiltration, flow, or irrigation methods. Treatment is provided by natural processes as the effluent flows through the soil and vegetation. With this method, organic material is removed within the top inch of soil. Nitrogen is removed by plants in the nitrogen cycle described in Chapter 4. Some of the wastewater is lost to evaporation and evapotranspiration, but most of it returns to surface waters or percolates through soil.

In one type of land treatment method, called the overland flow method, wastewater is applied to slightly sloping land and allowed to flow over the surface to a collection point at the bottom of the hill. After collection, it is disinfected and discharged to a water body. The treatment process that takes place involves suspended solids attaching themselves to surface vegetation and decomposing. Bacteria on the grass and in the soil consume organic matter, and nutrients are taken up by the vegetation. The resulting effluent from this type of system typically exceeds the reduction requirements of most secondary treatment systems. BOD and TSS are typically 85 to 92

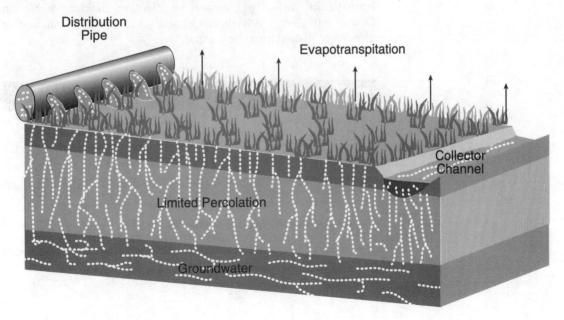

**Figure 7-25:** Land treatment method.

percent removed, nitrogen 60 to 80 percent removed, and phosphorus 20 to 50 percent removed. Obviously, this type of system provides a high level of treatment with a minimal use of equipment.

A variation of land treatment involves spray irrigation, where wastewater is pumped to various fields. Effluent is applied using sprinklers. Another variation applies pretreated wastewater to highly permeable soils. In this system, additional treatment occurs by filtration, adsorption, and microbial action as the wastewater percolates through the soil.

## Sludge Disposal

As mentioned earlier, most wastewater treatment processes generate sludge that must be properly treated prior to disposal. Prior to disposal, sludge must be stabilized, which will remove pathogens and reduce the organic content. After stabilization, some communities dewater the sludge to reduce its total volume. Many communities dispose of stabilized sludge to land.

Stabilization is accomplished by aerobic digestion or application of lime. For aerobic digestion, sludge is pumped into a digester, where it is typically retained for 20 to 30 days to reduce volatile suspended solids and pathogens. During this stabilization period the sludge is routinely aerated and mixed. In the lime application method, enough lime is added to either raw or digested sludges to raise the pH level to greater than 12. The high pH reduces pathogens.

Dewatering can be accomplished easily using uncovered sand drying beds. Some beds have liners, where the liquid collects and can be returned as **influent** to the wastewater treatment plant.

Some "clean" sludges can be used as soil conditioners since they typically contain nutrients such as nitrogen, phosphorus, and potassium. This makes certain sludges excellent supplements to commercial fertilizers and soil amendments. Liquid sludge can be applied beneath the surface of the land by injection, disking, or plowing. Dewatered sludge is typically spread over the land surface and plowed or disked into the ground.

## ■ Checking Your Understanding (7-3)

1. Describe the difference between aerobic, anaerobic, and facultative bacteria.

2. Explain how trickling filters achieve their purpose of reducing BOD and TSS.

3. Describe the purpose of lagoons in wastewater treatment.

4. Explain how land treatment results in treatment of wastewater.

# 8

# Hazardous Materials

## ■ Chapter Objectives

1. **Distinguish** between hazardous materials, hazardous substances, hazardous wastes, regulated substances, and hazardous chemicals as they are used in the federal regulatory structure.

2. **Relate** the purpose, basic provisions, and administering agencies for:

   ■ CERCLA
   ■ SARA Title III
   ■ HMTA
   ■ TSCA
   ■ FIFRA
   ■ RCRA, Subtitle I

3. **Describe** the special care that must be exercised in the transportation and storage of hazardous materials.

## ■ Terms and Concepts to Remember

Bactericides
Cathodic Protection
Chemical Manufacturers
  Association (CMA)
Community Awareness and
  Emergency Response (CAER)
Contingency Plan
Corrosive
Electrochemistry
Explosive
Extremely Hazardous
  Substances (EHS)
Feasibility Study (FS)
Federal Emergency
  Management Agency (FEMA)
Fire Hazard
Flammable
Fungicides
Hazard Classes
Hazard Ranking System (HRS)
Hazardous Substance
Health Hazard
Herbicides
Impressed Current
Insecticides
Labels
Local Emergency Planning
  Committee (LEPC)
Marine Pollutant
Markings
Material Safety Data Sheet
  (MSDS)
Monitoring Wells
National Contingency Plan
  (NCP)

National Priorities List (NPL)
National Response Center
Nematocides
Persistent
Placards
Polychlorinated Biphenyls
  (PCBs)
Potentially Responsible Party
  (PRP)
Preliminary Assessment
Radioactive Material
Reactive
Reactivity Hazard
Reduction
Regulated Substances
Remedial Action Plan (RAP)
Remedial Actions
Remedial Investigation (RI)
Remedial Investigation/
  Feasibility Study (RI/FS)
Removal Actions (RA)
Reportable Quantity (RQ)
Rodenticides
Sacrificial Anode
Secondary Containment
Shipping Papers
Site Investigation
Specific Hazard
State Emergency Response
  Commission (SERC)
Superfund Amendments and
  Reauthorization Act (SARA)
Toxic Chemical Release Form
Uniform Hazardous Waste
  Manifest

## ■ Chapter Introduction

The last chapter explained the many ways ground and surface water can become contaminated, including contamination from discharges and spills of industrial and household chemicals. Contamination of soil, air, and water is often the result of improper management of hazardous materials. This great potential for harm to public health and the environment necessitates strict regulations requiring hazardous materials management techniques to be used to prevent adverse effects.

This chapter covers the vast universe of hazardous materials and hazardous substances regulation. The profiled laws cover every aspect of hazardous materials/substance management, including their transportation, cleanup, and storage. These laws are designed to prevent environmental contamination and protect public health. From cleaning up old contaminated sites to transporting a load of hazardous materials, a number of agencies and overlapping laws regulate all the activity surrounding hazardous materials and hazardous substances.

# 8-1 Regulation of Hazardous Materials and Substances

## ■ Background and Overview

### Concepts

- Materials are hazardous when their chemical, physical, or biological nature poses an unacceptable risk if not controlled.

- Persistent materials do not readily break down in the environment.

- A release refers to material escaping from its container by spilling, leaking, or emitting toxic vapors.

A quick look under the sink or in the garage of your home can reveal the number of products we have to make tasks such as cleaning and polishing easier. During the first half of the twentieth century, Americans did not have the vast selection of consumer products containing chemicals that we now do. Think about the numerous items that perform useful services for us such as oven cleaners, toilet cleaners, floor strippers, flea and tick sprays for our pets, lighter fluid, and the list goes on and on. The surprising thing to many people is that a number of the materials we work with or use on a daily basis are hazardous. Many of these products make use of materials that don't exist in nature. In fact, the Department of Transportation estimates that over 1,000 new synthetic (man-made) chemicals enter commerce each year. As long as we use them properly, under normal circumstances, they should not pose health problems or environmental damage. But just what makes a material hazardous and what types of misuses can cause problems?

We have learned that hazardous materials, which can be solids, liquids, or gases, pose a potential risk to life, health,

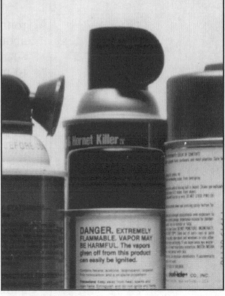

**Figure 8-1:** Products we use every day can be hazardous materials.

or property if they are released. A material's basic chemical, physical, or biological characteristics determine whether it is hazardous. If a material catches on fire easily or spontaneously under standard temperature and pressure conditions it is **flammable**. Paint thinners, gasoline, and charcoal lighter fluid are examples of flammable materials. A related category of hazardous materials, which assist in the production of fire by readily yielding oxygen, are called oxidizers. The chlorine released by the solid tablets used to chlorinate swimming pool water is an example of an oxidizer. A toxic material can affect health, causing permanent or temporary injury or incapacitation if it inhaled, swallowed, or if it comes in contact with your skin. If a substance reacts violently by catching on fire, exploding, or giving off fumes when it is exposed to water, air, or low heat, it is **reactive**. If it can corrode standard steel containers or burn the skin, eyes or other body tissue it is **corrosive**. Strong acids or bases are examples of corrosive materials. If a material releases pressure, gas, and heat suddenly when subjected to shock, heat, or high pressure, it is **explosive**. Lastly, a material is hazardous if it emits harmful radiation; we call it radioactive. There are still other ways a material can be hazardous under some laws, but they are not quite as common so we will cover them later in the chapter.

Many of the chemicals exhibiting these hazardous characteristics are not readily biodegradable, which means they are not able to be broken down easily into their component parts by microorganisms. These types of chemicals are said to be **persistent** in the environment. Once released, they tend to stay intact, possessing the same dangerous properties for long periods of time. Naturally occurring toxic substances can also pose problems. A wildlife refuge in California became contaminated by selenium, which is an element commonly found in high pH desert soil. The build-up of the selenium occurred because of the irrigation methods used by nearby farms. During irrigation, water carried dissolved selenium to the wildlife refuge. Deformities in developing embryos were found more frequently in area waterfowl that had ingested selenium.

Spills or releases of hazardous materials into the various environmental compartments are obviously the basis of our concern about their hazards. Hazardous materials releases have had their share of front page publicity over the years. This notoriety has served to make the public more aware of their hazards. One of the most tragic episodes occurred in Bhopal, India in 1984 when 44 tons of methyl isocyanate gas were released from an industrial operation in a densely populated community late at night while the local residents were sleeping. Over 3,000 deaths and injuries to hundreds of thousands more have been attributed to the release. In another case, 7,500 residents in a small Kentucky community were forced to evacuate when tank cars on a train line running right through the community derailed and burned. The fire itself created a column of smoke 3,000 feet high. Not only industry and trains have these problems. In Louisiana, close to 41,000 pounds of a concentrated acid polluted a part of the Mississippi River after it was released due to a collision of two ships. Stories are regularly published in newspapers reporting situations like the unfortunate fellow who caused a fire in his condominium when he was laying carpet with a flammable mastic and lit a cigarette.

When we speak of a release, we are referring to a material escaping its container by spilling, leaking, or emitting toxic vapors and creating a hazard or potential hazard by entering the environment. Sometimes the harm caused by hazardous materials on the loose is not so obvious. As we learned in Chapter 3, small concentrations of some chemicals in the water we drink or in the air we breath can lead to cancer and other health problems. Difficulty in pinpointing the cause of health problems complicates the situation even more. This is due to the fact that we are typically exposed to a great many chemicals in our lives, and the fact that many negative health effects of chemicals have long latency periods before they appear. One study conducted by the EPA to determine how many chemicals Americans are exposed to found over 200 industrial chemicals and pesticides in the systems of the sample group for the study. We have already discussed how difficult it would be to determine the specific exposure level for the multitude of chemicals available in commerce. The question, however, is not even purely scientific, but has social, political, and economic implications.

We know also that the effects of hazardous materials on health and the environment are not just an industry problem. They seriously affect us as consumers as well. We all use hazardous materials and add to the total problem at municipal landfills by our disposal practices for these chemicals. Our cars also emit pollutants from the use of gasoline, one of the most commonly used hazardous materials. The hazardous materials regulatory schemes that will be discussed in the following sections of this chapter have been designed and implemented with the goal of reducing the threat of many types of releases.

## ■ Underground Tanks

### Concepts

- Requirements for USTs containing regulated substances are found in Subtitle I of RCRA.

- Most USTs contain petroleum products.

- Regulations for USTs cover their installation, operation, and closure, as well as cleanup of any environmental contamination.

- UST systems leak due to corrosion, faulty installation, piping failure, and overfills.

**Figure 8-2:** Hazardous materials can be contained in many types of vessels.

Hazardous materials can be contained in many types of vessels including cans, drums, tanks on trucks or railcars, fiberboard boxes, above ground tanks, and underground tanks. Underground storage tank (UST) regulations are a unique category in that they are based on the vessel in which the hazardous materials are contained while most other regulations target the management of the hazardous material or substance itself regardless of its container. The reason for targeting underground tanks is that given the sheer numbers of them, leaks can be especially dangerous and costly. They can cause fires or explosions. In addition, leaking USTs can contaminate nearby groundwater. As we learned in earlier chapters, many of us depend on groundwater for the water we drink. UST regulations seek to safeguard public health and safety through protection of groundwater resources.

Federal underground storage tank requirements are found in *Subtitle I* of RCRA. Until *Subtitle I* was added, RCRA applied only to wastes. *Subtitle I* applies to **regulated substances**, which include substances defined as hazardous under the CERCLA of 1980 (with the exception of hazardous wastes) and petroleum. If the underground tanks contain hazardous waste, they are covered under *Subtitle C* of RCRA.

The EPA enforces the provisions of *Subtitle I* through regulations in 40 CFR Part 280. These rules went into effect in 1988. The regulations contain requirements for leak detection, leak prevention, financial responsibility, and correcting the effects of leaks. The definition of UST is important. It is defined as a tank system, including its piping, that has at least 10 percent of its volume underground. This means that USTs include both the UST vessel and the underground piping connected to it.

Developing regulations and standards for management of USTs was particularly challenging for the EPA because of the large numbers of tanks in operation. In 1988, when the rule was put into force, the EPA estimated that there were over 2 million UST systems at over 700,000 facilities. The vast majority of these tanks were used to store petroleum products for retail and industrial purposes. Less than 5 percent stored other hazardous materials. It was also estimated that over 75 percent of existing systems at the time were made of unprotected steel, which is the type of tank most likely to leak due to corrosion. These tanks were identified as those that created the greatest potential for health and environmental damage. Imagine, if only a small percentage of these tanks leaked, releases would occur at a great many sites nationwide. There was another difficult situation. It was found that most facilities were owned and operated by very small businesses not accustomed to dealing with complex regulatory requirements. In fact, a large portion of USTs were found to be owned by small businesses with $500,000 or less in total assets. A common example was the Mom and Pop gasoline service station and convenience store. These small entrepreneurs, used to operating their businesses with minimal regulation, found themselves significantly affected by environmental regulations for UST systems.

Studies by the EPA in the late 1980s discovered that well over 10 percent of UST facilities had noticeable or significant releases into the surrounding soil and groundwater. The EPA estimated that UST cleanups could cost $60 billion over 30 years. Further investigations pointed out that unprotected bare steel tanks begin leaking at a much greater frequency after about 12 years. In fact, depending on soil conditions, the critical age in a typical unprotected steel tank's life was determined to be between 10–20 years of age. This, along with the high estimated costs of cleanup, was an alarming discovery in light of the fact that the vast majority of existing tanks were installed during the 1940s through the 1960s when gas stations were popping up at every corner in an America moving to the suburbs.

Information obtained from studying UST system leaks was used to develop regulations. For instance, most releases were found not to result from the tank portion of the UST but from the piping. In fact, piping releases occurred twice as often as tank releases. Delivery piping has been determined to be the most significant source of reported releases. The two systems commonly used for delivery are suction piping and pressurized piping. Suction piping is normally used in low volume operations. It operates at less than atmospheric pressure so if a leak develops, air or groundwater is usually drawn into the pipe instead of product leaking out. Most retail motor fuel facilities, however, use pressurized piping. Pressurized piping is normally used in high volume applications where many dispensers are fed from one tank. If a line is breached, free product is released into the environment very rapidly. If no instrumentation is set up to detect loss of pressure, large volumes of product can be pushed out of the piping in a short time. These types of releases can be catastrophic in the absence of monitoring and automated pump flow restriction devices. The EPA has been made aware of incidents involving thousands of gallons of leaked materials. Piping failures are most often caused by corrosion, poor installation techniques and workmanship, accidents, and natural events such as frost heaves. Other common causes of releases were found to be spills and overfills.

Although older bare steel tanks failed primarily due to corrosion, the new USTs which are coated and protected from corrosion or are made of non-corrosive materials such as fiberglass, have nearly eliminated failures induced by external corrosion. These new technologies in tanks designed to minimize their failure began to appear in the 1970s in three basic forms: fiberglass reinforced plastic (FRP), steel with a corrosion-resistant coating and **cathodic protection**, and steel FRP composite. As of 1985, no tanks other than these were allowed in new installations. The most significant failure mode for these newer tanks is improper installation practices.

Under the federal standards, some tanks that would otherwise meet the definition of a UST are specifically excluded from being defined as such. Some of the more commonly excluded tanks are: residential and farm tanks of

**Figure 8-3:** A gasoline fueling station.

1,100 gallons or less capacity that store motor fuel for non-commercial purposes; tanks storing heating oil for consumptive use on the premises where stored; septic tanks; pipeline facilities; surface impoundments, pits, ponds, and lagoons; storm-water or waste water collection systems; flow-through process tanks; and storage tanks situated on or above the floor of underground areas such as basements and cellars. Specific state and local regulations, however, may cover these tanks.

Many states had established UST regulatory programs even before the federal EPA standard. Notable examples were New York, Florida, and California. Given the large number of UST facilities, tank systems, and potential cleanups needed, the EPA was convinced from the outset that the regulatory program would be most effectively carried out on the state level of government. In the preamble to the final rules, the EPA states that because there are so many UST sites nationwide, it would be very difficult to establish a credible federal presence through compliance monitoring and enforcement. The EPA therefore decided that a more realistic and effective approach was to provide support and guidance to state and local regulators so that they could improve their programs in order to achieve compliance. Many states currently use a permit program for USTs, requiring permits to operate existing tanks, install new tanks, and close tanks that are no longer used.

An early directive of the regulations required that UST system owners notify the appropriate state governmental agency of the existence of their tank(s). From this notification information, states were to compile inventories of all tanks for regulatory purposes. The EPA can authorize states to implement their own UST programs as long as their requirements are no less stringent than the federal standards. EPA regulations require the following:

- A leak detection system or comparable system designed to identify releases

- Maintenance of records of any release detected by detection system(s)

- Reporting of releases and any corrective action taken

- Appropriate corrective action in response to a release

- Closure of tanks to prevent future releases

- Performance standards for new UST systems including design, construction, installation, release detection, and compatibility standards.

The EPA's regulations establish comprehensive requirements for the management of a wide range of UST systems. They are designed to reduce the number of releases of petroleum and hazardous substances, increase the ability to quickly detect and minimize the contamination of soil and groundwater by such releases, and ensure adequate cleanup of contamination. To accomplish this, the standards affect every phase of the life cycle of a storage tank system, from its selection and installation to its operation and maintenance to its final closure and disposal, including any cleanup of the site that may be required.

New UST systems must be designed and constructed to retain their structural integrity for their operating life. All tanks and attached piping used to deliver the stored product must be protected from external corrosion, and any systems designed to protect against corrosion must be monitored and maintained.

Owners and operators of both new and existing UST systems must follow proper tank filling practices to prevent releases due to spills and overfills. In addition, owners and operators of either new or upgraded UST systems must use devices that prevent overfills and control or contain spills. Any tank repairs must be in accordance with nationally recognized industry codes. These codes include several tests that must be conducted to ensure quality repairs.

Any closure, including abandoning use or removal of an UST system, must follow industry-recommended practices. Federal regulations allow the UST system to be removed from the ground or left in place after removing all regulated substances and cleaning the tank, filling it with an inert substance, and closing it to all future outside access. State and local agencies may have more stringent requirements; for instance, requiring removal. In addition, owners and operators must perform an assessment at the time of UST closure to ensure that a release has not occurred at the site. If a release has occurred, then appropriate corrective action must be taken.

Release detection must be performed at all UST systems. All new or upgraded UST systems storing hazardous substances, however, are required to have **secondary containment** with interstitial monitoring, which is monitoring between the two containment systems. For petroleum systems, several methods are allowed, including inventory reconciliation, which means keeping accurate records on how much product was delivered and sold and matching these amounts with the amount of product in the tank each day. Petroleum systems are not required to have secondary containment with monitoring devices between the secondary container and the tank wall. Owners and operators must demonstrate to the implementing regulatory agency that their release detection method will detect releases of the stored substance in a manner no less stringent than the release detection methods allowed for petroleum USTs, and that a method of corrective action is available to clean up a release of the hazardous substance should one occur.

Release detection at existing UST systems was phased in over a five-year period based on the age of the tank. The oldest UST systems, which were usually unprotected from corrosion, were required to phase in release detection within 1 year, and the newest systems by the end of the five-year period. Release detection for all pressurized delivery lines was required to be retrofitted within two years.

At new or upgraded UST systems, periodic tank tightness testing, which tests whether the tank leaks, is required every five years if combined with monthly inventory control for ten years after new tank installation or existing tank upgrade. After ten years, monthly release detection is required.

Either monthly release detection or a combination of annual tank tightness testing with monthly inventory control is required of substandard existing USTs until they are upgraded. Existing UST

# Science & Technology

## What Do Leaking Underground Tanks and Batteries Have In Common?

We are all familiar with the dry-cell batteries that are commonly used to power our portable CD players, cellular phones, and lap-top computers. Have you ever had the unfortunate experience of biting down with one of your silver-amalgam fillings on a piece of aluminum foil stuck to your chewing gum? OUCH! What do these examples have in common with leaking underground tanks? The answer is **electrochemistry**!

In the 1830s, Michael Faraday, considered to be the father of electrochemistry, unraveled much of the mystery about the relationship between electricity and chemical processes. As explained today, all metal atoms would prefer to become charged particles called ions. For this to happen, it would mean that they must find something else that will take one or more of their electrons. This push-pull of electrons is nature's way of allowing an atom to gain a more stable electron arrangement. Just as physical strength varies among humans, not all elements have the same ability to give and take electrons.

Chemists chose the element hydrogen to be their standard for comparing all elements' ability to give and take electrons. If the element was better at giving away its electrons (oxidizing) than hydrogen, then it was given a positive number (voltage), and if it is not as strong as hydrogen then it was given a negative voltage. In general, metals tend to be better and non-metals tend to be poorer electron givers than hydrogen. By making many measurements, scientists were able to place all elements into a hierarchy from "best" to "poorest" electron givers, called an Electromotive Series. By using the information in the Electromotive Series, today we can now predict what will happen when any combination of metals and/or non-metals are put together.

Electrochemical reactions then are simply the result of one element's atoms, usually a metal, forcing another element's atoms or ions to accept its electrons. In every electrochemical reaction there must be one substance giving the electrons (oxidation) and one substance accepting electrons (**reduction**). This is why these reactions are also frequently referred to as simply "oxidation-reduction" or redox reactions.

Batteries are an example of a useful oxidation-reduction reaction. Batteries are constructed in such a way to keep the substance being oxidized separated from the substance being reduced. For metal atoms in the battery to become ions, the electrons must flow out of the minus (-) pole, through your CD player or flashlight bulb before they reach the positive (+) pole where the reduction will take place within the battery. We gain the benefit from that flow of electrons (called electricity) when we harness it and have it do our work.

Not all electrochemical reactions are useful. Consider the corrosion of an underground storage tank and its piping. The corrosion is the result of an oxidation-reduction reaction; however, it's the tank's and pipe's metal atoms that are changing into their corresponding ions. This re-arrangement of their electrons may make the metal atoms "happy," but the tank owner will be very unhappy when it results in a "rust" hole. The rate of rusting will depend on such soil characteristics as the presence of other metals nearby, the presence of underground electrical systems, the metals used for the tank and piping system, and the amount of water present in the surrounding soil.

To prevent corrosion of metal objects placed in the ground, cathodic protection systems are commonly used. The two types of cathodic protection generally used are the **impressed current** and **sacrificial anode** systems. In the impressed current system, an AC to DC rectifier is used to resist the flow of electrons from the tank's metal atoms out into the surrounding soil. This is accomplished by supplying a current from the rectifier to a metal anode placed nearby, through the soil, to the tank and pipes and back to the rectifier. To provide continued protection, the system must be constantly powered by the rectifier. In short, this system provides a backward flow of electrons that is strong enough to prevent the tank's metal atoms from becoming ions.

In the sacrificial anode system, the metal tank is connected to a metal anode, usually magnesium or zinc, that loses its electrons more easily than the tank's iron atoms. By connecting the tank to a metal with a greater tendency to lose electrons, the backward flow of electrons again prevents the tank's metal atoms from becoming ions. As the name suggests, the anode metal is sacrificed to protect the tank. The advantage of this system is that an external power supply is not required. The disadvantages of the system are that the anodes must be regularly monitored and replaced as they corrode away, and there are surrounding conditions where the anodes may not be able to generate an adequate flow of electrons to prevent the tank's corrosion.

systems must be upgraded or closed within ten years of the effective date of the final rule, or within one to five years if a release detection method is not available that can be applied during the required phase-in period for release detection. Upgrading of petroleum UST systems includes retrofitting of corrosion protection and both spill and overfill controls at all tanks.

Tank owners and operators must report suspected releases. Indications of a release must be reported to the implementing agency, including positive results from leak detection methods, unless the initial cause of the alarm has been immediately investigated and the alarm is found to be false. After reporting suspected releases, owners and operators must perform release investigation and confirmation tests and, where a release is confirmed, must begin corrective action.

Owners and operators of leaking UST systems must follow measures for corrective action. Immediate corrective action measures include mitigation of safety and fire hazards; removal of saturated soils and floating free product; and an assessment of the extent of further corrective action needed. A corrective action plan would be required for long-term cleanups addressing groundwater contamination, although these cleanups could begin upon notification of the implementing agency by the owner and operator. Cleanup levels are established on a site-by-site basis as approved by the implementing agency and are usually based on the risk posed by the actual or threatened release.

In 1988, the EPA also published final regulations concerning financial responsibilities of owners and operators of UST systems containing petroleum. Financial responsibility means that owners and operators of USTs must ensure, either through insurance or other approved means, that there will be money to help pay for the costs of third-party liability and corrective action caused by a leak from their tank. These costs could include cleaning up leaked petroleum, correcting environmental damage, supplying drinking water, and compensating people for personal injury or property damage. Financial responsibility regulations are also designed to protect the owner and/or operator. If a tank leaks, responsible individuals may be faced with high cleanup costs or with lawsuits brought by third parties. Having financial responsibility means that money will be available to meet these potentially high costs.

The financial responsibility regulations require owners or operators to show that they have either a minimum of $1 million to cover the costs of a leak or spill if they are a petroleum marketer, or at least $500,000 if they are not a marketer. These amounts of required financial responsibility do not, however, limit the total liability for damages caused by a leak.

# ■ Clues to Identifying Hazardous Materials

## Concepts

■ There are several clues that can be used to determine whether hazardous materials are present in an area or container.

■ The Hazardous Code Chart is one type of hazard identifier.

■ DOT identifies nine Hazard Classes.

A common thread in all hazardous materials regulations is the need to identify and communicate the specific hazards of the material in question. Identifiers and clues are necessary in order to determine whether a hazardous material is present. They are not only important for the day-to-day workers using the material but are especially important for emergency responders who otherwise would not be aware of the potential hazards. The basic clues used to identify a hazardous material include knowing the occupancy of a facility, referring to **labels** and **markings** on containers, identifying **placards** on trucks or railcars, and examining shipping papers or manifests delivered with a load. The occupancy of a facility is very revealing. For instance, one would expect a pool supply store to contain acids, disinfectants, and scale inhibitors. The right-to-know laws require this type of information to be made public, specifically making available both to the public and responding agencies lists of hazardous materials stored at business locations. Labels are used as warnings of the type of hazard the hazardous material poses and are placed on containers. For instance, a flammable liquid contained in a drum would require red flammable liquid labels that are 4-inch point-on-point diamonds. Markings are the other informational items that are required on a container of hazardous materials such as the proper name, name of manufacturer, and instructions/cautions. Placards are similar to labels except they are placed on the front, rear, and both sides of the vehicle transporting the hazardous material. Figure 8-4 shows labels and placards required by the DOT.

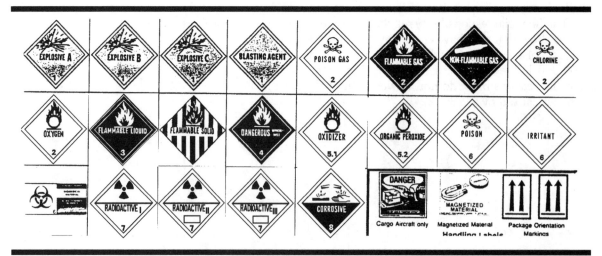

**Figure 8-4:** Labels and placards are required to identify hazardous substances.

A practice in many areas is to use a composite type of label called a Hazardous Code Chart on tanks and containers of hazardous materials. This type of labeling was developed by the National Fire Protection Association (NFPA) and is often referred to as the NFPA 704 System. The Hazardous Code Chart is a diamond-shaped identification display used to help fire departments or workers easily identify the hazards of a stored material. This is especially helpful in an emergency response situation where there has been a leak or spill of the material or if it is involved in fire. The diamond is divided into four segments, into which numbers or symbols are placed. The upper three segments show hazards pertaining to the health, fire, and reactivity characteristics of the stored material (Figure 8-5). The bottom segment explains specific hazards occurring with that material, such as corrosivity. The highest hazard for any category is designated by the number 4. The lowest hazard is designated by the number 0.

The four hazard areas shown by codes in the chart are designated by specific colors which mean the following:

**BLUE: Health Hazard**   The ability of a material to cause, either through direct or indirect exposure, temporary or permanent injury or incapacitation.

**RED: Fire Hazard**   The ability of a material to burn easily when exposed to a heat source.

**YELLOW: Reactivity Hazard**   The ability of a material to release energy when in contact with water. It can also be the tendency of a material, when in its pure state or as a commercially produced product, to vigorously polymerize, decompose or condense, or otherwise self-react and undergo violent chemical change.

**WHITE: Specific Hazard**   The term pertaining to a special hazard concerning the particular product or chemical, and not covered by the other labeled hazard code items. The special hazard must be written in the diamond.

The degree of hazard is expressed by the following numerical code:

4 = extreme hazard
3 = high hazard
2 = moderate hazard
1 = slight hazard
0 = no hazard

The chart depicted in Figure 8-5 informs us that the material in the container represents:

■ A moderate level of health hazard (code symbol 2 in left hand segment);

■ An extreme fire hazard (code 4, top segment);

■ A high reactivity hazard (code 3, right-hand segment); and

■ No specific hazards (bottom segment).

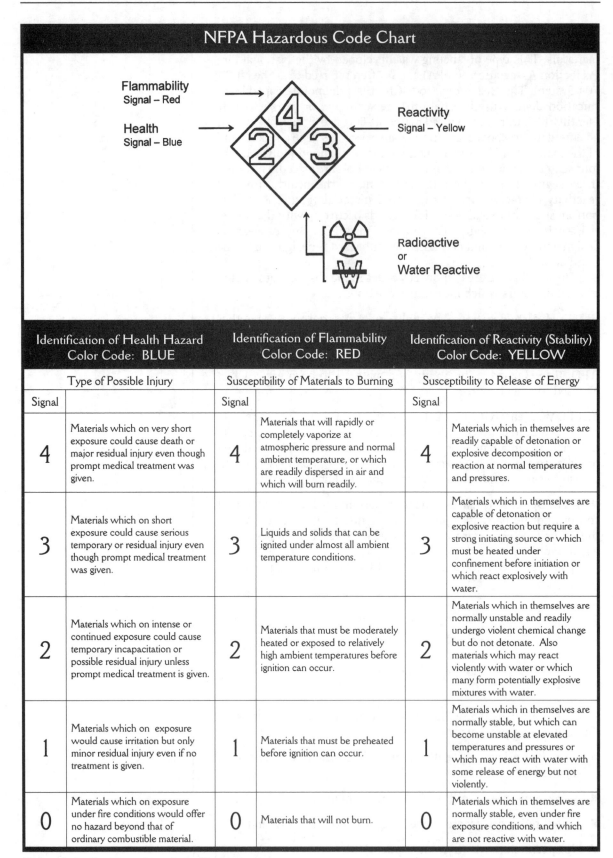

# NFPA Hazardous Code Chart

Flammability
Signal – Red

Health
Signal – Blue

Reactivity
Signal – Yellow

Radioactive
or
Water Reactive

| Identification of Health Hazard Color Code: BLUE | | Identification of Flammability Color Code: RED | | Identification of Reactivity (Stability) Color Code: YELLOW | |
|---|---|---|---|---|---|
| Type of Possible Injury | | Susceptibility of Materials to Burning | | Susceptibility to Release of Energy | |
| Signal | | Signal | | Signal | |
| 4 | Materials which on very short exposure could cause death or major residual injury even though prompt medical treatment was given. | 4 | Materials that will rapidly or completely vaporize at atmospheric pressure and normal ambient temperature, or which are readily dispersed in air and which will burn readily. | 4 | Materials which in themselves are readily capable of detonation or explosive decomposition or reaction at normal temperatures and pressures. |
| 3 | Materials which on short exposure could cause serious temporary or residual injury even though prompt medical treatment was given. | 3 | Liquids and solids that can be ignited under almost all ambient temperature conditions. | 3 | Materials which in themselves are capable of detonation or explosive reaction but require a strong initiating source or which must be heated under confinement before initiation or which react explosively with water. |
| 2 | Materials which on intense or continued exposure could cause temporary incapacitation or possible residual injury unless prompt medical treatment is given. | 2 | Materials that must be moderately heated or exposed to relatively high ambient temperatures before ignition can occur. | 2 | Materials which in themselves are normally unstable and readily undergo violent chemical change but do not detonate. Also materials which may react violently with water or which many form potentially explosive mixtures with water. |
| 1 | Materials which on exposure would cause irritation but only minor residual injury even if no treatment is given. | 1 | Materials that must be preheated before ignition can occur. | 1 | Materials which in themselves are normally stable, but which can become unstable at elevated temperatures and pressures or which may react with water with some release of energy but not violently. |
| 0 | Materials which on exposure under fire conditions would offer no hazard beyond that of ordinary combustible material. | 0 | Materials that will not burn. | 0 | Materials which in themselves are normally stable, even under fire exposure conditions, and which are not reactive with water. |

**Figure 8-5:** The Hazardous Code Chart.

Another way to classify hazardous materials was developed by the Department of Transportation. Under the DOT system, hazardous materials are divided into nine **hazard classes**. They are: explosives, gases, flammable liquids, flammable solids, oxidizers, poisons, **radioactive materials**, corrosives, and miscellaneous hazards. Section 8-3 on hazardous materials transportation will explain more about these hazard classes and how important they are in the labeling and transportation of hazardous materials.

When looking at the vast quantities of hazardous materials as well as their potential hazards that can severely impact public and environmental health, it is apparent that regulations are needed to ensure that they are handled, used, managed, and transported properly. It is important to note that there is no standard definition of hazardous materials. As such, be sure to check the specific regulations of the agencies involved (DOT, OSHA) to be assured of proper compliance. The rest of this chapter will highlight specific major hazardous materials laws and regulations that are designed to ensure that materials defined as hazardous are managed properly.

## ■ Checking Your Understanding (8-1)

1. List four ways in which a material can be hazardous.

2. Explain what we mean when we say a material is persistent in the environment.

3. Describe a hypothetical release into air, water, and soil.

4. List two reasons why hazardous materials stored in underground tanks can pose a greater public health and environmental risk than if stored in other ways.

5. Describe two main reasons for underground tank system leakage.

6. Define regulated substances.

7. List one technical requirement for USTs during their installation, use, and closure.

8. Define secondary containment.

9. Explain what is meant by financial responsibility.

10. Describe how corrosion takes place.

11. Contrast labels, markings, and placards.

12. Explain what the numbers and colors mean on the NFPA Hazardous Code Chart.

# 8-2 CERCLA and SARA

## ■ History and Overview

### Concepts

- CERCLA differs from RCRA in that it deals with past releases of hazardous substances.

- CERCLA provides money and procedures to identify, clean up and hold responsible parties liable for damages from hazardous substance contamination.

- Funding for CERCLA actions is from a large trust fund.

- SARA amended CERCLA, increased the Fund, and created Community Right-to-Know.

Congress passed the Comprehensive Emergency Response, Compensation, and Liability Act, known as CERCLA, or Superfund, in 1980 in response to the problem of severe contamination in the Love Canal area of New York. The Love Canal incident made government officials realize that RCRA did not address past contamination problems, which were starting to surface with alarming frequency. In fact, in the late 1970s a large number of abandoned and leaking hazardous waste dumps that were threatening human health and contaminating the environment were discovered throughout the country. Love Canal was just the most publicized. Early on, Hooker Chemical probably did not imagine that a law would be passed that would cause them to be listed as a responsible party in the Love Canal contamination. With this new law, however, they would continue to have liability in the case many years after they buried large amounts of hazardous waste in a canal originally designed to transport water. With only RCRA to turn to prior to 1980, there was no recourse for either the government to clean up contamination or to force clean up by responsible parties. Declaring the site a federal disaster area, which was what President Jimmy Carter did, was the only viable option available to the federal government. RCRA did not provide relief because the problem did not involve the current management of wastes. Legal actions against the responsible parties did not offer a solution because they were and still are too time-consuming. Clean-ups are often extremely expensive and beyond the resources of many small companies and individuals. Unfortunately, in the Love Canal incident it was discovered that the waste dump problem was far larger in scope, and the federal disaster relief option was soon deemed impractical. In late 1980, Congress passed CERCLA to address what they considered the ticking time bombs represented by thousands of potential Love Canals throughout the country.

**Figure 8-6:** The clean-up at Love Canal led to the enactment of CERCLA

CERCLA and RCRA are unique in that their primary purpose is to deal with dangers posed to the environment by chemicals. However, they deal with the problem from two very different perspectives. CERCLA is designed to remedy the mistakes in hazardous substance management made in the past, and uses a response focus. Whenever something has gone wrong in the substance management system, like a release, the statute allows for actions that will remedy the situation, including paying for them. RCRA is concerned with avoiding mistakes through proper management of hazardous waste in the present and future and therefore has a regulatory focus. RCRA mainly regulates how wastes should be managed from cradle to grave to avoid potential threats to human health and the environment. Chapter 10 covers the provisions of RCRA.

CERCLA implemented a five-year program carried out by the federal government to require performance of the following tasks:

■ Identify those sites where releases of hazardous substances had already occurred or might occur and posed a serious threat to human health, welfare, or the environment.

■ Take appropriate action to remedy those releases.

■ See that the parties responsible for the releases pay for the cleanup actions.

CERCLA's purpose was to fund cleanups and emergency response actions for some of the worst inactive or abandoned hazardous waste sites around the country. To accomplish this, CERCLA gave new cleanup authority to the federal government by creating a $1.6-billion trust fund to pay for cleanup of hazardous substances posing a threat to human health or the environment. This Superfund was financed primarily by imposing a tax on certain chemical and petroleum products, and was intended to allow the government to act quickly, delaying financial concerns. The fund was designed so that money is continually being infused for situations in which no responsible party can be found, or the responsible party does not have sufficient funds to undertake a cleanup.

The first five-year period of the original Superfund program revealed a lot about the problems of cleaning up hazardous waste sites. Congress and the EPA had to face two surprising facts: that the problem of abandoned hazardous waste sites was more extensive than originally thought, and that solutions would be more complex and time-consuming than had been imagined. During this initial time period, private parties agreed to conduct cleanups costing approximately $600 million, 580 removal actions were initiated, and 200 lawsuits were filed by the government. Litigation arose contesting the constitutionality of the statute and how it was to be implemented.

The **Superfund Amendments and Reauthorization Act of 1986 (SARA)** not only extended CERCLA for another five years, but increased the Fund five-fold: from $1.6 billion to $8.5 billion. New sources for funding were added, including a new environmental tax on corporations, $1.25 billion from general appropriations, and costs and damages recovered from responsible parties. SARA expanded on existing CERCLA programs by establishing new standards and schedules for site cleanup. It also created some new programs. The most important is the Community Right-to-Know program that informs the public of risks from hazardous substances in their community and prepares people for hazardous substance emergencies. SARA will be covered later in this section.

## ■ CERCLA

### Concepts

■ Hazardous substance is a more encompassing term than hazardous waste.

■ CERCLA authorizes the EPA to take removal actions and remedial actions.

■ The NPL lists sites slated for cleanup by rank.

■ Liability under CERCLA is strict and considered joint and several.

CERCLA response authorities are triggered by a release or a substantial threat of a release of dangerous substances into the environment (e.g. a chemical spill from a tank truck accident or a leak from a damaged drum). The release must involve either:

– A hazardous substance, as defined in the statute, or

– A pollutant or contaminant that may present an imminent or substantial danger to public health or welfare.

**Hazardous substance** is defined to include hazardous waste under RCRA, as well as substances regulated under the Clean Air Act, Clean Water Act, and Toxic Substances Control Act. Pollutant or contaminant is broadly defined to include any substance that is reasonably anticipated to cause illness or deformation in any organism. Both definitions specifically exclude petroleum and natural gas.

CERCLA authorizes two types of government response actions to remedy the release of a hazardous substance: **removal actions** and **remedial actions**. Removals are short-term cleanup actions that usually address problems only at the surface of a site. They are conducted in response to an emergency situation; for instance, to avert an explosion, to clean up a hazardous waste spill, or to stabilize a site until a permanent remedy can be found. Removal actions are limited to 12 months duration or $2 million in expenditures, although in certain cases these limits may be extended.

Remedial actions represent the final remedy for a site and generally are more expensive and longer in duration than removals. This is because the remedial actions are intended to provide permanent solutions to hazardous substance threats. Some current treatment remedies are estimated to cost an average of $16 million and to take approximately 10 years to complete. EPA can take remedial actions only at hazardous waste sites on the **National Priorities List (NPL).** Currently, there are over 1,100 sites either on the NPL or proposed for inclusion. Sites are placed on the NPL after being evaluated through the Hazard Ranking System (HRS). The HRS is a model that determines the relative risk to public health, the environment and soil. It is possible that both removal and remedial actions may be taken at the same site to deal with hazardous substances in groundwater, surface water, soil, and/or air.

As part of the rule-making process, EPA must implement and revise as needed the **National Contingency Plan (NCP),** 40 CFR §300, et seq. The NCP regulates how the EPA will use its authority and expend Superfund moneys. It details the process for investigating sites and cleaning them up.

Since CERCLA's enactment, the EPA has reported that Superfund has enabled the cleanup of nearly 250 contaminated sites, and some 700 others are in various stages of investigation and cleanup. In addition to remedial actions, removal actions have been taken at 2,800 different sites across the country. In the process of accomplishing both remedial and removal actions, Superfund has received negative publicity over flaws that are perceived to be in its structure. A common complaint is that the level of cleanup accomplished under the act is often inconsistent from one site to another. The extremely high costs of cleanups are attributed in part to a liability scheme that is generally considered unfair and cleanup levels that are sometimes overly stringent. State governments and local communities have complained that they do not have enough say in decisions that affect neighborhood jobs and economies.

The law offices of Sidley & Austin, a private law firm hired by industry, stated in a report on Superfund that the stringent liability provisions, exaggerated perceptions of risk, and burdensome corrective action requirements makes Superfund the most troublesome and potentially emotional regulatory program facing industry today. The liability issue has been extremely controversial from the outset.

## Box 8.1

### Diane M. Garcia

Diane M. Garcia is employed as an Environmental and Safety Specialist for a Santa Fe Springs, California-based manufacturer of tile, stone, and masonry care products. The 15-year-old company employs about 55 employees and is among the first in its industry to hire a full-time environmental technician to assist with the day-to-day management and implementation of business environmental compliance. Garcia coordinates all the company's environmental and safety activities including hazardous management, written compliance plans, filing of annual state and federal regulatory reports, preparation of material safety data sheets, application of DOT requirements, providing input for consumer product label cautionary language, training employees in hazard communication, and more.

Garcia, who is a single, returned to community college in her mid-thirties to continue her education and to make a career change. Although difficult at times, she says she realized that continuing to work on an education was a step that would result in more control over her own direction and future. It was both a motivating and a frightening thought. Garcia states, "I felt as if returning to school was a little risky, almost as if it was a waste of my time to be trying something new at my age. I couldn't help but wonder if, somehow, I was only deceiving myself; but I knew there was so much more that I wanted for myself and small child."

The fears soon gave way when, after completing general education requirements at Fullerton College, Garcia learned of the Environmental Hazardous Materials Technology program. She took an introductory course in an effort to get a glimpse of environmental engineering – her long term goal. Garcia found herself in an exciting world between business and environmentalism. She had long ago acknowledged two facts: (1) business and industry were never going to go away, and (2) environmentalists would not surrender their vision of a pollution-free world. Fifteen years ago, she did not see the possibility of any middle ground between the two. Today, Garcia finds herself working in the middle ground. It is an exhilarating place where global changes are occurring; a place where each individual is still very much a part of dynamic solutions; a place that is envisioning a higher environmental quality and demanding an end to industry's environmental abuses. It is a place that still needs all the creativity and hard work one can muster! "Sometimes progress is slow, sometimes rapid. Sometimes the solutions are right and sometimes they are wrong, but environmental technology is a place where every action taken helps shape our eco-industrial future. It's a place where the work I do makes a difference."

Garcia received an Associates Degree in Environmental Hazardous Materials Technology (EHMT) from Fullerton College. She also worked several years for the Business Environmental Assistance Center where she assisted numerous small businesses with environmental compliance issues before moving on to industry.

Eventually, Garcia would like to be in business for herself. She says, "This field is still emerging; it is in infancy stages. Opportunities here are still as big as your imagination. I, personally, would like to open an environmental practitioners firm . . . not a place of consulting, but a place of 'environmental re-infrastruction.' This is what I call the rebuilding of existing business infrastructures to include environmental components. In today's world, businesses require a revamping of their infrastructures, cornerstone by cornerstone. The trick is that while business wants to be revamped, it also insists that the rest of the structure be disturbed as little as possible. This is extremely difficult to accomplish, but I'm betting it can be done painlessly."

*Q: What do you do in your position as Environmental and Safety Specialist for your company?*

A: I do double-duty. I am responsible for environmental compliance functions, such as filing the company's Business Plan, annual SARA reports, evaluating which permits the company may or may not need, assist in determining proper treatment and disposal of hazardous and industrial wastes, and keeping abreast of regulations that impact the company in some way.

I am also responsible for several occupational safety programs, including writing and implementing an Injury and Illness Prevention Plan, Hazard Communication Standards, monitoring and coordinating respiratory protection programs, forklift programs, and more. In most industrial settings – particularly smaller companies – environmental compliance and health and safety get all lumped together. Contributing to this misconception is the notion, by those persons unfamiliar with the regulatory structure, that all regulatory agencies are synonymous with EPA.

# Box 8.1

**Q: What aspects of your work do you find most challenging?**

A: All of it. Make no mistake, it is not easy work and each task is not self-defined. I find that the field is still emerging and inventing itself. When I say this, I mean there are still many ways to accomplish a given task. While this gives you options, it also means that what are considered the best environmental practices keeps changing. There is a continuum of refining going on all the time, and that means the "chore" is never really done. This is very difficult in a world that thrives on definitive solutions.

There is one particularly rough spot – semi-isolation. If you work for a small company, chances are you are the only person working on your company's environmental and/or safety issues. You must be part entrepreneur and a self-starter at this stage of the game. You begin to realize that others in your company do not understand "hazmat lingo" nor the basic tenets of environmental compliance. You must rely on your own wits and your own resources.

**Q: What do you find most rewarding?**

A: All of it. I get to be creative, I work on the cutting edge of the eco-industrial future, I feel like I am making the community a better and safer place. I help educate people about workplace hazards and help them protect themselves against unnecessary or unlawful risks; I am compensated well for my work. And as corny as it may sound, I get to be part of the solution to pollution!

**Q: Did your course work prepare you for your position?**

A: Yes and no, but mostly yes. After taking only two courses, I found that I had new skills that someone needed. I found part-time work with a budding environmental assistance center that specialized in assisting small businesses with compliance. As I continued my course work, coupled with my work experience, I found that I was quite prepared to handle a variety of tasks, reports, and written plans that were expected of me. And most importantly, I had gained a broad base of resources and knowledge that told me how to go about and find the things I did not know. This is very important. Sometimes I am called upon to provide answers to issues I know nothing about, but I have learned how to research them, to deduce where they fit in the structure of environmental affairs, and then seek them out.

For example, my course work did not cover anything on Consumer Product Safety Commission regulations which govern, among other things, the types of hazard warnings a product must carry when it contains hazardous materials. I discovered the code of federal regulations section numbers and went to the law library and read up on the regulations. I then assisted my company with selecting appropriate wording for product labels.

**Q: What type of skills do you find you need on a day-to-day basis in your field?**

A: Besides the environmental compliance information base, computer skills are a must in business. Try to get experience with a range of software programs. Oral communication skills are important, too. I find that I must sometimes do internal employee training, make presentations at manager's meetings, board meetings, national sales meetings, and occasionally the boss volunteers me to give a lecture to a trade association! Oh, yes – practice smiling, and a good firm hand shake is an important skill to have, too.

**Q: How were your job expectations and job realities different?**

A: I expected to implement sweeping environmental changes in my first year in industry. I expected my company would have open arms and a fistful of dollars to implement product life cycle analysis projects, process changes, equipment retooling, and massive employee education programs. And I expected a support staff to assist me. But, change is slow in a "mom and pop" shop and most other shops, too. There is a learning curve that one must go through with any company, discovering how it operates, and what internal processes, politics, and protocols will allow. There are trust factors, there are budget constraints, space constraints, and personality constraints. And if that is not enough, there is often no clerical support.

**Q: What recommendations do you have for someone considering Environmental Technologies as a career choice?**

A: Come on down! Do it! Expect a rough time in this field for a few more years. Not everyone understands the industrial base has no alternative but to change to accommodate new generations that demand cleaner air, cleaner water, cleaner soil and cleaner manufacturing. Expect to do a little trailblazing. Expect change. Expect to be imaginative. Expect to work long hours. Expect to succeed.

CERCLA gives the EPA authority to require responsible parties to conduct cleanups and for the EPA to recover from the responsible parties any moneys the government has spent. Court precedents have interpreted the law, in accordance with Congressional intent, to be retroactive, which means the provisions apply to releases that occurred even before the law was enacted. The courts have also interpreted the liability to be strict and joint and several. Strict liability means that one can be liable even if not at fault. In other words, if you were to purchase property which you had no idea was contaminated, you could be held liable for cleanup, even though the contamination was obviously not your doing. Joint and several liability means that all **potentially responsible parties (PRPs)** could be held liable jointly or severally (which comes from the root word sever). What this means is that ten potentially responsible parties could be held jointly liable for some percentage each of the entire cleanup or one of the parties, usually the one with the greatest monetary resources, can be held liable for the whole amount. This liability provision is sometimes referred to as a deep pocket provision.

The list of potentially responsible parties that can be identified as responsible for financial liabilities at a Superfund site are many. They include the current owner or operators of the facility; owners or operators of the facility at the time the release took place; those who arranged for treatment or disposal of the substances at the facility; and those who accepted hazardous substances for transport to a facility that they selected. Individuals and corporate employees have been held liable for their hazardous waste management activities. Financial institutions that hold a trust deed on a contaminated site may also be liable if they foreclose on the site.

Based on years of feedback and experience, the EPA has acknowledged that problems do indeed exist with CERCLA. On February 3, 1994, EPA Administrator Carol Browner suggested a bill to amend the Superfund portions of CERCLA to remedy some of these problems. These amendments were not passed during the 1994 legislative session of Congress, so Superfund remains essentially the same at the time of this writing.

One very important change in the liability provisions of CERCLA was added by SARA. This provision has come to be known as the "innocent landowner defense." This defense allows a landowner to be shielded from liability for releases on property he or she purchased after the contamination of the property took place. In order to prove this, the innocent landowner must demonstrate that he or she took diligent precautions to establish that there was no reason to know of any hazardous substance on the property at the time of acquisition. In order to do this, careful record searches are needed to determine the prior uses of the property. The innocent landowner defense does not apply to releases caused by the current owner, nor can they be claimed if the owner knew of the release and then transferred the property without telling the new owner.

CERCLA requires spill reports and facility notifications to be made. Spill reporting is required for actual releases of hazardous substances in amounts equal to or greater than the CERCLA-listed **reportable quantity (RQ)** for that substance. Minimum reportable quantities range from 1 pound to 5,000 pounds per 24-hour day.

Spills or releases of this magnitude must be reported immediately to the **National Response Center**, which is located at US Coast Guard headquarters in Washington, DC. A facility notification is required of facilities where hazardous wastes have been disposed of and where releases might potentially occur. This information is used to learn the location of old disposal sites and to determine whether sites should be added to the NPL. Separate reporting under SARA is covered in the next section.

States may also have their own version of Superfund that may be more stringent than the federal law. In the case of releases, the designated state agency must be notified if the state has notification requirements in addition to the federal requirements. In fact, some states require notification for any quantity of hazardous substance released that may pose a threat to public or environmental health. Reporting to a state does not relieve facility personnel of the responsibility to report to the National Response Center, if warranted. Some states also have separate lists of sites prioritized for state cleanup in addition to the federal NPL list.

## ■ Site Characterization and Remediation

### Concepts

- There are a number of steps involved in cleaning up sites under CERCLA.

- Preliminary Assessments and Site Investigations provide information on the site and PRPs.

- A Hazard Ranking System is used to list sites on the NPL.

One of the goals of CERCLA is to identify contaminated sites that have high potential for harming public health and/or the environment. Contamination could have occurred through leaking underground storage tanks or drums, leaking waste disposal landfills or ponds, hazardous materials spills, or improper disposal of hazardous wastes. These uncontrolled releases can contaminate all the environmental compartments. The NPL prioritizes sites so that the most threatening will be acted upon first. This part of the chapter will discuss the mechanisms used to clean up a site.

As stated earlier, the National Contingency Plan, or NCP, describes the procedures for cleanup of contaminated sites. It does this by describing the various steps that must be taken in preparation, actual cleanup, and follow-up after the site has been cleaned. The basic steps include: **preliminary assessment**, **site investigation**, the **Hazard Ranking System**, the **National Priorities List**, and **Remedial Action**. In order for a site to be listed on the NPL it must meet one of three requirements: the U.S. Department of Health and Human Services has determined that people must stay off the site and has issued a health advisory to that effect; the EPA has determined that the site poses a significant threat to public health; or the EPA believes that it is more cost-effective to use its remedial authority rather than its removal authority to remedy the problem at the site.

Contaminated sites are discovered through facility notifications, reports by responsible parties, EPA investigations, referrals from other agencies, or reports by the public. Once a site is discovered, the severity of the problem must be evaluated. The first step for regulatory agencies made aware of a contaminated site is to conduct a preliminary assessment, which is a quick analysis to determine how serious the situation is and to identify all potentially responsible parties. The preliminary assessment utilizes readily available information from records, aerial photographs, and personnel interviews. Regulatory staff review existing files and records and may take minimal samples at the site. This background information is compiled into what is known as a Preliminary Site Assessment Report. The potentially responsible parties (PRPs) are also identified and a Community Relations Plan is prepared. Finding all responsible parties can prove difficult because as we learned before, responsible parties can be current or former owners, operators, users of the site, or originators of the products that contaminated the site. In many cases, more than one party may be responsible for a site's contamination. For example, the way the definition of responsible party is stated, if a drum-cleaning facility is found to be contaminated with hazardous waste, any business that sent waste-containing drums to the site may be a responsible party. The Community Relations Plan describes how the EPA will inform and involve the community in the cleanup process. These initial steps are time consuming and may take up to nine months.

The information obtained during the preliminary assessment is typically not sufficient to determine whether the site needs to be placed on the NPL. The site investigation is designed to provide additional site-specific information. The site investigation usually entails sampling. This information combined with the information from the preliminary assessment will allow ranking of the site using the Hazard Ranking System. The Hazard Ranking System is a mathematical model that uses sources, pathways, and receptor populations to attempt to rank the overall threat to public health and the environment.

The EPA is authorized to begin the remedial planning process and can expend Superfund money once a site is proposed for addition to the NPL. PRPs can, however, enter into agreements with the EPA to conduct the remediation, which, if approved may result in the site not being listed on the NPL.

The **Remedial Investigation (RI)** is a detailed investigation of the contaminated site. This study provides all the information necessary to characterize a site. Remedial Investigations involve defining the nature and extent of the contamination, and gathering any technical information needed to evaluate possible ways to clean up the site. Site conditions such as soil types, seasonal and annual rainfall, depth to aquifers, and proximity of humans and animals that could be affected by the contamination are examined. At sites where groundwater is contaminated, information is needed about the soil profile and the groundwater movement. The EPA may drill **monitoring wells** to gather this information.

The Remedial Investigation phase also entails collecting and analyzing data about the waste. Samples of soil, groundwater, surface water, or living organisms such as fish may be collected in the contaminated area. These samples will help to determine both what the contaminants are and at what levels they are found. Strategically locating the samples can show how far the contamination has spread. A risk assessment process similar to the one detailed in Chapter 3 is utilized. In this process, a number of determinations are made, including the possible routes of exposure, the potential levels of exposure, and the health effects which could result from specific exposures.

The next step is called the **Feasibility Study (FS)** in which the EPA identifies and evaluates site cleanup alternatives. Imagine, for instance, that a site has been contaminated because drums have leaked their contents onto the soil. The EPA could immediately remove the soil with the highest levels of contamination and send it to a permitted disposal facility. The area could then be capped with asphalt or cement to prevent the remaining contamination from spreading and/or it could be treated on-site to destroy the contamination. All potential alternatives are listed and evaluated for effectiveness, feasibility, and cost. The alternative that is chosen must effectively protect human health and the environment. The results of the **Remedial Investigation/Feasibility Study** (also known as **RI/FS**) are compiled into a report that is used to recommend a final cleanup plan. This step can take anywhere from four months to several years.

The **Remedial Action Plan (RAP)** recommends the actual cleanup procedure. At most sites, more than one method is typically used to perform cleanup. Before this plan is approved, it must be made available to other agencies, the responsible parties, and the public for a minimum 30-day comment period. After conducting public meetings and hearing comments, the EPA makes whatever changes it deems necessary and prepares the final Remedial Action Plan. This step usually takes three to six months.

After all of these preliminary steps, the approved remedial action can actually begin. This step may take from two months to two years. This, however, is still not the end of the process. After the remedial actions are complete or treatment systems are in place, the EPA will oversee the maintenance and operation of the site. At some sites, groundwater can be pumped and treated in aerators or adsorbers to remove or trap chemicals. The clean water can then be discharged to a stream or surface water if a NPDES permit allows it. Groundwater treatment systems may need to operate for 20 years or more to complete a cleanup. Other ongoing activities may include monitoring the effectiveness of the cleanup by regularly sampling water and/or soil and checking for cracks in soil caps. These activities typically continue for long periods of time – 15 years and upward.

The EPA can give responsible parties the opportunity to conduct the site cleanup. They must agree to meet specific time frames and gain approval for all steps before beginning. The EPA then oversees the work of cleanup. If responsible parties cannot or will not assume cleanup responsibilities, the EPA can conduct the investigation and ultimately attempt to regain cleanup costs through a legal provision that allows them to recoup treble damages, which means they can be awarded three times the actual cost of the cleanup.

The description of the process the government uses for cleanup actions allows us to see that several steps are involved: finding the responsible parties; addressing any immediate threats to public health or the environment through removal actions; informing and involving the community in the cleanup process; and finally, investigating and cleaning up the site. Cleaning up the site does not necessarily mean removing all the contamination. At some sites, the EPA may allow the contamination to be contained with underground barriers to prevent its spread or capped so that the threat is minimized. Some wastes lend themselves to being treated on the site to remove or lessen the contamination.

If an immediate health threat needs prompt action, the EPA can perform a removal action. For instance, if contaminated soils pose a threat to nearby residents, the soils can be removed. Removal actions allow the EPA to respond to immediate health threats before the lengthy process of site investigation is complete. These actions are temporary in nature, but must be consistent with the long-term goals of the cleanup.

# ■ SARA

## *Concepts*

- SARA Title III was passed to help communities prepare for chemical emergencies and to increase the public's knowledge of the presence and threat of hazardous chemicals.

- Title III has specific requirements for planning, notification, and reporting.

- SARA Title III was enacted as a result of the Bhopal incident.

The Superfund Amendments and Reauthorization Act of 1986, also known as SARA, was signed into law on October 17, 1986. The third part of SARA, *Title III*, is the Emergency Planning and Community Right-to-Know Act. Prior to this law, citizens had little or no legal backing in their attempts to obtain information about toxic releases from facilities in their own communities. As the public and its Congressional representatives became more aware of the increasing use of hazardous materials and the corresponding increase in the number of accidents, pressure grew for better information at the local level.

The single incident that is credited with raising the level of concern to the point that such a law could be passed was the release of methyl isocyanate in Bhopal, India which was described earlier in this chapter. To help reduce the likelihood that such a tragedy would occur in the United States, and also to increase a local government's ability to anticipate and plan for such a major emergency if one were to occur, *Title III* seeks to provide reliable information to those who would be most affected by an accidental release of this kind – the communities located in the immediate area of industrial plants. *Title III* has two primary purposes: to help communities prepare to respond in the event of a chemical emergency and to increase the public's knowledge of the presence and threat of hazardous chemicals. Toward this end, *Title III* requires the establishment of state and local committees to prepare communities for potential chemical emergencies. The focus of the preparation is a community emergency response plan that must: 1) identify the sources of potential emergencies, 2) develop procedures for responding to emergencies, and 3) designate who will coordinate the emergency response.

*Title III* also requires facilities to notify the appropriate state and local authorities if releases of certain chemicals occur. Facilities also must compile specified information about hazardous substances they have on-site and the threat posed by those substances. Some of this information must be provided to state and local authorities. More specific data must be made available upon request from those authorities or from the general public.

As used in SARA, the term "hazardous materials" refers to substances transported, used, and stored at petroleum refineries and natural gas facilities; hazardous chemicals such as PCBs and trichloroethylene (dry cleaning chemicals); acutely toxic chemicals; and fumes and dust from metals such as arsenic, lead, and cadmium. For the first time, the law even requires the agricultural industry to report production, use, storage, or release of certain chemicals. The EPA maintains an updated list that includes over 300 **extremely hazardous substances (EHS),** selected on the basis of their ability to pose an immediate threat to life and health. These EHS chemicals have been involved in some of the most serious accidents that have occurred in the U.S. to date.

*Title III* establishes requirements for federal, state, and local governments and industry regarding local hazardous materials, emergency planning, and reporting. It also provides a framework within which federal, state, and local governments are to work together with industry to reduce risks. *Title III* has four major sections: Emergency Planning; Emergency Notification; Community Right-to-Know Reporting Requirements; and Toxic Chemicals Release and Emissions Reporting.

The Emergency Planning portion of *Title III* requires that the governor of each state designate a **State Emergency Response Commission (SERC).** This commission generally includes representatives of public agencies and departments with expertise in environmental issues, natural resources, emergency services, public health, occupational safety, and transportation. Various public and private sector groups and associations with an interest in *Title III* issues may also be included in the state commission.

The SERC designates local emergency planning districts and appoints **Local Emergency Planning Committees (LEPCs)** within each of these districts. Each state determines the number of LEPCs that are appropriate to accomplish the goals of SARA's emergency planning provisions. Each SERC is responsible for supervising and coordinating the activities of the Local Emergency Planning Committees within the state, as well as for establishing procedures for receiving and processing public requests for information collected under other sections of *Title III*, and for reviewing plans generated by the LEPCs.

Each LEPC is expected to include broad representation, including: elected state and local officials; police, fire, civil defense, public health, environmental, hospital, and transportation officials; representatives of facilities subject to the emergency planning requirements; community groups; and the media. Public notice is given of meetings and activities, and procedures must be established for handling public requests for information. Citizens who want to help their community prevent and plan for hazardous materials emergencies should contact the LEPC.

The LEPC's primary responsibility is to develop and implement a local emergency response plan. In developing their plans, local committees analyzed local risks and evaluated resources available to their area that could help them to prepare for and respond to a hazardous materials accident. LEPCs are also encouraged to consider strategies for preventing or mitigating chemical emergencies: that is, to identify ways to prevent emergencies from happening, or at a minimum, to make their consequences less severe. Some examples that meet these goals include the installation of sprinklers in a chemical plant, or the routing of certain hazardous materials carriers away from highly populated or residential areas.

The **contingency plan** generated by the LEPC must include a list of hazardous materials facilities and the routes they use to transport their listed materials, emergency response procedures, and evacuation plans. Contingency plans are reviewed by the SERC and updated annually by the LEPC. Emergency plans must focus on the list of extremely hazardous substances published by the EPA, but are not limited to this list. Any facility that uses these substances in excess of specified threshold quantities is subject to emergency planning requirements.

The Emergency Notification section of *Title III* requires an industry to notify the Local Emergency Planning Committee, the state, and the National Response Center if there is a release of a listed hazardous substance that exceeds a certain quantity, as specified in the November 17, 1986 issue of the Federal Register. The emergency notification must include the name of the chemical released, the quantity involved, how and where it was released, and the health risks from exposure. This section of *Title III* also requires industry to submit reports to the state and the LEPC after the event that explain what actions were taken to control the release, and provide more data on health risks and any medical attention required for victims. This part of the law allows communities to learn if significant releases from hazardous materials facilities are occurring or are likely to occur, and whether state-of-the-art technology is being used by the plants to protect nearby communities from unnecessary adverse health effects.

Community Right-to-Know Reporting is a particularly important part of *Title III* because it gives citizens the right to obtain information on hazardous materials in their community. Environmental, health, and labor groups were instrumental in the passage of this law. *Title III* requires facilities to submit either **Material Safety Data Sheets (MSDSs)** or lists of specific hazardous chemicals present on sites in amounts over threshold quantities to the

LEPC, the SERC, and the local fire department. MSDSs are chemical information sheets provided by the chemical manufacturer that include information such as: chemical and physical characteristics; long and short-term health hazards; spill control procedures; personal protective equipment to be used when handling the chemical; reactivity with other chemicals; incompatibility with other chemicals; and manufacturer's name, address and phone number. MSDS formats vary considerably among providers, but all include vital information about the properties and effects of the hazardous material involved. Industry facilities must also submit inventories of the amounts and locations of these chemicals in their plants. OSHA and EPA rules specify which chemicals and quantities must be reported.

The Toxic Chemical Release and Emissions Inventory Reporting section of *Title III* requires hazardous materials facilities to inform the public about routine, day-to-day releases of chemicals. The intent is to provide information on the extent of the cumulative toxic chemical burden on the environment. Over 300 chemicals listed by EPA must be reported if they are emitted on a regular basis. Facilities must submit **toxic chemical release forms** for each of these chemicals. This requirement only applies to facilities that have ten or more full-time employees and that are in certain specified industries. Facilities that use less than 10,000 pounds of a listed chemical each year are currently exempted.

## ■ Federal, State, Local, Industry, and Citizen Roles Under SARA Title III

### Concepts

- The federal role under SARA Title III is one of support to states, which implement the law through SERCs.

- Local LEPCs develop emergency plans and collect information on hazardous materials facilities that is made available to the public.

- Industry must comply with the regulations and provide public and agency disclosure of their hazardous materials.

- Citizens can use SARA Title III to gain knowledge of community risks from hazardous materials.

SARA describes the federal role in reducing public risk from exposure to hazardous materials as providing technical guidance, developing standards and procedures, providing states access to data about chemical releases, and providing training. The **Federal Emergency Management Agency (FEMA)** has the lead role in coordinating civil emergency response planning and disaster response. For *Title III* reporting and enforcement activities, the EPA is the key federal agency; it maintains the National Toxic Chemical Inventory, publishes regulations concerning hazardous materials, reports on the status of various emergency systems, conducts training, and assists citizens with hazardous materials site identification and investigation.

As discussed in the previous section, SARA imposes major responsibilities on the states as well. Under *Title III*, each state governor must appoint a State Emergency Response Commission (SERC). These commissions are intended to ensure that an emergency planning and implementation structure is developed at the local level and to review plans developed by communities. In addition, SERCs provide training and technical assistance to local communities. In the case of an emergency which is too expensive for a local community to handle, the state may contribute resources. In general, the burden of funding for training and information management for *Title III* record keeping falls at the state and local level.

States may, of course, elect to exceed federal requirements for hazardous materials management. For example, over 30 states have enacted Right-to-Know laws similar to the federal law, some of which cover more chemicals and more potentially hazardous facilities with lower thresholds. States have added specific requirements to increase their protection from particular hazards. For example, California's Right-to-Know law has much lower thresholds for reporting than the federal version. California businesses that store or handle hazardous materials in the amounts of 55 gallons for a liquid, 500 pounds of a solid, or 200 cubic feet of a compressed gas, must submit a list or inventory of these chemicals to local authorities. States have broad authority to control how hazardous materials are stored, used, transported, and disposed of within their borders. For instance, states establish zoning control policies that determine where chemical plants may be located, and control siting for hazardous waste facilities and landfills. Pennsylvania and Connecticut currently have laws that deny permits to companies found in violation of environmental protection laws. Regulating transportation of hazardous materials within state borders is also a state responsibility.

Local communities play a key role under *Title III* to inform and protect citizens from hazardous materials. Local communities, represented by LEPCs, are responsible for developing an emergency plan for disasters involving hazardous substances. This includes identifying the resources that would be available in an emergency (such as trained personnel and specialized equipment) and ensuring coordination among responding groups. The LEPC also collects and stores information from hazardous materials facilities and makes that information available to the public.

Local officials have the lead role in responding to hazardous materials emergencies; usually, management of incidents is the specific responsibility of the local fire department. Communities also regulate the disposal of hazardous waste and inspect hazardous materials storage areas for violations of local codes. Many communities regulate hazardous materials traffic through specific zoning requirements.

Under *Title III*, facilities that use hazardous materials are responsible for complying with packaging, labeling, storage, transportation, and workplace safety regulations. Additionally, industry is required to furnish information about the quantities and health effects of materials used at their plants, and promptly notify local and state officials whenever a significant release of hazardous materials occurs. Small businesses and farmers are also included under the *Title III* umbrella if they use extremely hazardous substances, such as chlorine or anhydrous ammonia, in reportable quantities.

The role of industry in *Title III* is to inform the public and the proper agencies of their hazardous materials and emergency planning. The **Chemical Manufacturers Association (CMA)** has set up a voluntary, industry-wide **Community Awareness and Emergency Response** program **(CAER)**. This program encourages plant managers to listen to community concerns, participate in planning, and explain their plant's operations and policies. By working with the community to ensure safe handling, storage, transportation, and disposal of dangerous chemicals, industry can protect itself as well as the public from the high costs of chemical accidents.

Citizens can and should utilize the provisions of SARA *Title III* to gain an understanding of community risks from hazardous materials. Many federal and state agencies maintain hotlines for citizen inquiries and reports of violations. *Title III* has specific provisions that enable citizens to bring legal actions against facilities or industries that do not comply with its provisions. Litigation is a slow and costly process, and should be used only after discussions between the polluter and the enforcing agency have proven fruitless. However, lawsuits can force a government agency to act if it is shown to be:

- Violating normal agency procedures

- Violating a substantive statute or regulation

- Abusing its discretionary authority (that is, making a decision based on inadequate information or inappropriate standards)

- Violating legally required decision-making procedures

- Violating environmental impact review requirements

Under both federal and state environmental laws, citizens have the right to file a suit for an injunction (halt) to pollution if it can be shown that the defendant is in violation of the state law, or (in some states) if he or she is creating an imminent danger. However, it is only in extreme cases, when the potential damage is clear and irreparable, that a judge is likely to take short-term action before the full-scale legal process has come to its conclusion.

## ■ Checking Your Understanding (8-2)

1. List two of the main provisions of CERCLA.

2. Explain how SARA changed CERCLA.

3. Contrast the terms hazardous substance, hazardous material, and hazardous waste.

4. Explain how the National Priorities List is used.

5. Describe the purpose of the National Contingency Plan.

6. Define and give a hypothetical example of joint and several liability.

7. Explain how the innocent landowner provision could help a purchaser of property.

8. Contrast potentially responsible parties with guilty parties.

9. List and describe three basic steps in getting a site cleaned up under CERCLA.

10. Determine the shortest amount of time a complete cleanup may take under the NCP given the estimates for each of the steps in this chapter.

11. Contrast the duties and composition of SERCs and LEPCs.

12. List three types of information that can be obtained from a MSDS.

13. Describe how SARA Title III benefits citizens.

# 8-3 Hazardous Materials Transportation

## ■ History and Overview

### Concepts

■ Large quantities of hazardous materials are transported annually by land, sea, and air.

■ Regulation of hazardous materials transportation has a longer history than most environmental regulations.

■ Most accidents involving hazardous materials in transportation are caused by human error.

■ DOT, through RSPA, is responsible for most aspects of hazardous materials transportation.

Over one billion tons of hazardous materials are transported by land, sea, and air annually in the United States, not including pipeline transportation which would more than double the annual total. Even though most of these materials reach their destinations safely, there are times when an accident occurs and a hazardous material is released. Accidental releases from transportation mishaps can have grave consequences. For this reason, the occasional serious accident is frightening to the public and dramatically raises general uneasiness about hazardous materials. Transportation of hazardous materials is in fact one of the areas of hazardous materials management that arouses the highest level of public concern, and has for some time.

The first federal law to regulate the transportation of hazardous materials was passed in 1866 to address the movement of explosives and flammable materials such as nitroglycerin and certain oils. In 1871 a statute established criminal penalties for transporting specific hazardous materials on passenger vessels. At the time of the Civil War, rail shipments of explosives were regulated. These early statutes obligated the shipper to identify the hazards of dangerous cargo, use adequate packaging, and provide clear warnings of the shipments hazards. Regulations really haven't changed much from these early goals.

The Hazardous Materials Transportation Act, passed in 1975, is the primary federal law governing transportation of materials that are determined to present unreasonable risk if transported incorrectly. It is generally acknowledged that the law was passed as a result of the crash of a 707 cargo jet carrying several tons of hazardous materials in 1973. During the accident investigation, it was determined that there was lack of compliance with existing laws due to lack of communication between regulatory authorities, the complexity of the regulations, industry's lack of knowledge of the regulations, and inadequate government surveillance. The law was largely unchanged until 1990, when an amendment called the Hazardous Materials Transportation and Uniform Safety Act of 1990 revamped it almost entirely. These major changes were determined to be needed because of several deficiencies in the methods of regulation, as well as coordination and uniformity with states and other countries.

Reading any major newspaper will reveal stories of hair-raising transportation incidents. Houston residents, for instance, were shocked when a speeding truck carrying a tank of highly flammable methyl methacrylate hit the rail of an exit ramp. The driver was killed. As a result of the accident, the tank broke open, its contents ignited, and the resulting inferno destroyed part of the freeway and dropped burning debris on the street below. Fortunately, no one else was hurt, and the Houston Fire Department had a hazardous materials response team that was able to control the accident. Denver residents were similarly stunned when a truckload of Navy torpedoes overturned one Sunday morning on the exit loop of a city freeway. Again, fortunately, no one was injured, but hours passed before federal personnel with experience handling military weapons arrived. Worried state and local emergency response officials did not know whether the scattered weapons needed to be defused before cleanup could begin.

These are but two examples of the newsworthy and spectacular incidents that have a tendency to etch themselves upon the public consciousness because of their enormous potential for health and environmental damage. It is because of situations just like these that the public has demanded something be done to prevent these types of accidents from happening. Preparation, through planning and training of emergency response personnel, is important to handle emergency situations safely.

The types of vehicles carrying hazardous materials on the highway range from tank trucks to conventional tractor-trailers and flatbeds that carry large portable tank containers or non-bulk packages, such as cylinders, drums, and other small containers. Rail shipments are usually bulk commodities such as liquid or gaseous chemicals and fuels, which are carried in tank cars. Most hazardous materials transported by water are moved in bulk containers, such as tank ships or barges, while air shipments are typically small packages, often of high-value or time-critical material.

A majority of the accidents and injuries involving hazardous materials in transportation have been determined to have been caused by human error. In other words, reasons such as inadequately trained personnel, poor coordination and communication, and faulty judgment have been at the root of more problems than technological shortcomings such as bad brakes and poorly designed packaging. Knowing this, Congress revised HMTA to make it more comprehensive and include training requirements. The current law is estimated to address over 30,000 hazardous materials that may be transported in commerce.

The Federal Government has four roles related to hazardous materials transportation: regulation, enforcement, emergency response, and data collection and analysis. The DOT is the lead agency for establishing and enforcing regulations regarding safe transportation of hazardous materials. The DOT Research and Special Programs Administration (RSPA) has authority to issue regulations on many aspects of hazardous materials containers, except for bulk marine shipments, which are regulated by the U.S. Coast Guard. RSPA shares inspection and enforcement activities with the administrations that have authority over the vehicles or vessels for specific modes of transportation: the Federal Highway Administration, the Federal Railroad Administration (FRA), the Federal Aviation Administration, the National Highway Traffic Safety Administration, and the Coast Guard. RSPA is responsible for identification of hazardous materials as well as:

- regulation of hazardous materials containers, handling, and shipments;
- development of container standards and testing procedures;
- inspection and enforcement for shippers using more than one mode of transportation and container manufacturers; and
- data collection.

There is quite a bit of overlap between several federal agencies when it comes to hazardous materials regulation. Some other federal agencies that regulate aspects of hazardous materials transportation include: the Nuclear Regulatory Commission (NRC), the EPA, and the Occupational Safety and Health Administration (OSHA). The NRC has jurisdiction over high-level radioactive substances in the civil sector, the EPA has responsibilities for chemicals and hazardous non-nuclear wastes, and OSHA is concerned with worker safety. These agencies also undertake training activities and provide technical support for state and local governments.

Three additional agencies have non-regulatory functions related to the transportation of hazardous materials. The U.S. Department of Energy (DOE) is responsible for high-level nuclear waste movement, storage, and disposal under the Nuclear Waste Policy Act of 1982. The U.S. Department of Defense (DOD) transports many hazardous materials for military purposes. The Federal Emergency Management Agency (FEMA) is responsible for coordinating federal assistance, planning, and training activities for all types of emergency response with state and local governments.

The data collection function is similarly spread among federal agencies, most of which record accidents and spills and monitor compliance. RSPA is the principal agency collecting data on releases of hazardous materials during transportation, but every other federal entity keeps records pertaining to its area of interest. One complaint Congress has had with this system is that there is a lack of inter-agency coordination for record keeping on accidents and releases of hazardous materials. The division of responsibilities among multiple federal agencies and the DOT entities developed on the premise that hazardous wastes, radioactive materials, emergency response training, and the like should be handled by those with the appropriate expertise. This, however, has created a situation in which issues that require the coordinated effort of more than one agency to resolve may take years to do so. In an effort to improve communication between agencies, especially in emergencies, the federal government has a coordinating group called the National Response Team. The responsibilities of this team include emergency preparedness and coordination of response activities.

One improvement that the most recent amendments to the HMTA have made is to revise U.S. regulations to conform more closely with international codes. One example is the requirement to use United Nations identification numbers, which will be explained in the next section of the chapter.

Regulations contained in Title 49 of the Code of Federal Regulations do not apply to shipments that do not move between states. They only apply to interstate shipments. Many states have adopted 49 CFR wholly or in part, resulting in great variation among state regulations. It should be fairly obvious that ensuring the safe transportation of hazardous materials is a complex activity. If accidentally released, hazardous materials pose risks to human safety, property, and the environment. The next section will cover key components of the regulatory program designed to accomplish this.

## ■ Transportation Requirements

### Concepts

- DOT has different definitions for hazardous materials, hazardous substances, and marine pollutants than other agencies.

- Transporters must use authorized packaging and clearly communicate onboard hazards.

■ The Hazardous Materials Table provides necessary information on shipping name, hazard class, labels, and packaging.

■ Hazardous materials must be shipped accompanied by shipping papers; hazardous wastes, with the Uniform Hazardous Waste Manifest.

The current regulatory system governing the transportation of hazardous materials developed over the past century with industry involvement. Regulations are extensive and cover items ranging from detailed engineering specifications for containers, to steps that communicate the hazards of onboard materials such as placards and labeling, to handling and operating requirements for each mode of transport. Shippers and carriers must also comply with regulations pertaining to specific types of hazardous materials, worker safety, and environmental protection issued by other federal agencies.

Regulations issued by RSPA pertain to all four modes of transportation and consist of two basic types of requirements: use of authorized packaging and clear communication of the hazards. Regulations cover the activities of both shippers and carriers of hazardous materials.

A hazardous material is defined by DOT as a substance or material, including a hazardous substance, hazardous waste, marine pollutant, or elevated temperature material which has been determined by the Secretary of Transportation to be capable of posing an unreasonable risk to health, safety, and property when transported in commerce and which has been so designated. DOT also has their own definition for a hazardous substance (Figure 8-7). It is defined as a material, including mixtures and solutions, that is listed in Appendix A of the Hazardous Materials Table, is in a quantity in one

| Class 1 | Any substance, article, or device designed to function by explosion (extremely rapid release of gas and heat). |
| Class 2 | Flammable gas: ignitable at low concentrations (<13%). |
|  | Compressed gas: shipped at >41 psi. |
|  | Poisonous gas: toxic to humans or hazardous to health (or $LC_{50}$ of not more than 5000 ml/m$^3$ for laboratory animals). (Toxic in low concentrations) |
| Class 3 | Flammable liquid: flash point <141°F. |
|  | Combustible liquid: flash point > 141°F. (100°–200°F for domestic shipments) |
| Class 4 | Explosives shipped with sufficient wetting agent to suppress explosive properties. |
|  | Substance that can ignite if in contact with air < 5 minutes. |
|  | Substance that gives off flammable or toxic vapors or is spontaneously flammable upon contact with water. |
| Class 5 | A material that can cause or enhance the combustion of other materials (usually by giving up oxygen). |
| Class 6 | Toxic to humans, hazardous to human health or presumed toxic to humans based upon tests on laboratory animals. |
| Class 7 | Substance with specific activity >0.002 microcuries per gram. |
| Class 8 | Substance that causes a visible destruction or irreversible alteration in human skin tissue, or a liquid that has a severe corrosion rate on steel or aluminum. |
| Class 9 | Material with anesthetic, noxious, or similar property that could cause extreme annoyance or discomfort to flight crew and prevent performance of assigned duties. **Does not meet the definition of any other class.** |

**Figure 8-7:** DOT definitions for the nine Hazard Classes.

package which equals or exceeds the reportable quantity (RQ) listed in Appendix A to the Hazardous Materials Table, and is in concentrations that meet or exceed those listed on a hazardous substance reportable quantity (RQ) table. A **marine pollutant** is a material listed in Appendix B of the Hazardous Materials Table with a concentration which equals or exceeds 10 percent by weight, or 1 percent by weight for materials identified as severe marine pollutants.

Shippers begin the regulatory process by identifying the hazards of their cargo. Hazardous materials subject to RSPA regulations are listed in the Hazardous Materials Table in CFR 49 Part 172.101. This table not only lists specific chemicals but also categories and end-uses of chemicals, which allows it to encompass a much larger universe of chemicals than it would at first seem to encompass. The Hazardous Materials Table (Figure 8-8) indicates the hazard class to which each material belongs and references the packaging, labeling, and special requirements applicable to rail, air, and water transportation that must be met by shippers and carriers. Each column of the Hazardous Materials Table provides important information that is necessary in order to properly and legally ship hazardous materials.

Column 1 has codes for special situations. For instance, an A means that the material is subject to the regulations only if it will be transported by air. Column 2 lists the proper shipping name of hazardous material. No matter what you may call the chemical you will be shipping, the name on the shipping papers must match the proper shipping name shown in Roman type on the Hazardous Materials Table. Column 3 shows the numerical hazard class of the material that must be shown on the shipping papers.

| Descriptions and shipping names | Hazard | ID No. | PG | Label(s) |
|---|---|---|---|---|
| Corrosive liquids, which in contact with water emit flammable gases, n.o.s. | 8 | UN3094 | I | CORROSIVE DANGEROUS WHEN WET |
| | | | II | CORROSIVE DANGEROUS WHEN WET |
| Corrosive solids, flammable, n.o.s. | 8 | UN2921 | I | CORROSIVE FLAMMABLE SOLID |
| | | | II | CORROSIVE FLAMMABLE SOLID |
| Corrosive solids, n.o.s. | 8 | UN1759 | I | CORROSIVE |
| | | | II | CORROSIVE |
| | | | III | CORROSIVE |
| Corrosive solids, oxidizing, n.o.s. | 8 | UN3084 | I | CORROSIVE, OXIDIZER |
| | | | II | CORROSIVE, OXIDIZER |
| Corrosive solids, poison, n.o.s. | 8 | UN2923 | I | CORROSIVE, POISON |
| | | | II | CORROSIVE, POISON |
| Corrosive solids, self heating, n.o.s. | 8 | UN3095 | I | CORROSIVE, SPONTANEOUSLY COMBUSTIBLE |
| | | | II | CORROSIVE, SPONTANEOUSLY COMBUSTIBLE |
| Corrosive solids, which in contact with water emit flammable gases, n.o.s. | 8 | UN3095 | I | CORROSIVE DANGEROUS WHEN WET |
| | | | II | CORROSIVE DANGEROUS WHEN WET |

**Figure 8-8:** Hazardous Materials Table from DOT 49 CFR.

The regulations also require shippers and carriers to communicate the hazards of their cargo by providing **shipping papers**, markings, labels, and placards. These requirements are important because they are intended to furnish essential information about the cargo to emergency response personnel if accidents occur.

Shipments of hazardous materials must be accompanied by shipping papers that describe the hazardous material and contain a certification by the shipper that the material is offered for transport in accordance with applicable DOT regulations. DOT does not specify the use of a particular document. The information can be provided on a bill of lading, waybill, or similar document. The exception is a hazardous waste shipment which must be accompanied by a specific document called the **Uniform Hazardous Waste Manifest**. A manifest lists EPA identification numbers of the shipper, carrier, and the designated treatment, storage, and disposal facility, in addition to the standard information required by DOT. More information on manifests will be provided in Chapter 10.

Instructions for describing hazardous materials are provided in the regulations. These descriptions must include the quantity of the material, its shipping name and hazard class (taken from Hazardous Materials Table), the United Nations (UN) hazard identification number, and the packaging type and group. UN identification numbers, which also must be marked on packages and bulk containers, correspond to emergency response information provided in a guidebook that is published and distributed nationally by the DOT. The DOT Emergency Response Guidebook contains information on potential health, fire, or explosion hazards and basic emergency action instructions. Isolation and evacuation information is also provided for a limited number of highly hazardous substances.

In those instances where a specific technical name of a hazardous material is not listed in the Hazardous Material Table, a proper shipping name must be selected from general description and n.o.s. (not otherwise specified) entries corresponding to the hazard class of the material.

The DOT has established marking requirements for packages, freight containers, and transport vehicles. Shippers are required to mark all packages with a capacity of 110 gallons or less with the proper shipping name of the hazardous material, including its UN identification number. This is done so that the contents of a package can be identified if it is separated from its shipping papers. Requirements for portable tanks, highway cargo tanks, and rail tank cars specify that the UN identification number be displayed on a 12" placard or an orange rectangle panel.

Labels are symbolic representations of the hazards associated with a particular material. Refer to Figure 8-4 earlier in the chapter for depictions of labels and placards. Labels are required on most packages and must be printed or affixed near the marked shipping name. The Hazardous Materials Table indicates which materials require labels. Special labels, such as MAGNETIZED MATERIALS or CARGO AIRCRAFT ONLY, are required under specific circumstances. In addition, packages containing materials that meet more than one hazard class may require multiple labels.

Placards are symbols that are placed on the ends and sides of motor vehicles, railcars, and freight containers indicating the haz-

State of California—Health and Welfare Agency
Form Approved OMB No. 2050—0039 (Expires 9-30-91)
Please print or type. *(Form designed for use on elite (12-pitch typewriter).*

**See Instructions on Back of Page 6
and Front of Page 7**

Department of Health Services
Toxic Substances Control Division
Sacramento, California

**UNIFORM HAZARDOUS WASTE MANIFEST**

IN CASE OF AN EMERGENCY OR SPILL, CALL THE NATIONAL RESPONSE-CENTER 1-800-424-8802; WITHIN CALIFORNIA CALL 1-800-852-7550

89925845

GENERATOR

TRANSPORTER

FACILITY

1. Generator's US EPA ID No.

Manifest Document No.

2. Page 1 of

Information in the shaded areas is not required by Federal law.

3. Generator's Name and Mailing Address

A. State Manifest Document Number
**89925845**

B. State Generator's ID

4. Generator's Phone (    )

5. Transporter 1 Company Name

6.    US EPA ID Number

C. State Transporter's ID

D. Transporter's Phone

7. Transporter 2 Company Name

8.    US EPA ID Number

E. State Transporter's ID

F. Transporter's Phone

9. Designated Facility Name and Site Address

10.    US EPA ID Number

G. State Facility's ID

H. Facility's Phone

| 11. US DOT Description (Including Proper Shipping Name, Hazard Class, and ID Number) | 12. Containers | | 13. Total Quantity | 14. Unit Wt/Vol | I. Waste No. |
|---|---|---|---|---|---|
| | No. | Type | | | |
| a. | | | | | State |
| | | | | | EPA/Other |
| b. | | | | | State |
| | | | | | EPA/Other |
| c. | | | | | State |
| | | | | | EPA/Other |
| d. | | | | | State |
| | | | | | EPA/Other |

J. Additional Descriptions for Materials Listed Above

K. Handling Codes for Wastes Listed Above
a.        b.
c.        d.

15. Special Handling Instructions and Additional Information

16.

**GENERATOR'S CERTIFICATION:** I hereby declare that the contents of this consignment are fully and accurately described above by proper shipping name and are classified, packed, marked, and labeled, and are in all respects in proper condition for transport by highway according to applicable international and national government regulations.

If I am a large quantity generator, I certify that I have a program in place to reduce the volume and toxicity of waste generated to the degree I have determined to be economically practicable and that I have selected the practicable method of treatment, storage, or disposal currently available to me which minimizes the present and future threat to human health and the environment; OR, if I am a small quantity generator, I have made a good faith effort to minimize my waste generation and select the best waste management method that is available to me and that I can afford.

Printed/Typed Name        Signature        Month  Day  Year

17. Transporter 1 Acknowledgement of Receipt of Materials

Printed/Typed Name        Signature        Month  Day  Year

18. Transporter 2 Acknowledgement of Receipt of Materials

Printed/Typed Name        Signature        Month  Day  Year

19. Discrepancy Indication Space

20. Facility Owner or Operator Certification of receipt of hazardous materials covered by this manifest except as noted in Item 19.

Printed/Typed Name        Signature        Month  Day  Year

DHS 8022 A (1/88)
EPA 8700—22
(Rev. 9-88) Previous editions are obsolete.

Do Not Write Below This Line

White: TSDF SENDS THIS COPY TO DOHS WITHIN 30 DAYS
To: P.O. Box 3000, Sacramento, CA  95812

**Figure 8-9** A Uniform Hazardous Waste Manifest.

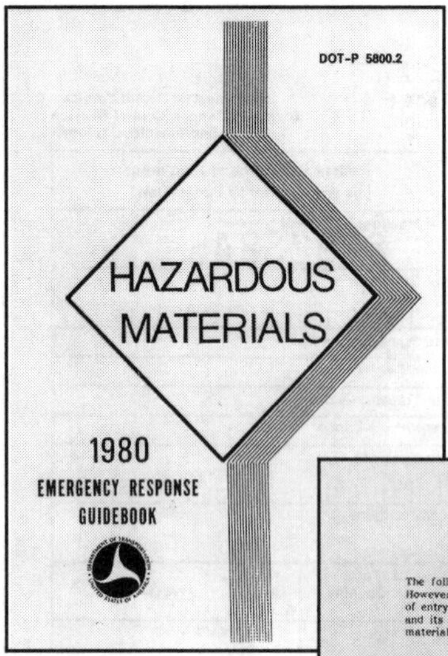

**Figure 8-10:** The DOT Emergency Response Guidebook contains information on potential hazards and emergency

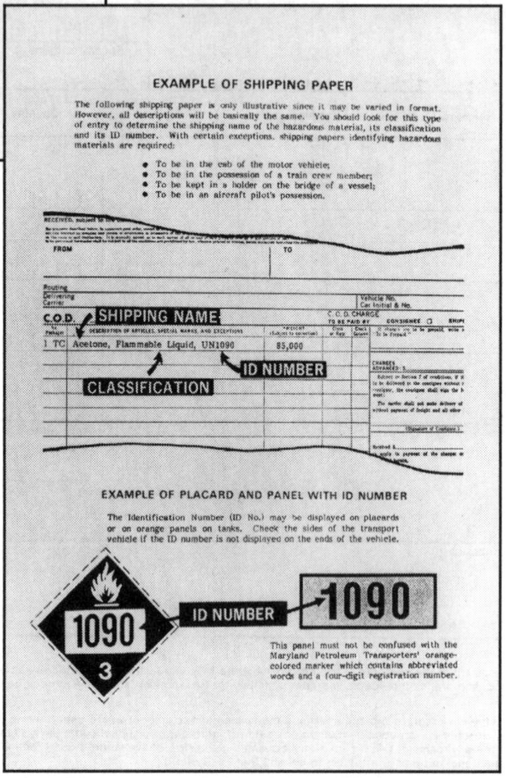

ards of the cargo. Recent amendments to the HMTA require that UN identification numbers be displayed on some placards. Placards are extremely important to emergency response personnel in the event of an accident because they are highly visible.

The DOT has developed tables, which are found in CFR 49, that indicate the placards required for each hazard class. Placarding is the joint responsibility of shippers and carriers. Placards are not required for all shipments of hazardous materials, such as infectious materials; materials classed as ORM-A, B, C, D, or E; or limited quantities of hazardous materials. Moreover, motor vehicles or freight containers transported by highway containing less than 1,000 pounds of certain types of hazardous materials do not have to be placarded.

The integrity of containers or packagings used for shipping is critical for preventing releases. Containers are required by regulation to be adequate to contain their contents during normal transport and must meet specific tests to demonstrate that they can withstand accident conditions without a dangerous release.

Hazardous materials, essential to the business and industrial economy of the United States, are shipped under regulations that reflect the history and different operating characteristics of the various modes of transportation. Industry has developed various forms of packaging that correlate the strength and integrity of the containers to the characteristics and hazards of the materials they are intended to contain. Packaging for hazardous materials during transportation is a major element of DOT's regulatory system. The Department, through RSPA, establishes technical standards for the design and testing of packages and associated transportation equipment for all hazardous materials and small quantities of radioactive materials.

The DOT requires packaging for shipping hazardous materials to be designed and constructed so that under typical transportation conditions:

– There will be no significant release of the hazardous material to the environment;

– The effectiveness of the packaging will not be substantially reduced;

– There will be no mixture of gases or vapors in the package which could, through any spontaneous increase of heat or pressure, or through an explosion, significantly reduce the effectiveness of the packaging.

In addition, packaging materials must be designed to ensure that there will be no significant chemical reaction among any of the materials in the package or between the packaging and its contents. Closures must prevent leakage, and gaskets must be used that will not be significantly deteriorated by the contents.

The EPA manages several programs that affect the transportation of certain hazardous materials. RCRA requires the EPA to establish requirements for transporters of hazardous wastes. The EPA has adopted DOT's regulations for hazard communication, packaging, and reporting discharges and has enacted additional notification, marking, manifest, and cleanup requirements.

## ■ Checking Your Understanding (8-3)

1. Explain how the basic regulatory requirements for hazardous materials transportation are expected to improve the human error problem which is at the root of the majority of transportation accidents.

2. List two agencies other than DOT that have overlapping regulatory responsibilities for hazardous materials transportation.

3. Contrast hazardous material, hazardous substance, and marine pollutant as used by DOT.

4. List four hazardous materials that you feel fit any four of the DOT hazard classes, based on the class definitions.

5. Explain the value of using uniform UN numbers worldwide.

6. Contrast use and placement of shipping papers, markings, labels, and placards.

7. Explain why container testing and specification is an important and stringent part of DOT regulations.

# 8-4 FIFRA and TSCA

## ■ FIFRA

### Concepts

■ FIFRA applies to several subgroups of pesticides.

■ FIFRA requires pesticide producers to register their product with the EPA.

- Registered pesticides are classified as general or restricted use.

- Applicators of restricted pesticides must be certified.

FIFRA, the Federal Insecticide, Fungicide, and Rodenticide Act was enacted in 1947. Originally, the U.S. Department of Agriculture was charged with its implementation and enforcement. The first version of this law was markedly different than the FIFRA we know today. In the early years it only applied to the labeling of pesticides that were involved in interstate commerce. A pesticide is defined in FIFRA as 1) any substance or mixture of substances intended for preventing, destroying, repelling, or mitigating any pest; and 2) any substance or mixture of substances intended for use in a plant regulator, defoliant or desiccant. An example of the type of violation that the law was originally designed to prevent was the defacing or destroying of pesticide labels. It is now much broader in scope, including programs requiring pesticide registration and controlling the production and application of pesticides. These changes are the result of FIFRA's many amendments over the years.

As stated in the definition above, pesticides are designed to kill pests. Pests are defined as insects, rodents, nematodes, fungi, weeds or any form of terrestrial or aquatic plant and animal life; viruses, bacteria or other microorganisms except those on or in living humans or other living animals which are injurious to health or the environment. Subgroups of pesticides identify which of these specific pests are the target: **insecticides** are designed to control insects; **rodenticides** kill rodents; **nematocides** kill nematodes; **fungicides** kill fungi; **herbicides** kill weeds; and **bactericides** kill bacteria.

Some of FIFRA's amendments over the years involved major revisions of the preceding version, and others resulted in only minor changes. A major revision in 1972 called the Federal Environmental Pesticide Control Act of 1972 (Public Law 92-516) essentially created the act as we know it today. This amendment provided for a strict registration program, a set of requirements for the use and management of pesticides, and an enforcement program.

FIFRA is currently administered by the Environmental Protection Agency, which received this program when the agency was formed in 1970. The program now has a more expansive regulatory structure in that it includes pesticides used throughout the United States, not just those involved in interstate commerce. It is important to note that FIFRA does not develop or enforce allowable pesticide residues on food. This public protection measure is covered by a pesticide amendment to the Federal Food, Drug, and Cosmetic Act (FFDCA).

The EPA has primary enforcement responsibility for the provisions of FIFRA. EPA can, however, enter into cooperative agreements with states to enforce pesticide use restrictions. A state will have primary enforcement responsibilities if it demonstrates to the EPA that it has adopted laws and regulations for pesticide use that are at least as stringent as FIFRA; has adequate procedures in place for enforcement of its pesticide laws; and will keep records and make reports showing compliance with the laws and regulations. States cannot, however, impose any requirements for labeling or packaging that are in addition to or different from those required by FIFRA. If the EPA determines that a state program is inadequate, it can take over enforcement responsibilities.

FIFRA requires that pesticides to be used in the United States must be registered with the EPA before they can be distributed, sold, offered for sale, shipped, received, or in any way offered for use. It is the responsibility of the producer to obtain this registration. The producer is individual, partnership, or corporation, who manufactures, prepares, compounds, propagates, or processes any pesticide or active ingredient used in producing a pesticide. With very few exceptions, unregistered pesticides may not be distributed, sold, shipped, or received by anyone.

The application sent to the EPA or to the authorized state to receive a pesticide registration must include specific information. The name and address of the applicant and the name of the pesticide must be listed. A copy of the intended label, which must include directions for use, must also be sent with the application. Complete and valid test data must be submitted that demonstrates to the EPA's satisfaction that the pesticide will be effective against the pests claimed and listed on the label; will not cause unreasonable adverse effects to people, animals, crops, or the environment; and will not exceed risk criteria associated with human dietary and animal feed exposure limits. Not every company must conduct research and come up with this data. In some cases the EPA will allow data they have already accepted from other producers to be used by a new producer.

The EPA will only register a pesticide if it determines that, if used in accordance with instructions and commonly recognized practices, it will meet the claims it makes, it is labeled in accordance with the law, and it will perform its intended function without unreasonable adverse effects on the environment. Registrations are only good for five years but can be renewed if proper notice is given to the EPA. A state can only register additional uses of federally-registered pesticides to meet special needs if the EPA has not previously denied, disapproved, or canceled such uses.

Once a pesticide is registered and allowed to enter commerce, additional specific compliance requirements of law must be followed. For instance, pesticide containers must meet strict labeling requirements including listing the EPA product registration number, the ingredients, the name and address of the registrant or producer, statements warning users of potential hazards, and precise directions on the pesticide's use. Pesticide producers are also required to register each establishment in which pesticides are produced. After reviewing establishment applications, the EPA registers the establishment and issues it an establishment number. All pesticide products must have the establishment registration number on the label or the container. Upon request, the producer must inform the EPA of the name and address of any recipient of pesticides produced at a specific establishment. Annual reports are required that detail the types and amounts of pesticides produced, sold, or distributed during the year. Any other records maintained by the producer must be accessible to the EPA upon request.

In order to enforce the provisions of FIFRA, EPA employees are authorized to enter, at reasonable times, establishments where pesticides or devices are held for distribution or sale, or are packaged,

**Figure 8-11:** Restricted pesticides are required to have a proper label.

labeled, or released for shipment. EPA employees are authorized to conduct inspections and obtain samples as long as they present appropriate credentials, explain the reason for the inspection, provide a receipt for any samples taken, and promptly furnish a copy of the results of any sample analysis.

FIFRA classifies pesticides as either general or restricted. General use pesticides can be found in any building supply store and easily purchased by anyone. These are pesticides that the EPA has determined will not generally cause unreasonable adverse effects to the environment. Restricted pesticides, on the other hand, are those that have been determined to cause unreasonable adverse environmental and health effects unless additional restrictions are imposed on them. Restricted use products are either highly toxic or require specialized knowledge in order to apply them. The statement "Restricted Use Pesticide" must be prominently noted on the label above the product name on these pesticides, and they can only be used and applied by certified operators.

The certification requirements for applicators are designed to ensure that they are sufficiently competent and knowledgeable in the safe use of pesticides. They must be able to handle restricted use pesticides in a manner that does not endanger themselves, the public, or the environment. Although the EPA oversees this program, the states have the primary responsibility for certification if their state plans for implementation of FIFRA have been approved by the EPA.

There are two types of certified applicators – private and commercial. A private applicator utilizes restricted use pesticides for agricultural purposes on his or her own property. A commercial applicator is someone who uses or supervises the use of restricted use pesticides for any purpose or on any property other than what is defined as private application. Commercial applicators can only use pesticides in categories for which they are certified. There are a number of categories for commercial applicators: agricultural pest control; forest pest control; ornamental pest control; seed treatment; aquatic pest control; right-of-way pest control; industrial, institutional, structural, and health-related pest control; public health pest control; regulatory pest control; and demonstration and research pest control.

An interesting provision of FIFRA is that the EPA must compensate damages, loss, or injury suffered by registrants or persons owning any quantity of pesticide whose registration has been canceled. The amount of payment is determined by the cost of the pesticide owned immediately before notice. Disposal of excess, off-specification, or canceled pesticides and containers with pesticide residues can be a major problem for pesticide producers, distributors, and users. Chapter 10 contains more information on the disposal of hazardous wastes, including pesticides.

Violations of the provisions of FIFRA carry both civil and criminal penalties. Civil penalties carry fines of up to $5,000 per offense, and criminal penalties can involve fines of up to $25,000 and imprisonment for up to one year.

## ■ TSCA

### Concepts

■ TSCA controls chemical substances entering commerce.

■ There are no provisions in TSCA for state authorization.

■ PCBs are notable chemical substances regulated by TSCA.

TSCA (pronounced Tosca), the Toxic Substances Control Act, was enacted in 1976 and is administered by the EPA. The purpose of TSCA is to control the risks of chemical substances that enter commerce. A chemical substance is defined as any organic or inorganic substance of a particular molecular identity, including: any combination of such substances occurring in whole or part as a result of a chemical reaction or occurring in nature, and any element or uncombined radical. The definition goes on to further exclude any mixture; any pesticide as defined in FIFRA, as long as it is used as a pesticide; tobacco or tobacco products; certain nuclear materials; firearms; and foods, food additives, drugs, cosmetics when covered under the provisions of the Federal Food, Drug, and Cosmetic Act. Essentially, TSCA prohibits or controls the manufacturing, importing, processing, use, and disposal of any chemical substance that presents an unreasonable risk of injury to human health or the environment.

TSCA differs from most other federal regulations because it does not contain provisions for state authorization. Because of this, the EPA has the sole responsibility for its implementation and enforcement. States can assist the EPA if there is mutual agreement between the agencies to do this. The one exception is related to **PCBs**, or **polychlorinated biphenyls.** PCBs are a group of organic compounds containing chlorine that were once widely used as liquid coolants and insulators in industrial equipment, especially in power transformers. Unfortunately, even though they did a superb job as insulators, they were later found to be dangerous environmental pollutants. Their manufacture was banned after 1978 in one of the first major bans of TSCA. Many states have implemented stricter regulations than those in TSCA for the regulation of PCBs.

Another difference between TSCA and RCRA, for instance, is that TSCA regulates individual chemical substances. RCRA, on the other hand, provides for the regulation of waste streams that may contain several chemicals. A controversial aspect of TSCA regulation involves the requirement that economic considerations be taken into account in its decision-making. TSCA actually requires the EPA to conduct economic cost/benefit analyses when determining unreasonable risk. RCRA has no parallel to this.

TSCA's main provisions involve toxicity testing, reporting, and listing of chemical substances. Existing as well as new chemical substances are covered by these standards. For instance, if the EPA can justify the need for more toxicological information on an existing chemical substance, it can require the chemical manufacturer, importer, or processor to test it and submit the data. In fact, the act actually calls for the formation of an Interagency Testing Committee whose responsibility includes establishing and updating a priority list of at least 50 substances recommended for testing. This committee is comprised of representatives from several federal agencies. The act also has a provision to allow citizens to petition the EPA for the regulation of any chemical substance.

TSCA allows the EPA to use a tool called the Pre-Manufacture Notification (PMN) to control the manufacture, processing, and use in commerce of new and existing chemical substances. Manufacturers or importers must submit a PMN to the EPA at least 90 days before manufacturing or importing a new chemical substance. Certain existing chemical substances must also submit a PMN prior to any significant new use.

Existing chemical substances are also required to be labeled. Some are even banned from being manufactured when new evidence shows they present unreasonable risks. The EPA is authorized by TSCA to take any action on any chemical substance if there is an imminent hazard.

The reporting aspects of TSCA are extensive. They include requirements for manufacturers, importers, and processors to report information on manufacturing, importing, processing, use, disposal, worker exposure, and production volumes for certain chemical substances. TSCA contains a list of all chemical substances that meet the definition of such that are in commerce in the United States. A substance is considered to be existing if it is on this list. If a substance is not on the list, it is considered to be new.

Parties covered under TSCA must retain records pertaining to adverse reactions alleged by workers or citizens to the chemical substances. They must also report any unpublished health and safety study data on specified chemicals. There is an overall umbrella clause in the act that states that any substantial risks to human health or the environment must be reported for any chemical substance.

As previously noted, one of the major groups of chemical substances regulated under TSCA are PCBs. The next section will cover these very important provisions.

## ■ PCBs

### Concepts

- PCBs are the only chemical substances regulated by name in TSCA and are not regulated at all by RCRA.

- PCBs belong to a family called chlorinated hydrocarbons.

- The primary use of PCBs has been in electrical transformers where the same chemical properties that make them so useful in industrial applications result in high environmental hazards.

- TSCA currently prohibits the manufacture of PCBs.

The term PCBs is short for polychlorinated biphenyls. PCBs belong to a broad family of organic chemicals known as chlorinated hydrocarbons. Although PCBs may be produced naturally in the environment, almost all PCBs in existence today have been synthetically manufactured.

PCBs are interesting in that they are conspicuously missing from the list of hazardous wastes in RCRA and are regulated by TSCA. PCBs, in fact, are the only chemical substances that are regulated specifically by name in TSCA. There have been a number of attempts to transfer the regulation of PCBs to RCRA but they have all been unsuccessful.

PCBs are designated as a hazardous substance under CERCLA due to the Clean Water Act. This means that any person identified as a responsible party in a release or threatened release of PCBs is liable for any and all costs incurred for cleanup. Under *Title III* of SARA, a spill of one pound of PCBs is a reportable quantity requiring immediate notification of the National Response Center.

Monsanto Corporation was the principal manufacturer of PCBs in the United States. It began production of PCBs in 1929; in 1977 production was voluntarily terminated because of the widespread environmental concerns about PCBs. The trade name of PCBs sold by Monsanto Corporation was Askarel. Companies who used PCBs in the manufacture of transformers and capacitors, and for other uses, often used other trade names such as Aroclor, Therminol, Pyroclor, Asbestol, No-Flamol, Saf-T-Kuhl, and Clorinol.

PCBs have a heavy liquid, oil-like consistency and weigh 10-13 pounds per gallon. The properties that made them commercially attractive include a high degree of chemical stability and resistance to degradation, low solubility in water and high solubility in fats/oils, low vapor pressure, low flammability, high heat capacity, and low electrical conductivity. Although these properties are ideal for industrial uses, they are also thought to be the cause of negative environmental effects. Since they easily dissolve fats, PCBs can bioaccumulate in the food chain. Low concentrations have been shown to be highly toxic to fish.

There are 209 possible PCB compounds depending on the arrangement of the chlorine on the molecule. The severity of their toxicological effects is dependent on the number and location of chlorine atoms.

The primary use of PCBs has been in closed or semi-closed systems in electrical transformers, capacitors, heat transfer systems, and hydraulic systems. PCBs have also been used in paints, adhesives, caulking compounds, plasticizers, inks, lubricants, carbonless copy paper, sealants, coatings, and dust control agents. The majority of PCB equipment marketed between 1929 and 1977 is still thought to be in service.

PCBs are harmful because they are persistent, and once released into the environment they do not break apart into new chemical arrangements; instead they bioaccumulate in organisms throughout the environment. In addition, PCBs biomagnify in the food chain; that is, they accumulate in the tissues of living organisms and as they move up the food chain towards humans their concentration increases. These facts are significant because PCBs have been shown to cause chronic toxic effects in many species even when exposed to very low concentrations.

There are well-documented tests which show PCBs cause, among other things, reproductive failure, gastric disorders, skin lesions, and tumors in laboratory animals. Studies of workers exposed to PCBs have shown a number of symptoms and adverse effects including, but not limited to, chloracne and other skin disorders, digestive disturbances, jaundice, impotence, throat and respiratory irritations, and severe headaches. In Times Beach, Missouri, PCBs were inadvertently used as a road cover. Research from this disaster indicates that the chemical is toxic to a wide variety of animals, especially fish, and may also affect animal reproduction. As the result of a PCB leak in Japan in 1968 that contaminated rice oil, 1,000 people developed a skin disease, and babies showed signs of poisoning.

TSCA requires EPA to establish rules to: (1) govern the disposal and marking of PCBs; and (2) prohibit, with certain exceptions, the manufacture, processing, distribution in commerce, and non-totally enclosed use of PCBs. The Final PCB Ban Rule appeared in the Federal Register on May 31, 1979. The manufacture of PCBs for reasons other than research purposes is now prohibited.

Petitions for exemption from the 1979 bans on processing and distribution in commerce of PCBs had to have been filed by July 1, 1979. In order to practically implement this rule (i.e., exceptions, disposal and marking requirements), EPA had to adopt a PCB concentration cut-off point for regulation. Therefore, the final rule applies to any substance, mixture, or item with 50 ppm or greater PCB; wherever the term PCB or PCBs is used in the rule, it means PCBs at a concentration of 50 ppm or greater, unless otherwise specified.

One notable exception to the 50 ppm rule is waste oil. Waste oil containing any detectable concentration of PCBs is forbidden from being used as a sealant, coating, or dust control agent. To permit the use of waste oil with any PCB contamination to be used in road oiling, pipe coating, or vegetation spraying would cause PCBs to directly enter the environmental compartments, which could introduce them into the food chain.

A distinction is made in the law between manufacturing of PCBs and processing of PCBs. The actual creation of the chemical substance PCB, or a substance contaminated with PCBs (e.g., PCBs as an impurity), is the manufacturing of PCBs. The production of *PCB Articles* and *PCB Equipment* is considered processing of PCBs. Processing PCBs includes such activities as placing manufactured PCBs into capacitors or transformers.

A PCB Article is defined in TSCA as an article whose surface is directly contacted by PCBs. Examples of articles include capacitors, transformers, electric motors, pumps, and pipes. Equipment whose surface is not directly contacted by PCBs, but contains a PCB article, is considered PCB Equipment. Examples include televisions, air conditioners, microwave ovens, electronic equipment, and fluorescent light ballasts and fixtures. PCB Item is a collective term used throughout the rule to refer to PCB Equipment, Articles, Containers, Article Containers.

After the July 2, 1979 ban, PCB Articles could no longer be produced because the production is not totally enclosed. In order to continue PCB Equipment production after that date, an exemption must have been obtained from the EPA. Totally enclosed manner is a term which refers to PCBs contained in a way that does not permit any detectable exposure to PCBs.

## ■ Checking Your Understanding (8-4)

1. Define pesticides and list three subcategories of pesticide.

2. Describe three ways that FIFRA differs from other TSCA.

3. Explain what it means to be registered under FIFRA.

4. Contrast general and restricted use pesticides.

5. Contrast private and commercial certified applicators.

6. Explain how TSCA's use of the term chemical substances differs from other similar terms in this chapter.

7. Describe the problems associated with use of PCBs.

8. Explain how PCBs are controlled by TSCA.

PRESERVING
THE LEGACY

# 9

# Occupational Safety and Health

## ■ Chapter Objectives

After completing this chapter, you will be able to:

1. **List** significant events leading to the passage of OSHA and other safety and health related standards.

2. **Define** the principle terms of occupational safety and health.

3. **Describe** the basic components of the Hazard Communication, Laboratory Safety, and Hazwoper standards.

4. **Define** the rights and responsibilities of an employer and employee, in regards to workplace health and safety.

5. **Outline** the required components of labels and Material Safety Data Sheets.

6. **Describe** common workplace hazards and practices to reduce them.

## ■ Terms and Concepts to Remember

Air-Purifying Respirator
Air-Supplying Respirator
American Council of
  Governmental Industrial
  Hygienists (ACGIH)
Blood Borne Pathogens
Boiling Point
Ceiling Limit (C)
Chemical Abstracts Service
  Number (CAS#)
Compliance Safety Health
  Officer (CSHO)
Fire Triangle
Flammable Range
Flash Point
General Duty Clause
Halogen
Hazard Communication
  Standard (HazCom)
Hazardous Waste Operations
  and Emergency Response
  Standard (Hazwoper)
Immediately Dangerous to Life
  or Health (IDLH)
Incident Command System
  (ICS)

Industrial Hygiene
Industrial Hygienist (IH)
Laboratory Safety Standard
Lower Explosive Limit (LEL)
Medical Monitoring
National Institute of
  Occupational Safety and
  Health (NIOSH)
Octet Theory
Oxygen Deficient Atmospheres
Performance Standards
Permissible Exposure Limit
  (PEL)
Recommended Exposure
  Limits (RELs)
Short Term Exposure Limit
  (STEL)
Specific Gravity
Specific Standards
Threshold Limit Value (TLV)
Time-Weighted Average (TWA)
Upper Explosive Limit (UEL)
Vapor Density
Vapor Pressure

## ■ Chapter Introduction

The last chapter covered the many regulatory requirements for facilities that transport, store, and clean up hazardous materials. These legal requirements are designed primarily for the protection of the environment and public health.

In this chapter, the primary focus is on protecting and maintaining the health and safety of workers. The Occupational Safety and Health Act specifies requirements that employers must follow to maintain a healthy and safe work environment for their workers. This chapter begins with an historical overview that shows how important these laws have been in improving workplace conditions. It also details three far-reaching OSHA standards pertaining to workers who handle hazardous chemicals and hazardous wastes.

# 9-1 History of Workplace Safety

## ■ OSHA History

### Concepts

- Union and public outcry over bad working conditions and resultant workplace disasters played a major role in improving conditions in the workplace.

- A performance-based standard is different from a hazard specific standard.

- The General Duty Clause expands OSHA's enforcement power.

*Even more criminal than the waste of material resources is the waste of human energy. The exploitation of human beings in factories and foundries, in mines and in the railway service, is no less ruthless than the destruction of our forests.*

from *The Industrial History of the United States*
by Katharine Coman, 1920

**Figure 9-1:** Before workplaces were regulated, workers were often unknowingly subjected to hazardous conditions.

Throughout the industrial revolution of the 1800s, working conditions were deplorable. Newly emerging factories, mills, processing and industrial plants, railroad yards, docks, mines and mechanized farms were notoriously hazardous places where workers physical safety and health often were in constant peril. Worker casualties per capita each year was higher in the United States than in any other industrialized country. With few exceptions, workplace safety was of little concern to most business owners. Workers were obligated as a condition of employment to assume the risk of physical injury or damaged health. Workers tolerated hazardous conditions, often not even realizing that they were in danger. Consequently, workplace injuries and fatalities were common. In 1908, for example, 10,000 of every 500,000 workers were either killed or suffered a major injury on the job.

Several major catastrophes, at the cost of many lives, occurred before occupational health and safety laws were drafted and enforced. On January 11, 1860, 600 workers were injured when the badly constructed Pemberton Textile Mill in New England collapsed. While rescue workers were digging out trapped workers, a fire broke out resulting in the death of 200 people. No action was ever taken against its owner. One of the worst tragedies occurred at New York City's Triangle Shirtwaist Company on March 25, 1911, which killed 146 workers, most of them women. A fire caused by faulty electrical wiring broke out in the loft of the building. Although there were fire escapes in the building, the doors were bolted shut to prevent workers from idling. Many of the trapped women jumped out windows, falling 10 stories to their deaths, while terrified onlookers watched, unable to stop the fire or assist the women. No action was taken against the company since they had broken no laws. The incident, however, focused public attention on working conditions and eventually led to health and safety regulatory reform.

Between 1880 and 1920, the American labor force grew from 28 million workers to 42 million workers with the vast majority working in manufacturing and heavy industries. Due to the large supply of labor, all workers, including children, were forced to work 14–18 hours per day to keep their job. These types of working conditions led to the organization of unions like the International Ladies' Garment Workers Union. Unions lobbied to improve working conditions and establish standard pay scales. As a result of the union's effort, the working day was decreased to 10 hours per day and pay scales were increased. Other agencies, such as the Workers Health Bureau, incorporated the research of health professionals into the trade union movement. As a result, significant changes in workplace safety took place. By 1920, all but six states had established some form of workers' compensation law, which paid employees for some part of the cost of injuries and in some cases of occupational disease. By 1930, all states had such laws.

The Social Security Act (1935) established a cooperative federal-state system of unemployment compensation and retirement insurance. According to the law, workers who paid social security taxes out of their wages would receive retirement benefits at age 65. In 1938 Congress passed the Fair Labor Standards Act (also called the Wages and Hours Act) to establish a minimum standard of living for all workers engaged directly or indirectly in interstate commerce.

The first minimum wage was 25 cents per hour for a 44-hour maximum work week. After 1942, the Wage and Hours Act was administered jointly with the Walsh-Healy Public Contracts Act of 1936, which requires that employers holding federal contracts in excess of $10,000 pay the prevailing minimum wage and maintain a minimum standard of safe and healthful conditions. The act also provided for an 8-hour workday and 40-hour week, with overtime pay at 1.5 times the base pay.

In the 1960s a movement occurred in the field of occupational health that paralleled the environmental movement. Strikes by asbestos workers focused on health and workplace hazards and similar issues emerged concerning the hazards of coal mining, especially in the increase of black lung disease. The Mine Safety Act of 1966 addressed issues not covered in previous labor legislation. It established standards for injury and illness record keeping. An explosion in the early hours of November 20, 1968 at the Mountaineer No. 9 coal mine of the Consolidated Coal Company in Farmington, West Virginia, that left 78 miners trapped, dramatically shifted the nation's and Congress' attention to environmental hazards in the mines and in other industries. In 1969 the Federal Coal Mine and Safety Act established regulations for airborne exposure to coal dust and quartz. The act mandated controls on important workplace safety hazards such as ventilation to reduce workers exposure to harmful dust.

Figure 9-2: Prior to the Mine Safety Act of 1966, many workers suffered from Black Lung disease due to a lack of ventilation and over-exposure to coal

Even while laws were put into place to protect workers from hazardous workplace exposures, and workplace fatalities were decreasing, the occupational illness and injury rate was increasing. This was particularly true among insulation workers and chemical manufacturers. The United States Congress determined that the workers' compensation system had not provided enough incentive to businesses to improve safety and health.

In 1970, the Congress passed the Occupational Safety and Health Act (OSH Act) over the veto of President Nixon. The OSH Act established the Occupational Safety and Health Administration (OSHA) to administer the act and "to assure so far as possible every working man and woman in the nation safe and healthful working conditions." Since its enactment, there has been an overall decrease in occupational injuries and fatalities.

The OSH Act regulatory program is organized into eight parts: universal coverage, employer obligations, enforcement, employee participation, checks and balances, information, training, and rights and responsibilities. It covers every employer, both private and public. This is known as universal coverage. The employer obligations include maintaining a safe and healthful workplace through performance of required activities such as training and self-inspections. Since OSHA is part of the Department of Labor, the responsibility for making sure the regulations are implemented, monitored, and enforced falls on the Secretary of Labor.

The **General Duty Clause** of the OSH Act is designed as a sort of safety net. The premise of this unique provision is that the overriding obligation of the employer is to provide a workplace "free from recognized hazards likely to cause death or serious physical

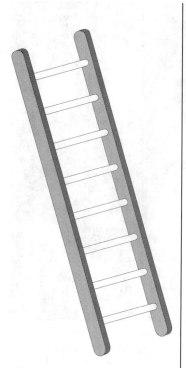

**Figure 9-3:** Specific standards explain how to comply with OSHA regulations. OSHA regulates the width of ladder steps and the distance between steps.

harm." That means that even if a hazard in the workplace is not specifically covered by a regulation, the employer must protect the employee anyway. Neither the Environmental Protection Agency nor the Department of Transportation have this type of authority.

OSHA regulations take two basic forms (as do most regulations): they are either **specific standards** or **performance standards**. Specific standards explain exactly how to comply. For example, the OSHA regulation covering ladders very specifically lists the minimum width of the ladder steps, distance between steps, etc. A performance standard lists the ultimate goal of compliance but does not explain exactly how to accomplish it. For instance, the general duty clause described above is a performance standard. It states that the employer must protect the health and safety of the employee even if there is no OSHA regulation currently written to cover the work activity in question. It does not, however, explain how to accomplish this; that is up to the employer.

## ■ Regulatory Framework

### Concepts

■ Accurate records on occupational injuries and illnesses are necessary in order to establish meaningful regulatory programs.

■ Both employers and employees have rights and responsibilities for workplace safety.

■ OSHA regulations include national consensus standards that are adopted by reference.

OSHA determines how well a program works based on injury data and insurance costs. In order to get accurate data upon which to base its judgment, OSHA requires extensive record keeping of written plans, injuries, illnesses, safety audits, inspections, corrections, and training. In theory, if a program is working effectively, then the injury data and insurance costs will reflect the performance. At the time of the OSH Act, very little information was available on occupational injuries and illnesses. This was a concern to Congress because without this information it was difficult to identify the workplace activities and industries most in need of regulation. For this reason, a part of the act mandated the compilation of information and data. This information is also used to determine the effectiveness of various programs. The responsibility for compiling the data falls to the Bureau of Labor Statistics. For example, a cornerstone of OSHAs record keeping requirements is the OSHA 200 Log. Information for this Log, which lists all workplace injuries except first aid treatments, must be submitted annually by employers.

Training is a major part of the OSH Act. Injury statistics have shown that newer employees without adequate training are far more likely to be injured on the job than those with more experience and training. Almost every regulation requires some sort of information and training. It makes sense to require training since it is difficult for an employee to protect him or herself if unaware of the hazards.

While the original OSH Act is a federal program, each state is allowed to develop its own OSHA program with 50 percent of the funding for the state provided by the federal OSHA program. State programs have to be "at least as effective as" the federal OSHA program; this is known as state authorization which has already been discussed for other statutes. Approximately half of the states have an occupational health and safety program authorized by OSHA.

Compliance with workplace safety regulations requires both the employer and employee to fulfill certain responsibilities which are specified in the OSH Act. Employees and employers may also exercise certain rights that are described in the Act. For instance, an employer has the *right* to: refuse an OSHA inspection unless it is preceded by an inspection or search warrant, request training and monitoring from OSHA, and accompany an inspector that is making an inspection. In addition, the employer has the *responsibility* to: maintain a workplace free from recognized hazards, monitor the effectiveness of health and safety programs, train all employees on job safety and enforce their safety program, including taking necessary disciplinary action(s) allowed by law.

**Figure 9-4:** Employers are required to train their employees about safety precautions and hazards in their workplace.

The employee has the right to: a workplace free of recognized hazards, anonymously request an inspection or audit by OSHA, refuse unsafe work as long as certain good faith conditions are met, be informed of hazards in the workplace, and request a copy of medical information pertaining to his or her health. The employee has the responsibility to obey the safety regulations implemented by the employer, to inform the employer of safety hazards in the workplace, and to perform his or her work in a safe manner.

The enforcement of the OSH Act is carried out through inspections, citations, and levying of civil penalties. These three increasingly punitive steps are designed to achieve a safe workplace by requiring the removal of hazards. Both requested and unannounced inspections can be conducted to determine if hazards exist. If hazardous situations are discovered, follow-up inspections assure that the appropriate corrections are made.

**Figure 9-5:** OSHA inspectors ensure that employers maintain a safe and healthful workplace for their employees.

Employee participation is a vital aspect of the OSHA program. An employee or his or her representative may request an inspection from OSHA with protection against reprisals from the employer. For example, a union representative may request an inspection from OSHA, may participate in the inspection, and participate in negotiating when the hazard will be removed.

OSHA investigates and writes citations based on inspections of the work site. An OSHA inspector may visit a site based on the following: an employee complaint; a report that an injury or fatality has occurred; or a random visit to a high-risk business. If the inspection uncovers one or more violations, the OSHA inspector who is titled the **Compliance Safety and Health Officer, CSHO,** provides an explanation on a written inspection report. The CSHO will discuss with the employer the nature of the violation, possible measures to take to correct the problem, and dates by which the hazard(s) must be controlled. Violations are grouped into several areas:

■ Willful: A violation in which the employer knew that a hazardous condition existed that violated a regulation of OSHA, but made no reasonable effort to eliminate it.

■ Serious: A violation that could cause serious harm or permanent injury to the employee, and the employer did not know or could not have known of the violation.

■ Repeated: A citation given to the employer for a previously documented violation. A citation is considered repeated if it occurs within three years of the first citation.

■ General: Inadequate or non-existent written programs, lack of training, training records, etc.

■ *de minimis*: A relatively minor deviation of regulations, generally in the area of interpretation of a regulation or a simple paper violation.

Any citation issued as a result of non-compliance must be posted in clear view near the place where the violation occurred, for three working days or until corrected, whichever is longer. An employer can take the following courses of action regarding citations: the employer can agree with the citation and fix the problem by the date given on the citation and pay any fines; or the employer can contest the citation, proposed penalty, or correction date, as long as it is done within 15 days of the date the matter in question is issued.

To contest a citation, OSHA must be provided with the appropriate paperwork indicating the employer's intent to do so. In addition, before filing an intent to contest, the employer may request an informal conference and settlement hearing. During this hearing, the employer can request a better explanation of the violations cited, as well as a more complete interpretation of the specific regulations that apply. The employer may also negotiate and enter into an informal settlement agreement.

When an employer contests a citation, and an informal hearing cannot resolve the issues, the next step is for the parties to meet with an administrative law judge or review commission. The judge

## 29 CFR 1910 Subparts A-Z

- Subpart A – General provides the provisions for OSHA's initial implementation of regulations.

- Subpart B – Adoption and Extension of Established Federal Standards explains which businesses are covered by the OSHA regulations. The construction industry, for instance, falls under 29 CFR 1926. Those industries that are not covered under a specific standard fall under the General Industry Safety Orders.

- Subpart C – General Safety and Health Provisions provides the right for an employee to gain access to exposure and medical records.

- SUBPART D – Walking and Working Surfaces establishes requirements for fixed and portable ladders, scaffolding, manually propelled ladder stands, and general walking surfaces.

- Subpart E – Means of Egress establishes general requirements for employee emergency plans and fire prevention plans.

- Subpart F – Powered Platforms, Man Lifts, and Vehicle-Mounted Work Platforms mandates the minimum requirements for an elevated safe work platform.

- Subpart G – Occupational Health and Environmental Control mandates engineering controls of physical hazards such as ventilation for dusts, noise, ionizing and non-ionizing radiation.

- Subpart H – Hazardous Materials provides requirements for the use, handling and storage of hazardous materials.

- Subpart I – Personal Protective Equipment provides general requirements for personal protective equipment.

- Subpart J – General Environmental Controls mandates the requirements for sanitation, accident prevention signs and tags, Confined Space Entry, Lockout/Tagout requirements for hazardous energy.

- Subpart K – Medical and First Aid requires that an employer provide first aid facilities or personnel trained in first aid to be at the facility.

- **Subpart L – Fire Protection** mandates portable or fixed fire suppression systems for work places.

- Subpart M – Compressed Gas and Compressed Air Equipment in this section the requirements for air receivers are presented.

- Subpart N – Materials Handling and Storage basically covers the use of mechanical lifting devices, changing a flat tire, forklifts, and helicopters.

- Subpart O – Machinery and Machine Guarding provides requirements for guarding rotating machinery.

- Subpart P – Hand and Portable Powered Tools and Other Hand-Held Equipment.

- Subpart Q – Welding, Cutting, and Brazing requires the use of eye protection, face shields with arc lenses, proper handling of oxygen and acetylene tanks.

- Subpart R – Special Industries in this section, special requirements for Textiles, Bakery equipment, Laundry machinery, Sawmills, Pulpwood logging, Grain handling, and Telecommunications are covered.

- Subpart S – Electrical requires the use of protection mechanisms for electrical installations.

- Subpart T – Commercial Diving Operation mandates requirements for the dive team.

- Subpart U – Y Not currently assigned.

- Subpart Z – Toxic and Hazardous Substances requires monitoring and protective methods for controlling hazardous airborne contaminants.

**Figure 9-6** Outline of Subparts of 29 CFR 1910, Subparts A–Z.

or commission listens to the evidence presented by the CSHO and counter-arguments by the employer and has the authority to make a binding decision. This process is similar to a court trial held without a jury.

When a citation is written, the burden of proof is on the inspector. The inspector must provide evidence using photos, videos, interviews, monitoring data or observations and cite the specific violation of a regulation. The regulations governing labor practices, including safety and health, are listed under Title 29 of the Code of Federal Regulations, with the occupational safety and health regulations found in parts 1900–1999.

The regulations are broken down into Parts and Subparts. Parts 1900–1909 establish the authority for the Secretary of Labor to write and implement occupational safety and health regulations. The administrative requirements for conducting inspections, writing citations, maintaining records, contesting citations, applying for variances, modifying rules, holding informal conferences, and authorizing state programs are included in this part. Essentially, anything you would like to know about how OSHA operates is found in these sections.

Part 1910 contains the actual workplace regulations and is divided into Subparts A to Z. Subpart A contains the general provisions for OSHA's initial implementation of regulations. When the program first began, the OSH Act allowed any national consensus standard to become a part of the Act. A National Consensus Standard is a performance objective, mechanical design, or procedure established by professional or industrial organizations for their members.

The National Consensus Standards were derived from groups such as the American National Standards Institute (ANSI), National Fire Protection Association (NFPA), American Society of Mechanical Engineers (ASME), American Society of Testing Materials (ASTM), American Conference of Governmental Industrial Hygienists (ACGIH) and the American Petroleum Institute (API). Included in subpart A are provisions for amendments, applicability, incorporation by reference, and definitions. Incorporation by reference means that OSHA inserts these standards into their regulations by stating, for instance, that safety shoes shall meet ANSI standards. OSHA does not approve standards of design; they merely mandate that materials, equipment, or performance standards meet the minimum requirements already established by a nationally recognized authority. Using the safety shoe example, there is no such thing as one that is "OSHA approved." However, there is an ANSI approved design for safety shoes, ANSI Z41.1. OSHA incorporates this design requirement by reference in 29 CFR 1910.132, Foot Protection.

## ■ Checking Your Understanding (9-1)

1. Describe the rights an employer has when an OSHA inspector shows up.

2. List two options that an employer can exercise after being cited for an OSHA violation.

3. How does the General Duty Clause differ from a specific hazard?

4. List two responsibilities of an employer and two of an employee in regards to workplace safety.

5. Define National Consensus Standard and give two examples.

6. Describe the types of workplace conditions and events that led to the enactment of health and safety laws.

7. Describe two pieces of legislation that preceded the OSH Act.

8. What is the difference between a performance-based standard and a hazard-specific standard?

# 9-2 Industrial Hygiene

## ■ Overview and Background

### Concepts

- Industrial hygiene is the application of science and technology to reduce workplace injuries.

- Industrial hygienists recognize, evaluate, and control physical, chemical, and biological hazards in the workplace.

- There are professional certifications for industrial hygienists and safety professionals.

Industrial hygiene is the application of scientific principles to reduce occupational exposure to potential hazards. An example of the practice of industrial hygiene would be determining the type of filter, either in the ventilation system or in an employee's respirator, that would reduce exposures in a work area where airborne particulates pose a threat. The person who does this, the practitioner of industrial hygiene, is called an **industrial hygienist** or **IH**. The basic goals of industrial hygiene are the recognition, evaluation, and control of physical, chemical, and biological hazards in the workplace.

The profession of industrial hygiene dates back to the fourth century B.C. when Hippocrates recognized that lead caused health problems in miners. A few centuries later, the Romans documented symptoms resulting from exposure to lead, zinc and sulfur. In addition, scholars of the time recommended a type of protective mask to be worn by workers when dealing with high concentrations of dust and metals.

Although science was considered "magic" throughout the Middle Ages and into the seventeenth century, there were some notable advances made during these time periods in the area of industrial hygiene. In 1556, a German scholar named Georgius Agricola described the hazards of mining and provided solutions to the problems he outlined. He published articles addressing respiratory protection for silicosis (a lung disease caused by inhalation of silica dust), ventilation procedures for mines, and observations of "trench foot" disease. Trench foot is a disease where the tissue of the toes and feet begins to die off due to a lack of circulation and bacterial growth. It is brought on by cold and wet feet. In addition, Agricola discussed mining injuries and fatalities and how to prevent their recurrence.

During the 1700s scientific studies on occupational health were published. Bernardo Ramazzini, an Italian physician, published a book called *De Morbis Artificium Diatiba* in which he described the health effects he noted when he did autopsies of workers. From the damaged lungs of miners, he described the effect silicosis had on health and outlined some precautionary measures for reducing the hazard. In his work, he also described how workers who had hammered copper damaged their hearing from excessive noise.

In England, Dr. Percival Pott studied chimney sweeps with the express purpose of finding out why they were dying at an early age. In his study, he correctly identified the soot from chimneys as the cause of scrotal cancer. It was another 200 years before benz(a)pyrene in the soot was identified as the cancer-causing agent. His study is credited as the first to identify a workplace carcinogen, and it became the primary force behind the British Parliament passing the Chimney-Sweepers Act of 1788. This act regulated how the cleaners used their tools and required rudimentary sanitary practices. For instance, it called for chimney sweeps to wear protective aprons while they were sweeping.

**Figure 9-7:** Chimney Sweeps developed cancer from the benz(a)pyrene in chimney soot.

In the early part of the twentieth century in the United States, Dr. Alice Hamilton led the way for the protection of workers' health. Dr. Hamilton studied various occupational diseases and identified causative agents. She presented scientific evidence for the relationship between occupational disease and industrial toxins. In addition, for each disease she presented solutions to reduce workers' exposures. Her primary goal was the implementation of engineering controls for the prevention of industrial illnesses. Her principles and priorities of prevention are best summed up in her words, "Whatever money is available for factory hygiene must be expended first on mechanisms to prevent poisoning the air, even if this means scanty equipment for the washrooms and lunchrooms. The physician will sometimes be told that certain processes cannot be carried on without contamination of the air, that the workman must be protected in some other way, by some sort of respirator or mask . . . but at present . . . no apparatus, respirator or army mask, through which a man can breathe with entire ease and comfort while doing heavy work, will serve to hold back all the poisonous dust or vapor in the air." This is a principle still employed in modern **industrial hygiene**.

Dr. Hamilton wrote numerous texts on industrial toxins while she was Assistant Professor of Industrial Medicine at Harvard Medical School. In 1918, Harvard was the first university to confer a master's degree in industrial hygiene, and Dr. Alice Hamilton was named to head up that program. She died in 1971, the year the OSH Act was passed.

Changes of energy create physical hazards that affect the worker. Energy sources include: pressure, temperature, sound waves, movement of machinery, ionizing and non-ionizing radiation, and vibration. This means that the workers position relative to the energy source influences the exposure. For instance, a worker standing too close to a piece of machinery that has a rotating arm runs the risk of injury from getting entangled in it. The IH would either recommend changing the position of the worker or shielding the moving part. Chemical hazards affect the worker through the interaction with body systems by chemical agents. Biological hazards are living organisms or their toxins that cause illness in workers.

A control measure is a method used to separate the hazard from the worker. The control measures and IH uses fall into three categories: engineering, administrative, and personal protective equipment (PPE). Engineering controls require physical changes to the facility or use of special equipment. A sound barrier to control noise is a type of engineering control as is ventilation to remove air contaminants. Engineering controls can be portable or fixed. With engineering controls, the employee is separated completely from the hazard but does not wear the control.

Administrative controls are policies and procedures, which if followed, are designed to reduce the likelihood of injury and illness. They do not actually separate the employee from the hazard. Examples of administrative controls include written programs, monitoring the work area, safety audits, and training. A list of general safe work practices and instructing workers on how to avoid specific hazards are excellent uses of administrative controls.

Personal Protective Equipment includes special clothing, devices, or equipment that the worker wears to provide protection to body systems from the hazardous environment. PPE is the least desirable form of controlling a hazard, because a worker is directly exposed to the hazard and if the PPE fails, the worker will come in contact with the hazard. A hard hat is an example of personal protective equipment designed to save the head from the hazard of falling objects.

Today, there are many universities offering undergraduate and graduate programs in Industrial Hygiene, Occupational Safety, or Environmental Health. In addition to formal academic degrees, the profession of industrial hygiene and safety has a voluntary certification process. The American Industrial Hygiene Association (AIHA)

**Figure 9-8:** Dr. Alice Hamilton.

is a professional society established in 1939 to provide professional status and improve the practice of industrial hygiene. One important way they do this is through a professional certification called the Certified Industrial Hygienist (CIH). To become a CIH one must possess specific educational and experiential qualifications and pass an exam. A similar application, qualification, and examination process is used by the American Society of Safety Engineers for their certification called the Certified Safety Professional (CSP) The examinations for both certifications are administered nationally on a semi-annual basis.

Another organization mentioned previously is the American Conference of Governmental Industrial Hygienists (ACGIH). ACGIH was organized in 1938 by a group of government IHs who wanted a way to promote standard techniques of industrial hygiene work and exchange ideas and data. An ACGIH committee is accepted for its expertise in the field of establishing safe and acceptable workplace levels for physical and chemical hazards.

The field of industrial hygiene is very different from the field of toxicology. A toxicologist is usually engaged in scientific research in pharmaceutical companies, chemical manufacturers, or in pathology labs. As we learned in Chapter 3, the primary function of a toxicologist is the study of chemical effects on humans and other animals. An industrial hygienist takes into account the job procedures and how the worker comes into contact with a chemical. In other words, the IH looks at the working relationship between occupational hazards and employees.

Interaction between the two professions occurs in the exchange of information. Industrial hygienists apply toxicological data to determine the risk of injury. Toxicologists use injury and illness records to refine their conclusions about the potency of chemicals and physical hazards.

Many large companies have in-house industrial hygienists as part of their environmental and safety staff. The responsibilities of the IH include monitoring workers' exposure to hazardous environments and working with engineering and facility design staff to reduce the likelihood that injury or illness will occur as the result of the set-up, layout, or use of facility equipment. An IH will also review operations that involve repetitive motions to make sure that measures are taken to reduce employees' risk of injuries. In order to make these determinations, an IH is likely to conduct employee interviews to determine the risks of various job duties.

## ■ Industrial Hygiene Terms

### Concepts

■ The PEL and TLV have the same definition, but different origins.

■ PELs and TLVs apply only to airborne contaminants; the route of entry being protected is inhalation.

- OSHA and ACGIH standards do not agree on allowable workplace exposures.

- Medical monitoring does not prevent injuries from occurring.

- Accurate baseline medical monitoring is important.

Industrial hygienists and safety professionals use terms specific to their professions that are important for understanding and communicating health hazards. In this section, we will introduce the most common terms, define them, and provide some additional information on how the terms are used.

The first terms we will cover identify allowable workplace exposures to physical and chemical hazards. Allowable means that most healthy adult workers would not be irreparably harmed if exposed at or below the established limits. A number of organizations develop and promote their own workplace exposure limits. They include OSHA, ACGIH and the **National Institute of Occupational Safety and Health** (**NIOSH**). Only OSHA-enforced exposure limits are legally mandated. These are called **Permissible Exposure Limits** (**PELs**) and are based on time weighted averages (TWA). So, to define it more completely, a PEL is a time weighted average concentration of an airborne contaminant that a healthy worker may be exposed to 8 hours per day or 40 hours per week without suffering any adverse health effects. It is established by legal means and is enforceable by OSHA.

Many readers may be unfamiliar with the concept of time weighted average. A **time-weighted average** is a mathematical average of exposure concentration over a specific time; for PELs the time is 8 hours. The time-weighted average is calculated by multiplying the concentration of the worker's exposure by the amount of time he or she is exposed, adding the totals for the day, and dividing by 8 hours [ $\Sigma$ (exposure in ppm $\times$ time in hours) $\div$ 8 hrs. = time weighted average in ppm]. It's easier if we look at an example. Jane works on a job where she has some exposure to a waste stream with benzene, which has a PEL of 1 ppm. Let's look at a typical day to get data on her exposure levels to benzene (an IH would use monitoring devices to get this information).

- <u>8 am – 8:30 am</u>: Arrives at work and fills out daily paperwork: zero exposure.

- <u>8:30 am – 9:30 am</u>: Drives to the remote station where she is responsible for the waste pumping units: zero exposure.

- <u>9:30 am – 10:30 am</u>: Checks and primes pumps: typical exposure level is 4 ppm.

- <u>10:30 am – Noon</u>: Makes rounds to check employee work on proper filing of reports on waste shipments: zero exposure.

- <u>1:00 pm – 3:00 pm</u>: Works in the waste treatment lab testing various treatment technologies: typical exposure is 2 ppm.

- <u>3:00 pm – 4:00 pm</u>: Drives back to central office: zero exposure.

- <u>4:00 pm – 5:00 pm</u>: Completes paperwork and governmental reports: zero exposure.

Although this is a simplistic scenario, it should help you to understand the concept of time-weighted average. Let's calculate her TWA exposure:

$$
\begin{array}{rclcl}
.5 \text{ hour} & \times & 0 \text{ ppm} & = & 0 \text{ hour ppm} \\
1 \text{ hour} & \times & 0 \text{ ppm} & = & 0 \text{ hour ppm} \\
1 \text{ hour} & \times & 4 \text{ ppm} & = & 4 \text{ hour ppm} \\
1.5 \text{ hour} & \times & 0 \text{ ppm} & = & 0 \text{ hour ppm} \\
2 \text{ hours} & \times & 2 \text{ ppm} & = & 4 \text{ hour ppm} \\
1 \text{ hour} & \times & 0 \text{ ppm} & = & 0 \text{ hour ppm} \\
1 \text{ hour} & \times & 0 \text{ ppm} & = & 0 \text{ hour ppm} \\
& & \text{Total} & = & 8 \text{ hour ppm} \\
& & \text{TWA} & = & 8 \text{ hour ppm} \div 8 \text{ hours} \\
& & \text{TWA} & = & 1 \text{ ppm}
\end{array}
$$

This means that her TWA exposure is 1 ppm. If we compare that with the PEL, which is the legal limit, we see that she has not exceeded it. This would be a legal exposure. It is important to note that even though she did receive exposures higher than 1 ppm, her time weighted average exposure was within the regulation.

**Threshold Limit Values** (TLVs) are based on exactly the same concept as PELs, except that TLVs do not have the force of governmental regulations behind them. They are recommended limits established and promoted by the **American Conference of Governmental Industrial Hygienists** (**ACGIH**). This professional organization constantly revises and updates their recommended limits in a booklet called *Threshold Limit Values for Chemical Substances and Physical Agents*, which is published every two years.

Another group that develops recommendations for exposure limits is the National Institute of Occupational Safety and Health (NIOSH). NIOSH is the primary federal agency that conducts research to eliminate workplace hazards. It also establishes **Recommended Exposure Limits** (**RELs**), which are reviewed by OSHA as input for revising the PELs.

In regards to our example of an 8 hour time-weighted average, you may have wondered how very high exposures for short periods of time are handled. Remember, if we use 8 hours as our averaging time, and the exposure lasts 5 minutes, a very high short-term exposure would still have a relatively low 8-hour time-weighted average. This is the reason **Short Term Exposure Limits** (**STELs**) were developed. They represent the time-weighted average concentration to which workers can be exposed continuously for a short period of time (typically 15 minutes) without suffering irritation, chronic or irreversible tissue damage, or impairment for self-rescue. Note that the STEL is a 15-minute time-weighted average. There can be no more than four STEL exposures per day, and there must be a minimum of one hour between exposures.

The last exposure limit used in the field of industrial hygiene is called the **Ceiling Limit** (**C**). This is truly an upper limit, not a time-weighted average. The ceiling represents the concentration that should not be exceeded during any part of the work day, even for an instant.

All the exposure limits discussed to this point are levels that are used together to identify safe exposures in working environments. Healthy workers should be able to work at these levels without respiratory protection and suffer no damaging effects. One other designation, however, represents a very dangerous condition. This is the **Immediately Dangerous to Life or Health (IDLH)** level. This is defined as the concentration or condition that poses an immediate threat to life or health. The worker must be able to escape within 30 minutes without losing his or her life or suffering permanent health damage.

For any of the exposure limits related to exposure to airborne contaminants, there may be a "skin" designation. When the word skin is noted next to the limit, this indicates that the chemical is known to absorb through the skin, mucus membranes or eyes.

It is important to review the difference between a **TLV** and a PEL. TLVs are not enforced by OSHA unless incorporated by reference into a regulation. Both of these terms, PEL and TLV, use the designations, such as TWA, STEL or C that specify the type of limit. For example, you may see airborne contaminants listed in the following ways: OSHA designations as PEL, PEL-TWA, PEL-STEL, PEL-C, or PEL-TWA(s), and the ACGIH designations as TLV-TWA, TLV-STEL, TLV-C or TLV-TWA(s). The important point to remember is you must know what the values refer to. In addition, not all airborne chemicals have an established allowable workplace exposure level. Exposures to more than one chemical raise other problems. OSHA has addressed these situations as additive exposures. It is beyond the scope of this text to examine the more complex calculations these types of situations require.

| | TLV | NIOSH REL | PEL |
|---|---|---|---|
| Acetaldehyde | 100 ppm | Carcinogen (no threshold) | 200 ppm |
| Acetone | 750 ppm | 250 ppm | 1000 ppm |
| Benzene | | 0.1 ppm | 1 ppm |
| 2-Butanone | 200 ppm | 200 ppm | 200 ppm |
| Methylene Chloride | 50 ppm | Carcinogen (no threshold) | 500 ppm |

**Figure 9-9:** Comparison of exposure limits.

The units for PEL and TLV are typically parts per million (ppm), milligrams per cubic meter ($mg/m^3$) or fibers per cubic centimeter (fibers/$cm^3$). Parts per million is a measure of the ratio of the chemical per million parts of air. Milligrams per cubic meter is a ratio of the mass of the chemical compared to a standardized breathing space, or mass per unit volume (density). Fibers per cubic centimeter is a measure of how many pieces of a solid material there are in 1 cubic centimeter of air space. This is how, for example, exposure limits for asbestos are described. Physical agents such as noise also have PELs. In this case, the limits are denoted in decibels. Noise as a physical hazard will be discussed in more detail in the next section.

There are a number of physical hazards that are defined and controlled by OSHA regulations. One important physical hazard for which a safe limit is set is an **oxygen deficient atmosphere**. An oxygen deficient atmosphere is legally defined as an atmosphere where the oxygen concentration is less than 19.5% by volume of air. In Chapter 3 we learned that air is made up of the following components in the approximate amounts listed: 78% nitrogen, 21% oxygen, and 1% of miscellaneous gases (carbon dioxide, argon, neon). When air is physically displaced from an environment, all of the components are displaced equally. Since oxygen is so vital to our survival, we start seeing the physiological effects of oxygen deficiency at about 16 percent. OSHA regulations have built-in safety margins. That is why the definition of oxygen deficient atmosphere was not set at 16 percent but at 19.5 percent. When the oxygen content in air is reduced below 19.5% by volume, OSHA considers the environment oxygen deficient and immediately dangerous to life or health, or IDLH. In these atmospheres the use of supplied air is required. ACGIH uses a value of 18% by volume to define an oxygen deficient atmosphere.

**Medical monitoring** consists of an initial medical exam of a worker followed by periodic exams. The purposes of medical monitoring are to assess workers' health, determine fitness to wear personal protective equipment, and maintain records of a person's health. Monitoring is also a procedure for assessing the effects of chemical or physical hazards and is one of the ways an Industrial Hygienist tracks the effects of workplace exposures. A primary goal of medical monitoring is to prevent illnesses related to workplace exposures. The method of prevention involves detecting the contaminant in the body or recognizing changes in health after an individual has been exposed. In theory, there is a threshold dose that the body can tolerate without adverse effects. We just learned that

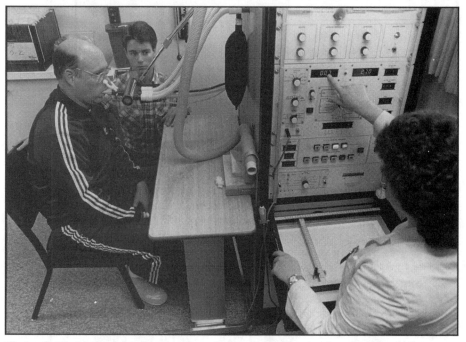

**Figure 9-10:** Medical monitoring of employees is a vital part of tracking workplace

the maximum allowable levels in the body are set by ACGIH, NIOSH, and OSHA. Many of the OSHA regulatory programs, for instance the asbestos regulations, require some type of medical monitoring of employees.

Various federal regulations requiring medical monitoring use similar wording. If an employee is exposed to a contaminant at or above the levels established by regulation or is injured at the workplace, the employer is required to provide medical treatment and follow up monitoring for the effects of exposure. In addition, if the employee works in an industry with an OSHA regulated substance, such as asbestos, then the employer must periodically monitor the person for damaging health effects as a result of exposure to the substance. Continuing our use of asbestos workers as examples, this means they must be given periodic chest X-rays and lung function tests.

Medical monitoring is specifically required for employees in the following situations:

1. Any employee who wears a respirator during any part of a day for a period of 30 days or more in a year or as required by the regulations relating to respiratory protection.

2. Any employee who is injured, becomes ill or develops signs or symptoms due to possible overexposure involving hazardous substances or health hazards from an emergency response or hazardous waste cleanup operation.

3. Members of hazardous materials spill cleanup teams.

For these employees, medical monitoring is to be conducted prior to their hazardous assignment and at least once every year, unless a physician believes that every two years is more appropriate. An examination prior to employment is usually called a baseline physical since it establishes the level of body function(s) prior to exposure. Employees must also be medically monitored when they terminate employment or are reassigned to an area where the medical monitoring requirements would not apply. In the course of their everyday work, they must be medically monitored as soon as possible if symptoms have developed indicating possible overexposure to a hazardous substance, or if it is believed that the employee may have been exposed above the PELs.

Because medical science does not have the technology to detect small amounts of damage to specific body parts, medical monitoring is not preventive. For instance, when a liver can no longer process a chemical at the original baseline rate, then damage to the liver has occurred. The manifestation of the impairment or injury is determined by a change from the baseline medical evaluation. Medical monitoring can catch acute injuries and illnesses, but chronic illnesses are difficult to detect until they are well along.

# ■ Typical Occupational Diseases and Their Control

## Concepts

- Occupational diseases result from repeated and long term exposure to physical and chemical hazards.

- Physical hazards come in many different forms.

- Respirators are of two general types: air-purifying and air-supplying.

- Safe work habits can be directly related to a person's attitude.

We already learned that physical hazards are the result of energy. Noise, radiation, heat, cold, vibration, repetitive motion, dust, and oxygen displacement are forms of physical hazards. Machines cause noise and have dangerous moving parts; nuclear energy or electricity causes ionizing or non-ionizing radiation; boilers cause heat; large packing refrigerators cause cold; typing is a repetitive motion; farming generates dusts; and the release of chemical vapors can cause oxygen displacement. This section will discuss some common occupational diseases that can result from exposure to these hazards.

Increased use of machines, steam, turbines, and automobiles has led to a workplace where noise is commonplace. The problem, of course, is that excessive noise causes hearing loss. This disease was so prevalent during the industrial revolution of the nineteenth century that it was called "boilermaker's disease. "

Noise is unwanted sound. It may do damage because of frequency or intensity. Frequency is pitch; a soprano or baritone. Intensity is loudness. The combination of frequency and intensity result in a sound pressure. If sound is excessive, then a worker's hearing may become damaged. For instance, attending a concert may cause your ears to hurt. The pain is caused by the sound pressure pushing on the ear drum in your ear canal. Too much pressure or pushing permanently destroys the sensory cells in the inner ear.

The most common methods for controlling excessive noise involve hearing protection devices such as ear plugs or muffs or sound deadening engineering controls such as covers on noisy printers. Noise levels must be monitored in the workplace. Even without precisely measuring the noise levels, if you have to raise your voice to speak to someone at arm's length, you are in an environment with excessive noise. Temporary hearing loss is also an indicator that you have been exposed to excessive noise.

The evaluation and control of noise is fairly straightforward. The first step is to measure the amount of sound employees are being exposed to. This is done with an instrument that measures sound pressure levels in decibels. If the monitoring indicates an employee exposure above 90 dB(A), then OSHA requires that a written Hearing Conservation Program be implemented.

The components of a written Hearing Conservation Program are found in 29 CFR 1910.95 and consist of the following:

| Common Equivalents of Sound Levels in Decibels (dB) | |
|---|---|
| Source Sound | Pressure Level |
| Jet Plane Gun Shot | 140 |
| Riveting (steel tank) | 130 |
| Auto Horn, Thunder | 120 |
| Power Saw Rock Band | 110 |
| Punch Press Garbage Truck | 100 |
| Subway Heavy Truck | 90 |
| Restaurant Alarm Clock | 80 |
| Conversation | 70 |
| Soft Whisper | 30 |

**Figure 9-11:** Comparison of common noises and their decibel levels. Exposure to noise with a loudness of 80 dB is annoying. Exposure to 90 dBA (A = time weighted average) can cause physical damage to the ear. At about 120 dBA, hearing actually becomes painful and damage to hearing is certain and rapid.

1. The permitted 8-hour time weighted average exposure is 90 dB(A); however, at a TWA of 90 dB(A), the employer must implement controls to reduce employee exposures to a TWA of 85 dB(A).

2. Employees can watch the monitoring procedures and must be notified of the results.

3. Employers must re-monitor exposures whenever a new employee joins the program or whenever job procedures or equipment use change.

4. Monitoring instruments must be calibrated.

5. Audiometric testing of employees to determine current hearing ability and annual hearing tests are required. This is a form of medical monitoring to check the performance of the employer's Hearing Conservation Program.

6. The use of engineering and hearing protectors must be provided to employees to reduce the noise exposure to a TWA of 85 dB(A).

Besides noise, dust is one of the most common causes of occupational diseases. Recorded problems of worker exposure to dust dates back to the early Bronze Age. Some of the occupational diseases caused by dust are silicosis, byssinosis, asbestosis, and pneumoconiosis. Dust is controlled through engineering controls such as dust capture devices or respiratory protection devices worn by the worker.

Silicosis is caused by inhalation of silica dust or quartz dust. Once in the lungs, the immune system attacks the particles to destroy them; however, fibrous tissue replaces the affected area of the lung. The result is a less elastic lung that does not exchange gases efficiently. Industries with silica hazards are mines, quarries, and tunnel excavators.

Byssinosis is caused by cotton, hemp, and flax dust. The acute effects are shortness of breath and reduction in lung capacity. The exact cause of byssinosis is not known, but it is believed that an active fungi on the product is part of the problem.

Asbestosis is a scarring of the lungs. Asbestos is a general term describing naturally occurring fibrous materials of mineral origin. Asbestos materials were traditionally mined to be used in brake linings, clutch disks, and in roofing, flooring, cement, and insulating materials. The latency period of asbestos-related lung cancers is estimated to be upwards of 20 years. Because of its toxicity, in 1989 the EPA (not OSHA), ordered a 94 percent reduction over a 7-year period, of the manufacturing, use and exportation of asbestos products.

Ergonomics refers to the motions a worker uses on the job and how physical conditions of the work station affect the ability to do the job. Lighting, temperature, work surface height, computer/operator distance, and computer

**Figure 9-12:** Workers removing asbestos from a building are required to wear air-supplying respirators.

screen coloring all affect a worker. If it is too cold a worker may not think clearly or move as rapidly; if it is too hot a worker may need more fluids or become easily tired. If the lighting is dim, a worker may miss a defect on a product. Ergonomics takes into account the person and his or her immediate work environment.

One important aspect of ergonomics is the reduction of repetitive stress injuries, or RSI. Typists, grocery clerks, production line or assembly line workers are all susceptible to RSI. One common example is Carpal Tunnel Syndrome. The symptoms are pain in the hands, arms, and sometimes numbing or partial loss of hand function. Changing the way employees position their hands while at a computer keyboard and providing more breaks to rest the hands can alleviate this problem.

Low frequency vibrations can cause your body to rattle. While this seems a minor problem, human bodies were not designed to tolerate continuous shaking, rattling and rolling. The effects of vibration can be muscle pain, joint fatigue, numbing of extremities, headaches, and pains in the midsection. Jack hammers, weed whackers, lawnmowers, saws, and production facilities may be contributors to this problem. "Kidney belts" are sometimes useful in reducing the effects by minimizing the amount of vibration carried through to the internal organs.

Non-ionizing radiation comes from electromechanical devices such as computers, microwaves, high tension power lines, and electrical generating equipment. There is a significant amount of controversy over the amount of non-ionizing radiation that may cause injury. The shielding built into the door of a microwave oven is an example of an engineering control to minimize exposure to non-ionizing radiation.

We learned in Chapter 3 that chemicals can cause a variety of injuries and illnesses through their toxicity, flammability, explosivity or corrosivity. Flammable and combustible chemicals are generally volatile; meaning they have low **boiling points**. The vapor given off by these materials is a fire hazard and an inhalation hazard. Chemicals that are organic solvents dissolve fats, proteins, and other organic material. Consequently, a solvent used with bare hands will have an additional route of entry into the body, through the skin. If a worker does not wash his or her hands, there may be a third route of entry, through ingestion. Some chemicals cause physical damage to the body very rapidly. Chemicals like sodium hydroxide (drain cleaner) or hydrochloric acid (pool acid) can cause immediate and permanent damage if eaten or by dissolving tissues of the body. The eyes are especially susceptible to injury by chemicals. The only natural protection the eyes have, tearing and closing the eyelids, are no match for workplace chemicals.

As a last resort in protecting a worker from occupational diseases, personal protective equipment is used (Figure 9-13). In general,

Apron, gloves, hardhat, faceshield, boot covers    Fully-encapsulating suit

**Figure 9-13:** PPE can include steel-toed shoes, safety glasses or goggles, face shields, ear plugs, welders' shields, gloves, chemical protective clothing, and respirators or supplied air breathing apparatus.

PPE is uncomfortable and increases the stress on a worker. When wearing PPE, the body's ability to cool is usually diminished; nevertheless, PPE is frequently required to reduce the risk of injury. PPE can include steel-toed shoes, safety glasses or goggles, face shields, ear plugs, welders' shields, gloves, chemical protective clothing, and respirators or supplied air breathing apparatus. Barrier creams and personal monitors are not considered PPE.

Respiratory protection equipment is commonly used because inhalation provides such rapid entry for volatile chemicals into the bloodstream. There are two general types of respiratory protection devices, called respirators: **air-purifying** and **air-supplying**. Both consist of a face piece connected to either an air source or an air-purifying device. The air-purifying respirator uses cartridges with filters to purify air before it is inhaled. This type of protection is not adequate in an oxygen deficient atmosphere. Respirators with their own air source are called supplied-air respirators. They consist of either a self-contained unit that the worker wears or a hose connecting a worker to a remote air source. See Figure 9-14 for examples of both types of respirator.

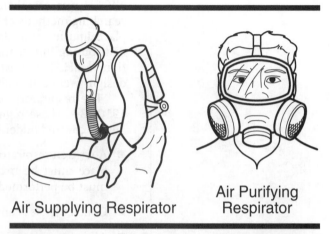

**Air Supplying Respirator**

**Air Purifying Respirator**

**Figure 9-14:** Respirators are either (a) air-supplying or (b) air-purifying.

OSHA has written regulations specifically to address the use of equipment designed to protect a worker's breathing. Respirator selection is a very technical procedure. Selecting a respirator requires the understanding of the chemical, the concentrations a worker is exposed to, and an understanding of how the respirator works. It is irresponsible to provide a respirator without knowing much about the hazard. Respiratory hazards are divided into two groupings: oxygen deficiency and toxic contaminants. If oxygen deficiency is the problem, then only approved supplied air respirators can be used. In addition, the IH must consider how the oxygen was displaced to provide complete protection to the worker.

If an air contaminant is at or above the IDLH value, then only supplied air respirators can be used. If the concentration is lower than the IDLH value, then the appropriate air purifying respirator must be selected. There are different cartridge filters for use with specific chemical families. Particulate and gas/vapor filters are also available, so it must be known whether the contaminant is a dust, fume, mist, or vapor.

**Figure 9-15:** A specific respiratory hazard determines which type of respirator is required.

Workers must be properly fit-tested to wear a respirator. Facial hair, scars, and temple bars on glasses interfere with the tight seal necessary for a respirator to function correctly. In addition, the worker must be physically fit to be able to wear a respirator. Air purifying respirators require the user to breathe through a filtering device that adds stress to the breathing system.

All respirators must meet NIOSH-MSHA design criteria. NIOSH was established by OSHA with the responsibility for researching methods of identifying, evaluating, and controlling workplace injuries and illnesses, training of health professionals, and distributing information. The Mine Safety and Health Administration (MSHA) was established to improve mine safety. Respirator designs were originally developed by MSHA.

The respiratory protection standard is found in Title 29 CFR 1910.134. These regulations require the following of those businesses covered under the standard:

■ A written respiratory protection plan for the hazards at the work place must be maintained at the facility. Hazard assessments must be performed prior to selecting PPE.

■ Contact lenses cannot be allowed to be used with a respirator.

■ All personnel using respirators must be properly trained and physically able to use the equipment. This means a worker must be medically certified by a doctor to use a respirator.

■ Personnel must be informed of the hazards and trained in the proper selection, use and limitations of respirators, including fit testing. Training requires certification.

■ Respiratory protection must not be obstructed by facial hair between the skin to mask seal.

■ Where practical, respirators are to be assigned to individual workers for their exclusive use.

■ Respirators must be regularly cleaned and disinfected. Those used by more than one worker are to be thoroughly cleaned and disinfected after each use. Respirators used routinely must be inspected and maintained.

■ Respirators must be stored in a convenient, clean and sanitary location and protected against dust, sunlight, heat, extreme cold, excessive moisture, or damaging chemicals.

■ Defective or damaged PPE cannot be used.

■ Respirators used for emergency use must be thoroughly inspected monthly and after each use, and such inspections properly documented.

The goal of OSHA is to ensure that employees are provided the appropriate equipment for the job and know how to use the equipment given to them. An important aspect of using PPE is that the wrong personal protective equipment may be more hazardous to the worker than no PPE at all. Regardless of the hazard, all worker exposures can be mitigated. Before selecting the PPE option, remember to consider the other options, such as removing the hazard or using engineering controls, or changing the job procedure.

As a final note, the best recognition and evaluation techniques and the finest control methods will not do any good unless the worker has a positive attitude toward his or her work. Safety and health is behavior related more than it is condition related. Understanding this concept is important. All injuries can be directly or indirectly related to an unsafe act or behavior. Comments like, "It won't happen to me," "It takes too long to do the job that way," "The equipment is uncomfortable, so why wear it?" "I always have to work on that part, so I leave the guard off," are all indicators of underlying attitude problems and/or a lack of knowledge.

A facility that has a spill, injury, or fatality has a negative impact not only on the person and his or her family, but the community as a whole. By performing our job only when we are properly trained and equipped to do it safely, we protect not only ourselves but our community and the ecosystem we live in.

So far we have outlined the general history and discussed the basic structure of the Occupational Safety and Health Administration. We have also reviewed the basic vocabulary used by Industrial Hygienists and Safety Professionals as well as common occupational diseases and their controls. The next two sections will apply these terms and concepts. The next section will take a close look at three hazard specific standards, three performance based standards, and one state standard.

## ■ Checking Your Understanding (9-2)

1. List two examples of work activities an IH would be involved in.

2. Give one example each of an engineering, administrative, and PPE control for a job hazard you are familiar with.

3. List two similarities and two differences between PELs and TLVs.

4. Compare physical, chemical, and biological hazards.

5. Draw a diagram explaining TWA.

6. Explain the difference between STELs, Ceiling Limits, and IDLH.

7. Define oxygen deficient atmosphere and explain what type of PPE is required.

8. List two common occupational diseases and their control.

9. Contrast air-purifying respirators and air-supplying respirators.

# 9-3 Health and Safety Regulations

## ■ Some Hazard Specific Regulations

### Concepts

■ OSHA has a list of allowable workplace air contaminant concentrations (PELs).

■ The original list of air contaminants was an incorporation of a national consensus standard; OSHA PELs do not keep up with current scientific data.

■ The original asbestos standard was 60 times higher than the current standard.

■ The blood borne pathogen standard affects any employee who administers health care and does not address colds, flu, or tuberculosis.

This section will cover some of the most prominent hazard specific occupational health standards that affect hazardous materials workers. The overview will begin with air contaminants regulations, which cover many airborne contaminants found in the workplace. Secondly, asbestos regulations will be discussed. Finally, standards that are of special note to health care providers who may be in contact with certain infectious materials will be outlined.

One of the first hazard specific regulations OSHA put forth was for air contaminants. Air contaminant regulations are found in Subpart Z of Title 29 CFR 1910.1000 and cover an employee's exposure to any substance for which OSHA has established a PEL. The original PEL list was derived from ACGIHs 1970 Threshold Limit Values. It has been expanded and modified since that time and is currently quite an extensive list. In addition to requiring monitoring of the workplace, this regulation requires the use of engineering, administrative, or personal protective equipment controls, in that order, to keep airborne exposures below the PEL.

The intent of regulating air contaminants was to reduce the number of acute and chronic occupational illnesses. Based on epidemiological data, illnesses were occurring due to worker exposures to airborne chemicals. For example, workers in the petrochemical industry were developing cancer due to exposure to benzene. Workers in the insulation industry were contracting lung cancers. In order to protect future workers, OSHA mandated allowable workplace exposure levels and the implementation of controls, such as ventilation or respiratory protection.

It is worth noting that while the ACGIH Threshold Limit Values are updated every two years, the OSHA air contaminant standard was not updated until 1989. In 1989, OSHA adopted the ACGIH standards again; however, OSHA was challenged in federal court for not having provided enough scientific evidence to warrant a sweeping change. The court ruled that the new standard would not go into effect until OSHA could produce the scientific evidence proving the need for the change. This occurred in 1992 and is an example of the checks and balances within the rule-making system.

Although asbestos is certainly an airborne contaminant, and even has a PEL listed on the air contaminants list, its regulation is much stricter. Asbestos actually has its own set of regulations, found in 29 CFR Section 1910.1001, which came about under great controversy. When OSHA was enacted in 1970, asbestos was known to be a carcinogen. Even though accurate scientific evidence did not exist regarding the threshold dose for asbestos to cause cancer (and remember, there are some scientists who believe there is no threshold for carcinogens), Congress and the medical community were very aware of the anecdotal association between asbestos and cancer. The potency of the fiber was not fully realized until more accurate scientific data became available, but the fact that 3.5 million workers were at risk of contracting asbestosis, pulmonary cancer, and mesothelioma was a concern to Congress in 1971.

OSHA adopted the Walsh-Healey Asbestos Standard in 1971, which required employers to reduce exposures to 12 fibers per cubic centimeter of air. This proved to be inadequate and OSHA issued an emergency temporary standard of 5 fibers per cubic centimeter of air in the same year. An emergency temporary standard is implemented when OSHA has sufficient evidence to warrant immediately requiring specific protections against a workplace hazard. Usually an emergency temporary standard allows OSHA and industry to gather more evidence to develop permanent regulations.

During the process of developing permanent regulations to reflect the 5 fibers per cubic centimeter of air limit, several court cases were initiated to dispute OSHAs rationale for the limit. The courts required better scientific data on the causal relationship between asbestos and cancer and stated that OSHA was to use economic and technical feasibility considerations when developing new standards. Economic *feasibility* refers to the financial burden placed on an industry to control the workplace hazard. In the case of asbestos, court decisions required OSHA to take into account the amount of money the new regulations would cost the industry.

In 1986, the asbestos hazard-specific standard was finalized and has since undergone some revisions. The current workplace exposure level for asbestos is 0.2 fibers per cubic centimeter of air, 60 times lower than the original standard. The current regulations pertaining specifically to asbestos are 41 pages in length! These regulations require medical monitoring, monitoring of the employee and work area, appropriate ventilation, and respiratory protection. They state that protective clothing must be worn in the work area. Hazard communication information must be provided to employees who work around asbestos-containing material (ACM), hygiene practices such as washing before eating and using the rest room, and decontamination procedures are also required.

Although this is a lengthy and comprehensive standard, it must be emphasized that federal OSHA does not mandate the removal of asbestos from the workplace (nor does the EPA); OSHA regulates worker exposures and control measures that must be taken to reduce

1910.1002 Coal tar pitch volatiles;
1910.1003 4-Nitrobiphenyl;
1910.1004 alpha-Naphthylamine;
1910.1005 (Reserved);
1910.1006 Methyl chloromethyl ether;
1910.1007 3,3'-Dichlorobenzidine (and its salts);
1910.1008 bis-Chloromethyl ether;
1910.1009 beta-Naphthylamine;
1910.1010 Benzidine;
1910.1011 4-Aminodiphenyl;
1910.1012 Ethyleneimine;
1910.1013 beta-Propiolactone;
1910.1014 2-Acetylaminofluorene;
1910.1015 4-Dimethylaminoazobenzene;
1910.1016 N-Nitrosodimethylamine;
1910.1017 Vinyl chloride;
1910.1018 Inorganic arsenic;
1910.1025 Lead;
1910.1028 Benzene;
1910.1029 Coke oven emissions;
1910.1043 Cotton dust;
1910.1044 1,2-dibromo-3-chloropropane;
1910.1045 Acrylonitrile;
1910.1047 Ethylene oxide;
1910.1048 Formaldehyde

**Figure 9-16:** OSHA Toxic Specific Standards

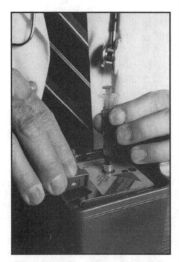

**Figure 9-17:** "Sharps" must be handled carefully because of the risk of infectious disease.

worker exposure. In addition to the asbestos standard, 1910.1001, there are toxic specific standards for a number of other chemicals. See Figure 9-16 for a listing.

One of the newest hazard-specific regulatory programs to be put into effect controls **blood borne pathogens**, which are disease-causing organisms that can be transmitted through blood. This standard was developed as a result of health care workers becoming infected with HIV and Hepatitis B (HBV). Prior to regulation, an estimated 6,000 employees were being infected with HBV annually. There have also been some cases of health care workers and emergency responders being exposed to HIV.

The blood borne pathogen regulations address procedures and work practices to minimize the number of needle stick injuries and blood splashing contamination episodes, ensure the appropriate labeling and packaging of specimens and regulated wastes, and decontaminate equipment and work areas contaminated with body fluids to prevent the transmission of blood borne pathogens (Figure 9-17). In addition, personal protective equipment must be provided to employees to protect them from contracting a disease during their normal job routine. Confidential medical monitoring and evaluation is addressed along with employee counseling.

All health care providers are covered by the standard, including employees who provide first aid. Employers are required to provide Hepatitis B vaccinations or receive a written declination (refusal) from the employee. The employee, however, has the option to take the vaccination at any time.

It is important to understand that the blood borne pathogen standard does not address airborne pathogens such as those that cause colds, flu, or tuberculosis. These viruses and bacteria are not blood borne pathogens; they act on the body in a different manner.

## ■ Workplace Hazard Communication Standards

### Concepts

- All businesses that use hazardous chemicals are covered by the HazCom standard.
- Labeling, MSDSs, and employee training are cornerstones of hazard communication programs.
- The Laboratory Safety Standard is a hazard communication program for laboratories.
- Hazwoper covers hazardous waste clean-up workers, TSD facilities, and emergency responders to chemical spills.

This section will cover three national standards that have essentially the same goals – that of communicating the hazards in the workplace to the employee.

The **Hazard Communication Standard (HazCom)**, which is found in 29 CFR 1910.1200, requires all employers to become aware of the chemical hazards in their workplace and relay that information to their employees. In addition, a contractor conducting work at the site of a client must provide chemical information to the client regarding the chemicals that are brought onto the work site. The standard requires an employer to do the following:

1. Develop an inventory or list of hazardous chemicals that employees may be exposed to in the workplace.

2. Have Material Safety Data Sheets for each hazardous chemical on the inventory and make them available to all employees.

3. Develop a written program describing the employers compliance plan wherein the person in charge of its implementation is clearly identified.

4. Provide training to all employees on chemical hazards, safety, and employer and employee rights and responsibilities.

5. Ensure that all hazardous substances are labeled correctly.

The cornerstone of the HazCom program is the Material Safety Data Sheet (MSDS), which is an informational sheet describing the hazards of a chemical. Once an employee learns how to read a MSDS, important information on how to work safely with the chemical will be readily available. Section 9-4 will give an overview of the information that can be found on a MSDS.

Before an employee can look up information on a MSDS, he or she must know the identity of the chemical. This is why proper labeling is so important. A distinction is made between user containers and storage containers for labeling purposes. A user container is one that holds the chemical for day-to-day use. User containers must have the name of the chemical clearly printed on them. If the container only holds enough for immediate use by the person intending to use the chemical, then the container is not required to be labeled. A storage container must have a label that lists the name of the manufacturer, manufacturers telephone number, chemical name, physical hazards, health hazards, and any special conditions of use.

The HazCom Standard pertains to workplaces where hazardous chemicals are used. A hazardous chemical is defined by OSHA as any chemical that presents a physical hazard or a health hazard; any chemical that is explosive, reactive, flammable, combustible, or causes toxic effects would be included. The exceptions to this standard are materials regulated by the Food and Drug Administration, tobacco and tobacco products, hazardous waste, and consumer commodities. Hazardous waste is an exception because it has a parallel standard that will be discussed later in this section.

The **Laboratory Safety Standard** is a more specific hazard communication program for laboratories. It can be found in 29 CFR 1910.1450. These regulations have far-reaching implications because they apply to any facility that engages in the small-scale use of hazardous chemicals that are not part of a production process. This includes all colleges, universities, hospitals, research facilities, environmental labs and product testing areas. For example, if an employer conducts a simple test to measure the water content of a product, the Lab Safety Standard would apply. These regulations are essentially a blend of hazard communication and emergency response for laboratories.

The cornerstone of the Lab Safety Standard is the requirement for a written Chemical Hygiene Plan. This plan must be capable of demonstrating protection of the employees from hazardous chemicals in the laboratory. You will see from the following list of required items that this plan is very similar to the HazCom program. It must contain:

1. An inventory of all chemicals stored or used in the lab.

2. MSDSs for each chemical.

3. Identity of the Chemical Hygiene Officer, who is responsible for safety in the laboratory.

4. Methods that the employer will use to reduce employee exposures.

5. Requirement that fume hoods and other protective equipment be functioning properly with specific measures to determine equipment performance.

6. Employee information and training.

7. Provisions for medical consultation with and medical examinations for employees.

8. Provisions for protecting employees from selected carcinogens.

The employee must be told of: the plan contents and its location, allowable workplace exposure levels for hazardous substances the employee works with, signs and symptoms of exposure, location of any referenced material, methods to detect exposures, physical and health hazards of the work area, measures to take to protect themselves, and the medical consultation and examination procedures. The standard is very thorough.

Earlier it was stated that HazCom exempted hazardous waste from the definition of hazardous chemical because there is a parallel standard for hazardous waste facilities. This standard is known as Hazwoper (pronounced haz-whopper), or **Hazardous Waste Operations and Emergency Response**. This standard essentially ties up the loose ends by covering just about everything HazCom doesn't. It covers any cleanup operation required by a governmental body, whether it is a federal, state, or local agency. An example would be an underground tank cleanup, where the hazardous material stored in the tank contaminated soil or groundwater. It covers site cleanups that result from illegal dumping, corrective actions required by RCRA, and even voluntary cleanup operations recognized by any agency. It specifies that hazardous waste treatment, storage, and disposal facilities (TSD facilities) as well as recycling facilities and landfills are covered. Finally, any emergency response actions for releases or substantial threats of releases of hazardous substances are covered by these regulations.

The Hazwoper standard is complex, but can be broken up into three major categories: Sections (a) through (o) cover hazardous waste operations including voluntary cleanup sites regulated or recognized by a government agency. Section (p) covers treatment, storage, and disposal facilities. Section (q) covers any employer whose employees are involved in emergency response activities.

One of the major components of the hazardous waste operations section is the requirement for a site-specific Safety and Health Program. This must be a written program that identifies all health and safety risks at the hazardous waste site. The person at the site who is responsible for health and safety must be identified and medical monitoring must be conducted. Training is a major component of the Hazwoper regulations. All employees at a hazardous waste site must have training certifications appropriate to the work they do. These regulations spell out the minimum amounts of training required for various site functions.

A Health and Safety Plan is comprehensive. It will identify, evaluate, and state the means of controlling safety and health hazards at the site. Site characterization and analysis is a means used to identify each physical or chemical hazard at the site. For example, unlabeled drums require a specific procedure to reduce the risk of injury to workers handling them. A site characterization would determine the condition of the drums and their contents. Work practices and standard operating procedures have to be written. Drum handling procedures and use of engineering controls must be implemented. Air monitoring must be conducted to determine the level of personal protective equipment required for each work area. Chemical protective clothing and decontamination procedures are required for each work area along with standard operating procedures for minimizing employee contamination. Illumination, sanitation, and potable water are required at each site.

All employees and contractors permanently or temporarily on site must complete 24 or 40 hours of training. In addition, managers and supervisors of the site must complete 8 hours of additional training in administrative controls measures. A written certification must be issued to each employee that receives training stating that he or she successfully completed the course. Furthermore, annual refresher training is required for each employee engaged in hazardous waste operations.

Section (p) covers hazardous waste treatment, storage, and disposal (TSD) facilities. Essentially the same requirements exist for a TSD as are mandated in the hazardous waste operations section. The primary difference is the fact that TSDs are permanent facilities and only require 24 hours of training for employees. This training must include how to implement the facility's emergency response plan. These requirements include an emergency alarm system, designated safe areas, notification procedures, and conducting regular drills to test the emergency response plan.

Section (q) covers any facility that has employees who respond to emergencies. This section requires a written Emergency Response Plan identifying pre-emergency actions, personnel roles and lines of authority, designated safe refuge areas, evacuation routes, site security and control measures, incident critique and followup procedures, available emergency response equipment, and emergency medical treatment.

In addition, Hazwoper requires the use of a standardized management system called the **Incident Command System**, or **ICS**. ICS is a structured method of managing any emergency, the intent of which is to minimize the number of bosses and maximize the number of workers, which helps keep communication and information moving in the right directions. There are seven basic group assignments used in ICS: safety, public information officer, liaison, financial, logistics, planning, and operations. Most importantly, regulations require that there be an Incident Commander who is responsible for the work being done in response to the emergency. Most communities have agreements between responding agencies as to who takes on this role. The system is used by almost all public safety agencies in the country and is a written standard of the National Fire Protection Association. The incorporation of ICS into the emergency response section was from the NFPA national consensus standard.

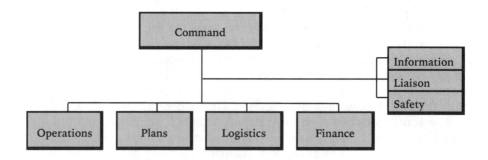

**Figure 9-18:** A sample Incident Command System (ICS).

The training requirements for emergency responders are also very strict, mandating a minimum time for each level of response. There are five levels of training. Levels 1 and 2 require 8 or more hours of training. Levels 3 and 4 require 24 or more hours, and Level 5 requires 40 or more hours of training. Employers must decide which levels are appropriate for their facilities.

■ **Level One – First Responder** *Awareness*; individuals who are likely to witness or discover a hazardous substance release and who are trained to initiate the notification procedure and secure the area.

■ **Level Two – First Responder** *Operations*; individuals who respond to releases in a defensive manner without actually trying to stop the release. These individuals must have knowledge of basic hazard and risk assessment, know how to select and use available personal protective equipment, know basic hazardous materials terms, know how to conduct basic control, containment and confinement procedures, and implement basic decontamination procedures.

■ **Level Three – Hazardous Materials** *Technicians* perform their duties in an offensive manner. Their job is to stop the release with specialized tools, plugs and patches. The requirements for HazMat Technician training is knowing how to implement the emergency response plan, classify and verify known and unknown materials using field instruments, having the ability to function within an assigned role of the Incident Command System, knowing how to select and use specialized personal protective equipment, implementing decontamination procedures, and understanding basic toxicology and chemistry.

■ **Level Four – Hazardous Material** *Specialists* are personnel who provide support to the incident. In addition to the technician level knowledge, these individuals know the local emergency response plan, understand in-depth hazard and risk assessment techniques, and have the ability to develop a site safety and control plan.

■ **Level Five – On Scene** *Incident Commanders* assume control of the overall operation of the incident. These individuals know how to implement and work within the employer emergency response plan, know the local and state emergency response plan, and know the importance of decontamination procedures.

The standard specifies the minimum requirements for emergency responders. Its original intent was to regulate the minimum training and knowledge requirements for responders working at unregulated or uncontrolled waste sites, such as areas like Love Canal or those on the national priority list. OSHA expanded the rule to include almost every employer after emergency responders at businesses suffered similar injuries and fatalities as responders at uncontrolled hazardous waste sites.

Many states have enacted even more stringent hazard communication regulations. For instance, in 1991, California OSHA enacted an innovative standard which required all employers to have and maintain an **injury and illness prevention plan**, or **IIPP** (CCR, Title 8, Subsection 3203). The plan requires the documentation of all administrative controls that the employer uses to maintain a safe workplace. The IIPP is designed to enhance management commitment to safety. Management must, for instance, implement a program where safety issues can be communicated back and forth between supervisory staff and their employees. They must have a system to ensure employee compliance with safe work practices. Scheduled inspections, an accident evaluation and investigation system, and procedures for correcting unsafe or unhealthy conditions must be demonstrated and documented. Of course, safety training and instruction and record keeping are emphasized; in fact, all employers are required to conduct at least monthly safety meetings. The intent of this regulation was to capture all employers without exception.

## ■ Checking Your Understanding (9-3)

1. Explain why PELs are less likely to be up-to-date than TLVs.

2. List two required protections for workers working in environments with airborne contaminants.

3. List two required protections for workers working in environments with asbestos.

4. List two required protections for workers working in environments with blood borne pathogens.

5. Explain how the MSDS and labels are used in hazard communication standards.

6. Explain how Hazwoper helps fill the gap left by HazCom.

7. Contrast the five levels of emergency responder.

# 9-4 Understanding Material Safety Data Sheets

## *Concepts*

■ There are 18 points of information required on a MSDS.

■ There are 8 sections to a MSDS.

■ A fire can change a non-toxic material into a toxic one.

■ Flash point, Lower Explosive Limit and the Fire Triangle are interrelated concepts.

In this section, we will take a close look at the different sections and terminology used on a MSDS. At a minimum, the Material Safety Data Sheet must have the following 18 points of information:

1. The specific chemical identification and its components (if over 1% or 0.1% if a carcinogen).

2. The **Chemical Abstracts Service (CAS#) Number** of the substance or each component that have been determined to cause health hazards. The CAS# is the specific identification number for chemicals. A product with more than one component will have a CAS# for each component.

3. The potential for fire or explosion.

4. The physical properties of the substance.

5. Conditions that may cause a reaction.

6. Acute and chronic health risks.

7. Medical conditions to avoid or symptoms of over-exposure.

8. The Permissible Exposure Limit and any other published exposure limit recommended by the preparer.

9. The primary routes of entry into the body.

10. Whether the substance or any of its components (greater than 0.1%) are known as carcinogens.

11. The recommended level of personal protective equipment to handle the substance.

12. Emergency and first aid procedures to follow for each route of entry into the body.

13. Guidance for spill cleanup and disposal.

14. Conditions for fire fighters.

15. The date of preparation and the date of the latest revision.

16. The identification and location of the preparer.

17. Potential health risks of the material explained in terms that a person without medical knowledge can understand.

18. General precautions for using and handling the substance.

At the present time, no standardized format exists. As long as the 18 points mentioned above are included on the MSDS, the chemical manufacturer and the user are in compliance. This issue has been debated since MSDSs were being developed. Figure 9-19(a) and (b) illustrate an OSHA recommended format for an MSDS.

Let's take a close look at the different sections of an MSDS:

**Section 1.** Contains the material and manufacturer identification. The name, address, and telephone number of the manufacturer must be identified. The identity of the chemical, as it is found on a label and hazardous substance inventory, must be noted. The date the MSDS was prepared or updated must be clearly indicated.

**Section 2.** Contains information about the ingredients and published exposure limits. In this section, the components of the chemical that are considered hazardous are listed along with the percentage by weight of each component. Any ingredient exceeding 1% by weight, or 0.1% by weight if it is a carcinogen, must be listed. The Chemical Abstract Service number is not required by the federal standard, but may be listed. The common names or synonyms for the chemical would be found in this section as well. The OSHA PEL or ACGIH TLV must be listed for each ingredient.

**Section 3.** Contains the physical data of the chemical including the appearance and odor of the material at 70° Fahrenheit and one atmosphere of pressure. In this section, information about the physical characteristics of the material are found. For instance, the boiling point, which is the temperature in degrees Fahrenheit, at which a liquid changes state into a gas, would be found in this section. Water boils at 212° Fahrenheit; at 212.1° Fahrenheit, liquid water would rapidly change to steam. The vapor pressure of a material is also an important piece of information found in Section 3. The **vapor pressure** tells us how much pressure, in millimeters of mercury, that the vapor above a liquid exerts in a closed container. This helps us to understand how much a liquid wants to be a gas, or, stated another way, how easily it evaporates. The higher a liquids vapor pressure, the more its molecules want to escape to the gas form, hence evaporate. For example, water has a vapor pressure of 24 mm of mercury at standard temperature and pressure and 50 % humidity. Ethanol has a vapor pressure of 50 mm of mercury under the same conditions. Which chemical evaporates faster and by relatively how much faster?

The **specific gravity** refers to the ratio of the weight of the liquid material compared to the weight of an equal volume of water. The specific gravity of water is defined as 1.0. Materials with a specific gravity greater than one will generally sink (the exception is when the material entirely dissolves or mixes with water), and those chemicals with a specific gravity less than one will generally float (with the same exception noted above). A material that floats will generally spread out on and cover the surface of the liquid it is floating on. This is why using water to try to put out a fire of a flammable material that floats on water only serves to spread the fire.

Vapor density or vapor specific gravity refers to the ratio of the weight of a gas compared to the weight of air. The vapor density of air is given as 1.0. Any gas or vapor with a vapor density greater than 1 would tend to sink to low areas. Those with vapor densities of less than 1 would tend to rise in air. Liquids must change to a vapor or gas for the vapor density to have meaning.

**Section 4.** Contains information on fire and explosion hazards. The flash point, flammable limits (flammable range) and special fire fighting procedures are presented. It is important to note that for a flammable liquid, it is not the liquid that burns, but its vapors. In order to understand the fire-related characteristics terms that would be found in this section, it is important to review some basic fire chemistry principles.

When a material burns, it is actually undergoing a rapid chemical reaction that breaks the bonds of the flammable material. The **fire triangle** is a concept used by safety trainers to help people understand the components needed to start a fire. The three components are fuel-heat-oxygen. Fuel can take the more common forms were familiar with, such as wood, paper, or coal, or it can be a flammable liquid such as gasoline, acetone, or another solvent. We generally think of heat sources as ignition sources, like matches, but they can also be a chemical reaction or due to friction. The heat (or ignition) source must provide enough energy to start the chain reaction or fire. For most combustible materials this is approximately 400° Fahrenheit, and enough oxygen must be present to sustain the burning. The volume of oxygen varies from chemical to chemical, but in general approximately 6 to 10 percent oxygen by volume is enough to sustain a fire. Take away one part of the fire triangle, and the fire will go out.

Recall that chemical reactions can provide sufficient heat to start fires. This is actually fairly common with flammables and oxidizers, which, when mixed together can immediately burst into flames. This is, in fact, one of the leading causes of fires in chemical storage areas. When the oxidizer (recall that an oxidizer is a chemical that yields oxygen readily) and fuel react, the reaction itself provides the necessary heat for the fire, and of course, the oxidizer helps out by providing even more oxygen for burning the fuel. It is important to remember that in all cases of burning, the three components must be in the right quantities or the chain reaction does not take place.

Fire needs heat to get started, but once the process of burning is started, a large quantity of energy in the form of heat and light is released. The energy evolved is the difference between breaking and reforming bonds for such things as $CO_2$ and $H_2O$. So, once a fire is started, the heat takes care of itself as long as there is sufficient fuel and oxygen to sustain the fire. Remember that the correct ratio of fuel to oxygen is needed for this to occur. Another way of saying this is that the fuel must be within its flammable range. The **flammable range** of a fuel represents the entire range of percentages of that fuel in air that will burn in the presence of an ignition source (heat). The flammable limits are listed as percent by volume in air. The **LEL**, or **lower explosive limit**, is the minimum concentration of a flammable gas in air required for ignition in the presence of an ignition source. The **UEL** or **upper explosive limit**, is the maximum concentration of a flammable gas in air required for ignition in the presence of an ignition source. The flammable range is the concentration between and including the LEL and UEL. Materials with wide flammable ranges such as acetylene are difficult to control.

The **flash point** is the only relative indicator of a materials flammability. It is defined as the minimum temperature at which a flammable liquid will produce sufficient vapors to cause a temporary flame, when exposed to an ignition source, across the surface of the liquid. If we apply the fire triangle, lower explosive limit and flash point, we can visualize the interrelationship between these concepts.

Let's use gasoline as an example. Gasoline has a flash point of approximately 45° Fahrenheit. At room temperature, the gasoline may produce vapors in sufficient quantity to develop a lower explosive limit concentration in the immediate area. If we add a heat source, like a cigarette, then we have completed the fire triangle, and an instantaneous fire will start. With a slight breeze, the fuel source (gasoline vapors) may be below the LEL.

A flammable atmosphere is another OSHA-regulated condition that presents a common physical hazard in the workplace. It is interesting to think back to our discussion of PELs and look at how this concentration relates in magnitude to the flammable range numbers. The minimum amount of gasoline vapor needed for a fire exceeds the PEL by a factor of 47. The PEL for gasoline is 300 parts per million. The Lower Explosive Limit is 1.4% by volume or 14,000 parts per million. 14,000 ppm divided by 300 ppm equals about 47. This shows that concentrating on the health hazards will minimize the fire danger.

**Section 5.** Presents general information about health hazards. Any typical acute or chronic symptoms from exposure should be provided. Remember that acute symptoms arise immediately after a

## MATERIAL SAFETY DATA SHEET

### SECTION I  NAME

| | | | |
|---|---|---|---|
| Product | Cyclohexene | **CHEMTREC** | |
| | | 800-424-9300 | **Health** 2 |
| Chemical Synonyms | | | **FLAMMABILITY** 4 |
| | | Day 312-468-3700 | **REACTIVITY** 0 |
| Formula | $C_6H_{10}$ | **HAZARD RATING** | |
| | | LEAST  SLIGHT  MODERATE  HIGH  EXTREME | |
| RETECS # | GW2500000 | 0    1    2    3    4 | |
| C.A.S. No. | 110-83-8 | | |

### SECTION II  HAZARDOUS INGREDIENTS OF MIXTURES

| Principal Hazardous Component(s) | % | TLV Units |
|---|---|---|
| | | |
| NA = Not Applicable | | |
| NE = None Established | | |

### SECTION III  PHYSICAL DATA

| | | | |
|---|---|---|---|
| **Melting Point (°F)** | -104 °C | Specific Gravity (H₂O=1) | 0.81 |
| **Boiling Point (°F)** | 83 °C | Percent Volatile by Volume (%) | 100 |
| **Vapor Pressure** (mm Hg) | 160 @ 38 °C | Evaporation Rate ( =1) | NA |
| **Vapor Density** (Air = 1) | 2.8 | | |
| **Solubility in Water** | Insoluble | | |
| **Appearance & Odor** | Clear colorless liquid, etheral odor | | |

### SECTION IV  FIRE AND EXPLOSION HAZARD DATA

| | | | Lower | Upper |
|---|---|---|---|---|
| Flash Point | -20 °C | Flammable Limits in | | |
| (Method Used) | Closed Cup | % by volume | 1.5 | 6.6 |
| Extinguisher Media | Dry chemical, CO₂ or foam. | | | |
| **SPECIAL FIRE FIGHTING PROCEDURES** | In event of a fire, wear full protective clothing and NIOSH-approved self-contained breathing apparatus with a full facepiece operated in the pressure demand or other positive pressure mode. | | | |
| **UNUSUAL FIRE AND EXPLOSION HAZARDS** | Flammable Liquid. Autoignition temperature 244 °C. Above flash point, vapor-air mixtures are explosive within flammable limits noted above. Vapors can flow along surfaces to distant ignition source and flash back. | | | |

**NFPA Hazard Rating**   1 / 3 / 0

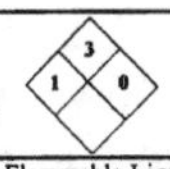

| D.O.T. | Flammable Liquid NOS   UN1993   Cyhclohexene |
|---|---|

Approved by US Department of Labor "essentially similar" to form OSHA-174

**Figure 9-19(a):** A Material Safety Data Sheet (part one).

high exposure. Chronic symptoms arise after a low exposure over a long period of time. The chemical's toxicity will be noted with an $LD_{50}$ value. Recall that this information is typically from epidemiological data from an animal study. Caution should be exercised in using this $LD_{50}$ information because, given the many animal studies conducted, the value that reflects the lowest toxicity is usually selected. Also, the extrapolation from animals to humans has always been a source of dispute amongst scientists.

In addition, all routes of entry (Recall from Chapter 3 the routes of entry) into an unprotected body are identified. Any medical

**SECTION V   HEALTH HAZARD DATA**

| | |
|---|---|
| Threshold Limited Value | 300 ppm TWA |
| Effects of Overexposure | Absorbed thru skin. **Acute:** Irritation to skin, eyes, lungs and GI tract. headache, nausea, CNS depression, dizziness, narcosis. **Cronic:** Dermatitis, respiratory sensitization. |
| Emergency and<br><br>First Aid Procedures | **Skin:** Remove any contaminated clothing. Wash skin with soap or mild detergent and water for at least 15 minutes. Get medical attention if irritation develops or persists. **Eyes:** Wash eyes with plenty of water for at least 15 minutes, lifting lower and upper eyelids occasionally. Get medical attention immediately. **Inhalation:** Remove to fresh air. If not breathing, give artificial respiration. If breathing is difficult, give oxygen. Call a physician. **Ingestion:** Give several glasses of water or milk to drink. Vomiting may occur spontaneously, but DO NOT INDUCE! Never given anything by mouth to an unconscious person. Get medical attention immediately. |

**SECTION VI   REACTIVITY DATA**

| Stability | Unstable | | Stable   X | Conditions to Avoid   Heat, ignition sources |
|---|---|---|---|---|
| **Incompatibility**<br>**(Materials to avoid)** | | Oxidizing Agents | | |
| **Hazardous**<br>**Decomposition Products** | | Carbon oxides | | |
| **Hazardous Polymerization** | | | **Conditions to Avoid**   None | |
| May Occur | Will Not Occur<br>   X | | | |

**SECTION VII   SPILL OR LEAK PROCEDURES**

| | |
|---|---|
| **Steps to be taken in case material is released or spilled** | Ventilate area of leak or spill. Remove all sources of ignition. Clean-up personnel require protective clothing and respiratory protection |
| from vapors. Small spills may be absorbed on paper towels and evaporated in a fume hood. Allow enough time for fumes to clear hood, then ignite paper in a sutiable location away from combustible materials. Contain and recover liquid when possible. | |
| **Waste Disposal Method** | Can be liquid incinerated in RCRA-approved combustible chamber or absorbed with vermiculite, dry sand, earth, or similar material. Scoop up with non-sparking tools and place |
| in a closed container, and dispose in a RCRA-approved facility. Do not flush to the sewer. Discharge, treatment, or disposal may be subject to federal, state, or local laws. | |

**SECTION VIII   SPECIAL PROTECTION INFORMATION**

| **Respiratory Protection**<br>(Specify Type) | Organic cartridge or self-contained | | | |
|---|---|---|---|---|
| **Ventilation** | Local Exhaust | X | **Special** | |
| | Mechanical (General) | X | **Other** | |
| **Protective Gloves** | Vinyl or Rubber | | **Eye Protection** | Goggles |
| **Other Protective Equipment** | | Lab Coat | | |

**SECTION IX   SPECIAL PRECAUTIONS**

| | |
|---|---|
| **Precautions to be Taken in Handling & Storing** | Store in a cool well ventilated place |
| Keep container closed when not in use | |
| **Other Precautions** | Read Label on Container before using.<br>Conditions Aggravated/Targen Organs: Persons with preexisting skin, eye, respiratory, or kidney liver disorders may be more susceptible |
| For laboratory use only. Not for drug, food or household use. Keep out of reach of children. | |

| Rev. No.   2 | Date 6/2/96 | Approved | I. M. O'Kaye | Chemical Safety Coordinator |
|---|---|---|---|---|

The information contained herein is furnished without warranty of any kind. Employers should use the information only as a supplement of other information gathered by them and must make independent determinations of suitability and completeness of information from all sources to assure proper use of these materials and the safety and health of employees.

**Figure 9-19(b):** A Material Safety Data Sheet (part two).

conditions that are aggravated by exposure are listed. The emergency first aid procedures for each route of entry identified are discussed.

**Section 6.**   Contains information on the stability and reactivity of the chemical. Stability is defined as the susceptibility of the material to decomposition. Knowing whether a chemical is stable will help answer the question as to what will happen to it, over time, if it is left alone in its container or out in the open. Will it form new substances that can explode or spontaneously erupt into flames, or will it stay just as it was if left for a long period of time? Even if the material is marked as stable, there may be certain conditions which should be avoided, e.g. temperature extremes, incompatible chemicals, or storage conditions. Reactivity tells you when, under what conditions, and with which chemicals the material will cause a reaction.

**Section 7.** Provides information on precautions for safe handling and use. This section also describes the steps that should be taken in case of material spillage or release. Any precautions for handling or storage, such as keeping the material away from sparks or static grounding requirements, are also noted. This information is general in nature and is not meant to replace federal, state, or local regulations governing hazardous waste disposal or spill cleanup.

**Section 8.** Under this section, special protection information and control measures are described. This section provides a recommendation for the type of personal protective equipment that should be used when handling the equipment under normal conditions. In addition, the classification for electrical equipment use will be stated, such as "use only non-sparking fans and motors."

MSDSs contain a great deal of information and are difficult to read even for a trained professional. The important point to remember is that the MSDS is only one source of information. Additional sources can be used to help employees understand the hazards they work around. When training employees on a MSDS, it is imperative that employees take the time to learn and understand the new terms and ask questions on items unknown to them. With this in mind, annual refresher training is not mandatory, but it is highly recommended.

## Some Chemical Safety Rules of Thumb

- The lower the boiling point, the higher the vapor pressure.
- A material with a high vapor pressure will evaporate rapidly, producing a vapor cloud.
- Keeping a material below its flash point will prevent the material from releasing a flammable vapor; it will reduce the chances of starting a fire.
- Flammable gases, vapors, and solids such as dusts burn; liquids do not burn, but remember, most flammable liquids evaporate very rapidly.
- If a worker concentrates on preventing health exposures, then the fire potential is reduced dramatically.
- Permissible Exposure Limits and Threshold Limit Values are established for and apply only to airborne hazards and are established for healthy workers only.
- Inhalation is the greatest health risk to a worker, but skin contact with chemicals is the most common and causes the most injuries.
- The lower the PEL or TLV the greater the potential harm of the chemical in the workplace.
- The flash point is the only relative indicator of flammability; the lower the flash point the greater the fire danger.
- 1 percent by volume is equal to 10,000 parts per million.

## Science & Technology

# The Chemical and Physical Properties of Elements and Compounds

To a chemist, matter is all the "stuff" around us. Over the years chemists have sorted it into two groups – elements and compounds. There are only about 89 naturally occurring elements and a few others that have been "man made." The number of compounds that can be made from these elements, however, number in the millions, with new ones being made or discovered every day.

Each element and compound has a unique set of physical and chemical properties, much as we do. Everyone has a skin, hair, and eye color, a height and a weight. These are examples of some of our physical properties. We also have personalities. Some of us show little emotion and are slow to anger, while others of us might be labeled as "hot heads." It's how we get along with others; some would call it our personal chemistry. In a similar fashion, some elements and compounds are very unemotional; in fact, they are considered to be inert, like Teflon and helium. Others, like the alkali metals and **halogens** are known for their fiery reactions with most other substances.

OSHA now requires manufacturers to distribute a Material Safety Data Sheet (MSDS) with each of its products. Included in the 18 items required in its eight sections is information about both the substance's physical and chemical properties.

Today we believe that most of the chemical and physical properties have to do with how the substance's electrons are arranged. For elements, it is a matter of how many electrons it has in its outer energy level. According to the **Octet Theory**, those having eight electrons tend to be very calm and do not readily react with other elements. Those having just less than eight turn out to be some of the most active elements. These elements will gladly combine with almost any others in an attempt improve their electron arrangements. This may be accomplished by either giving up or taking each other's electrons or by sharing them.

The compounds resulting from giving and taking electrons are called ionic compounds. Their physical properties generally include being water soluble; not soluble in organic solvents; and being composed of hard crystals with both high melting and boiling points. Table salt (sodium chloride, $NaCl$) is a typical example of this type of substance. Chemically, most inorganic compounds do not burn. Their other chemical properties are dependent on the nature of both their positively and negatively charged ions. Their chemical characteristics can range from essential, as in the case of the ferrous ion ($Fe^{2+}$) used in making hemoglobin, to deadly, as in the case of the cyanide ion ($CN^-$).

For those compounds that form shared (covalent) bonds, their physical properties vary greatly. If the electrons are shared equally, called nonpolar covalent, their physical properties will include, being insoluble in water, but soluble in organic solvents. They will also be composed of soft, waxy or greasy crystals that have low melting and boiling points. Lard is a good example of a nonpolar organic compound. Most tend to be organic compounds that burn readily when mixed with air. They, too, exhibit a range of chemical characteristics from essential, as in the case of vitamins, to deadly, as in the case of dioxin.

The other type of covalent compound results from an unequal sharing of electrons. These are called polar covalent compounds. The term "polar" was borrowed from our understanding of magnets, each with its own magnetic North and South pole. In this instance, it results from the electrons not being equally spaced around the two nuclei. Much like when you drop two eggs into a skillet, the yolks remain separate, but the whites blend together. However, as the whites blend, the yolks do not stay in the exact center of the blended whites. It is this lopsidedness of the electrons around the two nuclei that results in the polar covalent bond. Having this lopsidedness causes the resulting molecule to have North and South poles, just like a magnet.

When two polar molecules are close together, there is an attraction between these North and South poles, much like two small magnets would seek the other's opposite end. When substances composed of polar organic molecules are put together there is also an attraction between the differing molecules. When hydrogen is involved in one of the attracting molecules, then this is referred to as a hydrogen bond. Hydrogen bonding is not a strong bond like ionic or covalent bonding, but rather the weak attraction between polar covalent molecules.

Since water itself is one of these polar covalent substances, other polar covalent substances are highly attracted to it. The hydrogen bonds formed

## Science & Technology

result in some significant changes in the physical properties of the substance. These substances may range from partially to completely soluble, depending on the size of its molecules. This is why alcohols, like sugar and antifreeze, can be completely dissolved in water.

It also changes the substance's melting and boiling points, making it harder (higher temperature required) to both melt and boil and reach its vapor pressure. For a substance to evaporate, molecules must leave the surface and go flying off into space. If all the neighboring molecules in the substance are pulling on the molecule trying to escape, it will take more energy for them to get away.

As you can see, understanding just a few basic chemistry principles can greatly help explain much of the information contained on the MSDS. As your knowledge of chemistry and chemical substances grow, you too will be better prepared to understand the chemical and physical properties of a wide range of substances.

## ■ Checking Your Understanding (9-4)

Answer the following questions using the sample MSDS.

1. Describe this chemical's potential for fire or explosion.

2. List the acute and chronic health effects that may result if workers are exposed to this material.

3. Describe the personal protective equipment and clothing that should be used when handling this material.

4. Explain what the listed exposure level means as it relates to workers that work with this chemical every day.

5. Describe the steps that should be taken in case of an emergency spill or release of this chemical.

# 10

# Nonhazardous, Hazardous, and Nuclear Waste

## ■ Chapter Objectives

1. **Describe** the basic provisions of RCRA as they pertain to the regulation of solid waste, hazardous waste, and medical waste.

2. **List** major regulatory responsibilities for hazardous waste generators, transporters, and TSD facilities.

3. **Explain** the impact of both consumers and industry on the generation of solid and hazardous waste.

4. **Describe** the types of wastes generated by medical facilities and the potential problems they pose for handling and disposal.

5. **List** the range of sources that contribute to nuclear waste.

6. **Describe** the three classifications of nuclear waste and particular challenges for disposal of each.

7. **Define** basic terms related to radioactivity.

## ■ Terms and Concepts to Remember

Alpha Particles
Background Radiation
Beta (β) particles
Characteristic Waste
Delist
Disintegration Series
Free Radicals
Gamma Rays
Generation
Half-life
Hazardous and Solid Waste
  Amendments (HSWA)
High-Level Waste
Ignitable
Incineration
Indoor Air Pollution
Ion
Ionization
Ionizing radiation
Land Ban
Listed Waste
Low-Level Waste
Metric ton
Moderator

Non-ionizing Radiation
Nuclear Fission
Nuclear Regulatory
  Commission (NRC)
Plutonium
Radiation
Radiation Absorbed Dose (Rad)
Radioactive Decay
Radioisotopes
Radionuclide
Radiation Equivalent Man
  (Rem)
Recycle
Repository
Reprocessing
Small Quantity Generator
  (SQG)
Solid Waste
Spent Nuclear Fuel
Transuranic waste
Treatment, Storage, and
  Disposal Facility (TSD)
Uranium
Waste Treatment Plant

## ■ Chapter Introduction

In the first chapter of this text, we introduced the term hazardous waste and defined it as a material that has no intended use and has at least one of four hazardous characteristics – ignitability, toxicity, corrosivity, or reactivity. A hazardous material is essentially the same, only it has a useful purpose. A waste is anything we no longer need. If it is not hazardous, it is **solid waste**, i.e., garbage.

And garbage is piling up! Over the years we have realized that we need better solid and hazardous waste management systems to deal with the huge volumes of waste being generated. The Resource Conservation and Recovery Act (RCRA) was established in 1976 to do this.

In this chapter, the legal definitions and finer distinctions between different types of wastes are covered. The basic categories that are discussed include nonhazardous solid wastes, hazardous wastes, nuclear wastes, and medical wastes.

# 10-1 RCRA and Solid Waste

## ■ History and Overview

### Concepts

■ Generation of wastes, both hazardous and nonhazardous, continues to increase.

■ RCRA, as amended by HSWA, is the federal law that regulates waste disposal and conservation of resources.

■ Subtitles C, D, I and J of RCRA cover requirements for management of wastes as well as underground tanks.

Before the nineteenth century, the amount of waste produced was small and its impact on the environment relatively minor. Conventional wisdom at the turn of the century held that a river purified itself every ten miles. This is just one example of the misconception that the environment could take care of all our waste on its own. We now know that the ability of natural as well as human-made systems to deal with the huge volume of wastes this nation produces is seriously inadequate.

**Figure 10-1:** The amount of solid waste generated in the United States has more than tripled since the 1950s.

Starting during the period of the industrial revolution in the mid-1800s, this nation experienced phenomenal growth. An explosion of new products hit a consumer market that has continued to grow. This growth, however, was not entirely positive. These new goods resulted in more waste, both hazardous and nonhazardous. "Miracle" products, such as plastics, nylon stockings, and coated paper goods, found a ready market as soon as industry introduced them in the 1950s. Our appetite for material goods created a problem: how to manage the increasing amount of waste produced by industry and consumers. To put this growth in perspective, industries in this country generated about 500,000 **metric tons** (550,000 tons) of hazardous waste per year in the years just after World War II. Just 40 years later, the EPA estimates that about 275,000,000 metric tons (302,500,000 tons) of hazardous waste were generated nationwide per year!

Progress in managing all of the wastes produced by industry and consumers had not matched the rate at which we were generating increased wastes. Much of the waste found its way into the land, air, and water where it now poses a serious threat to public health and the environment.

By the second half of this century it was obvious that garbage and other wastes were causing serious problems and we needed better disposal methods for their growing volumes. To remedy this, in 1965 Congress passed the Solid Waste Disposal Act, the first federal law to require safeguards and encourage environmentally sound methods for disposal of household, municipal, commercial, and industrial wastes. It established grant programs so that states could develop solid waste management plans. In the mid-1970s it became clear that a more comprehensive nationwide program was required to make sure that solid and hazardous wastes were disposed of properly. The Resource Conservation and Recovery Act (RCRA) is the result. It was enacted in 1976, and amended the original Solid Waste Disposal Act. One significant change from the original law was that RCRA expanded requirements for *hazardous waste* management. RCRA has been amended several times since 1976, most significantly in 1984. The 1984 amendments, called the **Hazardous and Solid Waste Amendments (HSWA),** again significantly expanded the scope and requirements of RCRA. The Act will no doubt continue to evolve as Congress amends it to reflect our changing needs. This section will highlight the basic provisions of RCRA, as amended.

The primary goals of RCRA are:

■ To protect human health and the environment from the potential hazards of waste disposal.

■ To conserve energy and natural resources.

■ To reduce the amount of waste generated, including hazardous waste, as quickly as possible.

■ To ensure that wastes are managed in an environmentally sound manner.

To achieve these goals, four programs were brought together: a *solid waste* management program, a *hazardous waste* management program, an *underground tank* management program, and a *medical waste* program. These four distinct, yet interrelated programs, make up RCRA.

RCRA is currently divided into ten subtitles, with the letter designations A through J. Subtitles A, B, E, F, G, and H outline general provisions including the authorities of the Administrator and federal responsibilities. Subtitles C, D, I, and J spell out the framework of the four programs that establish management procedures for wastes and underground tanks. See Figure 10-2 for an outline of RCRA. RCRA requires the EPA to develop regulations to implement its provisions.

The solid waste program is found in Subtitle D of the Act. Solid waste is defined in RCRA as "garbage, refuse, or sludge or any other waste material." Subtitle D addresses solid waste management planning by the states, which are required to develop plans intended to promote recycling and reduction of solid wastes. Subtitle D also requires that all environmentally unsound solid waste dumps be closed or upgraded. It includes very specific requirements for protection of groundwater from the leachate that may result from the wastes buried in a municipal landfill. This subtitle requires lining and leachate collection systems in any landfill that is to accept solid waste.

Subtitle C of the Act addresses the management of hazardous waste from "cradle to grave." RCRA defines hazardous waste as a solid waste that "because of its quantity, concentration, or physical, chemical, or infectious characteristics, may cause, or significantly contribute to, an increase in mortality or an increase in serious irreversible, or incapacitating reversible illness; or pose a substantial present or potential hazard to human health and the environment when improperly treated, stored, transported, or disposed of, or otherwise managed." A more complete definition is covered in the following section on hazardous waste management under RCRA. The ultimate goal of this subtitle is to ensure that hazardous waste is managed in a manner that protects both human health and the environment. To accomplish this, EPA regulations address the **generation**, transportation, treatment, storage, and disposal of hazardous wastes. To generate means to originate or produce. The EPA estimates that there are nearly 200,000 large and small quantity generators of hazardous waste in this country (Figure 10-3).

Subtitle C is considered one of the most comprehensive laws that Congress has ever developed. It starts by defining and classifying a hazardous waste then goes on to establish management requirements for generators, transporters, and owners and operators of **treatment, storage, and disposal facilities (TSD facilities)**. It also sets technical standards

| Outline of RCRA | |
|---|---|
| Subtitle | Provisions |
| A | General Provisions |
| B | Office of Solid Waste; Authorities of the Administrator |
| C | Hazardous Waste Management |
| D | State or Regional Solid Waste Plans |
| E | Duties of the Secretary of Commerce in Resource and Recovery |
| F | Federal Responsibilities |
| G | Miscellaneous Provisions |
| H | Research, Development, Demonstration, and Information |
| I | Regulation of Underground Storage Tanks |

**Figure 10-2:** Outline of RCRA Subtitles.

**Figure 10-3:** A dry cleaner is a typical small quantity generator.

**Figure 10-4:** Medical waste washing up on Atlantic beaches publicized the problem of proper waste disposal.

for the design and safe operation of TSD facilities as well as the specific requirements for necessary permits. The issuance of permits is one of the critical components of the Subtitle C program since it is through permits that the EPA or a state actually applies the technical standards to facilities.

It is important to note that the main difference between Subtitle C and Subtitle D is the type of waste that each addresses. Subtitle C addresses only activities relating to managing hazardous waste, which is a subset of solid waste. Subtitle D addresses nonhazardous solid waste.

Subtitle I was introduced in Chapter 8 on hazardous materials. The reason for this is obvious – Subtitle I is unique in that it does not cover management of a waste; instead it covers a variety of substances that are contained in underground tanks. Specifically, Subtitle I addresses petroleum products and hazardous substances (as defined in CERCLA), which are stored in underground tanks. Subtitle C covers underground tanks containing hazardous waste, but tanks containing hazardous substances far outnumber those that contain hazardous waste.

The goal of Subtitle I is to prevent leakage from tanks into groundwater and to clean up past releases. This is another program, like Subtitle C, that may be delegated to the states. Under this program, the EPA has developed performance standards for new tanks and regulations for leak detection, prevention, closure, financial responsibility, and corrective action at all underground tank sites.

Subtitle J, the most recent addition to RCRA, was added to deal with the problem of mismanagement of medical waste. In the summer of 1988, there were several instances of medical waste washing up on Atlantic coast beaches causing widespread concern. These events served to highlight the inadequacy of medical waste management practices. Subtitle J was designed as a demonstration program to study the problem further and report back to Congress with suggestions for improvement.

The regulations that implement RCRA are found in 40 CFR, Parts 240–280. In Chapter 2, we learned that regulations specify implementation details that expand on the intent of the statute. This relationship is a little different in the case of the Hazardous and Solid Waste Amendments (HSWA) because Congress placed explicit requirements in the statute in addition to instructing the EPA to develop regulations. In fact, many of these requirements are so detailed that the EPA incorporated them directly into the regulations. HSWA also contains very ambitious deadlines and hammer provisions for implementing the Acts requirements. Hammer provisions are statutory requirements that go into effect automatically as regulations if the EPA fails to issue regulations by specific dates.

The EPA also issues guidance and policy documents to help with the implementation of the statutes and regulations. Guidance documents, which are not laws, simply serve to elaborate on and provide direction for implementing regulations. Essentially, they explain how to do something. For example, the regulations in 40 CFR Part 270 require a hazardous waste management facility to obtain a permit. The guidance document for this Part gives instructions on how to evaluate a permit application to ensure that everything has been included.

Policy statements, which also are not laws, specify operating procedures that should be followed. They are used by EPA offices to outline the manner in which they will carry out the provisions of the RCRA program. For example, as we saw in Chapter 2, an enforcement office may issue a policy outlining what enforcement action must be taken if a specific groundwater violation is found. In most cases, policy statements are addressed to the regulatory staff working on implementation of the law and regulations.

There are many guidance and policy documents that have been developed to help implement the RCRA program. The Office of Solid Waste and Emergency Response is the office within the EPA that is responsible for developing and implementing regulations that address the provisions of RCRA. This office also maintains a list of RCRA-related documents, including memoranda, guidance documents and policy documents that are available from the EPA. These lists can be obtained from Regional Policy Directives Coordinators.

The EPA can delegate implementation of RCRA provisions to states that meet certain criteria, allowing them to be authorized. An authorized state may be provided with technical and financial assistance. The general public is provided with many opportunities to comment and provide input throughout all stages of state or federal program development and implementation.

Unfortunately, this nations experience with hazardous waste site cleanups has not been good. Chapter 8 highlighted the high costs and dismal track record of CERCLA in providing satisfactory cleanup of the many contaminated sites that have been identified. None of the states have a much better track record. These negative clean-up experiences along with the rising costs of disposal, limited permitted disposal sites, and the difficulty in siting new facilities to meet future needs has made it clear that the best way to manage waste is not to generate it in the first place. Congress noted this and made waste minimization a major goal of Subtitle C. Waste minimization means the reduction, to the extent feasible, of hazardous waste generated prior to any treatment, storage, or disposal of the waste. It is a source reduction or recycling activity that results in either reduction of the total volume of hazardous waste or reduction of toxicity of hazardous waste, or both.

Although RCRA creates programs requiring the proper management of hazardous waste, it does not deal with problems related to past sins such as contamination at inactive or abandoned disposal sites or those resulting from spills that require emergency response. Problems associated with past mismanagement of hazardous wastes are covered by CERCLA which was covered in Chapter 8. The one exception is Subtitle I which enabled the EPA to establish what is known as the Leaking Underground Storage Tank Trust Fund Program to clean up certain underground storage tank leaks that threaten human health and the environment.

The next sections will describe the problems associated with management and disposal of all types of wastes. We will start with the most common to all of us, municipal solid waste, and work our way through industrial hazardous waste, medical waste, and nuclear waste.

## ■ The Solid Waste Problem

### Concepts

■ Our per capita waste generation is over double that of any other country.

■ Current methods for managing and disposing of solid waste are not expected to be adequate for anticipated future volumes.

■ The primary methods for managing solid waste are landfilling, recycling, and incineration.

> *What we throw away may be the closest we come to our pollution of Earth.*
>
> from *Saving the Earth, A Citizen's Guide to Environmental Action* by Will Steger

Throughout the world, solid waste woes make the news on a regular basis. Developing countries aspire to the standard of living enjoyed in the United States while we struggle to find ways to deal with the ever-increasing waste our lifestyle creates. Consumers in this coun-

| How Our Per Capita Garbage is Changing (in pounds per day, with projections to 2000) | | | | | | |
|---|---|---|---|---|---|---|
| Waste Materials | 1960 | 1970 | 1980 | 1990 | 1995 | 2000 |
| Total nonfood product wastes | 1.65 | 2.25 | 2.57 | 2.94 | 3.18 | 3.38 |
| Paper and paperboard | 0.91 | 1.19 | 1.32 | 1.60 | 1.80 | 1.96 |
| Glass | 0.20 | 0.34 | 0.36 | 0.28 | 0.23 | 0.21 |
| Metals | 0.32 | 0.30 | 0.35 | 0.34 | 0.34 | 0.35 |
| Plastics | 0.01 | 0.08 | 0.19 | 0.32 | 0.39 | 0.43 |
| Rubber and leather | 0.06 | 0.09 | 0.10 | 0.10 | 0.10 | 0.11 |
| Textiles | 0.05 | 0.05 | 0.06 | 0.09 | 0.09 | 0.09 |
| Wood | 0.09 | 0.11 | 0.12 | 0.14 | 0.16 | 0.17 |
| Other | 0.00 | 0.02 | 0.07 | 0.07 | 0.06 | 0.06 |
| Other wastes | | | | | | |
| Food wastes | 0.37 | 0.34 | 0.32 | 0.29 | 0.28 | 0.27 |
| Yard wastes | 0.61 | 0.62 | 0.66 | 0.70 | 0.70 | 0.70 |
| Miscellaneous inorganic wastes | 0.04 | 0.05 | 0.05 | 0.06 | 0.06 | 0.06 |
| Total waste generated | 2.66 | 3.27 | 3.61 | 4.00 | 4.21 | 4.41 |

**Figure 10-5:** Changes in amount of garbage generated.

try are presented with choices of consumer products that grow at a staggering rate. Cynthia Pollock, of the Worldwatch Institute, has stated that our consumer habits are fed by industries that design goods "for a one-night stand." Marketing and merchandising strategies have shaped a market that buys for the short term. How often have you heard someone remark that it is cheaper to toss out the VCR and buy a new one than repair the old one? This is why more than one commentator has called us the "throw-away society."

In order for mass produced goods to make their way from the factory to the consumer, they must be differentiated from other similar mass produced goods. Slick and colorful and carefully designed packaging generally amounts to nearly half of the volume of household waste. Take stock of the volume of paper in the form of newspapers, magazines, repeated draft print-outs on the computer, and junk mail that you receive or generate as waste on a weekly basis. Its staggering. In fact, our generation of waste is growing at such a pace that a study conducted in Montgomery County, Maryland indicated that in 20 years, even if 30 percent of the waste is recycled, we will still be generating more waste than we generate today (Figure 10-5). Each person in the United States produces more solid waste than persons anywhere else in the world. The average amount of waste generated per person in this country, about four pounds per day, is over twice that of the per capita waste generation of any other country. That adds up to over half a ton a year! In the 15 years between 1970 and 1985, while the population increased about 18 percent, our waste output increased 25 percent. Figure 10-6 shows waste generation in selected cities.

Even though these figures represent a lot of waste, still by far the largest generators of solid waste are industrial, agricultural, and mining sources (Figure 10-7). Mining alone generates 3.6 billion tons of waste a year. Agricultural and mining wastes are typically disposed of on the same land where they were generated. This

| Waste in Major World Cities (in pounds per capita) | |
|---|---|
| City | Waste |
| Los Angeles | 6.4 |
| Philadelphia | 5.8 |
| Chicago | 5.0 |
| New York | 4.0 |
| Tokyo | 3.0 |
| Paris | 2.4 |
| Toronto | 2.4 |
| Hamburg | 1.9 |
| Rome | 1.5 |

**Figure 10-6:** Waste generated in major world

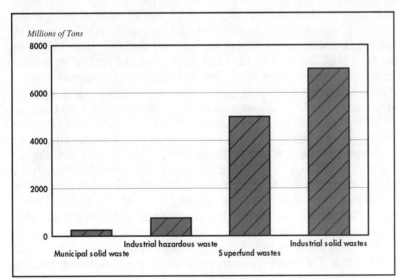

**Figure 10-7:** Sources of solid waste in the U.S. (Source: Council on Environmental Quality, United States of America National Report [prepared for the United Nations Conference on Environment and Development, 1992])

**Figure 10-8:** The garbage barge Mobro 4000 was not allowed to dock.

creates the potential for water pollution from runoff in these areas. Agricultural waste, which includes animal wastes, presents another source of water pollution. When animals are confined in relatively small areas, such as in feedlots, their wastes tend to concentrate and leach through the soil into groundwater, or runoff carries them to surface water sources. Before the era of vast agri-business enterprises, animal wastes were valued as excellent fertilizers since they contain valuable nutrients for plant growth. This is just another example of too much of a good thing not being better in the long run.

An unusual situation occurred in 1987 that served to focus our attention on the waste problems we had previously ignored. A floating garbage barge called the Mobro 4000 was carrying over 3,000 tons of garbage destined for a landfill in Islip, New York but was turned away when it arrived there (Figure 10-8). It then proceeded to find a home for the garbage in North Carolina, Florida, Alabama, Mississippi, Louisiana, and even Mexico, the Bahamas, and Belize. All rejected it. The Mobro's disastrous voyage, which made international news, finally ended 6,000 miles later at an incinerator at the same landfill where it had started. The Mobros voyage has since become a symbol of the dilemma surrounding the fact that we are running out of places to put our garbage.

Government documents use the term municipal solid waste to describe what everyone else generally calls garbage. The EPA reports that in 1990 this country generated 11 billion tons of solid waste including agricultural, mining, oil and gas, industrial, as well as municipal solid wastes. These wastes were found to be managed in an estimated 227,000 storage and disposal units, including surface impoundments, waste piles, landfills, and land treatment units. Approximately 160 million tons of municipal solid waste are generated each year, 131 million tons of which are landfilled. By the year 2000 we are projected to generate 190 million tons per year. One thing we know for sure is that the methods we currently employ to dispose of waste will not be adequate for these volumes. Eighty percent of garbage is disposed of in landfills. Landfill capacity in some places has fallen dangerously low and finding new landfills is becoming more and more difficult due to public resistance, commonly known as the **Not in My Back Yard** (**NIMBY**) syndrome. This resistance is based on the perception that solid waste management brings environmental damage, unpleasant smells, noise, and truck traffic. This public resistance is not targeted only at municipal landfills, even materials recovery facilities and recycling centers can be difficult to site.

It is estimated that more than one-third of the nation's landfills will be full by the end of the century. This is a national dilemma. Even though managing solid waste falls in the hands of local agencies, the problems it poses are national in scope. That is why a national strategy was developed to address the multitude of problems presented by solid waste generation.

Even though hazardous wastes are a subset of solid waste, the term solid waste is often used to refer to nonhazardous solid waste. Don't be mislead by the word *solid* because solid waste is defined in the law to mean solid, semi-solid, liquid, or compressed gas waste. This is just another example of why legal definitions are critical; many are not what they seem to be and do not necessarily match our common use definitions.

Besides landfilling, there are two other common methods for managing municipal solid waste: **recycling** and **incineration**. Recycling refers to using the waste as raw material for new products that may or may not resemble the original material. Incineration is controlled burning of the material. This country has overwhelmingly opted for landfilling most solid waste because in the past it has been the least expensive method. As older landfills reach capacity and new landfills become difficult to site, we must rely on other methods.

By weight, the most prevalent component of municipal solid waste is paper products, comprising roughly one-third of the waste stream. Yard wastes come in second representing about one-fifth of the total. These two types of waste, as well as metals, glass, and plastics that make up another one-third, are almost entirely recoverable by recycling. In 1992, EPA data showed that only about 14 percent of these wastes were recycled, and 16 percent were incinerated. That left 70 percent going to landfills.

We can certainly do much better with recycling than we have in the past. Many cities are implementing curbside recycling programs and there has been a greater educational effort, especially for school children, to learn about the need for and benefits of recycling. Realistically, however, as much as we'd like to think that recycling will solve our municipal solid waste problem, it can not. The EPA looks to Japan to offer a model in this regard because the Japanese have one of the most highly sophisticated programs for source reduction and recycling among industrialized countries. Even with many years of experience, and with their highly efficient program, the Japanese are able to recycle only about 50 percent of their waste, with little hope to achieve higher percentages. Fifty percent, however, is a lot more than our 14 percent!

An example of a common recyclable material is newspaper. Newspapers make up about seven percent of municipal solid waste by weight. In order to recycle newspapers, the fibers are first separated by beating the old papers into a pulp in water. Chemicals are added to remove inks and the de-inked pulp is washed and cleaned to remove traces of contaminants. The water is then removed so the pulp can be pressed, dried, and rolled into usable paper. Each time this process is repeated, the fibers break down to a greater extent until they are not strong enough to be reused as paper. This is why in most paper recycling processes some unrecycled, new fiber must be added.

**Figure 10-9:** Many communities are making it easier to recycle by developing curbside recycling programs.

**Figure 10-10:** Recycling glass is more cost effective than producing new glass.

Another common recyclable waste is aluminum cans. Of the metals, aluminum is second only to iron in total amounts used. Savings in energy use is one of the added benefits of recycling aluminum. The original production of aluminum from bauxite ore has the highest energy usage of any major metal; recycling aluminum requires about 95 percent less energy than producing it from scratch. Another bonus is that recycling also reduces air and water pollution.

Although less commonly recycled than aluminum cans, tin and bi-metal cans can also be recycled. Because of the way they are made, recycling these types of cans is more complicated since tin must be separated from the high quality steel, which can also be recycled. Currently, over one-fourth of the U.S. tin supply comes from used material. What we do not produce from used material must be purchased from other countries since we do not have tin resources in this country. The steel in cans currently makes up only a small fraction of steel scrap. Steel scrap is a very important industry – the U.S. exports close to 12 million tons of iron and steel scrap each year.

Glass and plastic are other common recyclable wastes. Over two billion pounds of glass containers are collected annually in this country. Most of these containers are melted down in furnaces to produce new containers. Not only are raw materials saved, but energy consumption is reduced because the furnaces used to re-melt glass make use of lower temperatures than those used to make new glass.

Plastics are deceptive in waste statistics because most of these statistics are given by weight, and plastics are light (have low densities). So, even though plastics only account for about 7 percent by weight of municipal solid waste, by volume they constitute up to about one-third of the total. The difficulty in recycling plastics is that there are so many varieties. Problems arise just identifying the type for proper segregation. To address this problem, many plastic containers now have a coding system that enables recyclers to better separate them. Many states have adopted and require this coding system for beverage containers. Recycled plastics are now used in a wide variety of new products, including construction materials such as recycled plastic 2 x 4 boards or patio furniture that was once your plastic milk carton.

A final, easily recycled, large quantity municipal solid waste category is yard wastes. Yard wastes include lawn clippings, tree limbs and leaves. Its a shame that so much of this type of waste has found its way into the precious space of landfills, because it can be so easily composted to provide nutrients and soil conditioners for yards. Many counties and cities now operate composting centers that either sell or donate the compost to gardeners and landscapers. Christmas trees are a great example of a high volume waste that can be easily chipped down to serve as a wonderful soil amendment, saving a large amount of landfill space.

Incineration was the other method identified to help solve the growing waste crisis. Historically, the burning of waste has probably been the most common method used to deal with wastes, and was common in this country in open burn dumps up until the 1960s. Open burning was banned by RCRA. Incineration is not open burning. Chapter 11 will explain in more detail the process used for incinerating wastes.

Probably the most exciting advance in the management of solid waste is the extraction of energy from it. It is estimated that about 75 percent of municipal solid waste is combustible and can yield high levels of energy. It has about one-third the fuel value of coal. Energy is generally recovered from waste by incineration to make steam which in turn is used to run electric generators. The simplest method involves removing non-combustible materials from the ashes of combustible wastes. Another method of separating the fuel components from other components is to shred all the waste and introduce it into a column in which a strong air current separates the lighter pieces, such a paper scraps, from the heavier pieces, such as glass and metals. The lighter pieces become fuel and the other pieces are reclaimed. If the combustible portions of the waste are pre-separated, they can be incinerated to produce high pressure gases to generate electricity. Finally, lighter organic wastes can be decomposed chemically in a high temperature, low-oxygen process called pyrolysis to produce oil or gas fuels.

With the increased cost of energy and landfills, the use of solid waste energy recovery using the thermal processes described above is being given more and more attention. Several cities have received funding to construct waste-to-energy (WTE) facilities. There are currently nearly 250 waste-to-energy plants providing electricity to over a million homes. They also reduce the volume of waste 80 to 90 percent (the ash must still be disposed of). The Integrated Wastes Services Association estimates that the equivalent energy output would require nearly 30 million barrels of crude oil.

# ■ Solid Waste Management Under Subtitle D

## Concepts

- Solid wastes include nonhazardous and hazardous wastes.

- There are two main components of Subtitle D: State solid waste management plans and Subtitle D Criteria.

- Landfills have strict design, construction, and operation requirements to protect the environment.

The primary goals of Subtitle D as stated in Section 4001 of RCRA "are to assist in developing and encouraging methods for the disposal of solid waste which are environmentally sound and which maximize the utilization of valuable resources including energy and materials, which are recoverable from solid waste and to encourage resource conservation." To do this, the EPA established standards for determining whether or not solid waste disposal facilities and practices pose adverse effects on human health and the environment. Facilities failing to meet the criteria were designated as "open dumps." The EPA also established guidelines for solid waste management plans to be developed by each state. The EPA provided technical assistance and approved state plans that complied with the requirements. Each state, however, is primarily responsible for developing and implementing its own plan.

The term *solid waste* is used very broadly to include not only the traditional nonhazardous solid wastes, like municipal garbage, but also hazardous solid wastes. Some examples of solid waste as defined in the Act include: garbage, such as food packages and coffee grounds; refuse, such as wall board, metal scrap, and empty containers; sludge from a wastewater treatment plant or scrubber sludges; and other discarded material, including solid, semisolid, liquid, or contained gaseous material resulting from industrial, commercial, mining, agricultural, and community activities, such as fly ash. There are a number of exceptions to the definition of solid waste. They include: domestic sewage; industrial wastewater discharges regulated under the Clean Water Act; irrigation return flows; and nuclear materials or by-products.

There are two main components to Subtitle D. The first, which is commonly referred to as the Subtitle D Criteria, classifies solid waste disposal facilities and determines which have the potential to harm the environment. Any facility that does not satisfy the criteria for safe and proper solid waste disposal is considered an open dump for the purposes of state solid waste management planning and must be upgraded or closed.

The Subtitle D Criteria include performance standards for eight major topics: floodplains, endangered species, surface water, groundwater, land application, disease, air, and safety. Solid waste disposal facilities cannot be sited in areas where there is the potential for washout of the waste to occur such as in a flood plain. A solid waste facility cannot cause destruction of critical habitats that would harm endangered or threatened species. No solid waste disposal facility can cause the discharge of pollutants or fill material to surface waters or cause Maximum Contaminant Levels (MCLs) to be exceeded in underground sources of water. Periodic cover material must be used to prevent diseases caused by vectors such as flies and mosquitoes. There is also a requirement that sewage sludges and septic tank pumpings be treated to reduce disease-causing organisms

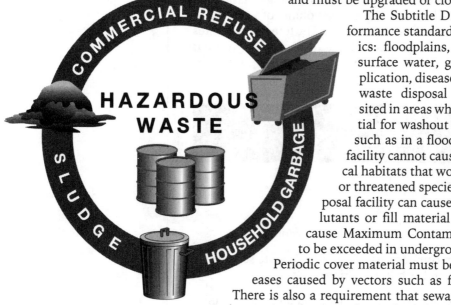

**Figure 10-11** The solid waste universe.

before they are applied to the land. Air quality is protected by the abolishing of open burning of solid waste at disposal facilities as well as specifying that applicable requirements of the SIPs developed under the Clean Air Act must be met. Finally, safety provisions require control of explosive gases, fires, bird hazards to aircraft, and public access to the facility.

The Subtitle D Criteria also include minimum technical standards for solid waste disposal facilities. States have the option of developing criteria more stringent than those in Subtitle D. Compliance with this part and the ban on dumping can be enforced through citizen suits or by the states. In addition to providing minimum technical standards, the Subtitle D Criteria are used to identify open dumps. An open dump is defined as a disposal facility that does not comply with one or more of the criteria. Each state compared its disposal facilities against the standards in Subtitle D and identified those that were open dumps. Open dumps are not allowed and were therefore closed or upgraded to meet the new standards.

The second major part of the management standards for solid waste are the guidelines for state solid waste management plans. The purpose of these guidelines is to assist states in developing and implementing EPA-approved solid waste management plans. These plans are designed to ensure environmentally sound solid waste management and disposal, resource conservation, and maximum utilization of valuable resources. The guidelines outline the minimum requirements for state plans and explain how the plans are approved by the EPA. Each state plan must include: identification of the responsibilities of state, local, and regional authorities in implementing the plan; a program that prohibits new open dumps and closes or upgrades existing open dumps through appropriate state powers; a strategy for encouraging resource recovery and conservation activities; methods to ensure that adequate facility capacity exists to dispose of solid waste in an environmentally sound manner; coordination with other environmental programs; and assurances of public participation in all stages of planning and management activities. Once a state develops and adopts a plan, it must be submitted to the EPA for approval. The Administrator must approve or disapprove the plan within six months of submittal.

As usual, there are some exceptions to the Subtitle D Criteria. Some examples are: agricultural wastes used as fertilizers or soil conditioners; domestic sewage applied to the land; hazardous waste disposal facilities regulated under Subtitle C of RCRA; industrial discharges that are point sources and subject to permits under the Clean Water Act.

Some important additions were made to Subtitle D under the Hazardous and Solid Waste Amendments. Congress was concerned about the adequacy of RCRA to protect groundwater from contamination, so groundwater monitoring as a requirement and more detailed standards on determining acceptable locations for new facilities were added. Also, the EPA revised the Subtitle D Criteria to address facilities receiving household hazardous waste or hazardous waste from small quantity generators, which under RCRA are defined as generators who produce less than 100 kilograms of hazardous waste per month or who produce less than 1 kilogram of acutely hazardous waste per month. States must establish permit programs or equivalent systems of approval for facilities receiving household hazardous waste or hazardous waste from small quantity generators.

In the past, municipal landfills were essentially pits in the ground where garbage was dumped; hence the term by which many people know them – garbage dump. They were not covered until the whole pit was filled in and oftentimes the garbage was burned. Now landfills require extensive environmental controls to prevent the types of damage we have seen from past practices. The strict requirements regarding landfills have caused a decrease in the number of landfills being built. This, in addition to the fact that existing landfills are rapidly filling to capacity, has led to decline in the total number of landfills currently in operation to the lowest level in the last several decades. There were over 8,000 landfills in operation in 1988, but only 5,812 at the end of 1991. However, even though there are fewer landfills, their average size is now larger.

In a modern landfill, solid wastes are spread in thin layers and each layer is compacted. After waste layers accumulate to the depth of about ten feet, an earth layer is added and compacted. This design incorporates pollution prevention through the use of impermeable liners made of clay and/or plastic systems to collect and treat any leachate, monitoring groundwater and surface water for indications of infiltration by harmful chemicals, diverting drainage away from the fill, monitoring the landfill surface for escaping methane gas, and siting the landfill in an area not subject to flooding or high groundwater. Gases are naturally generated from waste through bacterial decomposition. It is common to find methane gas at landfills as a byproduct of anaerobic decomposition of the waste. This gas can be explosive if found in significant amounts. Some landfills use systems to capture the gas and burn it to produce energy.

Existing landfills not meeting all of the requirements described above must either upgrade or close. EPA estimates that the cost of building and maintaining a landfill meeting the strict Subtitle C requirements is currently about $125 million.

**Figure 10-12:** A modern landfill.

## ■ Checking Your Understanding (10-1)

1. Contrast the definition of solid waste with that of hazardous waste.

2. List and describe three major regulatory components of RCRA.

3. Discuss the NIMBY syndrome in relation to activities in your community.

4. Describe your community's recycling efforts and discuss where you think these efforts might be improved.

# 10-2 Hazardous and Medical Waste

## ■ Hazardous Waste Generation

### Concepts

■ A great majority of industrial processes generate hazardous wastes.

■ Hazardous wastes are a subset of solid wastes.

■ Hazardous wastes are designated as such because they either possess one or more of four characteristics or are listed.

In the past, waste disposal was relatively inexpensive, and for the most part unregulated. Industry did not consider the costs of generation of waste in its production costs. Society has since determined that generation of waste, especially hazardous waste, is indeed costly, both in terms of health and environmental damage. In fact, we have already discussed several examples where society is paying extraordinarily high costs for past hazardous waste disposal practices.

Many people believe that only a few industries generate hazardous waste. The impression is that only large chemical manufacturers produce hazardous waste as a result of their industrial processes. This is not the case. A business is likely to produce hazardous wastes if it uses:

– oils or other petroleum products;
– dyes, paints, printing inks, thinner, solvents, cleaning fluids;
– pesticides or other poisonous chemicals;
– materials that dissolve metals, wood, paper, or clothing (for instance, acids and caustics); ignitable materials;
– materials that burn or itch upon contact with skin;

- materials that bubble or fume upon contact with water;
- metal-bearing solutions; or
- products that are accompanied by a MSDS or a label indicating that the product is hazardous.

Think about it. This long list includes – as waste generators – food processing companies, building and cleaning maintenance processes, formulating cosmetics, construction, educational institutions, equipment repair, funeral services, furniture manufacturing and finishing, health care facilities, laboratories, laundries/dry cleaners, metal manufacturers, motor freight terminals, printing, vehicle maintenance, wood preserving, and the list goes on. The EPA estimates that the amount of industrial hazardous waste generated each year is more than double the amount of municipal solid waste generated.

We already discovered that hazardous wastes are a subset of solid waste and can be liquids, solids, or sludges and are capable of causing greater harm if mismanaged. Hazardous wastes come in all shapes and forms. They may be a result of manufacturing processes, such as those named above, or may simply be commercial products that have been discarded, such as household cleaning fluids or battery acid. Whatever their form, they must first be classified as a waste, which means they have no use and are to be discarded. A discarded material, 40 CFR states, "is one that is relinquished, recycled, or considered inherently wastelike."

The RCRA definition of the term *hazardous waste* was provided earlier in the chapter. The EPA felt that this definition of hazardous wastes was too broad to apply in the field. In order to regulate hazardous wastes, the EPA first had to determine which specific wastes were hazardous. This was not a simple task, and the EPA realized that the decision would have major economic consequences for the generators of tens of thousands of wastes that could end up on the hazardous list for a variety of different reasons. Congress intended that only those wastes fitting EPA interpretation of the basic definition of hazardous waste would be subject to RCRA's Subtitle C requirements, which are far more stringent than the Subtitle D solid waste requirements.

Initially, the EPA spent many months in consultation with industry and the public to develop a workable definition of hazardous waste for its regulations. As a result of this work, regulations implementing RCRA identify hazardous wastes based on their characteristics and provide a list of specific hazardous wastes.

**Characteristic Wastes** possess one or more of four characteristics:

■ **Ignitability** (Waste Code D001). Ignitable wastes easily catch on fire under certain conditions. They have a flash point of less than 140°F (60°C) or less or they could be a solid that ignites as a result of friction when rubbing against another solid. Examples include liquids such as solvents that readily catch fire, and friction-sensitive substances.

■ **Corrosivity** (Waste Code D002). Corrosive wastes include those that are strongly acidic or basic with a pH 2.0 or 12.5. They are capable of corroding metal, such as tanks, containers, drums, and barrels, at a rate of .25 inch per year and damaging human tissue.

■ **Reactivity** (Waste Code D003). Reactive wastes are unstable under normal conditions. They can create explosions and/or toxic fumes, gases, and vapors when mixed with water. (Laboratory tests evaluate the generation of $H_2S$ or HCN gas from an acidified sample.)

■ **Toxicity** (Waste Code D004 – 043). Toxic wastes are harmful or fatal when ingested or absorbed. When toxic wastes are disposed of on land, contaminated liquid may leach from the waste and pollute groundwater. The EPA has developed a list of elements and compounds that could pose a hazard to health at certain concentrations in water. Toxicity is identified through a laboratory procedure called the Toxicity Characteristic Leaching Procedure (TCLP). This procedure attempts to mimic the situation that would arise in the land disposal of wastes. Rain water, which we have already learned is slightly acidic, would form leachate and carry contaminants through the soil and into groundwater. This test uses a weak acid to leach metal, organic chemicals, and pesticides from the sample. If predetermined levels of certain chemicals are found in the leachate, the waste is hazardous under the toxicity characteristic.

| Hazardous Waste # | Hazardous constituents for which listed | Hazardous Waste # | Hazardous constituents for which listed |
|---|---|---|---|
| F035 | Arsenic, chromium, lead | K038 | Phorate, formaldehyde, phosphorodithioic acid esters |
| F037 | Benzene, benzo(a)pyrene, chrysene, lead, chromium | K039 | Phosphorodithioic and phosphorothioic acids |
| F038 | Benzene, benzo(a)pyrene, chrysene, lead, chromium | K040 | Phorate, formaldehyde, phosphorodithioic and phosphorothioic acid esters |
| F039 | All constituents which treatments standards are specified for multi-source leachate (wastewaters and nonwaste-waters) under Section 66268,43(a), Table CCW | K041 | Toxaphene |
| | | K042 | Hexachlorobenzene, ortho-dichlorobenzene |
| K001 | Pentachlorophenol, phenol, 2-chlorophenol, p-chloro-m-cresol, 2,4-dimethylphenyl, 2,4-dinitrophenol, trichlorophenols, tetrachlorophenols, creosote, chrysene, naphthalene, flouranthene, benzo(b)fluroanthene, benzo(a)pyrene, . . . | K043 | 2,4-dichlorophenol, 2,6-dichlorophenol, 2,4,6-trichlorophenol |
| | | K044 | N.A. |
| | | K045 | N.A. |
| K002 | Hexavalent chromium, lead | K046 | Lead |
| K003 | Hexavalent chromium, lead | K048 | Hexavalent chromium, lead |
| K004 | Hexavalent chromium | K049 | Hexavalent chromium, lead |
| K005 | Hexavalent chromium, lead | K050 | Hexavalent chromium |
| K006 | Hexavalent chromium | K051 | Hexavalent chromium, lead |
| K007 | Cyanide (complexed), hexavalent chromium | K052 | Lead |

**Figure 10-13:** Examples of listed wastes by category.

Besides the characteristic wastes, the EPA has determined that some specific wastes are hazardous. These wastes are now incorporated into lists published by the EPA, hence their designation as **listed wastes**. Generators that produce any of the listed wastes must also abide by RCRA's Subtitle C hazardous waste requirements. The lists are organized into three categories:

■ **Source-Specific Wastes.** This list (K List) includes wastes from specific industries such as petroleum refining and wood preserving. Sludges and wastewaters from treatment and production processes in these industries are examples of source-specific wastes.

- **Generic Wastes.** This list (F List) identifies wastes from common manufacturing and industrial processes. Generic wastes include solvents that have been used in degreasing operations in an industry.

- **Commercial Chemical Products.** These lists (P and U Lists) include specific commercial chemical products such as creosote and some pesticides.

All listed wastes are presumed to be hazardous regardless of their concentrations. However, if a company can demonstrate that its specific waste is not hazardous, the waste may be **delisted** and is then no longer subject to Subtitle C requirements. A delisted waste, however, is still covered by Subtitle D solid waste management requirements.

Determining which wastes are hazardous is an ever-changing process influenced by new concerns, research data, and test development. The EPA is now adding certain types and classes of wastes to its hazardous waste lists, and is deciding whether to identify additional hazardous characteristics.

There are a number of categories of waste that are specifically excluded from the definition of hazardous waste. They are:

- Domestic sewage.
- Irrigation waters or industrial discharges permitted under the Federal Water Pollution Control Act.
- Certain nuclear material as defined by the Atomic Energy Act.
- Household wastes, including toxic and hazardous waste.
- Certain mining wastes.
- Agricultural wastes, excluding some pesticides.
- Small quantity wastes (that is, wastes from businesses generating fewer than 220 pounds of hazardous waste per month).

According to EPA estimates, of the six billion tons of industrial, agricultural, commercial, and domestic wastes we generate annually, about 275 million tons are hazardous as defined by RCRA regulations.

## ■ Hazardous Waste Management Under Subtitle C

### Concepts

- The generator is responsible for determining whether wastes are hazardous and for proper hazardous waste management from "cradle to grave."

- Hazardous wastes are tracked using a uniform hazardous waste manifest.

- A system of generator identification numbers and permits allow the EPA to monitor hazardous waste.

- Untreated hazardous waste is restricted from land disposal.

**Figure 10-14:** Illegal waste disposal has tragic consequences in all the environmental compartments.

The tragic consequences of past hazardous waste mismanagement are reflected in the long lists of sites where pollution has affected one or more of the environmental compartments. Improper disposal of hazardous waste has been linked to elevated levels of toxic contaminants in humans, aquatic species, and livestock. Illegal dumping of hazardous waste on roadsides or in open fields has resulted in explosions, fires, contamination of groundwater, and generation of toxic vapors.

To prevent these types of negative impacts, the EPA designed the hazardous waste regulations to ensure proper management of hazardous waste from the moment the waste is generated until its ultimate disposal. This step-by-step management approach enables the EPA and the states to monitor and control hazardous waste at every point in the waste cycle. Congress felt that this approach would offer the highest level of protection of human health and natural resources from the dangers posed by hazardous wastes. This approach has three key elements:

- A tracking system requiring a uniform hazardous waste manifest document to accompany any transported hazardous waste from the point of generation to the point of final disposal.

- An identification and permitting system enabling the EPA and the states to monitor the generation of waste and the operation of facilities involved in the treatment, storage, and disposal of hazardous waste.

- A system of restrictions and controls on the placement of hazardous waste on or into the land.

The first step in the waste cycle is its generation. The person who actually produces the waste or first causes it to become a waste subject to the RCRA requirements. As we learned earlier, generators include large industries, small businesses, universities, and hospitals. The EPA regulations put the responsibility on waste generators to evaluate their wastes to determine whether the wastes are listed or if they exhibit any of the four hazardous characteristics. The

**Figure 10-15:** Hazardous wastes must be properly labeled for transport.

critical point is that it is the generator's responsibility to determine whether the waste is hazardous and to oversee its ultimate fate.

Once a generator determines that a waste is hazardous, they must obtain an EPA identification number for each site. According to EPA estimates, generators treat or dispose of about 96 percent of the nation's hazardous waste onsite. Onsite treatment, storage, and disposal facilities generally are found at larger businesses that can afford treatment equipment and that possess the necessary space for storage and disposal. Smaller firms and those in crowded urban locations are more likely to transport their waste offsite where it is managed by a commercial firm or a publicly owned and operated facility. RCRA applies to both onsite and offsite facilities.

If the generator chooses to dispose of the hazardous waste offsite, the generator must package and label the waste properly for transportation. Proper packaging helps to ensure that no hazardous waste leaks from containers during transport. Labeling enables transporters and public officials, including those who respond to emergencies, to rapidly identify the waste and its hazards.

Although only a small percentage of the nation's hazardous waste is transported offsite to treatment, storage, or disposal facilities, its volume is still substantial: approximately 12 million tons per year. To track transported waste, the EPA requires generators to prepare a Uniform Hazardous Waste Manifest. This form, with copies for everyone involved in the shipment, identifies the type and quantity of waste, the generator, the transporter, and the facility to which the waste is being shipped. Large quantity generators (LQGs) generate over 1,000 kg (2,200 lb.) of hazardous waste per month. They must certify on the manifest that they are minimizing the amount and toxicity of their waste, and that the method of treatment, storage, or disposal they have chosen will minimize the risk to human health and the environment. Small quantity generators (SQGs) generate up to 1,000 kg (2,200 lb.) of hazardous waste per month. Although they do not have to certify that they are minimizing their waste production, they do have to state that they are doing the best they can to minimize the amount generated. The certification requirement was added by the 1984 RCRA amendments. It reflects RCRAs new emphasis on reducing the volume of waste.

The manifest must accompany the waste wherever it travels. Each individual handler of the waste must sign the manifest and keep one copy (see Figure 10-16). When the waste reaches its destination, the owner of that facility returns a copy of the manifest to the

Storage Facility

Transporter

Transporter

MANIFEST

NNEPA or State Agency

Generator

Treatment Facility

Transporter

Disposal Facility

**Figure 10-16:** A copy of the Hazardous Waste Manifest must be kept by each handler of hazardous waste during its transport.

generator to confirm that the waste arrived. If the waste does not arrive as scheduled, generators must immediately notify the EPA or the authorized state environmental agency so that they can investigate and take appropriate action. Generators must retain copies of the manifest for three years after shipment. Every other year, generators must provide information on their activities to their authorized state agency or to the EPA in a biennial report. Some businesses that generate small quantities of hazardous waste are subject to fewer paperwork requirements.

Transporters pick up properly packaged and labeled hazardous waste from generators and transport it to the permitted facilities that treat, store, or dispose of the waste. Transporters must carry copies of the completed manifests and must put proper placards on the transport vehicle to identify the type of waste being transported. These placards, like the labels on the hazardous waste containers, enable fire fighters, police, and other officials to immediately identify potential hazards in case of an emergency. Transporters, like generators, are also required to obtain an EPA identification number. Because an accident involving hazardous waste could create very serious problems, the EPA regulations also require transporters to comply with procedures for hazardous waste spill cleanup.

The treatment, storage, and disposal (TSD) facilities that receive hazardous waste from the transporter must be permitted by the EPA. Treatment facilities use various processes to alter the character or composition of a hazardous waste. Some treatment processes enable waste to be recovered and reused in manufacturing, while other treatment processes dramatically reduce the volume of waste to be disposed of. Storage facilities hold hazardous waste temporarily pending its treatment or disposal. Historically, most disposal facilities buried hazardous waste or piled it on the land.

Like generators and transporters, TSD facilities must obtain an EPA identification number. They must also obtain a permit to operate which ensures safe operation. The permit system requires that facilities meet the standards established by the RCRA program for proper waste management. Many of these standards are designed to protect groundwater. For example, permitted TSD facilities must:

**Figure 10-17:** A treatment-storage-disposal (TSD) facility.

- Analyze and identify wastes prior to treatment, storage, or disposal.

- Prevent the entry of unauthorized personnel into the facility by installing fences and surveillance systems and by posting warning signs.

- Periodically inspect the facility to determine if there are any problems.

- Train employees.

- Prepare a contingency plan for emergencies and establish other emergency response procedures.
- Comply with the manifest system and with various reporting and record keeping requirements.
- Comply with technology requirements, such as installing double liners and leachate detection and collection systems.

Permits also contain requirements specific to the individual facility. Approximately 4,000 hazardous waste TSD facilities are currently authorized to operate under RCRA.

Air emissions released during the treatment, storage, and disposal of hazardous waste can present risks to human health and the environment. To control these emissions and to limit the emissions from all hazardous waste treatment, storage, and disposal facilities, the EPA has established standards for hazardous waste incinerators. Chapter 11 will provide more information on hazardous waste treatment technologies.

As their capacity is used up and as new stringent operating requirements are imposed, many hazardous waste disposal facilities are closing. What happens to the facility after it closes is a long-term concern, with substantial health and environmental implications. For example, at Love Canal, disturbance of the hazardous waste landfill after the site had been closed is believed to have caused the subsequent leakage of hazardous chemicals into the basements of nearby homes.

EPA regulations are designed to prevent post-closure problems by requiring TSD facility owners to prepare carefully for the time when their facility will close. Owners must:

- Acquire sufficient financial assurance mechanisms (such as trust funds, surety bonds, or letters of credit) to pay for completion of all operations.
- Be prepared to pay for 30 years of groundwater monitoring, waste system maintenance, and security measures after the facility closes.
- Obtain liability insurance to cover third party damages that may arise from accidents or waste mismanagement.

The tracking and permitting system for hazardous waste, from generation through treatment, storage, and disposal, applies to all hazardous waste, regardless of the disposal method employed. Most hazardous waste, about 80 percent, has been disposed of into or onto the land at a variety of disposal locations including landfills, surface impoundments, waste piles, lagoons, and underground injection wells. Improper land disposal practices in the past have endangered public health and the environment and continue to pose a threat, particularly to groundwater. To prevent these threats, RCRA requires:

- Minimization of wastes by reducing, recycling, and treatment.
- A ban on unsafe, untreated wastes from land disposal.
- That land disposal facilities be designed, constructed, and operated according to stringent standards.
- Corrective action for releases of hazardous waste into the environment.

Large Quantity Generators (LQGs) must certify that they have taken steps to reduce the volume of the hazardous waste they generate. Generators may reduce their waste volume in several ways including manufacturing process changes, source separation, recycling, raw material substitution, and product substitution. In many cases, waste reduction can benefit industry by decreasing the costs of waste management. When companies produce less waste, their disposal costs are lower. Further, companies may sell recovered materials at a profit. Chapter 11 will provide more specifics on waste minimization techniques.

In addition to reducing the volume of waste generated, companies with proper permits may treat waste prior to disposal to reduce the waste volume or eliminate the wastes hazardous constituents. Several different types of treatment involving physical, chemical, thermal and/or biological processes are available. One or more processes may be used, depending on the characteristics of the waste. Nearly all treatment processes produce residues that ultimately require disposal.

RCRA standards for the operation of land disposal facilities have become increasingly stringent in response to the growing concern about adverse environmental impacts. While EPA regulations strongly discourage land disposal of most hazardous wastes, it is very likely that there will always be some wastes that must be disposed of on land. RCRA provides several restrictions and standards for land disposal facilities to ensure more thorough protection of the environment, particularly groundwater. These include:

- Banning all liquids from landfills.
- Banning underground injection of hazardous waste within one-quarter mile of a drinking water well.
- Requiring more stringent structural and design conditions for landfills and surface impoundments, including two or more liners, leachate collection systems above and between the liners, and groundwater monitoring.
- Requiring cleanup or corrective action if hazardous waste leaks from a facility.
- Requiring information from disposal facilities on pathways of potential human exposure to hazardous substances.
- Requiring location standards that are protective of human health and the environment; for example, allowing disposal facilities to be constructed only in suitable hydrogeologic settings.

Note that there is a ban on untreated hazardous wastes and liquids from land disposal. This feature of RCRA is commonly called the **Land Ban**. The land ban provisions have served to jump-start efforts to develop more economic and more effective means of treating waste. As a result, treatment technologies are being improved rapidly. The EPA is now sponsoring research on: new treatment technologies to destroy, detoxify, or incinerate hazardous waste; ways to recover and reuse hazardous waste; and methods to reduce the volume of hazardous waste requiring treatment or disposal.

Land disposal of untreated hazardous waste is currently prohibited unless the EPA makes a finding that there will be "no migration of hazardous constituents . . . for as long as the wastes remain hazardous." Only waste that is determined not to pose a health or environmental threat may continue to be disposed of on land.

Hazardous wastes banned from land disposal (most of which are), must be treated and rendered less hazardous before they can be disposed of on the land. EPA established treatment standards for banned wastes. These standards specify either a level that the hazardous components must be reduced to or methods of treatment which substantially reduce the toxicity or mobility of the hazardous constituents (see Figure 10-18 for examples of treatment standards). This "Land Ban" is designed to minimize long-term threats to human health and the environment. The 1984 RCRA amendments set strict deadlines (most of which have passed) for determining whether wastes should be banned from land disposal and for developing treatment standards.

| Waste Code | Waste Description and/or Treatment Subcategory | CAS No. for Regulated Hazardous Constituents | Technology Code | |
|---|---|---|---|---|
| | | | Wastewaters | Nonwastewaters |
| P003 | Acrolein | 107-02-8 | (WETOX or CHOXO) lb. CARBN; or INCIN | FSUBS or INCIN |
| P005 | Allyl alcohol | 107-18-8 | (WETOX or CHOXO) lb. CARBN; or INCIN | FSUBS or INCIN |
| P006 | Aluminum phosphide | 20658-73-8 | CHOXD; CHRED; or INCIN | CHOXD; CHRED; or INCIN |
| P007 | 5-Aminoethyl 3-isoxazolol | 2763-96-4 | (WETOX or CHOXO) lb. CARBN; or INCIN | INCIN |
| P008 | 4-Aminopyridine | 504-24-5 | (WETOX or CHOXO) lb. CARBN; or INCIN | INCIN |
| P014 | Thiolphenol (Benzene thiol) | 108-98-5 | (WETOX or CHOXO) lb. CARBN; or INCIN | INCIN |
| P015 | Berylium dust | 7440-41-7 | NA | RMETL; or RHTRM |
| P016 | Bis(chlormethyl)ether | 542-88-1 | (WETOX or CHOXO) lb. CARBN; or INCIN | INCIN |
| P017 | Bromoacetone | 598-31-2 | (WETOX or CHOXO) lb. CARBN; or INCIN | INCIN |

**Figure 10-18:** Treatment standards.

RCRA requires that all TSD facilities take corrective action for any release of hazardous waste into the environment. To enforce this requirement, the EPA or an authorized state can issue an administrative order to require corrective action (such as repairing liners or pumping to remove a plume of contamination at a facility). The EPA or the state may require corrective action beyond the facility boundary. They also may require corrective action regardless of when the waste was placed at the facility. That means that this mechanism may be used to clean up past problems. TSD facilities are also required to provide financial assurance that they can complete the corrective action.

When the EPA first issued hazardous waste regulations under RCRA, the Agency focused on those companies generating the largest amounts of waste – more than 1,000 kg (2,200 pounds or about five full 55-gallon drums) per month. As previously stated, these generators are called LQGs. This regulated community included about 15,000 companies that produced approximately 90 percent of the nations hazardous waste. **Small Quantity Generators (SQGs)** were exempted from most of the federal hazardous waste requirements. Some states, however, had more stringent regulations that did cover small quantity generators. For instance, federal law restricts LQGs to a 90-day accumulation period for their hazardous wastes. Hazardous wastes stored beyond this 90-day limit require a TSD permit. Some states have also imposed this 90-day accumulation period on SQGs.

In the 1984 RCRA amendments, Congress closed this gap by requiring the EPA to regulate those small quantity generators who produce between 220 and 2,200 pounds (100 to 1,000 kilograms) of hazardous waste in a calendar month. There are about 100,000 of these small generators, including businesses such as vehicle repair shops, metal manufacturing and finishing operations, laboratories, printers, laundries, and dry cleaners. *Conditionally-exempt* generators – businesses that generate less than 220 pounds per month – will remain exempt from most of the hazardous waste requirements on a federal level. Many states, however, have more stringent regulations that include these smaller quantity generators in their regulations.

As of 1986, 220 to 2,200 pounds per month generators were also required to meet most of the key requirements of the RCRA program for hazardous waste management. For example, they must:

– Obtain an EPA identification number.

– Use the fully completed manifest when shipping waste offsite.

– Use only hazardous waste transporters and authorized facilities with EPA identification numbers to transport, treat, store, or dispose of their hazardous wastes.

– Not store hazardous waste onsite for longer than 180 days (270 days if the waste is to be shipped more than 200 miles) without a permit.

Prior to 1986, utilities, industries, commercial facilities, schools, government institutions, and apartment and home owners across the country burned about one to two million tons of hazardous wastes and about 500 to 600 million gallons of used oils per year for energy recovery purposes. Used oils include automotive crankcase oil and oils used in industrial processes. Used oils are often contaminated with toxic metals (such as cadmium, chromium, lead), the metalloid arsenic, and chlorinated organic compounds (such as cleaning solvents).

Facilities often mixed hazardous waste or used oil with fuels to increase the heating value of the fuels and to reduce the costs of waste disposal. When burned, fuels containing used oil contaminated with toxic constituents or hazardous waste emit toxic fumes. In one incident, residents of a building in New York City complained of respiratory problems, headaches, nausea, and digestive problems. An investigation revealed that a fuel-blending facility had mixed hazardous wastes containing PCBs and other chlorinated solvents with the buildings heating fuel supply. Incidents such as this prompted Congress to require the EPA to regulate the burning of hazardous waste and used oils.

The EPA's RCRA regulations now prohibit industrial and non-industrial facilities, such as homes and apartment buildings, from burning hazardous waste fuel or used oil that contains significant levels of hazardous contaminants unless they have special permits. In addition, the EPA requires all producers, distributors, and marketers of contaminated used oil fuel to place a label on the bill of sale indicating that the fuel is subject to EPA regulations. Marketers of contaminated used oil fuel also must notify the EPA of their activities, obtain EPA identification numbers, and keep certain records. Facilities that blend hazardous waste fuel must also comply with these notification and record keeping requirements, comply with hazardous waste storage and facility standards, and fill out a manifest form if they ship the waste fuel.

# ■ Subtitle J: Managing Medical Waste

## Concepts

- Medical facilities present unique workplace hazards due to the types of materials and equipment routinely used by workers.

- Infectious agents are a severe potential contamination problem at health care facilities.

- Although there are currently no federal regulations for managing medical waste, the states have adopted versions of previously enacted federal regulations.

Before discussing the management of medical waste, it is important to know about the types of materials medical facility workers routinely use or come in contact with that generate these harmful wastes. Health care facilities present a unique, complex setting for potential health and safety hazards because of the infectious and hazardous materials they routinely have onsite. In fact, compared with the total civilian workforce, hospital workers have a greater percentage of workers' compensation claims, including those related to handling medical wastes such as acquiring infectious and parasitic diseases, dermatitis, hepatitis, and influenza. The list of potential hazards also includes radiation, toxic chemicals, and infectious materials.

Maintenance workers are potentially exposed to solvents, asbestos, and electrical hazards. Persons working around boiler rooms are regularly exposed to high levels of noise and heat. Housekeeping staff are exposed to detergents and disinfectants that can cause skin rashes and eye and throat irritation. They risk exposure to hepatitis and other diseases from hypodermic needles that have not been properly discarded. Nurses and doctors confront such potential problems as exposure to infectious diseases and toxic substances, back injury, and radiation exposure. Radiology technicians are exposed to radiation from X-rays and other radioactive sources. Operating room workers may face increased risk of reproductive problems as a result of exposure to waste anesthetic gases. They are also subject to cuts and puncture wounds, infection, radiation, and electrical hazards.

The **Nuclear Regulatory Commission (NRC)** adopts and enforces standards for departments of nuclear medicine in hospitals, although some states have agreements with the federal government to assume these responsibilities. The NRC regulates all radioactive sources except naturally occurring materials such as radium or radon, which are regulated by the Food and Drug Administration (FDA). Section 10-3 of this chapter covers radiation and nuclear waste.

The widespread use and storage of flammable and combustible liquids presents a major fire hazard in all medical facilities. Common flammable liquids found in medical facilities include: alcohols, ketones (acetone), hydrocarbons (toluene, benzene, xylene) and ethers such as ethylene oxide, and diethyl ether. Some of these are toxic as well. Other toxic agents commonly used include: ammonia, chlorine, and corrosive drain cleaners.

Microorganisms such as bacteria and viruses found in medical facilities can be inhaled, ingested, or injected through the skin. This type of "contamination" can cause serious infectious diseases. Two of the most common infectious diseases that present a problem with the handling of medical facility waste are hepatitis and tuberculosis. Currently there is great concern over the HIV virus. The wastes most commonly contaminated with disease-causing organisms are used hypodermic needles, which can easily puncture the skin if not properly managed (see Figure 10-19). Syringes can also leak infectious materials.

Other exposures can occur from the sprays or aerosols that result from spilling or breaking containers of infectious material. Aerosols are airborne droplets of infectious material that may be generated by: opening containers; blowing out pipettes; mixing test tube contents; opening freeze-dried cultures; pouring liquids; mixing with high-speed blenders; and spilling liquids. Small aerosol particles dry almost instantly and remain suspended in the air for long periods. When inhaled, they penetrate deep into the lung and may cause infections. Larger and heavier particles settle slowly on surfaces and workers' skin. They may enter the body through contaminated foods, contaminated skin, or objects that touch the eyes or mouth. Even such common items as laundry and soiled linens can contaminate workers.

Precautions for safe handling of infectious and hazardous materials in medical facilities have been established because of their potential for serious harm. During the summer of 1988, it became apparent that these potential health and safety issues were not restricted to the environment of medical facilities. Widespread mismanagement led to medical waste washing up on the Atlantic seaboard. In response to this problem, Subtitle J was added to RCRA. Subtitle J required the EPA to develop a two year demonstration program to track

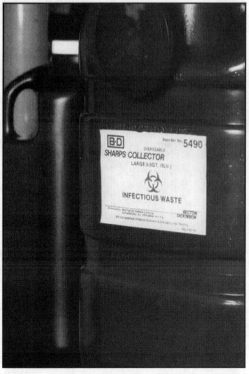

**Figure 10-19:** Disposal of "sharps" is strictly regulated in medical facilities.

medical waste from generation to disposal in the areas that had chosen to participate in the program (Connecticut, New Jersey, New York, Rhode Island, and Puerto Rico). The idea was that after completion of the demonstration program in 1991, the Agency would report its findings on the program to Congress, which would consider the merits of establishing nationwide medical waste requirements.

In June of 1991, the interim regulations pertaining to medical waste expired. There are currently no federal medical waste regulations. Fortunately, most of the states adopted their own versions of the federal regulations as they were originally written. Many states, in fact, have more stringent regulations concerning medical waste than the original federal regulations. Because the federal regulations provided a common basis for states to develop their own medical waste management programs, this section will concentrate on the original federal regulations. The student is advised to refer to specific state regulations for more specific compliance details.

There are a number of agencies and regulatory schemes other than the EPA's RCRA program that affect facilities generating medical waste. One example is OSHA's blood borne pathogen standard, which was covered in Chapter 9. DOT also regulates the transport of medical wastes under the hazardous materials transportation regulations. The Postal Service regulations and the Clean Air Act are other examples of laws that deal with specific medical waste issues.

The EPA enacted interim final regulations pertaining to the management of medical waste in March, 1989. These regulations were placed in 40 CFR Part 259 and established requirements for medical waste generators, transporters, and treatment, destruction, and disposal facilities (TDDs). This original EPA framework for medical waste management included:

- Medical Waste Identification
- Medical Waste Generator Requirements
- Medical Waste Transporter Requirements
- Medical Waste Treatment, Destruction, and Disposal Facility Requirements

Medical waste is defined as "any solid waste which is generated in the diagnosis, treatment, or immunization of human beings or animals, in related research, biologicals production, or testing." Regulated medical wastes are a subset of all medical wastes and include seven distinct categories:

- Cultures and stocks of infectious agents
- Human wastes related to diseases (e.g., tissues, body parts)
- Human blood and blood products
- Sharps (e.g., hypodermic needles and syringes used in animal or human patient care)
- Certain animal wastes
- Certain isolation wastes (e.g., wastes from patients with highly communicable diseases)
- Unused sharps (e.g., suture needles, scalpel blades, hypodermic needles)

In addition, mixtures of solid waste and medical waste are considered medical waste. Mixtures of hazardous and regulated medical waste were subject to the medical waste requirements only if the shipment was not subject to hazardous waste manifesting (e.g., the hazardous waste is generated by a conditionally exempt generator).

The definition of *medical waste* excludes any hazardous waste identified or listed under 40 CFR Part 261 or any household waste defined in 40 CFR 261.4(b)(2). In addition, residues from treatment and destruction processes or from the incineration of regulated medical wastes were excluded from the requirements, as were human remains intended for burial or cremation. Infectious agents being shipped pursuant to other federal regulations and samples of regulated medical waste shipped for enforcement purposes were also exempt from the requirements.

The federal regulations required segregation, where practical, of regulated medical wastes at the point of generation. The tracking requirements for generators of less than 50 pounds per month were more flexible than those for larger generators. All generators, however, were subject to the same waste management requirements.

Generators (including transporters who repackage shipments) were identified as being responsible for properly handling medical waste before shipping it off-site. The waste was required to be segregated into sharps, fluids, and other wastes and then packaged in rigid, leak-resistant containers. Additionally, sharps were to be packaged in puncture resistant, and fluids in leak resistant, containers. Untreated medical waste required a water resistant label on the outside of the packaging identifying it as "infectious waste" or "medical waste" or displaying the bio-hazard symbol. Regulated medical waste shipments had to be marked with information identifying the generator and each individual transporter.

A tracking form was developed similar to the uniform hazardous waste manifest. It was designed to track medical waste from "cradle to grave," or from generation through disposal. As in the case of the uniform hazardous waste manifest, copies of the form stayed with the generator, each transporter, and each facility handling the waste.

Transporters of medical waste were required to obtain a medical waste identification number that was to be used on all paperwork, including the tracking form. Transporters could only accept medical waste that was properly packaged, labeled, and marked, and was accompanied by a properly completed tracking form.

Facilities accepting the medical waste had to sign and return tracking forms to the generator. These facilities had to check the waste they received to make sure it matched what was on the tracking form. If it did not match, they had specific reporting requirements. If they incinerated the waste, they had to submit regular reports detailing the types and amounts of medical waste that were burned.

**Figure 10-20:** Infectious wastes must be properly labeled before transport.

Although these regulations have been described in the past tense because they are no longer on the federal books, the states have for the most part instituted similar regulatory programs. The federal government is still looking at the need for federal regulations and may at some future date reinstate similar provisions.

## ■ Checking Your Understanding (10-2)

1. Explain the difference between a small quantity generator and a generator under RCRA.

2. Describe how you would determine whether a sample of waste was hazardous waste.

3. Explain the role of the Uniform Hazardous Waste Manifest in "cradle to grave" management of hazardous wastes.

4. Describe the basic provisions of the Land Ban.

5. Discuss the potential hazards of mismanaging medical wastes.

# 10-3 Managing Nuclear Waste

## ■ Radioactive Materials and Waste

### Concepts

■ Nuclear energy and medical facilities are the largest users of radioactive materials.

■ Practically every industry uses radioisotopes in some form.

■ Both the DOE and the EPA are involved in radioactive material regulation.

■ The main law regulating the management and disposal of radioactive materials and waste is the Atomic Energy Act.

Radioactive materials are widely used in our society. They are used in food irradiation, generating electrical energy, medical treatment and diagnosis, weapons development, industrial applications, product sterilization (for bacteria), detection of buried pipelines, research in many fields, and consumer products. They are also by-products of certain mining operations. The large amounts of radioactive wastes resulting from these activities create a potential for exposure of the population to levels of radiation that are well above naturally occurring levels. It is well known that exposure to radiation increases the risk of cancer and genetic damage.

The public tends to view **radiation** as one of the most frightening risks modern society faces. The very thought of invisible particles "zapping" a person makes the concept that much more alarming. Many people overlook medical facilities and tend to associate radiation only with nuclear energy plants, which are very significant sources of radioactive substances. Another major source, and one that is often overlooked, is medical facilities. Memorable disasters, such as the 1986 accident at the Chernobyl power plant that caused a lethal radioactive release contribute to the public's apprehensions about the hazards of radioactive substances. Unfortunately, these types of disasters and the damage inflicted on people and the environment have overshadowed in many people's minds the benefits to society of many uses of radioactive materials.

To gain an understanding of the widespread use of radioactive materials, we need to look at the DOE estimate of over 3 million packages of radioactive materials shipped each year in the United States. These radioactive materials are routinely used for food preservation, medical diagnosis, agriculture, power generation, as well

**Figure 10-21:** Nuclear power plants are widely recognized users of

# Science & Technology

## What is Radiation and How Does It Harm Us?

In Chapter 1 we learned that atoms are composed of protons, neutrons and electrons. In this model of the atom, the protons and neutrons are located in the nucleus and are surrounded by the electrons in energy levels. We also learned that all atoms of the same element have the same number of protons, but differing numbers of neutrons. What we did not explain is what effect this may have on the stability of the atom's nucleus. When it is important to know the number of neutrons present, the atom's atomic mass number is included on the upper left-hand side of the symbol, and its atomic number may or may not also be included in the lower left-hand side , e.g., $^{238}_{92}U$ or $^{238}U$. When speaking or writing this isotope, it is also common to refer to it as **uranium**-238.

As we also learned in Chapter 1, when two or more atoms of the same element have differing numbers of neutrons they are called isotopes. Of the 100 or more natural and manmade elements, all have at least one isotope. A common example would be the isotopes of carbon, $^{12}_{6}C$ (carbon-12) and $^{14}_{6}C$ (carbon-14), whose ratio of isotopes is compared to determine the age of artifacts. Hydrogen is another example. If we look at the Periodic Chart, we see that all hydrogen atoms have one proton in their nucleus. They can, however, have either zero, one, or two neutrons.

When some combination of hydrogen's two heavier isotopes, $^{2}_{1}H$, or $^{3}_{1}H$ combines with oxygen to form water, it is referred to as "heavy" water. Heavy water is used as a **moderator** in some types of nuclear reactors. Moderators are substances that have little tendency to absorb, but are used to reduce the speed of neutrons given off by radioactive isotopes.

Not all isotopes are radioactive. Radioactive isotopes are those that have an unstable arrangement of protons and neutrons in their nucleus. Isotopes of elements such as uranium (U), radium (Ra), and thorium (Th) are highly unstable. No one knows exactly, however, when a nucleus will split. This splitting, or **radioactive decay**, is also known as **nuclear fission** and includes the throwing off of a variety of subatomic particles and is always accompanied by the release of energy. Those isotopes that undergo this type of decay are called **radioisotopes** or **radionuclides**.

One theory of nuclear stability is that the farther the ratio of neutrons to protons is from one ($8n^o/6p^+ = 1.33$), the more likely it will be a radioisotope. This theory is strengthened by the fact that whenever an unstable isotope undergoes a series of nuclear decays, it always results in an improvement of the proton to neutron ratio. The energy and/or particles released from radioactive decay are generally referred to collectively as **radiation**. In each step, as the radioactive nucleus changes, the new nucleus may be either stable or radioactive itself. It is common for radioactive nuclei to undergo several transformations before they convert to stable nuclei.

| | | | | | |
|---|---|---|---|---|---|
| $^{238}_{92}U$ | $\rightarrow$ | $^{234}_{90}Th$ | $+$ | $^{4}_{2}He$ |
| $^{234}_{90}Th$ | $\rightarrow$ | $^{234}_{91}Pa$ | $+$ | $^{0}_{-1}e$ |
| $^{234}_{91}Pa$ | $\rightarrow$ | $^{234}_{92}U$ | $+$ | $^{0}_{-1}e$ |
| $^{234}_{92}U$ | $\rightarrow$ | $^{230}_{90}Th$ | $+$ | $^{4}_{2}He$ |
| $^{230}_{90}Th$ | $\rightarrow$ | $^{226}_{88}Ra$ | $+$ | $^{4}_{2}He$ |
| $^{226}_{88}Ra$ | $\rightarrow$ | $^{222}_{86}Rn$ | $+$ | $^{4}_{2}He$ |
| $^{222}_{86}Rn$ | $\rightarrow$ | $^{218}_{84}Po$ | $+$ | $^{4}_{2}He$ |
| $^{218}_{84}Po$ | $\rightarrow$ | $^{214}_{82}Pb$ | $+$ | $^{4}_{2}He$ |
| $^{214}_{82}Pb$ | $\rightarrow$ | $^{214}_{83}Bi$ | $+$ | $^{0}_{-1}e$ |
| $^{214}_{84}Bi$ | $\rightarrow$ | $^{214}_{84}Po$ | $+$ | $^{0}_{-1}e$ |
| $^{214}_{84}Po$ | $\rightarrow$ | $^{210}_{82}Pb$ | $+$ | $^{4}_{2}He$ |
| $^{210}_{82}Pb$ | $\rightarrow$ | $^{210}_{83}Bi$ | $+$ | $^{4}_{2}He$ |
| $^{210}_{83}Bi$ | $\rightarrow$ | $^{210}_{84}Po$ | $+$ | $^{4}_{2}He$ |
| $^{210}_{84}Po$ | $\rightarrow$ | $^{206}_{82}Pb$ | $+$ | $^{4}_{2}He$ |

It was previously noted these transformations are accompanied by the release of energy and a variety of subatomic particles other than just protons and neutrons. The other two types of particles are alpha ($^{4}_{2}He$) and beta ($^{0}_{-1}e$) particles. An **alpha** particle ($\alpha^{++}$) is two protons and two neutrons that are being ejected from the nucleus as a single particle. This particle lacks electrons, so its composition would be the same as a positively charged helium nuclei, $He^{2+}$. The energy from these relatively slow speed positively charged alpha particles is easily absorbed by thin pieces of most forms of matter (such as this page or your skin). Alpha particles are most hazardous when they are inhaled or ingested, reaching sensitive internal organs. Radon gas is an alpha particle emitter and is believed to be one of the primary causes of lung cancer in miners.

## Science & Technology

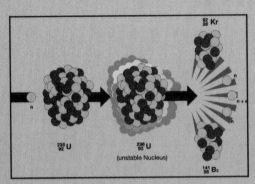

**Figure 10-22:** Nuclear fission.

**Beta** particles ($\beta^-$) are high speed electrons ejected from the nucleus when a neutron is converted to a proton. They can penetrate short distances into solid material and can cause skin damage, but are stopped before reaching internal organs. Heavy clothing can protect the skin from damage from beta particles. If ingested or inhaled, these particles can reach internal organs and cause severe damage. Consider the following disintegration series where the radioactive $^{238}_{92}U$ is converted to the stable $^{206}_{82}Pb$ isotope, through a series of alpha and beta particle emissions.

The most penetrating emission during a fission is energy in the form of powerful gamma rays ($\gamma$). **Gamma rays** are a form of electromagnetic radiation, which also includes X-rays, visible light and radio waves. X-rays have more energy than ordinary light, but gamma rays are the most powerful. The hazards and the behavior of gamma and X-rays are much the same. These rays can penetrate deeply into solid materials thereby requiring heavy materials such as lead or concrete to stop them. Exposure to these rays is extremely hazardous, requiring strict adherence to occupational exposure limits.

Because of the amount of energy transmitted when alpha, beta, and gamma radiation strike atoms, the result is the loss of one or more of their electrons. This process is called **ionization** and results in the formation of an **ion** and a free electron. An ion is a charged particle produced during an ionizing collision. These forms of radiation can also result in the formation of **free radicals** which are highly reactive uncharged molecular fragments (uncharged pieces of a molecule with an unpaired electron).

Water molecules are the most abundant molecules in cells. Interactions of the water in a cell with radiation can result in both ionization and free-radical formation.

$$H_2O + radiation \rightarrow H_2O^+ + e^-$$

$$H_2O^+ \rightarrow OH\bullet + H^+$$

The highly reactive OH• species, with its unpaired electron, can react with a host of other molecules to produce new free radicals. It is this ionizing and free-radical formation ability that makes radiation harmful to living matter.

The amount of damage caused by the various forms of radiation varies inversely to its penetrating power. Since gamma radiation is the most penetrating, it does the least tissue harm. Due to its mass and slower speed, the alpha particle has the least penetrating power, but delivers up to 10 times the ionizing power of either beta or gamma radiation.

When can one expect a radioactive nuclei to undergo the release of this damaging radiation? This important question can only be answered by understanding the concept of half-life. **Half-life** is defined as the time it takes for one-half of the nuclei in a sample to undergo radioactive decay. Half-lives of different isotopes may vary tremendously, as is demonstrated below:

| Isotope | Half-life |
|---------|-----------|
| $^{238}_{92}U$ | $4.46 \times 10^9$ years |
| $^{234}_{91}Pa$ | 6.69 hours |
| $^{226}_{88}Ra$ | 1,600 years |
| $^{234}_{92}U$ | $2.45 \times 10^5$ years |
| $^{212}_{84}Po$ | $2.98 \times 10^{-7}$ sec |
| $^{210}_{84}Po$ | 138 days |

What this means is that in the first half-life, one-half of the nuclei would undergo disintegration. In the second half-life, one-half of the remaining nuclei would disintegrate. This process repeats over and over, resulting in $\frac{1}{2}$ to $\frac{1}{4}$, $\frac{1}{8}$, $\frac{1}{16}$, etc. of the nuclei remaining. In other words, something radioactive will always remain radioactive.

There are also lower forms of radiation called **non-ionizing radiation**. These forms of radiation do not release enough energy when striking atoms to result in their ionization. Some examples of nonionizing radiation include: radiant heat, light and radio waves. It is generally easy to control exposure to these forms of radiation with simple

## PERIODIC TABLE OF THE ELEMENTS

| IA | IIA | IIIB | IVB | VB | VIB | VIIB | ←──VIIIB──→ | | | IB | IIB | IIIA | IVA | VA | VIA | VIIA | VIIIA |
|---|---|---|---|---|---|---|---|---|---|---|---|---|---|---|---|---|---|
| 1<br>H<br>1.008 | | | | | | | | | | | | | | | | | 2<br>He<br>4.003 |
| 3<br>Li<br>6.941 | 4<br>Be<br>9.012 | | | | | | | | | | | 5<br>B<br>10.811 | 6<br>C<br>12.011 | 7<br>N<br>14.007 | 8<br>O<br>15.999 | 9<br>F<br>18.998 | 10<br>Ne<br>20.180 |
| 11<br>Na<br>22.990 | 12<br>Mg<br>24.305 | | | | | | | | | | | 13<br>Al<br>26.982 | 14<br>Si<br>28.086 | 15<br>P<br>30.974 | 16<br>S<br>32.066 | 17<br>Cl<br>35.453 | 18<br>Ar<br>39.948 |
| 19<br>K<br>39.098 | 20<br>Ca<br>40.08 | 21<br>Sc<br>44.956 | 22<br>Ti<br>47.88 | 23<br>V<br>50.942 | 24<br>Cr<br>51.996 | 25<br>Mn<br>54.9380 | 26<br>Fe<br>55.847 | 27<br>Co<br>58.933 | 28<br>Ni<br>58.69 | 29<br>Cu<br>63.546 | 30<br>Zn<br>65.39 | 31<br>Ga<br>69.723 | 32<br>Ge<br>72.61 | 33<br>As<br>74.922 | 34<br>Se<br>78.96 | 35<br>Br<br>79.904 | 36<br>Kr<br>83.80 |
| 37<br>Rb<br>85.468 | 38<br>Sr<br>87.62 | 39<br>Y<br>88.906 | 40<br>Zr<br>91.224 | 41<br>Nb<br>92.906 | 42<br>Mo<br>95.94 | 43<br>Tc<br>98.9062 | 44<br>Ru<br>101.07 | 45<br>Rh<br>102.906 | 46<br>Pd<br>106.42 | 47<br>Ag<br>107.868 | 48<br>Cd<br>112.41 | 49<br>In<br>114.82 | 50<br>Sn<br>118.71 | 51<br>Sb<br>121.75 | 52<br>Te<br>127.60 | 53<br>I<br>126.904 | 54<br>Xe<br>131.3 |
| 55<br>Cs<br>132.905 | 56<br>Ba<br>137.33 | 57<br>La<br>138.906 | 72<br>Hf<br>178.49 | 73<br>Ta<br>180.948 | 74<br>W<br>183.85 | 75<br>Re<br>186.21 | 76<br>Os<br>190.23 | 77<br>Ir<br>192.22 | 78<br>Pt<br>195.08 | 79<br>Au<br>196.966 | 80<br>Hg<br>200.59 | 81<br>Tl<br>204.38 | 82<br>Pb<br>207.2 | 83<br>Bi<br>208.98 | 84<br>Po<br>(209) | 85<br>At<br>(210) | 86<br>Rn<br>(222) |
| 87<br>Fr<br>(223) | 88<br>Ra<br>226.025 | 89<br>Ac<br>227.03 | 104<br>Unq*<br>(261) | 105<br>Unp*<br>(262) | 106<br>Unh*<br>(263) | 107<br>Uns*<br>(262) | 108<br>Uno*<br>(265) | 109<br>Une*<br>(266) | | | | | | | | | |

| 58<br>Ce<br>140.11 | 59<br>Pr<br>140.91 | 60<br>Nd<br>144.24 | 61<br>Pm<br>(145) | 62<br>Sm<br>150.36 | 63<br>Eu<br>151.96 | 64<br>Gd<br>157.25 | 65<br>Tb<br>158.92 | 66<br>Dy<br>162.50 | 67<br>Ho<br>164.93 | 68<br>Er<br>167.26 | 69<br>Tm<br>168.93 | 70<br>Yb<br>173.04 | 71<br>Lu<br>174.97 |
|---|---|---|---|---|---|---|---|---|---|---|---|---|---|
| 90<br>Th<br>232.04 | 91<br>Pa<br>231.04 | 92<br>U<br>238.03 | 93<br>Np<br>237.05 | 94<br>Pu<br>244 | 95<br>Am<br>(243) | 96<br>Cm<br>(247) | 97<br>Bk<br>(247) | 98<br>Cf<br>(251) | 99<br>Es<br>(252) | 100<br>Fm<br>(257) | 101<br>Md<br>(258) | 102<br>No<br>(259) | 103<br>Lr<br>(260) |

*Names not officially assigned

**Figure 10-23:** The Periodic Table of the Elements.

as the many applications listed above. Many people are not aware that radioactive materials are used in consumer goods; for example, smoke detectors found in millions of American homes use americium-241. Radioactive materials are also used in research, historical dating and food preservation.

This growth in uses and amounts of radioactive materials has caused concern regarding their disposal. The public not only fears the potential contamination of soil and groundwater at the site of disposal but also the potential for transportation accidents as the materials move from point of use to disposal sites.

The radioactive material used in hospitals and research facilities includes many **radioisotopes**. The DOE estimates that about one-third of all patients admitted to U.S. hospitals are diagnosed or treated using them. One of the most common treatments for cancer, for instance, uses cobalt-60. Iodine-131 is used in the treatment of brain tumors. Strontium-89 is used to reduce pain from prostate and breast cancer. Technetium is used for early detection of AIDS-related abnormalities in the lungs of patients infected with HIV. Some heart pacemakers provide regular stimuli to defective hearts with power from small **plutonium**-238 radioisotope generators.

Medical use of radioisotopes is not limited to diagnosis and treatment of diseases. They are also used to sterilize medical equipment. Supplies that might be damaged in other methods of sterilization include syringes, surgical instruments, and pharmaceuticals. These can be sterilized by use of radiation from radioisotopes.

The DOE estimates that about 62 percent of all shipments of radioactive materials are to medical institutions. The remaining 38 percent is divided amongst research facilities, industrial facilities, nuclear power plants, and waste storage and disposal sites.

Radioactive materials are used in agriculture for a wide range of applications, including production of higher-yielding food crops through efficient use of fertilizers. Using fertilizers combined with phosphorus-32 provides a means of determining how much of the fertilizer is taken up by the plant and how much is lost to the environment. Effects of placement, timing, and amounts of fertilizer can also be determined using this method.

A major achievement in the use of radioisotopes has been in the control of pests and insects. Cesium-137 and cobalt-60 are used to develop seeds with improved resistance and product yields and extend the shelf life of certain foods by destroying spoiling bacteria. In addition, radioactive material is used to produce shrink-wrap food packaging.

Although nuclear power plants raise safety concerns, we now see that alternatives also have their downsides. In Chapter 4 we learned that the burning of fossil fuels for energy is being implicated in the potential warming of the planet. Nuclear energy is often touted as a cleaner source of energy. However, siting nuclear electric plants is a long, involved, and difficult process. A critical concern is the safe disposal of waste from the plant. Even highly respected scientists in the field can take opposite sides of the issue.

Both DOE and the EPA are involved in nuclear waste regulation. DOE is responsible for disposal of nuclear wastes related to nuclear power plant use and weapons production. All these wastes are currently temporarily contained in tanks, casks, or cooling pools. The

# Box 10.1

## Risk Communication

This section explains that the general public views the use and management of radioactive materials as a very high risk activity. The question is, how well does the public gauge the relative risk of these sorts of things? In Chapter 3 we learned about the accepted method for assessing toxicological risks, which involves a number of steps that utilize scientific data and dispassionate analysis to ultimately arrive at a numerical representation of risk. Scientists' efforts at educating the public often have become confused and frustrated because of the public's reactions to environmental risk. They note how tempers flare at a public meeting concerning a risk that they have estimated might cause considerably fewer than one-in-a-million increased cancer deaths. Yet people will smoke during the break and drive home without seat belts – risks far greater than those discussed at the public meeting. When scientists point to this apparent contradiction, people become even angrier. Conversely, risks that the scientists see as serious – naturally occurring radon gas in homes, for example – often can be met with relative indifference by the public.

Agencies sometimes respond to unexpected angry community reactions by dismissing them as irrational and concluding that the public is unable to understand the scientific aspects of risk. But when agencies make decisions that affect communities without involving those communities, they often elicit even angrier responses. Unfortunately, too often, the regulatory community assumes that because communities don't agree with an agency action, it is because they don't understand it.

People commonly overestimate the frequency and seriousness of dramatic, sensational, dreaded, well-publicized causes of death and underestimate the risks from more familiar, accepted causes that claim lives one by one. It has been shown that risk estimates by "experts" and the public on many key environmental problems differ significantly.

This situation and the reasons for it are extremely important since in our society the public generally does not trust experts to make important risk decisions alone. As a former EPA Administrator William D. Ruckelshaus said, "We have decided, in an unprecedented way, that the decision-making responsibility involving risk issues must be shared with the American people, and we are very unlikely to back away from that decision. The policy questions at stake are critical, affecting not only public and ecological health and welfare, but also massive amounts of public and private resources."

The EPA actually commissioned a study to compare the risks of major environmental problems with the idea that since the resources of the agency were limited, they should "give priority attention to those pollutants and problems that pose the greatest risks to our society." A special task force was created to compare problems. Their final result was a ranking of 31 environmental problems in the areas of cancer risk, non-cancer health risks, ecological effects, and welfare effects (i.e. materials damage).

The public's risk rankings for the same problems were also compiled. Comparing the public's ranking with the task force's ranking raised some interesting issues. Analysis of both sets of data coupled with review of legislation revealed that EPA's *actual* priorities and legislative actions correspond more closely with public opinion than they do with the EPA task force's estimates of the relative risk. The most significant differences concern hazardous waste and chemical plant accidents (high public concern, medium/low risk ranking by the task force) and pesticides, **indoor air pollution**, consumer product exposure, worker exposure to chemicals, and global warming (medium/low public concern, relatively high risk ranking by the task force).

The most obvious explanation for the difference is that the general public simply did not have all the information that was available to the task force experts. The areas of risk are vast and it is difficult for anyone to have full knowledge of the information. Indeed, the experts themselves had to expend considerable effort to develop their rankings, and all of them were surprised by at least some of the findings.

It is also important to note that the experts and the public were looking at the question of risk in somewhat different ways. The EPA task force purposely dealt with a limited number of dimensions of risks, ignoring most of the intangible aspects that are of great value to the public: the degree to which risks are familiar, generally accepted, voluntary, controllable by the individual, etc. These differences reflect a more general pattern of experts taking a societal (macro) perspective, while the public usually takes a more individual or personal (micro) perspective.

The most obvious message for those involved in environmental problems – representatives of government, industry, public interest groups, and the scientific community – is to recognize how people may react to the risks, to understand why

# Box 10.1

the risks have been assessed technically as high or low, and to tailor policies and communications to accommodate differing perspectives. Issues of high risk/high public concern and low risk/low public concern are issues of general agreement. But the high/low combinations can present challenges to leadership, values, and ethics to all involved.

Because outbursts of citizen anger make agencies understandably uncomfortable, they also tend to forget that public outrage can be extremely positive. In fact, most environmental agencies and a significant number of the laws they enforce are the results of citizen campaigns, fueled by anger over environmental degradation or an incident that endangered lives. Funding for these laws, and consequently for agency staff, also depends in some cases on tough legislative battles fought by citizens. In addition, most agencies can admit to a number of environmental problems that wouldn't have been uncovered were it not for community action. On the other hand, agencies particularly resent anger directed at them rather than at the environmental problem.

As we have seen, public fears are often not well-correlated with agency assessments. While agencies focus on data gathered from hazard evaluations, monitoring, and risk assessments, the public takes into account many other factors besides scientific data. Collectively, it is helpful to think of these non-technical factors as the "outrage" dimension of risk, as opposed to the "hazard" dimension more familiar to environmental professionals. Because the public pays more attention to outrage than the experts do, public risk assessments are likely to be very different from agency risk assessments.

Merely hammering away at the scientific information will rarely change this. While it may be tempting to conclude from this that the public cannot understand risk assessment data, research in the field of risk perception strongly suggests that other factors are at work. Below are some of the key variables that underlie community perception of risk.

■ Voluntary risks are accepted more readily than those that are imposed.

■ Risks under individual control are accepted more readily than those under government control.

■ Risks that seem fair are more acceptable than those that seem unfair.

■ Risks that are ethically objectionable will seem more risky than those that aren't.

■ Risk information that comes from trustworthy sources is more readily believed than information form untrustworthy sources.

■ Natural risks seem more acceptable than artificial risks. Exotic risks seem more risky than familiar risks.

■ Risks that are associated with other, memorable events are considered more risky.

The greater the number and seriousness of these factors, the greater the likelihood of public concern about the risk, regardless of the scientific data. As government agencies have seen many times, the risks that elicit public concern may not be the same ones that scientists have identified as most dangerous to health. When officials dismiss the public's concern as misguided, the result is controversy, anger, distrust, and still greater concern. None of this is meant to suggest that people disregard scientific information and make decisions based only on the other variables. It does suggest, however, that outrage also matters, and that by ignoring the outrage factors, agencies can cause people to become still more outraged.

Agency representatives and environmental managers sometimes believe that if they could only find a way to explain the data more clearly, communities would accept the risks scientists define as minimal and take seriously the risks scientists see as serious. However, simply finding ways to explain the numbers more clearly is not the simple solution practitioners might hope for. Agencies now realize this and are training their staff in *risk communication*. Guidelines provided to those who must explain risk include the following:

■ Consider the outrage factors when explaining risk.

■ Find out what risk information people want and in what form.

■ Anticipate and respond to people's concerns about their personal risk.

■ Take care to give adequate background when explaining risk numbers.

■ Take care when comparing environmental risks to other risks.

■ Acknowledge uncertainty.

■ Recognize that communities determine what is acceptable to them, not agencies nor industry.

■ Take even greater care presenting technical information than presenting other information.

EPA addresses radiation problems in four primary areas: radiation from nuclear accidents, radon emissions, land disposal of radioactive waste, and radiation in groundwater and drinking water. The EPA is responsible for setting certain radiation standards and for developing guidance to be implemented by other federal agencies such as DOE and the Nuclear Regulatory Commission (NRC).

The statute that governs the control of hazards associated with radioactive materials is the Atomic Energy Act of 1954. The administering agency for this act is the U.S. Nuclear Regulatory Commission (USNRC). This commission licenses and oversees the construction and operation of any facility that uses intensely radioactive materials as well as the disposal of the resulting wastes.

## ■ Radioactive Waste

### Concepts

■ There are three classifications of nuclear waste.

■ Siting a permanent repository for spent fuel has been a difficult proposition.

■ Radioactive waste is characterized by radioactivity, volume, and half-life.

Radioactive or nuclear waste is the unusable byproduct of nuclear activities. These wastes are classified as either high-level waste (HLW), transuranic (TRU) waste, or low-level waste (LLW). Each of these types of nuclear wastes presents its own challenges for proper management.

The three criteria used to categorize the level of radioactive waste are: radioactivity, volume, and half-life. These criteria determine the level of hazard of the waste. For instance:

– high-level waste is characterized by high radioactivity, a long half-life, and relatively low volume;

– low-level waste is characterized by low radioactivity, a relatively short half-life, and high volume;

– transuranic waste is characterized by low radioactivity, a long half-life, and low volume.

**High-level waste (HLW)** is the highly radioactive waste resulting from reprocessing spent fuel, which is the fuel that has been irradiated in a nuclear reactor. Currently, in the U.S., reprocessing involves removing the plutonium and uranium from spent fuel generated by the USDOE nuclear reactors. The plutonium and uranium is recycled for use in defense programs. What remains after reprocessing is highly radioactive waste that must be remotely handled behind heavy protective shielding. This waste requires long-term isolation, typically in an underground repository, while it stabilizes. Shipping HLW presents a unique hazard, so it is packaged in heavily shielded containers for storage and transport.

The spent fuel resulting from the generation of electricity from U.S. commercial nuclear power plants is currently not being reprocessed. The Nuclear Waste Policy Act of 1982 is the federal statute that creates the framework for managing nuclear waste. Under this act, the nuclear wastes from nuclear power plants are to be placed in an underground repository for long-term isolation. The act set forth ambitious deadlines for siting, constructing, and operating two geologic repositories. Five years later, Congress passed the Nuclear Waste Policy Amendments Act of 1987. These amendments built on the previous direction of the underground repository program and in fact required that DOE focus on one site – Yucca Mountain, in Nevada. Two repositories are currently being developed – Waste Isolation Pilot Plant (WIPPs) in New Mexico for waste generated by the DOE, and Yucca Mountain for waste generated by commercial power plants.

According to a federal commission charged with overseeing and reviewing the storage of nuclear waste, about 20,000 metric tons of spent fuel had accumulated from the beginning of the production and use of nuclear power to 1990 – about 50 years. This accumulation came from the more than 70 nuclear power plants in the U.S. Since all nuclear power plants are licensed, the commission also estimated that by the end of the current license period of all existing nuclear power plants, the amount of nuclear waste is expected to be nearly 90,000 metric tons.

Currently, practically all spent fuel is stored in water-filled pools at reactor sites. There is not enough space to store the expected quantities of fuel in this manner. Since current U.S. policy is to ultimately deposit spent fuel in a permanent underground geologic repository, the repository would have to contain the radioactive waste for the next 10,000 years and beyond. Congress felt the Yucca Mountain site in Nevada was the best candidate site to achieve this. This site is on federal land about 90 miles northwest of Las Vegas, Nevada. As can be imagined, progress on the project has been slow and fraught with many setbacks.

The DOE initially announced that the opening of the site was slated to be 2003. They were rather optimistic in this estimate given that the design and operation of such a repository is so technically and politically complex. In general, local public sentiment for a permanent repository has been negative. There has been tremendous controversy over just about every aspect of the project. This has caused the estimated opening to be revised upwards several times. The project is currently mired in lawsuits and red tape. Many experts believe that if the project is ever built, it will be well into the next century. Just the studies required to site and determine compliance standards for the facility as required by the various regulatory agencies have already cost $6 billion. That is twice the estimate of what it would cost to construct the actual facility. In the meantime, while waiting for a permanent repository, the radioactive waste continues to be stockpiled.

Given the national policy to dispose of spent fuel from civilian reactors in a permanent geologic repository, the question is, what is the best way to store the fuel that accumulates before the repository opens? Many options have been discussed over the years, including 1) continuing to store all spent fuel at the reactors where it is generated, and 2) store some spent fuel at a central location or locations as well as at some reactors.

Managing and disposing of **spent nuclear fuel** in a safe and efficient manner has proven to be a difficult public policy issue. Scientifically, there are many uncertainties and experts themselves differ in opinion over the best approaches. Few, if any, public policy issues require such long-range planning as nuclear waste disposal. Just think, the 10,000 year span that was eluded to earlier is longer than currently recorded history! Planning anything for this length means that geologic stability must be understood completely. Just choosing a location requires extensive study and evaluation that may take years, complicated by political and policy issues. In spite of considerable time and money expended on a permanent repository, to date none has been sited and no one can predict what date one will be available.

The nuclear waste disposal debate has fueled the more fundamental debate as to whether nuclear power is appropriate. Critics of nuclear power point out that it should not be used until an effective solution to the waste disposal problem is found. Environmental groups and concerned citizens view nuclear waste disposal as a major moral obligation facing this generation. Many believe that this generation has an awesome ethical obligation to assure that our radioactive waste will not harm future generations. Groups opposing nuclear power explain that the risks are too great because of the extremely long half-life of the radioactive materials involved and the staggering costs of isolating them from the biosphere.

Groups that are pro-nuclear power believe that wastes can be managed and disposed of safely. Even these groups are frustrated, however, because of the large sums of money, time, and resources already used with no apparent progress.

We have over 80 years of experience with radioactivity to help in accomplishing safe working conditions in occupations where exposure may occur. According to the Society for Nuclear Medicine, "More is known about radiation effects than about any other hazard to humans." In fact, no other standards have been so extensively reviewed and agreed upon by international experts, and ultimately adopted by so many countries (Figure 10-24).

Nuclear power plants generated about 19.5 percent of all electricity produced in the U.S. in 1988. The generation of energy at nuclear power plants is accomplished through fission. Fission is the splitting of atoms through a nuclear reaction. In nuclear power plants the heavy nucleus of uranium-235 is split, usually into two fragments of comparable masses and several free neutrons. These types of splits generate heat that produces steam to generate electricity. The neutrons cause the chain reaction that over time becomes less and less efficient. The uranium-235 is contained in zirconium alloy rods and it usually takes three to four years for the fuel in a rod to become so depleted that they must be removed from the reactor and replaced. Even though we use the term *spent* to

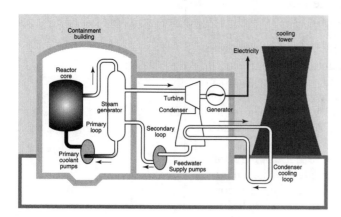

**Figure 10-24:** Nuclear power plant operation.

describe this condition of the fuel, it is important to remember that the fuel still emits a large amount of heat and radiation.

Government high level waste (HLW) is stored temporarily in underground tanks and vaults at government sites. This is a temporary measure until it is solidified and packaged in stainless steel canisters and inserted into heavily shielded containers for transport to a permanent disposal site. We have already stated that permanent disposal is not available at this time because siting these disposal areas is very difficult and takes many years. The public needs assurance that there will be permanent protection of society and the environment at these repositories. Geologic features such as faults, volcanoes or the potential for developing these features over geologic time make it difficult to assure the public that release of radioactive material will not occur.

**Transuranic (TRU) wastes** contain man-made elements heavier than uranium, which is why it is called transuranic (beyond uranium). Figure 10-22 is a periodic chart of the elements that shows the transuranic elements. Although most transuranic waste is no more radioactive than low-level waste, it decays so slowly that it is radioactive for a long time. Even though it does not require as much shielding, it requires the same sort of long-term isolation as high-level waste. TRU is generated as a byproduct of the reprocessing of spent fuel from government reactors. **Reprocessing** removes usable plutonium and uranium for use in defense programs as well as other isotopes useful in a variety of commercial applications. TRU waste can also include contaminated protective clothing, tools, glassware, and equipment. TRU waste is being stored at government sites throughout the U.S. while waiting to be shipped to a research and development facility in New Mexico that will dispose of it in deep salt beds.

**Low-level waste (LLW)** is radioactive material resulting from nuclear-related research, medical, and industrial processes. The radiation level of these materials ranges from natural background levels to very radioactive. LLW usually consists of rags, papers, filters, tools, equipment, and discarded protective clothing contaminated with radionuclides. These wastes are generated from research, operations, housekeeping, and maintenance activities at medical, research, commercial, industrial, or other nuclear facilities. Typically, LLW contains small amounts of radioactive material dispersed in large volumes of other material and generally poses little potential hazard. However, small volumes of very radioactive LLW material do require protective shielding during handling and transportation activities. All LLW is disposed of at specially licensed facilities.

Low-level wastes (LLW) are managed by the states. The Low-Level Radioactive Waste Policy Act of 1980 directs each state to be responsible for disposing of its own LLW by either licensing a facility serving the state, or states can team up for a regional approach. Commercial LLW is currently shipped to the existing commercial disposal sites. The bad news is there are currently only two, which service 19 states. Only California has licensed a new facility. Even this had its political pitfalls. Lawsuits were filed on several issues including the threatened desert tortoise. The application for the facility was 7,000 pages long. So far, no schedule issued by the government regarding nuclear waste disposal has been met. Government LLW is disposed of either at the generating facility or shipped to another DOE disposal site.

Uranium tailings are radioactive rock and soil that are the byproducts of uranium mining and milling. Tailings contain small amounts of radium that decay and emit a radioactive gas, radon. When radon gas is released into the atmosphere, it disperses harmlessly, but this gas would be harmful if a person were exposed to it in high concentrations for long periods of time.

## ■ Biological Effects of Radiation and Employee Exposure

### Concepts

■ Most of our exposure to radiation is from background sources.

■ The three key ways to minimize exposure to radiation are time, distance, and shielding.

■ Ionizing radiation is alpha particles, beta particles, or gamma/X-rays.

■ Rad and Rem are two exposure units for radiation.

In the last chapter, the concept of exposure limits was introduced. Radiation exposures are also governed by workplace standards. The primary recommendations for radiation exposure come from the International Commission on Radiation Protection and the U.S. National Council on Radiation Protection and Measurements. These organizations include physicians, radiologists, and scientists who are experts on the biological effects of radiation. Reviews of these recommendations are conducted by the U.S. National Academy of Sciences and the United Nations Scientific Committee on the Effects of Atomic Radiation.

Using these groups' recommendations, actual basic safety standards for radiation protection are issued by the International Atomic Energy Agency (IAEA). The IAEAs standards have been adopted by the United Nations and the U.S. Department of Transportation. The U.S. Nuclear Regulatory Commission has based its regulations on relevant portions of the IAEA regulations.

**Ionizing radiation** comes from natural and manmade sources. Three-fourths of the average American's annual exposure to radiation comes from natural background sources (Figure 10-25). Natural background radiation includes cosmic rays from the sun and from other sources in space. Background radiation also comes from the earth's rocks and soil, which contain naturally occurring radioactive isotopes such as radon, thorium, uranium, and radium. Most of the rest, 23 percent, comes from medical and dental radiation. Other uses of nuclear energy account for the remaining 2 percent.

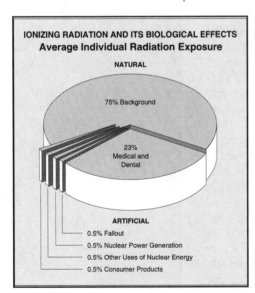

**Figure 10-25:** Average individual radiation exposure.

There are three keys to minimizing external exposure to radiation: time, distance, and shielding. The longer the time spent working near a radiation source, the greater the exposure. Minimizing on-the-job radiation exposure includes controlling the amount of time spent working near a radiation source. Distance is a simple but very effective way to reduce exposures. Radiation follows what is known as the *inverse square law*. That means that doubling the distance between you and the source actually reduces the dose to one quarter (the inverse of 2 is $\frac{1}{2}$, which squared is $\frac{1}{4}$). Shielding removes the pathway from the radiation to the worker. In areas where there may be low-level contamination, the use of protective clothing and respirators may be required.

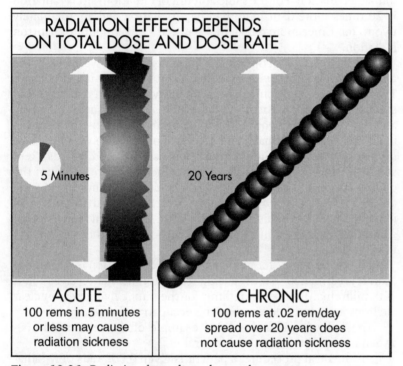

**Figure 10-26:** Radiation depends on dose and exposure.

The four types of ionizing radiation described earlier, gamma/X-ray, neutron, alpha, and beta, cause damage to living tissue. Each of these types of radiation cause different amounts of biological damage due to the difference in the amount of energy they deposit as they travel through body tissue. Interestingly, the length of the path over which the energy is deposited determines how much damage occurs. A large amount of energy deposited over a small distance is more damaging than the same amount of energy deposited over a larger distance.

Gamma rays and high energy X-rays are essentially the same type of radiation emitted from different sources. They travel very rapidly through both the air and through the body, losing only a very small amount of energy as they travel. Beta particles also travel quickly and lose only a small part of their energy during their travel. Beta radiation only penetrates a short distance through tissue, which explains why the primary external effect of beta radiation is on the skin. Neutron radiation can penetrate the body. Neutrons deposit more energy as they pass through the body than do gamma or beta radiation. Gamma, beta, and neutron radiation are sources of external radiation.

Alpha particles are heavy and very energetic. They lose energy rapidly as they travel. Alpha radiation does not penetrate unbroken skin. When alpha-emitting materials are taken into the body, all of the energy is given up in a short distance directly to the body tissues. Beta-emitting materials taken into the body also deposit energy directly to the tissues. Alpha and beta radiation are the principal sources of internal radiation exposure.

The quality factor is a number assignment used to compare the relative ability to cause biological damage for the different types of ionizing radiation. For the same amount of radiation, alpha radiation is 20 times more damaging than beta radiation and approximately two to ten times more damaging than the same amount of neutron radiation.

When we speak of dose for radiation exposures we speak of the amount of energy that is deposited in a material when it is irradiated. Two terms are commonly used to quantify dose: **rad** and **rem**. Rad stands for **radiation absorbed dose** and is the measurement of the amount of energy deposited by ionizing radiation in any material, including people, animals, and objects. For instance, when a piece of fruit is irradiated, the absorbed dose is measured in rads. Rem stands for **radiation equivalent man** which takes into account the quality factor of the dose. Rem is calculated by multiplying the absorbed dose in rads by the quality factor for the particular type of ionizing radiation. This is called a dose equivalent and is most commonly expressed in millirems, which are one-thousandth of a rem.

The current limit for worker exposure to radiation is 5 rems per year. It is estimated that the average radiation worker receives about 300 millirems. The exposure limit for the public living near nuclear facilities is 50 times lower than the occupational standard.

The following is a simplified example of the relationship between rad, quality factor, and millirem:

In a medical procedure, a patient is given 0.002 rads of X-radiation. X-rays, like gamma radiation, have a quality factor of 1, therefore the dose equivalent is 2 mrem (0.002 rad × 1 = 0.002 rem = 2 mrem). If the same procedure had used alpha radiation at 0.002 rads then the dose equivalent would be 40 mrem (0.002 rad × 20 = 0.04 rem = 40 mrem). This represents the greater power for damage that alpha radiation poses.

Although they are not used to measure radiation dose, two other radiation units are commonly used for other purposes. The curie (Ci) is a measurement of the radioactivity of a material in disintegrations per second. The Roentgen (R) is a measure of radiation intensity used to show the amount of ionization caused in air by gamma or X-rays.

Radiation can do direct damage to cells by breaking the membranes and by damaging the chromosomes. It can cause indirect damage by interacting with the water that makes up almost 90 percent of a cell. Cells are the basic unit of structure and function in the body. Most cells have a nucleus surrounded by cytoplasm. There is a membrane around the nucleus and another membrane around the whole cell. The cellular functions corresponding to breathing, eating and digesting, and getting rid of wastes take place in the cytoplasm.

The nucleus is the control center of the cell and also contains the genetic blueprint of the organism, its genes and chromosomes. The nucleus controls the cycle of cell division for both normal growth and damage repair. Many body cells normally have a short life span. They divide at certain stages in their lives and the daughter cells perform the same functions as the cells they came from. This is possible because all of the information needed for a cell to function is present on the chromosomes. When it is time for the cell to divide, each chromosome in the nucleus makes an identical copy of itself. The cell divides and each daughter cell ends up exactly like the parent cell. This process is called mitosis. It is only when egg and sperm cells are formed that each daughter cell contains one-half of the genetic material. Then, when the egg cell is fertilized, the two halves combine to give the new individual the whole number of chromosomes.

When radiation interacts with the water in the cell, very reactive free radicals (molecular fragments with unpaired electrons) are formed. In microseconds these free radicals can combine to produce harmless hydrogen gas, ore water, or the harmful chemical hydrogen peroxide. Hydrogen peroxide is a poison to the cell.

The body has systems to repair much of the direct and indirect damage caused by radiation. Some damage to a cell can be repaired by the cell itself. Badly injured or dead cells can be replaced by the division of nearby healthy cells.

The nucleus of a cell is more sensitive to radiation damage than the cytoplasm. Similarly, not all cells and tissues in the body have the same sensitivity to radiation. The most radiosensitive tissues are those that:

– Contain cells that are dividing at the time they are exposed to radiation,

– Contain cells that normally divide very often, and

– Are unspecialized in structure and function.

The most radiosensitive human tissues are the blood-forming tissues, bone marrow, spleen, and lymph nodes; the tissue which forms sperm cells; the lining of the intestines; and the blood lymphocytes. Lymphocytes are sensitive to radiation because they are almost all nucleus. Highly specialized tissues that are more resistant to the effects of radiation include bones, liver, kidney, muscles, and the brain and nervous system.

Adults are more resistant to the effects of radiation than are children, who are still growing and developing. A developing embryo is very sensitive to the effects of radiation because most of its cells are dividing rapidly to form the organ systems and to increase its size.

There are two types of effects: genetic and somatic. Genetic effects affect the descendants of the exposed individual while somatic effects are effects that occur in the exposed individual. Prompt somatic effects are those that occur almost immediately or within a few days or weeks after the radiation exposure. One example of a prompt somatic effect is the temporary hair loss that occurs about three weeks after a dose of 400 rem to the scalp. New hair is expected to grow within two months after the exposure, although the color and texture may be different. If cancer should develop in the irradiated area 30 years later, it would be an example of delayed somatic effect – one occurring many years after the radiation exposure.

The genetic effect of acute doses of radiation may be mutations that occur in the future children of the exposed person. Mutations, as we learned in Chapter 3, are changes in chromosomes and genes that are passed on to future generations. Some conditions and diseases resulting from gene mutations include asthma, diabetes, anemia, and Down's Syndrome. Mutations happen naturally in our population at a rate of one out of every ten births.

Experts estimate that 50 to 250 rem of radiation received at a constant low level of exposure before conception is the dose needed to double the mutation rate. It is difficult to estimate the genetic risk of ionizing radiation to people because there is no population chronically exposed to enough radiation to give data on actual genetic effects of radiation on their descendants.

The effects of radiation upon biological systems depend on the total radiation dose and the rate at which the dose is received. A whole body radiation dose of 100 rems received in five minutes or less is an acute exposure and may cause mild radiation sickness. A chronic radiation exposure in many small doses, such as two hundredths of a rem (0.02 rem) per day for 50 working weeks each year for 20 years, is also a total dose of 100 rems, but does not cause radiation sickness.

Radiation sickness is the result of an acute whole body dose of 100 to 400 rem of ionizing radiation. Low-level chronic radiation exposures at or below the occupational exposure limit have not been shown to result in damage to biological systems. The maximum averaged daily radiation dose for most workers is 10 to 50 percent of the 0.02 rem. Figure 10-27 shows the effects for various levels of acute exposure and Figure 10-28 for low level chronic effects.

| Acute Exposure Effects | |
| --- | --- |
| 20–50 rem | Acute whole body doses of 25–50 rem can be detected by changes in the number of blood cells. |
| 50–600 rem | An acute dose of 50–200 rem to the sex organs will cause temporary sterility in both men and women. Permanent sterility will follow a 350–400 rem dose to the ovaries of females and a 400–600 rem dose to the testes of males. |
| 400 rem | Without medical treatment, a whole body dose of 400 rem is likely to result in death for half of the exposed persons because of damage to the blood-forming cells in the bone marrow. |
| 600–900 rem | An acute dose of 600-900 rem delivered to the eye will probably result in the formation of cataracts. |
| 1,000 rem | An acute dose of 1,000 rem to an area of the skin has an effect much like a second degree burn. |
| 2,000–5,000 rem | Acute whole body doses of 2,000–5,000 rem result in death within 3–10 days because of damage to the lining of the small intestine. There is no medical treatment to prevent death after this radiation dose to the whole body. When kept in a small area of the body, radiation doses in this range are used to treat cancer. |
| 10,000 rem | Acute whole body doses of 10,000 rem or more cause death within 48 hours from damage to the central nervous system. |

Figure 10-27: Biological effects of acute exposure to ionizing radiation. (Source: DOE)

| Low Level Chronic Exposure Effects | |
|---|---|
| 5 rem | There are no detectable biological effects following an acute or chronic whole body radiation dose of 5 rem. |
| 5 rem per year | The upper occupational exposure limit for radiation workers is 5 rem per year |
| 0.1 – 0.25 rem per year | Normal administrative working levels range about 10–50% of the 5 rem per year occupational exposure limit. |

**Figure 10-28:** Biological effects of low level chronic exposure to ionizing radiation. (Source: DOE)

## ■ Checking Your Understanding (10-3)

1. Explain the difference between an isotope and an ion.

2. Define half-life.

3. Explain how the outrage factor affects perception of risk.

4. Contrast the definitions of high level, transuranic, and low level wastes.

5. Describe the potential genetic effects of radiation exposure.

# 11

# Pollution Prevention and Waste Minimization

## ■ Terms and Concepts to Remember

Bioremediation

Carbon Adsorption

Closed-Loop Recycling

Dechlorination

Distillation

Environmental Audit

Flotation

ISO 14000

Neutralization

Reclamation

Reuse

Solute

Solvent

Source or Waste Reduction

Sump

Thermal Treatment

Waste Minimization

## ■ Chapter Introduction

In the last chapter we learned how rapidly we have increased our generation of wastes and how we are now challenged to find ways to manage them safely and effectively. We also learned that recycling, although a waste management technique that is highly favored by the public, is currently used to manage only a relatively small percentage of this nation's waste. Its success is dependent on separation of recyclable materials by generators and good collection mechanisms as well as favorable market prices for recyclable materials.

This chapter will explore various ways to reduce the generation of wastes and minimize their effects on the environment. Waste reduction technologies will be discussed for both nonhazardous and hazardous solid wastes.

The Congress hereby declares it to be national policy of the United States that, wherever feasible, the generation of hazardous waste is to be reduced or eliminated as expeditiously as possible. Waste nevertheless generated should be treated, stored, or disposed of so as to minimize the present and future threat to human health and the environment.

RCRA, 1976

# 11-1 From Pollution to Prevention

## ▪ Introduction to Pollution Prevention

### Concepts

▪ Pollution prevention is necessary for environmental protection.

▪ Source reduction is one aspect of waste minimization.

▪ Source reduction is preferred over other methods of dealing with waste.

We begin the final chapter with a note of optimism and opportunity. There is a growing recognition that we can protect our environment while reducing costs to industry by practicing what is known as **pollution prevention.** Although it seems obvious, the concept has only recently been embraced by government agencies, industry, and citizens. Congressional mandates state that we are now committed to "preventing rather than controlling pollutants and wastes."

This is a direct turnabout from the conventional "command and control" approach which guided environmental protection since formation of the EPA two decades ago. With the command and control approach, regulations imposed strict controls that were enforced more firmly over time. While this approach resulted in improvement in many areas of environmental protection, there are some members of Congress who feel that the improvement in the environment has not met expectations. They recognize that land disposal of hazardous wastes is inexpensive and effective, but is not necessarily the best option for dealing with hazardous waste.

Other disadvantages of command and control strategies include increased government spending to enforce regulations and the added burden of record keeping, reporting, and control equipment to industry. These requirements may contribute to the competitive disadvantage of U.S. industries compared to foreign competitors, since American industries must spend more for pollution control. It's easy to see that reducing the generation of all wastes, particularly hazardous wastes, at their sources makes good sense both for business and the environment.

According to the EPA, source or waste reduction is the environmental option of choice; this option brings unique and undisputed environmental and economic benefits. **Waste reduction** means decreasing the generation of waste to avoid its handling, treatment, or disposal. Waste reduction includes in-plant practices that reduce or avoid the generation of waste. This greatly reduces risks to health and the environment.

It is important to understand the distinction between the terms waste minimization and waste reduction. **Waste minimization** is a broad umbrella term that includes waste reduction, recycling, and waste treatment processes such as incineration. Actions taken to

In the 1970s, Dr. Joseph T. Ling of 3M Corporation stated the case for pollution prevention through source reduction:

"Pollution controls solve no problem; they only alter the problem, shifting it from one form to another, contrary to this immutable law of nature: the form of matter may be changed, but matter does not disappear... It is apparent that conventional controls, at some point, create more pollution than they remove and consume resources out of proportion to the benefits derived. . . . What emerges is an environmental paradox. It takes resources to remove pollution; pollution removal generates residue; it takes more resources to dispose of this residue and disposal of residue also produces pollution."

reduce the volume or toxicity, or both, of a waste either *before* or *after* it is generated are considered waste minimization. Techniques that reduce the amount of waste after it has been produced are not considered source reduction. This includes waste recycling or treatment of wastes after they are generated. Also, an action that merely concentrates the hazardous content of a waste to reduce waste volume or dilutes it to reduce the degree of hazard is not considered waste reduction. By defining source reduction in this narrow way, the EPA emphasizes its goal of preventing the generation of waste at its source rather than controlling, treating, or managing waste after its generation.

The progress toward achieving higher levels of waste reduction has been slow but steady. The technology is available but the process of converting to new methods that generate less waste is inhibited by both regulatory and industrial obstacles. Industry must still commit significant resources to regulatory compliance. Regulatory agencies continue to press companies to fix the mistakes of the past, often placing little emphasis on prevention programs that would eliminate problems in the future. Furthermore, because of cost, industries that are marginally profitable are the least likely to examine and implement waste reduction measures, even though they may need them most.

The following sections will describe various waste minimization and source reduction techniques. Hazardous waste treatment methods will also be highlighted.

---

## Box 1.1

## Ways Companies Can Promote Waste Reduction

■ Conduct a waste reduction audit to provide information about: 1) types, amounts, and level of hazard of wastes generated; 2) sources of those wastes within the production operation; and 3) feasible reduction techniques for those wastes.

■ **Revise accounting methods** so that both short and long term costs of managing wastes, including liabilities, are charged to the departments and individuals responsible for the processes and operations that generate the waste.

■ **Involve all employees** in waste reduction planning and implementation. Waste reduction must be seen as the responsibility of all workers and managers involved in production, rather than just the responsibility of those who deal with pollution control and compliance.

■ **Motivate employees** and focus attention on waste reduction by setting goals and rewarding employees' suggestions that lead to successful waste reduction. Special education and training can help all types of employees identify waste reduction opportunities at all levels of operation and production.

■ **Transfer knowledge throughout the company** so that waste-reducing techniques implemented in one part of the company can benefit all divisions and plants. This is particularly important in large companies. Newsletters and company meetings can be helpful tools for disseminating information about waste reduction opportunities.

■ **Seek technical assistance** from outside sources. This may be particularly useful for smaller companies with limited technical resources. Sources of outside assistance include state programs, colleges and universities, and professional consultants.

(Source: EPA)

## ■ Integrated Waste Management

### Concepts

■ The hierarchy for waste management has four levels.

■ Integrated waste management brings together a blend of waste management options.

■ There are a number of relatively simple things both consumers and industry can do to reduce waste.

Although most of this chapter will deal with hazardous waste reduction issues, it is important to remember that municipal waste is also a major concern. The previous chapter touched on the two social forces that have created the serious and growing waste problem in America. One is the throwaway mentality of manufacturers and consumers who have become accustomed to goods that are packaged for convenience and that are disposal. The second is the NIMBY (Not In My Back Yard) syndrome. People who manage solid waste are fond of explaining NIMBY as the "First Law of Garbage:" "Everybody wants us to pick it up, and nobody wants us to put it down." These factors are preventing many cities from finding acceptable sites for their wastes, even while running out of space in existing landfills. Siting new combustors or incinerators can be equally difficult. The problem has gotten so severe that some cities are paying premium prices to have their trash shipped to other counties, states, and even foreign countries.

**Figure 11-1:** A sign protesting the siting of a landfill.

The EPA's Municipal Solid Waste Task Force stated in its "Solutions for the 90s Report" that "there is no single solution to this complex problem. A myriad of activities must be implemented, both in the short and long term, by all of us in order to solve the current and future problems with municipal solid waste." Some of the key suggestions to come out of this task force include:

■ Manufacturing products with consideration for their ultimate management as wastes

■ Encouraging, producing, and buying products that are made from recycled or recyclable materials

■ Separating bottles, cans and paper and turning them in for recycling

■ Improving the safety and efficiency of landfills and combustors

■ Wherever practical, choosing source reduction and recycling over landfilling and combustion for managing municipal solid waste

In response to the solid waste dilemma, many states, localities, and concerned individuals have stepped-up their recycling activities and initiated comprehensive waste management programs. Some localities have progressive programs while others lag far behind and may not even recognize or anticipate problems. The private sector has also recognized the benefits of waste minimization and recycling and has successfully implemented programs. In response to these pressures, the EPA has, over the last decade, revised minimum standards for operating municipal landfills; issued procurement guidelines for some recycled goods; promoted used oil recycling; and conducted a toxicity study on municipal waste combustor ash.

Most regulatory agencies favor using integrated systems for solid waste management. This is a holistic approach that is designed so that some or all of four basic waste management options are used to complement one another. The four options for municipal solid waste management are: source reduction (including reuse of products), recycling of materials (including composting), waste combustion (with energy recovery), and landfilling. A key aspect of integrated waste management is the hierarchy, which favors source reduction (including reuse) to first decrease the volume and toxicity and increase the useful life of products in order to reduce the volume and toxicity of *waste*.

Source reduction may occur through the design and manufacture of products and packaging with minimum toxic content, minimum volume of material, and/or a longer useful life. Individuals or companies can practice source reduction by altering their buying habits. Source reduction has many benefits. It slows down the depletion of our resources, prolongs the life of landfill space, and makes combustion and landfilling of wastes safer by removing toxic constituents.

The second level on the hierarchy is recycling materials, including composting food and yard waste. Recycling is high on the hierarchy because it prevents potentially useful materials from being combusted or landfilled. This preserves precious landfill space. Recycling can also save energy and natural resources, provide useful products from discarded materials, and even make a profit.

Although lower on the hierarchy, waste combustion is useful to reduce the bulk of municipal waste and can have the added benefit of energy production. A state-of-the-art combustor that is well maintained and operated should not present a significant risk to human health and the environment.

Landfilling is the lowest on the hierarchy but it is essential to handle nonrecyclable and noncombustible wastes and the ash from combustion. Landfills can also provide the benefit of energy production from the capture of methane gas. Landfills will still be necessary to handle a significant portion of our wastes, so they must be built and operated in the safest manner.

Integrated waste management systems combine the options best suited for the local environmental, economic, and institutional needs. The emphasis is always shifted toward source reduction and recycling. Major reductions in the wastes going to landfills will occur through increased recycling and composting of yard wastes. Many localities already have special "green waste" pickup containers to address this concern.

Reducing our current per capita municipal waste generation rate is very important. It involves educating the public as well as developing a willingness within industry to produce longer-lasting products. Industry must also work toward reducing the volume and toxicity of products and packaging that will ultimately require disposal. Consumers must be willing to change their throwaway mindset and begin to buy longer-lasting goods and support recycling. We are all responsible and have a role in the reduction, reuse, and recycling of the materials we buy.

As was stated in the previous chapter, Americans generate more waste than residents of equally industrialized nations, such as West Germany. Studies have traced much of this difference to the fact that U.S. industries produce more products and that U.S. citizens have higher consumption rates and a greater desire for convenience. In addition, American consumers generally have no incentive to limit their waste generation because disposal is not charged according to the amount of waste produced, like water, gas and electric services. There also are few incentives for manufacturers to design their products and packaging in a way that takes into account effective management when the item is discarded.

Some manufacturers are now looking at "life cycle" management by designing their products with recycling in mind. Building a product with snap-together pieces of like material, for example, enables the pieces to be easily interchanged or recycled.

## ■ Hazardous Waste Reduction

### Concepts

- Source reduction is any action that causes net reduction of waste and is taken before it is generated.

- The EPA has established a hazardous waste management hierarchy with four levels.

- Source reduction is pollution prevention.

We now know that the management of hazardous wastes and other contaminants exclusively through "end-of-pipe" pollution control approaches has proved inadequate. Despite strict laws and regulations and extensive spending on treatment and disposal of pollutants, dangerous and ecologically damaging chemicals have been and continue to be discharged into the environment.

In 1976 the EPA established, through a policy statement, a hazardous waste management hierarchy. This hierarchy is very similar to the hierarchy discussed for municipal solid waste. It is important to note that the hierarchy represents an attempt to build proactive environmental protection into the industrial waste management process. It encourages industries to reduce their hazardous wastes *at the source* and to recycle rather than treat and/or dispose of wastes to land, air and water. This hierarchy of hazardous waste management is presented in Figure 11-2.

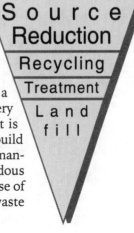

**Figure 11-2:** Hierarchy of hazardous waste management.

Source Reduction

Recycling

Treatment

Land fill

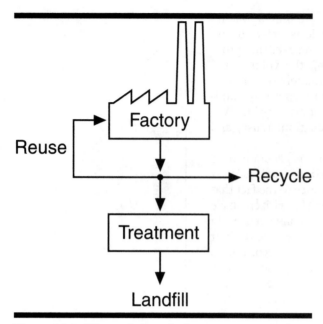

Reuse

Recycle

**Figure 11-3:** Wastes from a single waste stream can be reused in the same process, recycled and used in other products, or treated to reduce volume or toxicity and then disposed of to landfills.

Waste minimization effectively reduces the amount of hazardous material that leaves the production process as waste. It was the intention of the EPA that industry implement source reduction and recycling before resorting to treatment and/or disposal. Until recently, however, this intent has not been widely implemented.

Source reduction is pollution prevention, and for hazardous waste is generally considered any action:

– that causes a net reduction in the generation of hazardous waste.

– taken before the hazardous waste is generated that results in a lessening of the properties that cause it to be classified as a hazardous waste.

Waste minimization and source reduction have been identified in various, often contradictory ways, in government and private publications. Recently, it has been generally agreed that certain actions *do not* represent source reduction:

■ Actions taken *after* hazardous waste is generated.

This clearly excludes any form of treatment such as detoxification, incineration, thermal, chemical, or biological decomposition, stabilization through solidification, or embedding/encapsulation.

■ Actions that merely *concentrate* the constituents of a hazardous waste to reduce its volume.

Volume reduction operations that simply reduce the total volume without reducing the toxicity of a waste do not qualify as source reduction. An example would be dewatering a heavy metal sludge by pressure filtration and drying. The same number of atoms of toxic metal remain in the sludge. In fact, the concentration of the metals has been increased. Although dewatering is an important treatment that reduces both handling costs and the waste's ability to migrate, it does not prevent waste generation or reduce the amount or toxicity of waste being generated.

■ Actions that *dilute* hazardous waste after generation to reduce its hazardous characteristics.

Dilution is the opposite of volume reduction. As with volume reduction, waste dilution is treatment applied to the waste stream after generation. It does nothing to reduce the amount or hazard of the waste at its source of production.

■ Actions that merely *shift* hazardous waste from one environmental compartment to another.

Many waste management, treatment and control practices have simply collected pollutants and moved them from one environmental medium to another, often to one that is less regulated. Two examples are:

1. Collection of pollutants from air and water using pollution control devices and then legally disposing of them in land disposal sites.

2. Transfer of volatile pollutants from surface impoundments, landfills, water treatment units, groundwater air stripping operations, etc., to the air through evaporation.

## ■ Incentives and Impediments

### *Concepts*

- There are economic incentives to reduce waste.

- There are important legal reasons to reduce waste.

- Waste reduction provides a positive image for companies as well as benefiting the environment.

Economic incentives may reward those who reduce their wastes. These incentives can be current real savings, which can be clearly measured, or savings estimates based on avoiding future costs. In either case, the economic incentive is the potential for increased competitive advantage through lower production costs. Economic incentives can be grouped into four basic categories:

- Reduced waste management costs

  Waste management costs are lower when there is less waste to manage. Depending on the individual site and its operations, these savings might be achieved through one or more of the following:

  1. Lower on-site handling costs.
  2. Less waste storage area, leaving more space available for production.
  3. Lower offsite transportation and disposal costs.
  4. Lower paperwork and record-keeping costs.
  5. Change from treatment, storage, and disposal facility (TSD) status to non-TSD status, with associated lower costs.
  6. Reduced waste end-tax obligations.

Companies have already documented millions of dollars in savings as a result of waste minimization efforts. In New York State, each hazardous waste generator's waste end-tax obligations can be reduced by as much as $40,000 per year, depending on the amount of hazardous waste generated. In essence, generators can then double-up their savings – not only through reduced waste management costs, but through reducing taxes as well.

■ Improved operations

As wastes are reduced, the proportion of raw materials being converted to desired end-products increases. In this way, waste minimization leads directly to increased production. Many simple investments to reduce waste can pay for themselves in only a few months. These savings can be even more impressive when calculated on a long-term basis, which includes minimizing future liability costs. According to the EPA, annual costs for waste treatment and disposal can be reduced by 30 to 50 percent.

■ Reduced liability

Reducing the amount and toxicity of wastes being handled on-site also reduces the likelihood that violations of hazardous waste laws will occur. Civil and criminal penalties are often vigorously pursued. Criminal penalties are more severe for violations of hazardous waste regulations than other regulatory crimes. Some convictions relating to the mismanagement of hazardous wastes have occurred without a showing of criminal intent leading to the violation.

Another source of liability is worker exposure to hazardous materials. Reducing the amount of hazardous materials handled and their level of toxicity can have the added benefit of improving worker safety and morale and reducing the risk of lawsuits that may come with injury.

The liability risk is greatest with treatment and disposal activities; the second greatest is with recycling and resource recovery; and the least is with source reduction.

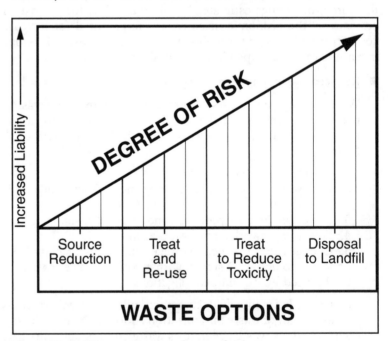

**Figure 11-4:** Waste options vs. degree of risk.

■ Increased competitive advantage

All these economic strategies result in increased competitive advantage leading to greater profits, which is what business is all about. By reducing costs, increasing productivity, and reducing long-term liabilities, organizations have the ability to increase revenues and profitability.

There are also very important legal reasons to minimize waste. These include compliance with federal and state regulations. The Land Disposal Ban, discussed in the previous chapter, is one example of a federal regulation with a major impact. The prohibitions on land disposal of untreated hazardous waste are a powerful incentive to reduce the generation of hazardous waste. Another example is Section 224(s) of HSWA, which has the following specific requirements for generators related to waste reduction:

■ Manifest certifications

Each Uniform Hazardous Waste Manifest requires generators to certify that they have ". . . a program in place to reduce the volume or quantity and toxicity of waste to the degree determined by the generator to be economically practicable. . . ." A false certification is a direct violation of RCRA, possibly leading to civil and criminal proceedings against the generator.

■ Permit certifications

Each generator obtaining a RCRA permit for the treatment, storage, or disposal of hazardous waste must certify, as a permit requirement, that he or she has a waste minimization program in place. A false certification carries the same potential liabilities as mentioned above.

■ Reporting of waste minimization activity

All hazardous waste generators are required, as part of their biennial EPA report or applicable state report, to describe their waste minimization activities during the report years. This would include any efforts to reduce the volume and toxicity of the waste generated and comparison of the actual waste generated over the years covered by the report. Reporting waste minimization activity is also required for various external federal and state surveys and questionnaires. In addition, mandatory federal and state survey questionnaires often inquire about waste minimization activities.

While economic and legal incentives should be sufficiently compelling reasons for manufacturers to reduce their wastes, there are additional incentives:

■ Positive public image

Waste minimization is very attractive to the public. It is strongly supported and any effort involving it projects a positive image for an organization; the organization is viewed as having concern for the community and the environment. Waste minimization provides firms with unique opportunities to: 1) meet regulatory requirements, 2) realize economic benefits, and 3) obtain a positive public image – all at the same time, and for the same effort.

**Figure 11-5:** Some companies tout their environmental friendliness in order to gain a competitive edge in the marketplace.

Notes on waste minimization in an organization's annual report or quotes in newspapers, magazines, and journals all enhance the public's perception of the company as a good corporate citizen.

As with other programs, increased personnel interaction and involvement are important incentives for companies to participate in waste minimization activities. Waste minimization projects can give employees a sense of individual contribution not only to the company but to society and their community. This, in turn, can improve employee morale.

■ Healthy environment

The quality of life depends on a healthy and sustainable environment. Preventing the deterioration of the environment is a responsibility shared by both producers and consumers.

■ Product quality improvement

Improving product quality can be a natural outgrowth of waste minimization evaluation. Any time a process or unit is investigated closely in a new manner, by a variety of people, opportunities emerge for improvements that might not otherwise have been discovered. In one case, waste minimization efforts created a new process technology that brought new life to a business that was near closure.

■ Government assistance programs

Several forms of assistance exist, and more are under development to help industries adopt waste minimization practices. The economic incentive programs include grants, loans, and tax and fee incentives.

Although these incentives are more difficult to measure in value than either regulatory or economic incentives, their value should not be underestimated.

Although a rosy picture has been painted with all these incentives, the move toward waste minimization is not without its obstacles. The recognition of the hazards, liabilities, and costs associated with hazardous waste generation is relatively new to the industrial community. The barriers to waste minimization are similar to those that delay other major changes in the way businesses operate. They include limitations on available financial resources, information, and technical capabilities.

Waste minimization can be further limited by the "end-of-pipe" outlook, which persists in many industries and regulatory agencies. It is sometimes difficult, particularly for producers and regulators that have invested years of attention, time and financial resources in strategies designed to cope with waste after it is generated, to find any value in seeking ways of decreasing the amount of waste generated.

Often the success story of a competitor or a larger company as "role model" is the most effective means of changing the way other companies look at waste minimization. The fact that many industries have not only discovered but publicized that pollution prevention can save money, improve worker safety, reduce environmental pollution and enhance public image has inspired others to follow suit. For instance, in summarizing its waste reduction efforts since 1975, the 3M Company, an industry leader in waste minimization, reported the following:

The combined total of almost 1,900 projects has resulted in eliminating annually the discharge of almost 110,000 tons of air pollutants, over 13,000 tons of water pollutants, and over 260,000 tons of sludge, over 18,000 tons of which are hazardous – along with the prevention of approximately 1.6 billion gallons of wastewater. Cost savings to 3M total more than $292 million.

3M reported the following sources of their savings:

1. Pollution control facilities that did not have to be built.
2. Reduced pollution control operating costs.
3. Reduced manufacturing costs.
4. Retained sales of products that might have been taken off the market as environmentally unacceptable.

The experiences of 3M and other businesses that have implemented waste reduction programs have shown that waste minimization succeeds when it is incorporated into the everyday awareness of all workers and managers involved with production who are aware of opportunities for waste reduction. This often requires a major change in outlook in companies where waste management is managed by those responsible for end-of-pipe pollution control regulations.

Source reduction requires a fundamentally different approach from practices that manage hazardous waste after it has been generated. Successful source reduction depends partly on technological changes, but also requires changes in attitude. For this reason greater involvement of employees is essential.

Source reduction techniques vary greatly from one industry to another, many are very process-specific. There are, nevertheless, some elements common to all types of businesses. Four general source reduction categories can be identified:

– Good Operating Practices
– Changes in Technology
– Changes in Input or Feedstock Materials
– Changes in Product

The next section covers each of these categories in detail.

## ■ Checking Your Understanding (11-1)

1. Contrast the terms "pollution prevention," "source reduction," and "waste minimization."

2. Explain how you would use the information in each of the six bullets from "Ways Companies Can Promote Waste Reduction" to promote waste reduction in your home.

3. List three incentives for reducing waste in your home and three incentives for businesses to reduce waste.

# 11-2 Approaches to Waste Reduction

## ■ Good Operating Practices

### Concepts

■ Good operating practices are procedural changes that minimize waste.

■ Cost-accounting, inventory control, and material handling improvements can result in immediate and direct reductions in waste and costs.

■ Waste stream segregation is the cheapest and easiest method of reducing the volume of hazardous waste.

■ Educating employees on waste minimization is critical to success.

Good operating practices include many procedural changes that can be implemented in many areas of plant operation. Since they are procedural, many effective operating practices can be implemented at low cost and give a good return on the investment in a relatively short time. Specific operating practices that will be discussed include: waste management cost-accounting; inventory management; procedural scheduling; material handling improvements and loss prevention; waste stream segregation; and personnel education, communication and involvement.

Waste management cost-accounting takes into consideration the actual cost of managing waste. This is an especially important practice for dealing with major hazardous waste streams and is one of the earliest assessments that should be conducted. It will be impossible to perform meaningful process review and evaluation required for other approaches to waste minimization without this information. Often the costs of generating hazardous wastes are unknown to facility managers. Complete and accurate record keeping is needed of all waste management costs including fees, storage, treatment, transportation, disposal, tracking, management overhead, insurance, energy, and raw material expenses. The rates of generation of existing hazardous wastes should also be documented.

A business may benefit from inventory management, which involves developing a review procedure for identifying incoming hazardous materials. The availability of non-hazardous alternative materials should be explored with suppliers.

Inventory management will result in reduction of waste when purchases are made of only the required amount of hazardous material needed for specific jobs at specific times. Buying hazardous materials in bulk may appear to be a cost savings, but may be deceiving if handling and disposal costs are not considered. Unused hazardous materials may represent a significant liability. The materials become hazardous waste when they no longer meet specifications because of prolonged storage or due to contamination, and their handling and disposal costs must be added to their original purchase price. Potential new product lines should also be reviewed with respect to their potential for increasing the use of hazardous materials.

Procedural scheduling is sometimes called "just-in-time" manufacturing because raw materials are delivered to the plant a short time before they are to be used. This approach is also called production scheduling and has been extensively field-tested by large companies such as 3M. This approach minimizes many problems associated with storage and product line changes. As an example, parts sitting exposed to the elements for long periods of time require cleaning before they can be used. The cleaning process often involves hazardous solvents.

In batch production operations, the amount of cleaning waste generated is directly related to the frequency of cleaning. Coordinating scheduled equipment usage to reduce the need for clean-out or wash-down between batches can reduce the amount of hazardous waste generated.

Material handling improvements and loss prevention are procedural changes that can result in immediate and direct reductions in waste and costs. This approach can involve taking steps as basic as compliance with existing hazardous material and hazardous waste management regulations. Often, with a little review, methods of handling hazardous materials and waste can be tremendously improved. Preventive measures can include:

1. Training employees in the operation, capacity, and capabilities of each type of equipment used to transfer loads (forklifts, dollies, conveyers, etc.) to prevent spillage.

2. Providing for regular monitoring of storage areas.

3. Allowing adequate, clear space between rows of drums so that each container can be visually inspected for corrosion and leaks.

4. Stacking containers in a way that minimizes the chance of tipping, tearing, puncturing, or breaking.

5. Raising drums off the floor to prevent corrosion from spills or "sweating" concrete.

6. Maintaining adequate distance between different chemicals to prevent cross-contamination or chemical reactions in case of spills.

7. Keeping containers closed except when material is being removed and providing funnels and other appropriate transfer equipment to reduce losses of material during transfer.

8. Providing adequate lighting in storage areas.

9. Maintaining a clean, even surface in areas traveled by personnel and equipment.

10. Curbing and diking storage and process areas where hazardous substances and/or wastes are present to contain any spillage.

11. Using reusable containers that are appropriate for the job. Many are equipped for top and/or bottom discharge, cleaning access, and easy transporting.

12. Implementing a strict and thorough maintenance program that stresses corrective and preventive maintenance. This will prevent loss of hazardous material due to equipment failure.

*Waste stream segregation*, which is also called source separation (or segregation) keeps hazardous waste from contaminating nonhazardous waste. This is accomplished through management practices designed to prevent the wastes from coming into contact with one another. This is the cheapest and easiest method of reducing the volume of hazardous waste to be disposed of, and is widely used by industry.

In addition to reducing disposal costs, source separation reduces handling and transportation costs. Keeping nonhazardous and hazardous waste separate lowers costs for disposal since a mixed stream must be managed as hazardous. For example, hosing down or sweeping ordinary "household" dust and debris into a **sump** designed to collect hazardous waste wash-down might be avoided by vacuuming the nonhazardous dust. This decreases the total volume of hazardous waste generated.

Separation of waste streams also increases opportunities for recycling and reuse. With solvents, segregation is essential in order for recycling to occur. Some recyclers cannot accept organic solvents that are contaminated with even minute amounts of chlorinated solvents. In addition, the quality of chlorinated solvents will deteriorate rapidly if they have been contaminated with water or other solvents.

Lastly, *personnel education, communication & involvement* in waste minimization programs is necessary if the people who can do the most good in this endeavor are to play the vital role required of them. These activities can vary from simple pollution awareness programs, where managers and employees are asked to identify ways of reducing the generation of waste, to complex programs that extend to worldwide operations. In either case, employee involvement is a key to success.

An effective campaign to reduce the amount of waste generated must incorporate an effective employee training program that teaches employees how to detect spills, leaks, and releases of material. Process operators and maintenance personnel should be given training that stresses waste minimization methods. Incentive programs that reward employees for waste reduction ideas with incentives and gifts ranging from T-shirts and jackets to weekend trips for two have been particularly popular.

Many source reduction options require coordination and cooperation of employees at various points in the overall operation of the plant. This means that information on the benefits of reducing hazardous material use and hazardous waste being generated needs to be communicated to *all* employees. As many employees as possible should also be involved in the search for waste minimization opportunities.

# ■ Changes in Technology

## Concepts

- Process changes alter a process to generate less waste.

- Equipment, piping, or layout changes replace or modify equipment.

- Additional automation is a costly but effective means to reduce waste.

Manufacturing process changes consist of either eliminating a process that produces a hazardous waste or altering the process so that it no longer produces the waste. Source reduction strategies involving changes in technology tend to be industry and process specific and require a good understanding of the details of a facility's operation. A few examples from specific industrial processes are provided below. Some categories of technological changes to achieve source reduction are: process changes; equipment, piping, or layout changes; and additional automation.

*Process changes* involve developing an alternate process that generates less waste. The challenge is achieving this change without changing product specifications. The way to start is by a careful review of the production process for ways to improve efficiency. Following this review, an examination of specific steps within the process may be conducted to identify ways to reduce either the quantity or toxicity, or both, of the wastes being generated.

A parts cleaning operation might, for example, determine just how much cleaning is needed. It may be decided that the amount of required cleaning can be lessened by reducing the time the parts are stored before cleaning. Another option may be to replace protective oil coatings with protective peel coatings. The question to be answered is whether there are alternative methods or materials available that result in less waste. In the case of a parts cleaning operation, some alternative methods might include mechanical brushing, sandblasting, plastic bead blasting, agitation, sonication (using sound waves), or nonhalogenated or aqueous solvents.

*Equipment, piping, or layout changes* replace or modify equipment that is operating inefficiently and/or generating excess amounts of waste. An example would be replacing a vapor degreaser designed to use a chlorinated solvent with one designed to used alkaline materials.

Replacement of equipment does not always need to occur; modifications are also effective. Attaching covers or spill guards to vats containing volatile liquids reduces losses and worker exposure. Mechanical wipers can be used to wipe out mixing vats or pipes between operations. Pipes and valves can be made more leak-proof by using double mechanical seals. Heat exchangers, which are used in a great majority of industrial operations, can be modified and maintained to reduce plugging and fouling. These generate less waste when clean. In general, more efficient equipment may reduce the number of rejected products and significantly reduce hazardous waste generation.

*Additional automation*, although very costly initially, can reduce waste. This approach replaces workers with automated systems in areas where it is feasible. Although there are many operations machines can't do, there are tasks they can perform more efficiently than humans. In operations requiring precise timing and close monitoring to minimize waste, automatic systems are usually more effective. For example, machines continually monitoring a plating bath can ensure that maximum use is made of solutions and can trigger quick responses when changes occur.

In many cases, automated systems can replace manual handling of substances. A closed, automated transfer system in bulk operations can reduce error and the likelihood of spills and emissions. Painting operations often use robotics, which have resulted in reducing paint waste as well as enhancing the uniformity of the paint thickness on the products. Due to the high costs of installing automated systems, this alternative is usually more feasible for larger companies.

## ■ Substitution of Raw Materials

### Concepts

■ Substitution of input materials can substantially reduce waste volume.

■ Material purification reduces certain hazardous waste byproducts.

Substitution of raw, or input, materials may offer the greatest opportunity for waste reduction. Manufacturers can substantially reduce waste volume by replacing a raw material that generates a large amount of hazardous waste with one that generates little or no hazardous waste. There are two general categories of changes in input material: material purification and material substitution.

*Material purification* removes contaminants or impurities resulting in reduction of certain hazardous waste by-products. Hazardous material may enter the process as a contaminant in raw material or feedstock. If raw material with a lower concentration of hazardous contaminants is available, it may be possible to reduce these by-products. One example of material purification occurs in industries that use acids and organic liquids in their processes. Using these materials with lower levels of metal impurities reduces the metal waste generated through reaction processes.

*Material substitution* has had great success in reducing hazardous waste generation. One example is the printing industry's change from solvent-based inks to water-based inks. Yet another example is the replacement of hexavalent chrome-plating baths with less toxic, lower concentrations of trivalent chrome in the production of plumbing fixtures.

## ■ Changes in Product

### Concepts

- Product substitutions can eliminate the use of a hazardous material.

- Product conservation extends the life of a product.

- Changing a product's composition reduces its hazardous components.

We don't often think about reducing hazardous waste by making changes in a product or in how it is used and maintained by the consumer. There are three categories of changes that can result in waste reduction: one product can be substituted for another, the product can be conserved, or the composition of the product can be altered.

*Substitution of products* may eliminate use of a hazardous material. For example, by substituting concrete posts for creosote-preserved wood posts in construction operations, builders can remove any possibility that the hazardous creosote will leach from the posts and contaminate underlying groundwater or surrounding soil.

Another excellent example of product substitution is the replacement of chrome bumpers on automobiles with plastic bumpers. As we learned earlier in the text, CFCs have been banned as aerosol propellants in the U.S.. Since the ban, a number of substitutes have emerged.

*Product conservation* refers to using the end product in a manner that extends its life. For example, improving maintenance of equipment can lengthen its useful life thereby reducing the need to replace it. This reduces the waste generated in producing the new component or equipment item. Consumers can also maintain products and derive the same results.

*Changes in product composition* involve manufacturing a product with reduced or no hazardous components. An example of this is using a nonhalogenated solvent in place of a halogenated solvent in a chemical formulation. Organic pigments have been used in place of heavy metal pigments in the manufacture of paint.

Since waste reduction cannot eliminate all waste, the small amount that is generated needs to be rendered less hazardous for proper management. Several processes exist for accomplishing this. The next section will describe several methods.

## ■ Checking Your Understanding (1-2)

1. Describe how you would use at least two of the "good operating practices" to minimize waste at home.

2. Provide one example, not described in this chapter, of a change in technology and of a substitution of raw materials that would result in generating less waste.

# 11-3 Hazardous Waste Recycling and Treatment Technologies

## ■ Introduction to Recycling and Treatment Technologies

### Concepts

- Hazardous waste treatment can be accomplished through physical, chemical, biological, or thermal means.

- Reclamation and reuse are subsets of recycling.

- Recycling is waste minimization and can be done on-site or off-site.

Chapter 7 covered the basic classification of treatments for wastewater. This section will build upon those principles to explain how similar technologies are used for hazardous waste treatment. Treatment technologies are commonly divided into four categories: physical, chemical, thermal, and biological.

Physical treatments for hazardous waste generally separate or concentrate components of the waste. Many of these methods separate solids from liquids. Thermal technologies rely on heat to destroy certain constituents of hazardous waste. Chemical treatments are chemical reactions that take advantage of different chemical properties. Biological treatment uses living organisms to consume or concentrate the waste constituents.

Before we explore each of the treatment categories in more detail, a discussion of recycling is in order. Recycling, which is also referred to as recovery and reuse, is a widely used approach to managing waste. Recycling removes a substance from a waste and returns it to productive use. Generators of wastes commonly recycle solvents, acids, and metals. Besides source reduction, it is the preferred waste minimization method. Reuse and reclamation are subsets of recycling and will be discussed below.

In general, recycling is not considered waste reduction because it involves the use, reuse, or reclamation of a waste *after* it has been generated. Through recycling and reuse, hazardous "waste" is returned to a production process rather than being released to the environment or disposed of as a waste. There are some situations, however, where recycling/reuse may be considered source reduction. For example, if the return of materials is within the existing process before they have actually become wastes, it is considered source reduction. This is often called a **"closed loop" system.**

**Reclamation** is usually considered to be a part of recycling because it recovers valuable *raw material* for reuse. Reclamation differs from use and reuse in that the recovered material is sold to another entity rather than being used in the facility in which it was generated. The processes used to reclaim useful materials from waste often generate hazardous wastes of their own. In addition, the transportation of waste to off-site facilities can also increase a generator's potential liability for spills and accidents. This is an example of the trade-offs that must be examined when assessing waste minimization options.

The most effective recycling requires minimal management and provides a high percentage of the material for reuse. Recycling is advantageous in two ways; it reduces costs because the need for raw materials is reduced, and it helps the environment because it conserves natural resources and avoids treatment and disposal risks.

Recycling can occur on-site or off-site. On-site recycling uses the waste materials at the site where they are generated. They can either be used in the same or different processes. The reuse of cleaning wastewaters is one example of a waste stream commonly recycled on-site. Off-site recycling involves transporting the waste to a commercial recycler who is permitted for this activity. The recycler processes the material and either returns it to the generator or sells it. Products commonly recycled off-site include batteries, antifreeze, freon, and used oil.

Recycling of specific wastes depends on the purity of the waste, its concentration, and sometimes its chemical form. For this reason some wastes must be treated before being recycled. This additional step is less desirable than direct in-process recycling because treatment involves additional handling that can lead to added environmental and safety risks, and the treatment process itself can generate waste. Always keep in mind that the methods that generate the lowest quantity and toxicity of hazardous waste should be the methods that are selected.

In-process and other on-site recycling options can be improved by using source reduction strategies up front. For instance, segregating wastes to cut down on contamination of materials such as solvents, baghouse dusts, lubricating oils, or rinse waters can make later recycling much easier.

If a waste cannot be used in the original process in which it was generated, it is sometimes possible to find an alternative process within the plant or with a recycler. Certain processes have less rigid raw material specifications than others. For instance, the electronics industry requires very pure raw materials. Solvents that have been used to clean circuit boards may have very low concentrations of contaminants and could easily be reused as cleaning agents in degreasing operations or as a thinner in paints.

An increasingly popular approach to recycling is the use of a "waste exchange." These exchanges, which go by various names, basically act as clearinghouses for wastes that may have value to industrial users. If a waste producer can find another industry that needs the waste material, disposal costs are eliminated, and in fact the waste may generate income. (See Figure 11-6 for a sample page from the *California Waste Exchange*.)

**Reuse** implies using the waste as a valuable raw material. The most common form of reuse is using the waste as a fuel for energy production. Several processes such as cement kilns and asphalt plants can use waste materials as fuels. Some incinerators can be

# CATALOG

## ITEMS AVAILABLE

### ACIDS

**Sulfuric Acid**                CA:A01/0001                                CA/408

Dilute sulfuric acid containing 72% water, 23% sulfuric acid, and about 1% of 4-methyl-3-nitroacetophenone, 2-amino-5-nitrobenzoic acid, and oxamide; generated from production vessel cleanup; 6,450 gallons, one time only; sample and analysis available.

San Benito County

**Sulfuric Acid**                CA:A01/2002                                CA/415

Spent sulfuric acid; contaminants: zinc and heavy metals; 4,800 gallons/3 months, continuously in tank trucks; sample available.

San Francisco County

**Acid Waste Streams**                CA:A01/2003                                CA/213

- 28-30% hydrochloric acid
- 8% nitric acid with 2% Amchem 414
- 25% sulfuric acid with 2.7% sodium dichromate
- 15% sulfuric acid with 2.2 to 3.3% sulfuric acid
- 1-1.4% chromic acid, 8-12 oz./gallon sulfuric acid
- 2-4 oz./gallon hydrofluoric acid, 38-52 oz./ gallon nitric acid

1-100 tons/month in tank truck; sample and some analyses available.

Los Angeles County

**Nitric Acid**                CA:A01/5001                                CA/916

Waste nitric acid; 10-20% nitric acid; 3,000 gallons/month, continuously, in tank cars; sample and analysis available.

Sacramento County

**Figure 11-6:** A page from the *California Waste Exchange*.

used as energy recovery systems. The characteristics of the waste, such as its chlorine content or ash composition, may restrict use of the waste as a fuel.

# ■ Some Physical and Chemical Hazardous Waste Treatment Technologies

## Concepts

■ Flotation separates particles by density.

■ Membrane technologies remove solvents from waste solutions.

■ Fractional distillation separates liquids with differing volatilities.

Table 11-7 lists some of the common physical treatments for hazardous waste that have been identified by the EPA. This list is rather extensive so only a handful of the treatment methods will be chosen for further discussion. Many of these methods have been covered in Chapter 7 since they are commonly used in the wastewater treatment field as well.

A number of **flotation** methods are commonly used in industry. Air flotation separates droplets of solids or liquids from a liquid waste according to the differences in their densities. Bubbles attach to the particles or droplets and buoy them to the surface. Electroflotation uses hydrogen and oxygen gas generated at the surface of electrodes to buoy the waste particles. Both of these flotation methods are commonly used to separate oil and latex emulsions.

Sedimentation also separates specific components of wastestreams based on density differences. Given sufficient time, dense solids will settle out of a liquid. The clarified liquid can then be poured off, leaving the settled solids for further treatment or disposal. One of the Superfund cleanups in California, the Stringfellow Acid Pits, operates a treatment plant in which sedimentation is used in conjunction with flocculation and precipitation to remove heavy metals from contaminated groundwater.

Centrifugation will also separate substances of differing densities. In this method, the waste is whirled in a device that causes the heavier particles to be pushed against the sides first, separating them from the less dense portion. Centrifuges can also include filters that further separate the finer particles from the liquids. Since centrifugation requires specialized equipment, it is more expensive than other methods of separation. However, it has the advantage of speed.

Filtration is another common physical treatment method. Filters of different pore size will capture different sized particles. The most nagging problem associated with this method is the clogging of the filter.

Ultrafiltration and reverse osmosis use membranes to remove solvent from a waste solution. The solvent passes through the membrane and the solute is retained in a more concentrated form. A **solvent** is a substance in which another substance is dissolved. A

## Physical Treatments

(the codes next to each treatment are EPA codes that are required on reports)

(1) Separation of Components

| | |
|---|---|
| T35 | Centrifugation |
| T36 | Clarification |
| T37 | Coagulation |
| T38 | Decanting |
| T39 | Encapsulation |
| T40 | Filtration |
| T41 | Flocculation |
| T42 | Flotation |
| T43 | Foaming |
| T44 | Sedimentation |
| T45 | Thickening |
| T46 | Ultrafiltration |

(2) Removal of Specific Components

| | |
|---|---|
| T48 | Adsorption – molecular sieve |
| T49 | Activated carbon |
| T50 | Blending |
| T51 | Catalysis |
| T52 | Crystallization |
| T53 | Dialysis |
| T54 | Distillation |
| T55 | Electrodialysis |
| T56 | Electrolysis |
| T57 | Evaporation |
| T58 | High gradient magnetic separation |
| T59 | Leaching |
| T60 | Liquid ion exchange |
| T61 | Liquid-liquid extraction |
| T62 | Reverse osmosis |
| T63 | Solvent recovery |
| T64 | Stripping |
| T65 | Sand filter |

**Figure 11-7:** Physical Treatments.

solute is a substance dissolved in another substance, and is usually the constituent in the smaller concentration. The processes of ultrafiltration and reverse osmosis differ only in the size of the particles each can separate. Ultrafiltration can separate only relatively large solute particles whereas reverse osmosis is applicable to atomic and molecular sized solutes. A series of membranes of differing pore sizes could even be used to separate a complex mixture of chemicals.

**Distillation** uses the difference in boiling points as a means of separation. When a mixture of two or more liquids with differing boiling points is separated, it is called "fractional distillation." The liquid with the lowest boiling point will volatize first, and the liquid with the highest boiling point will volatize last.

**Carbon adsorption** is a process in which substances adhere to the surface of specially treated (activated) carbon. This method is particularly effective in removing organic compounds from liquid or gaseous waste streams.

**Neutralization** is the process of combining acidic and basic substances to produce a nonhazardous (or reduced hazard) substance. This is accomplished by adding an alkaline substance to an acid or an acidic substance to a basic material.

**Dechlorination** removes chlorine from a substance by chemically replacing it with hydrogen or hydroxide ions. This process is used to detoxify chlorinated substances. Dechlorination will be discussed later in the chapter.

**Oxidation** occurs when a substance gains oxygen, loses hydrogen or electrons in a chemical reaction. When a substance loses oxygen or gains hydrogen or electrons it is reduced. Oxidation and reduction always happen at the same time because when one substance loses something the other must gain it. This process is used to treat wastes such as cyanides, phenols, and organic sulfur compounds.

**Solidification** and **stabilization** remove the aqueous portion of a waste or change it chemically, thereby making it less soluble and less susceptible to transport by water. Since liquid wastes are no longer allowed to be disposed of on land, solidification and stabilization are required prior to land disposal.

**Biological treatment** uses microorganisms to degrade waste organic compounds. Since it has had broad application in hazardous waste site cleanup, it will be discussed in detail in the next section.

# Box 11.2

## A Common Industrial Process Generating Hazardous Wastes: Electroplating

Electroplating is a finishing process that applies a metal coating to a surface. It is used to put chrome on bumpers of classic cars, tin on steel cans, and to put the copper tracks on printed circuit boards used in manufacturing electronics. Very basically, it is accomplished by suspending the object that is to be coated in a plating solution of the desired metal and giving the object a negative electrical charge. The positively charged metal ions in the solution move to the object's surface, where they form the metal coating.

Electroplating typically generates a variety of hazardous wastes. The plating solution can contain large concentrations of hazardous metals. It can also contain cyanides, strong acids, and other hazardous additives. As the solution is used, the concentrations of contaminants increase and reduce its effectiveness for use in plating. When this occurs, the material becomes a waste.

There are processes that remove contaminants from plating solutions that greatly extend the life of the solutions. Recovering the metals and other valuable constituents from the used plating solution and adding them to the fresh solution provides ongoing waste reduction.

Another waste that presents a problem in plating operations is the rinse solution. Typically, several rinses are needed to remove all the plating solution from the object being electroplated. As the object is moved from one bath to another, it carries liquid from the previous bath; this liquid is commonly referred to as *dragout*. Over time, the rinse baths become so contaminated that they lose effectiveness. The contaminants from these baths can also be recovered and used in new plating solutions or to replenish existing solutions.

A commonly used waste reduction measure for rinse baths is known as counter-current rinsing. In this approach, fresh water is only added to the final rinse. This keeps the final bath relatively clean. Figure 11-8 shows how counter-current rinsing keeps the baths replenished by using the following bath. This method results in only the first bath being a waste.

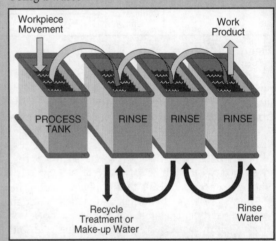

Figure 11-8: A counter-current rinse.

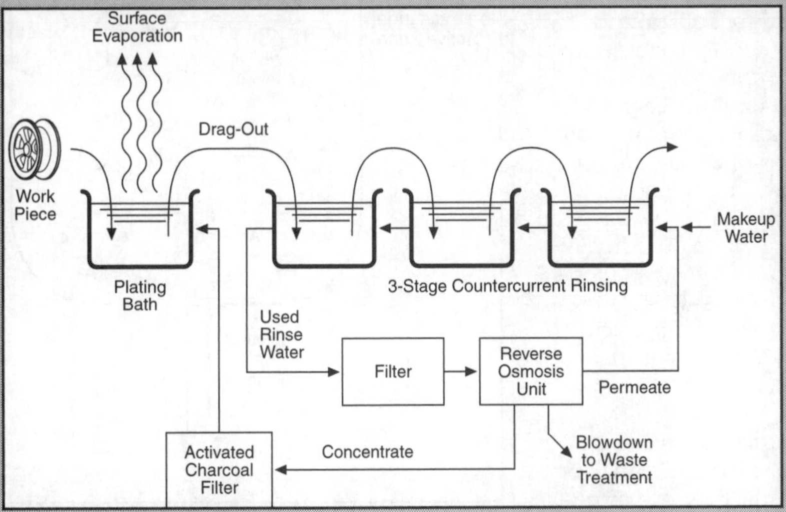

Figure 11-9 shows how a recovery process can result in a system that generates little, if any, waste rinse water. The EPA estimates that "under very favorable conditions, a recovery system can achieve almost zero effluent discharge to sewers." Some of the recovery technologies commonly used in the metal finishing industry follow.

Figure 11-9: The reverse osmosis process generates very little waste rinse water.

## Box 11.2

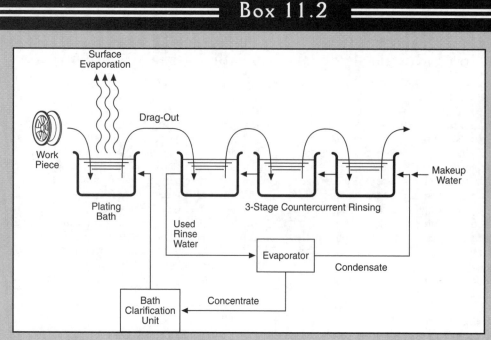

**Figure 11-10:** An evaporation system.

Reverse osmosis is a technology used successfully for recovery of electroplating wastes. The waste solution is pumped under pressure into a compartment with a semi-permeable membrane. Only the water passes through the membrane, leaving behind the salts and dissolved metals. When applied to wastewater from the rinse baths, there are two resultant waste streams – one with very concentrated metals that can be reused in the plating solution and another that is clean enough to be reused or discharged. The process does not work well on wastewater from chromic acid and high pH cyanide baths. The major disadvantage to these units is cost, which can easily run in the range of hundreds of thousands of dollars.

Evaporation is a simple way to concentrate components of a solution. In this case, water is evaporated from the solution until the chemicals remaining in the wastewater are at a concentration high enough to reuse in the plating solution. Figure 11-10 also shows the use of evaporation to recover metals from electroplating rinse water.

Ion exchange units remove cations (positively charged ions) or anions (negatively charged ions) from a waste stream. In these units, wastes flow over a resin bed that exchanges ions. The hazardous ions from the waste bind to the bed material, displacing less hazardous ions. Figure 11-11 shows an ion exchange column.

The disadvantages of ion exchange include the initial high costs for the system and the high cost of its maintenance. The resin beds must be routinely chemically treated to remove the absorbed

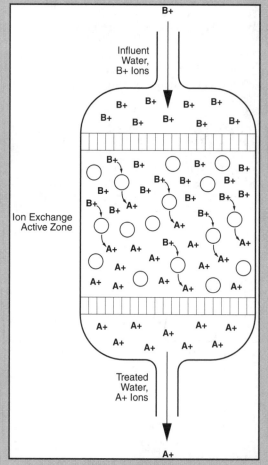

**Figure 11-11:** An ion exchange column.

## Box 11.2

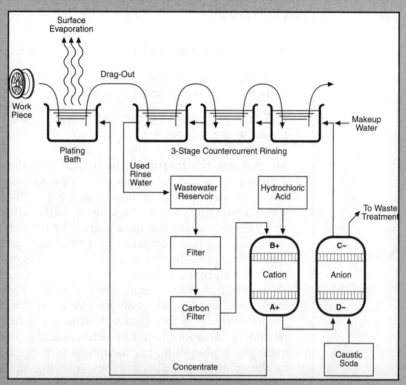

**Figure 11-12:** An ion exchange unit that recovers and reuses absorbed

waste. This involves labor and chemical costs. Fortunately, the recovered material can often be reused in the plating solutions. Ion exchange has been extensively used for the recovery of aluminum, arsenic, cadmium, chromium (hexavalent and trivalent), copper, cyanide, gold, iron, lead, manganese, nickel, selenium, silver, tin, and zinc. Figure 11-12 shows an ion exchange unit used in a treatment system.

Electrodialysis is also used in a number of waste recovery processes, including electroplating. This technology uses a series of alternating cation and anion permeable membranes that are located between two electrodes. Figure 11-13 shows an electrodialysis apparatus. Waste water passes between the membranes, and the ions migrate towards the positively charged electrodes. The concentrated streams are sent back to the process tank and the dilute stream goes to the rinse tank. Electrodialysis is effective for a number of wastestreams including those containing nickel, copper, chromic acid, iron, and zinc. It is beneficial for producing a highly concentrated stream but requires careful operation and maintenance. A difficulty with this process is that the membranes are prone to clogging and leakage.

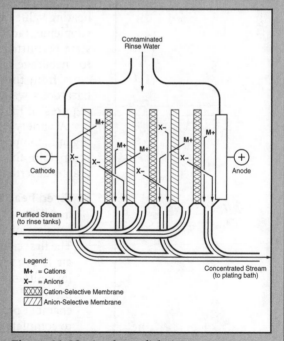

**Figure 11-13:** An electrodialysis apparatus.

## ■ Thermal Treatment

### Concepts

- Incineration destroys hazardous wastes by burning.

- Temperature, turbulence and residence time are the three important factors in efficient burning.

- There are a number of incinerator types that are used for hazardous waste treatment.

All **thermal treatment** methods use a combination of thermal decomposition and air oxidation to destroy waste materials. The most commonly used thermal treatment methods for hazardous waste treatment is some type of incineration, which either destroys or makes the waste less hazardous. The operating variables of an incinerator include temperature, turbulence, and residence time. The correct mix of these variables is required to minimize hazardous emissions and ensure destruction efficiency. In addition, installation of air pollution control equipment is usually required to meet the local, state, and federal regulations for allowable emissions.

Incineration is an effective destruction technology because the remaining hazardous material (ash) is small in volume and is easily managed. It also shifts the hazardous waste generator designation to the treatment facility thereby relieving the hazardous waste generator from many of the liability risks of disposal.

Incineration is frequently used to destroy high Btu content organic wastes. At present, the EPA allows hazardous wastes having heating values greater than 5,000 Btus per pound to be burned as supplementary fuel in on-site boilers. They may also be burned in state-permitted off-site incinerators for resource recovery. In fact, to encourage resource recovery, these incinerators are partially exempt from the EPA's RCRA regulations. Off-site incineration of hazardous wastes having values lower than 5,000 Btu per pound requires an EPA Hazardous Waste Facility Permit (TSD).

A variety of incinerator types are used to destroy hazardous organic compounds. They include: fixed hearth, liquid injection, rotary kiln, fluidized bed, and multiple hearth. The following paragraphs describe these types of incinerators.

- **Fixed hearth incinerators** usually have small capacities and can handle both liquid and solid wastes. The incinerator typically consists of two chambers: a primary and a secondary chamber. The first chamber operates in a starved oxygen mode (not enough air is present for complete combustion). Vortex-type burners inject liquid waste mixed with air into this chamber. Solids can be added through grates. Combustion products from the first chamber proceed to the second chamber where more air is added to complete the combustion. These incinerators can only handle low volumes of waste, and they are limited in their ability to destroy more stable compounds such as chlorinated organic liquid wastes.

■ **Liquid injection incinerators** decompose combustible liquid waste that is injected into the incinerator. The liquid waste is usually blended and filtered before it is injected. Burning takes place in a combustion chamber where the fuel and air are mixed. The injected waste is limited to pumpable liquids and slurries that can be readily atomized. Liquid halogenated hydrocarbons can be successfully destroyed in this type of incinerator at temperatures above 1,800°F.

■ **Rotary kilns** typically have a cylindrical, refractory-lined shell mounted at a slight slant. Rotation of the shell enhances mixing of solid wastes with the combustion air and provides for transport of the waste through the kiln. Rotary kilns usually have a secondary combustion chamber following the kiln.

Most organic wastes, including solids, sludges, and slurries can be burned in rotary kilns. The rotary kiln is one of the most versatile types of incinerators for hazardous wastes because of its capacity to handle large volumes of waste. Cement kilns, which are rotary kilns, have been used to destroy chlorinated organic compounds in Sweden, Canada and the U.S.

■ **Fluidized bed incinerators** are refractory-lined vessels containing a bed of inert granular material. Combustion air is blown through the bed to make the granular material "fluid." Waste is injected into the reactor either above or within the bed. Fluidized bed incinerators are effective on a wide variety of wastes, including solid wastes. Solid combustible materials remain in the bed until they become small and light enough to be carried off with the flue gas. Particulates are recovered by air pollution control equipment. Lime or limestone is added to neutralize acidic gases. The most common applications of fluidized bed incinerators are in the petroleum, paper and sewage disposal industries.

■ **Multiple hearth furnaces** typically include a refractory-lined steel shell, a vertical central rotating shaft, a series of vertically stacked flat hearths, and a series of mechanical arms that plow the waste material successively across the hearths. Waste enters the top of the unit and ash is discharged at the bottom. Additional liquid and gaseous wastes can be injected through the side ports. Multiple hearth furnaces can incinerate gases and liquids as well as sludges and solid wastes. They have been used mainly for sewage sludges and municipal wastes.

## ■ Checking Your Understanding (11-3)

1. Contrast "reclamation" and "reuse."

2. Define two hazardous waste treatment technologies.

3. Explain how counter-current rinsing reduces waste.

# Box 11.3

## Treating PCB-Containing Wastes

Polychlorinated biphenyls, commonly called PCBs, were discussed in Chapter 3. They represent a group of chlorinated organic compounds that have been in wide use as coolants and insulation fluids over the years. PCBs were produced in the United States from 1929 to 1979. Concern over PCBs arose because of their toxicity, their stability in the environment, and their tendency to concentrate in the tissues of animals at the top of the food chain.

Although no longer produced, PCBs remain in active use. PCBs and PCB-containing oils are used in: hydraulic and heat-transfer systems, roofing tar, asphalt, "carbonless" carbon paper, lubricants, paints, plastics, televisions, air conditioners, and fluorescent and mercury vapor lights. Because of their stability under a wide range of temperatures and their low electrical conductivity, PCBs have been widely used in electrical transformers and capacitors.

Laboratory studies have suggested a link between PCBs and reproductive problems, birth defects, gastric disorders, skin lesions, swollen limbs, cancers, tumors, eye problems, liver disorders, and other health problems. In 1968, approximately 1,300 people in Yusho, Japan, used rice oil that had been accidentally contaminated with PCBs. Five people died from this incident and the remaining victims developed a variety of ailments known as "Yusho Disease." They include skin lesions, eye discharges, abdominal pain, menstrual irregularity, fatigue, cough, disorders of the nervous system, and hyperpigmentation of the skin, nails, and mucous membranes.

The health threat from PCBs is compounded by their ability to bioaccumulate. Many microorganisms tolerate PCBs and concentrate them internally. Animals accumulate PCBs in fat cells. In this way, organisms highest in the food chain can consume considerable concentrations of PCBs in their diets.

PCBs are regulated under the Federal Toxic Substances Control Act (TSCA) and not under RCRA. Under the federal standards, liquids containing PCBs are regulated as hazardous at 50 mg/l. Recycling of PCBs is strictly prohibited by TSCA.

Currently, land disposal of liquids containing up to 500 ppm PCBs is allowed by the EPA. There are several methods to treat PCB wastes to reduce the hazardous constituents to below regulatory limits. The most widely used technique is dechlorination.

The original PCB-dechlorination technique was developed by Goodyear Tire and Rubber Company. This process uses a metallic sodium reagent (sodium napthalene tetrahydrofuran) to strip away chlorine atoms from the stable PCB molecule, thus reducing its toxicity.

Goodyear Company made the details of this dechlorination process public. Several companies have developed similar processes that do not employ a napthalene-based compound. This substitution is necessary because the EPA has classified napthalene as a restricted compound. The equipment required for these dechlorination processes is generally mobile, and can be transported on semi-trailers.

The utility company Sunohio has developed what they call the "PCBX" process. The PCBX process replaces the napthalene used in the original Goodyear Company process with a proprietary compound. The Sunohio mobile PCBX unit is a self-contained, continuous-flow unit. It is designed and equipped to treat transformer oil by removing moisture and acids, as well as removing and destroying PCBs.

Another process for PCB destruction is high-temperature incineration. A description of incineration technology was presented earlier. The incineration of PCBs is regulated under TSCA and is subject to very stringent performance standards. In order to receive a permit, the operator of a PCB incinerator must conduct a trial burn with a DRE (Destruction Rate Efficiency) of 99.9999% or greater. This requirement for destruction efficiency is often called "6-Nines." TSCA also mandates minimum temperatures and residence times for PCB incinerators; $1,200^0$F and 2 seconds or $1,600^0$F and 1.5 seconds.

At present, there are only a few incinerators permitted for disposal of PCBs in concentrations greater than 50 ppm. The EPA has issued permits to two off-site rotary kilns, four on-site rotary kilns, and several power generation boilers for PCB incineration. Currently, PCB-contaminated solids and sludges can only be treated at approved incineration facilities meeting the criteria described above.

# 11-4 Hazardous Waste Site Cleanup

## ■ Waste Site Cleanup

### Concepts

■ The site cleanup process is involved and may take many years to complete.

■ The main goals of site cleanup are to: find the responsible parties, address immediate threats, inform and involve the community, and investigate and clean up the site.

In the previous chapter we learned that hazardous waste is generated in many ways. Some manufacturing processes, such as those used in electronics and petroleum industries, produce hazardous waste. Other waste is generated through the use of products containing hazardous chemicals. Examples include solvents used in dry cleaning, plate cleaners used in printing, pesticides used in agriculture, and household paint thinners. Some naturally-occurring chemicals, such as selenium, can also build up to hazardous levels through runoff from soils.

At various sites throughout the country, hazardous wastes have not been properly stored or disposed of and have escaped into the environment. Hazardous waste releases can occur through leaking underground storage tanks or drums, leaking waste disposal landfills or ponds, hazardous waste spills, or other improper disposal of waste. Uncontrolled wastes can potentially contaminate groundwater, surface water (lakes, streams, bays), soils, or air. The federal government is investigating thousands of potential hazardous waste sites nationwide to assure that those likely to affect public health are addressed. It has targeted over 1,000 for federal cleanup under CERCLA. Many states are investigating other sites within their borders. The site cleanup process has several goals:

– Find the responsible parties
– Address any immediate threats to public health or the environment
– Inform and involve the community in the cleanup process
– Investigate and clean up the site

The federal "Superfund" law authorizes the Environmental Protection Agency to supervise cleanup of sites on or proposed for the National Priorities List, which is a ranking of the nation's contaminated sites. The parties responsible for generating the hazardous waste are also responsible for conducting the site cleanup. If the responsible parties cannot be found or will not comply with the law, federal EPA assumes the cleanup activities and cost. Expenses can later be recovered from the responsible parties, including all investigation and cleanup costs.

There are nine stages to a Superfund Cleanup:
- Site Identification
- Preliminary Investigation
- Short Term Cleanup Actions
- Remedial Investigation/Feasibility Study
- Draft Remedial Action Plan
- Remedial Action Plan
- Remedial Design
- Remedial Action
- Operation and Maintenance

Sites are identified through complaints and referrals to regulatory agencies as well as in the course of routine inspections conducted by officials. After a site is identified, the EPA conducts a *preliminary investigation*. During this stage, the EPA attempts to identify all potentially responsible parties who can perform and pay for the investigation and cleanup. Finding all responsible parties is usually difficult. Responsible parties can be current or former owners, operators, users of the site, or originators of products that contaminated the site. In many cases, more than one party may be responsible for a site's contamination. For instance, at a drum cleaning facility contaminated with hazardous waste, any business that sent waste-containing drums may be considered a responsible party. The responsible parties have the opportunity to assume site cleanup responsibility. If the responsible parties agree to comply with an order to conduct an investigation, EPA oversees their work. If the responsible parties do not comply with the order, EPA conducts the investigation.

The preliminary investigation, which may require up to nine months, assesses the potential severity of the contamination using public records and minimal sampling at the site. At the end of the investigation, the EPA issues another order requiring the responsible parties to perform the site cleanup. Again, if they do not comply, the EPA conducts the cleanup. The background information from the preliminary investigation is compiled into a report. The EPA also plans how it will inform and involve the community in the cleanup process.

If an immediate health threat exists, the EPA or the responsible party can take short term actions. For instance, if the public was exposed to soil contaminants at dangerous levels, the soils could be immediately removed. These short-term actions allow the EPA to address immediate health threats before the lengthy site investigation is complete. These actions are temporary measures, but taken with the overall long-term plans for the site in mind.

The remedial investigation involves defining the nature and extent of the contamination, and gathering the technical information needed to evaluate possible ways to clean up the site. Site conditions that affect the spread of contamination, and therefore possible cleanup actions, must be established. At sites where groundwater is involved, information about the site's underground geology and groundwater movement is needed. The EPA may use wells or deep bore holes to gather this information.

During the remedial investigation, the project team also collects and analyzes data about waste characteristics. Samples of soil, groundwater, surface water, or living organisms (such as fish or shellfish) are tested to determine the type and levels of contaminants that are present. Samples taken from different locations can show how far the contamination has spread. At groundwater sites, the EPA may drill monitoring wells in several places on or near the site.

At this point in the investigation, the EPA examines the site, looking for ways that living organisms might be exposed to contaminants and details the possible routes of exposure and how much exposure might occur. The health effects that could result from specific exposures are then identified. The risk assessment is used to insure that cleanup actions adequately protect public heath and the environment.

During the *remedial investigation/feasibility study* (RI/FS), the EPA identifies and evaluates site cleanup alternatives. In most cases there is a range of available alternatives. For instance, at a site where leaking drums have contaminated the soil, the soil could be removed to an appropriate disposal location, it could be capped to prevent contamination from spreading off-site, or treated on-site to destroy the contamination. After identifying the alternatives, EPA evaluates each one by considering effectiveness, feasibility, and cost. The chosen cleanup must effectively protect public health. Both long-term and short-term costs are considered. The results of the remedial investigation/feasibility study (RI/FS) are compiled into a report that is used to recommend a final cleanup plan. This step can last from four months to several years.

The *draft remedial action plan* describes the site conditions, the alternatives for cleanup that were considered, and a recommended cleanup plan. At most sites, the final cleanup involves a combination of alternatives. During the required comment period, the draft remedial action plan is available to other agencies, the responsible parties, and the public for review. Additionally, the EPA holds a public meeting. After receiving and considering comments, the EPA prepares the final remedial action plan. Development and review of this plan usually requires three to six months.

During the *remedial design* phase, the EPA or the responsible party designs the actual cleanup. For instance, if the cleanup involves removing contaminated soil from a site, the soil must be identified by location and depth; and provisions made for getting heavy equipment to and from the site, for decontaminating equipment, for disposal of the soil, and determining what to replace the removed soil with. The design step can last from two to ten months.

Following a public notification, the actual *remediation action*, or site cleanup begins. For example, contaminated soils might be excavated and disposed of appropriately. The excavated area could then be refilled with clean soil. Remedial action can take from two months to many years to carry out, depending on the cleanup plan and the extent of the contamination.

Cleaning up a site does not necessarily mean removing the contamination from the site. At some sites the EPA may allow containment of the contamination with underground barriers that prevent its spread or threat to human health or the environment.

Contaminated areas can also be covered or capped. Some wastes can be treated on the site to remove or destroy the contamination.

After the remedial actions are complete or treatment systems are in place, the EPA oversees *maintenance and operation* of the site. At some sites, groundwater can be pumped and treated in aerators or adsorbers to remove or trap chemicals. The clean water can then be discharged to a stream or other surface water (with the appropriate NPDES Permit). The EPA may have to operate a groundwater treatment system for 20 years to complete a cleanup. Examples or other site activities include:

– Sampling the site to check the cleanup's effectiveness

– Operating on-site incinerators

– Checking for cracks in soil caps

These activities may continue for 10 to 15 years. Operation and maintenance is likely to continue even longer at very complex sites with extensive contamination.

Other agencies besides the EPA work on site cleanup. These agencies include the U.S. Fish and Wildlife Service and the U.S. Army Corps of Engineers, county health departments, and city hazardous materials units.

## ■ Bioremediation

### Concepts

■ Bioremediation has been used successfully to cleanup hazardous waste sites

■ Bioremediation is a promising technology because it is an ecologically sound process

**Bioremediation** is a process that uses microorganisms to transform harmful substances to nontoxic or less harmful compounds. It has been one of the most promising new technologies for treating chemical spills and hazardous waste problems. In order to improve the technology and better understand its capabilities, the EPA has been encouraging bioremediation for waste site cleanups.

Bioremediation uses naturally occurring microorganisms, such as bacteria or fungi, to degrade harmful chemicals. Microorganisms need nutrients (such as nitrogen, phosphate, and trace metals), carbon, and energy to survive. They break down a wide variety of organic compounds found in nature to obtain energy for their growth. Many species of soil bacteria, for example, use petroleum hydrocarbons as a food and energy source, transforming them into harmless substances consisting mainly of carbon dioxide, water, and fatty acids. Bioremediation enhances this natural process by promoting the growth of microorganisms that can degrade contaminants.

Microorganisms exposed to contaminants develop an increased ability to degrade those substances. For example, when soil bacteria are exposed to organic contaminants, new strains of bacteria often naturally appear that can break down these substances to obtain energy. At many hazardous waste sites, microorganisms that degrade organic wastes, including some synthetic organic chemicals such as pesticides and solvents, have developed over time.

There are a number of environmental conditions that can slow down or stop the biodegradation process:

■ The concentration of the chemical may be so high that it is toxic to the microorganisms.

■ The number or type of microorganisms may be inadequate for biodegradation.

■ Conditions may be too acidic or alkaline.

■ The microorganisms may lack specific nutrients which they need in addition to the chemical as a food source.

■ Moisture conditions may be unfavorable (too wet or too dry).

■ The microorganisms may lack the oxygen, nitrate, or sulfate they need to use the chemical as an energy source.

In many instances, environmental conditions can be altered to enhance the biodegradation process. Samples are collected at the site and analyzed to determine what types of microorganisms are present and what nutrients and environmental conditions (such as pH, moisture, temperature, and oxygen levels) can enhance microbial degradation. For example, if adequate levels of nitrogen or phosphorous are not available, these nutrients can be added to enhance the growth of the microorganisms. If the concentration of the waste is too high, other chemicals or uncontaminated soil can be added to reduce toxicity so that biodegradation can occur. Bioremediation can be an attractive option for several reasons:

■ It is an ecologically sound process. Existing microorganisms can increase in number when the contaminant is present. When the contaminant that is their main food source is degraded, the microbial population naturally dies off. The residues from the biological treatment are usually harmless products (such as carbon dioxide, water, and fatty acids). However, the bioremediation process must be carefully monitored to avoid the possibility of creating a product that is more toxic than the original pollutant.

■ Instead of merely transferring contaminants from one environmental compartment to another, bioremediation destroys the target chemicals.

■ Bioremediation is usually less expensive than other technologies often used to clean up hazardous waste.

■ Bioremediation can often be accomplished on-site. This eliminates the need to transport large quantities of contaminated waste off-site and the potential threats to human health and the environment that can arise during such transportation.

## Example: Creosote

Creosote is a mixture of over 200 individual chemical compounds (including some substances known to cause cancer). It is used to treat wood to prevent rotting and termite attack. Creosote sometimes leaks from holding tanks and wood treatment areas. It can then seep into the soil and underlying ground water. Microorganisms that degrade creosote are currently the focus of extensive research efforts by the EPA. For example, EPA researchers have found that the bacterium *Pseudomonas paucimobilis* can break down many of the compounds found in creosote. The white-rot fungus *Phanerochaete chrysosporium* has been shown to degrade pentachlorophenol, another common soil contaminant at wood preserving facilities.

EPA studies have uncovered several limitations that have prevented more widespread use of bioremediation as a cleanup technology:

■ Research is needed to develop and engineer bioremediation technologies that are appropriate for sites with complex mixtures of contaminants.

■ Cleanup using bioremediation often takes longer than other remedial actions, such as excavation and removal of soil or incineration.

■ In some cases, depending on the parent compound, by-products may be formed. Some of these by-products may be toxic. The process must be carefully monitored to ensure the effectiveness of degradation.

Bioremediation techniques can be grouped into two general categories: *in situ* techniques, which treat wastes without removing the contaminated soil or water, and aboveground techniques, which treat soil or water in a vessel, reactor, or compost pile.

*In situ* treatment includes both surface remediation and subsurface remediation. In surface remediation, oxygen is available directly from the atmosphere, while in subsurface remediation, oxygen must be supplied by physically delivering water (containing dissolved oxygen) or air to the contaminated material.

Contamination in the top 6 to 12 inches of soil can be treated by tilling the soil to provide aeration and by adding nutrients and water to stimulate bacterial growth. Naturally occurring microorganisms can degrade contaminants that may volatize. Enhancing the growth of these microorganisms under controlled conditions can result in the removal of the pollutant, thus preventing it from volatizing.

Bioremediation techniques are also proving effective for cleaning up beaches contaminated by oil spills. When nutrients are applied to the oil, microorganisms can significantly reduce its concentration. Bioremediation can be more effective than physical cleaning methods for treating oil on the shoreline. It was used successfully to clean oily beaches in Prince William Sound after the Exxon *Valdez* spill.

In many cases, bacteria that can degrade contaminants are already present in the soil and groundwater. These bacteria may begin to degrade the contaminant, but they soon use up the available oxygen. To prevent this, oxygen must be delivered to the contaminated region. Nutrients can also be added to make environmental conditions more suitable for microbial activity.

Oxygen can be delivered to groundwater and to contaminated soils below the water table by withdrawing the groundwater, adding an oxygen source (such as air, pure oxygen, hydrogen peroxide, or ozone), and re-injecting the water using injection wells or trenches. Subsurface drains can also deliver nutrients and oxygen to depths of up to 40 feet.

Contaminants above the water table can be biodegraded *in situ* if the soil is relatively porous and permeable to air and water. A treatment solution containing nutrients can be added to the surface using spray irrigation, flooding, or ditches. Venting wells can be installed at intervals throughout the contaminated soil area to deliver oxygen to the contaminated areas.

A variety of aboveground biological treatment processes can effectively treat soil and water contaminated with organic chemicals. Composting is one method for treating soil containing hazardous organic compounds. Highly biodegradable materials, such as wood chips, are combined with a small percentage of biodegradable waste materials. Air can be provided by mixing the compost material or by forced air systems. Composting can also occur in closed bioreactors.

Many hazardous waste sites contain complex mixtures of organic and inorganic chemicals that are not readily biodegraded and that can be cleaned up only by a combination of treatment techniques. For example, a chemical dechlorination process can remove chlorine from PCBs. Then, biological treatment, which is more effective after dechlorination, can be used to complete the cleanup. The EPA is working to develop methods combining biological, chemical, and physical treatment processes to handle a variety of contaminants at hazardous waste sites.

Ultimately, the questions that must be answered to evaluate bioremediation as a remediation technology for the site include the following:

■ Are the chemicals at the site potentially biodegradable?

■ Are any of the contaminants potentially toxic to microbial degradation processes? If so, will another type of treatment prior to bioremediation change this?

■ To what level should the contaminants be reduced to meet the cleanup goals for the site?

■ What are the microbiological characteristics (e.g. mostly anaerobic or aerobic?) of the environment at the site?

■ Is the environment appropriate for bioremediation or can environmental conditions be adjusted to make it more appropriate for biological treatment?

■ What are the microbiological needs of the site ( e.g. do nutrients or bacteria need to be added)?

Bioremediation is a technology with a promising future. It can be a nondisruptive, cost-effective, and efficient means of destroying harmful chemicals at many chemical spill and hazardous waste sites. As scientists learn more about its capabilities and develop practical techniques to biodegrade an increasing number of wastes, bioremediation is likely to be used more and more to clean up the environment.

## ■ Epilogue – ISO 14000

It is appropriate to end this text with an exciting glimpse of the future of environmental health and safety protection. International Standards Organization  is developing an environmental counterpart to the ISO 9000 international quality standards that many companies have been working with for the past several years. **ISO 14000** is a single environmental standard that helps companies meet environmental goals and regulations regardless of their location. The standard is designed to eliminate potential non-tariff trade barriers resulting from multiple and sometimes conflicting environmental regulations and standards found throughout the world. Designed for international trade, the ISO 14000 standard is impacting companies doing business nationally, regionally and locally.

In sharp contrast to most environmental regulations, ISO 14000 centers on pollution prevention and does not include specific environmental goals. ISO 14000 focuses on an organization's operations and procedures. It promotes adoption of an environmental management system that brings companies into compliance with government regulations and voluntary codes of practice, and promotes continual improvement in environmental management. Companies that demonstrate compliance with ISO 14000 to external auditors will be able to have that compliance "certified."

The ISO 14000 standard includes the development of and implementation of an "Environmental Management System" with the following elements:

- Environmental Auditing;
- Performance Evaluation;
- Product Labeling; and
- Life-Cycle Assessments

The Environmental Management System is the only portion of the ISO 14000 for which companies can be "certified." The standard for corporate environmental management systems includes the following requirements:

- A commitment from the top management of the organization to improve the environmental performance of the organization in managing its activities, products and services;

- Established procedures to identify the environmental aspects of all of the organization's activities and to analyze which have significant impacts on the environment;

- Completion of an initial environmental review to identify the current position of the organization with respect to the environment including regulatory requirements, environmental impacts and liabilities, and contracting activities;

- Establishment of an environmental policy that commits the organization to compliance with applicable legal requirements, continual improvements and prevention of pollution;

- Development of internal environmental performance criteria and environmental objectives;

- Establishment and maintenance of procedures to identify all legal and other requirements applicable to the environmental aspects of the company's activities, products and services;

- Establishment and maintenance of operational procedures and controls, procedures for training of all personnel whose work may significantly affect the environment, and procedures for monitoring and evaluating environmental performance to ensure that the level of performance is consistent with the organization's policies, objectives, and targets;

- Documentation of all operational processes and procedures, including evaluations and monitoring of the organization's environmental performance and the processes used to advance the organization's environmental objectives;

- Establishment of emergency plans and procedures to ensure an appropriate response to unexpected or accidental incidents; and

- A periodic review of the organization's management of the Environmental Management System to ensure its continuing suitability and effectiveness with the goal of improving its overall environmental performance.

The ISO 14000 standard also addresses product labeling for international commerce and provides a basis for life-cycle assessments. A life-cycle assessment is an analysis of the impacts of the company's products and services at all stages, from design through manufacture, packaging, use and disposal.

At present the ISO 14000 standard effectively has become a condition of doing business in Europe. Countries such as the United Kingdom are adopting ISO 14000 as an equivalent for their own environmental standards. Other countries without developed environmental standards are considering adopting ISO 14000 as a requirement to foster internal environmental stewardship and regulation.

Similarly, lenders and insurers who need information on a company's environmental performance record, many of whom have supported the proposed standard from its inception, may require companies to become certified. Furthermore, some states and federal agencies are providing incentives in the form of reduced compliance inspections for certified companies. Finally, even if companies are not required to comply with ISO 14000, uncertified companies may be at a competitive disadvantage in the market place because their competitors are promoting themselves as being in compliance with ISO 14000.

## ■ Checking Your Understanding (11-4)

1. Contrast "short-term actions" with "remedial design" in the cleanup of a hazardous waste site.

2. Explain how Superfund cleanups are paid for.

3. Define "bioremediation."

4. Explain the importance of ISO 14000.

# Glossary

μg/g – Microgram per gram, one part per million (ppm)

μg/l – Microgram per liter, one part per billion (ppb)

**Abatement** – Reducing the degree of or eliminating pollution.

**Above Ground Tank** – A storage tank situated so that the entire surface area of the tank is completely above the plane of the surrounding ground surface and the entire surface area of the tank is able to be visually inspected.

**Absorption** – In toxicology, the ability of a chemical to find its way into the bloodstream. Physically, the passage of one substance into or through another; for example, a process where one or more soluble components of a gas mixture are dissolved in a liquid.

**Accumulation** – The build-up of dangerous chemicals in the body due to repeated exposures.

**ACGIH** – American Conference of Governmental Industrial Hygienists.

**Action Levels** – Under CERCLA, the existence of a contaminant concentration in the environment high enough to warrant action or trigger a response under SARA and the National Oil and Hazardous Substances Contingency Plan. The term is used similarly in other regulatory programs, for example, in asbestos regulation.

**Activated Sludge** – Sludge resulting from the mixing of primary effluent with bacteria-laden sludge and then agitated and aerated to promote biological treatment. This speeds breakdown of organic matter in raw sewage undergoing secondary waste treatment.

**Acute Toxicity** – A short-term exposure that causes effects felt at the time of exposure or shortly after.

**Additive Effect** – The combined effect of two or more chemicals is equal to the sum of the effect of each agent alone. The opposite of antagonism.

**Administrative Order** – A legal document signed by a governmental agency directing an individual, business, or other entity to take corrective action or refrain from an activity. It describes the violations and actions to be taken, and can be enforced in court.

**Adsorption** – The attraction and accumulation of one substance on the surface of another. One of the physical treatment methods. Activated carbon is commonly used to remove organic matter from wastewater in this manner.

**AEA** – See Atomic Energy Act

**AEC** – See Atomic Energy Commission.

**Aeration** – A physical treatment method that promotes biological degradation of organic matter. The process may be passive (when waste is exposed to air), or active (when a mixing or bubbling device introduces the air).

**Aerobic Bacteria** – A type of bacteria that requires free oxygen to carry on metabolism.

**AHERA** – See Asbestos Hazard Emergency Response Act.

**Air-Purifying Respirator** – A respiratory protection device composed of a face piece that uses cartridges with filters to purify air before it is inhaled. This type of respirator is not adequate in an oxygen deficient atmosphere.

**Air-Supplying Respirator** – A respiratory protection device composed of a face piece that is connected to its own air source. These devices consist of either a self-contained unit that the worker wears or a hose connecting a worker to a remote air source.

**Air Contaminant** – Any particulate matter, gas, or combination, other than water vapor or natural air.

**Air Pollution** – The term used to refer to the accumulation of substances in the atmosphere that can cause harmful health effects to living things or negatively affect the public welfare.

**Air Pollution Control District** – A county, or several counties that have joined to form a single district, responsible for the implementation of air quality control plans.

**Air Quality Control Region** – An area, designated by the federal government, in which communities share a common air pollution problem. Sometimes several states are involved.

**Air Quality Standards** – The concentration of pollutants allowed by regulations that may not be exceeded during a specified time in a defined area.

**Airborne Particulates** – Total suspended particulate matter found in the atmosphere as solid particles or liquid droplets. Airborne particulates include dust, emissions from industrial processes, smoke from burning, and motor-vehicle exhaust.

**Alpha Particles** – A relatively slow moving subatomic particle, designated by either $^4_2\text{He}$ or symbol $\alpha^{++}$. It is a helium nucleus composed to two protons and two neutrons, but without any surrounding electrons. The most damaging subatomic particle to human tissue due to its high energy.

**Alternative Fuels** – Fuels that can replace ordinary gasoline due to their desirable energy efficiency and pollution reduction features. Alternative fuels include compressed natural gas, alcohols, liquefied petroleum gas (LPG), and electricity.

**Ambient Air Quality** – A term defined in EPAs NAAQS air quality program designed to reduce outdoor pollution. Ambient air is any unconfined portion of the atmosphere, including open air and surrounding air.

**American Council of Governmental Industrial Hygienists** – A professional organization of industrial hygienists who establish guidelines (TLVs) for workplace exposures to air contaminants.

**Anaerobic Bacteria** – A type of bacteria that cannot tolerate free oxygen, using food sources in metabolism for both oxidizing and reducing agents.

**Antagonism** – This is a subtractive effect, where one substance interferes with the action of another. The opposite of additive effect.

**Anthropogenic Sources** – Pollutants that are generated by human activities, such as driving a car.

**Aquifer** – A permeable geologic unit with the ability to store, transmit, and yield fresh water in usable quantities.

**Aquitard** – A confining layer having little to no water permeability.

**Asbestos** – A mineral fiber that can pollute air or water and cause cancer or asbestosis when inhaled. EPA has banned or restricted its use in manufacturing and construction.

**Asbestos Hazard Emergency Response Act** – AHERA is the federal law that requires certain procedures to be followed for asbestos abatement in school buildings.

**Ash** – The mineral content of a product remaining after complete combustion, such as incineration.

**Atmosphere** – The gaseous envelope held close to the earth by gravity.

**Atom** – The smallest part of an element that can exist; atoms consist of a nucleus of protons and neutrons surrounded by orbiting electrons.

**Atomic Energy Act** – The federal law that provide controls over the possession, development, and use of radioactive materials.

**Atomic Energy Commission** – Its Functions have been assumed by DOE and NRC.

**Atomic Mass** The combined sum of an atoms proton and neutron masses.

**Atomic Number** – The number of protons in the nucleus of an atom.

**Atomic Theory** – The theory that proposes that atoms are composed of a nucleus, surrounded by a cloud of electrons.

**Attainment Area** – An area considered to have air quality meeting the national ambient air quality standards as defined in the Clean Air Act. An area may be an attainment area for one pollutant, and a nonattainment area for others.

**Avogadro's Law** – A scientific law that states that equal volumes of gasses at the same temperature and pressure contain equal numbers of particles.

**Background Radiation** – Nuclear radiation due to the natural environment and to naturally occurring radioactivity within the body.

**BACT** – See Best Available Control Technology

**Bactericides** – A subgroup of pesticides, it is any material that kills bacteria.

**Benign Tumors** – Well-defined tumors that do not tend to invade surrounding tissue or spread to additional distant locations in the body.

**Best Available Control Technology** – An air pollution emission limitation based on the maximum degree of emission reduction that, considering energy, environmental, and economic impacts and other costs, is achievable.

**Beta Particles** – A high speed, but low mass electron, designated by $_{-1}^{0}e$, ejected from an atoms nucleus. A charged particle that is emitted by certain radioactive materials that is physically identical with an electron.

**Bill** – A bill is a proposal to change, amend, repeal, or add to existing law.

**Bioaccumulation** – The accumulation of a toxin within an individual organism.

**Bioamplification** – The multiplying factor that results from concentrating the toxin at higher and higher levels within a food chain.

**Bioassay** – A part of biomonitoring where controlled laboratory experiments determine the effect of exposing test organisms to water containing a specific chemical or a complex mixture of chemicals.

**Biochemical Oxygen Demand** – The dissolved oxygen required to decompose organic matter in water; a numerical estimate of contamination in water expressed in milligrams per liter of dissolved oxygen.

**Biogenic Sources** – Sources of pollutants that are living organisms, such as pine trees, which emit volatile organic compounds.

**Biological Diversity** – The variety and variability among living organisms and the ecological complexes in which they occur.

**Biological Treatment** – Processes that use living organisms to bring about chemical changes. Biological treatment can be viewed as a subset of chemical treatment.

**Biome** – The largest terrestrial ecosystem. They are distinct regional ecosystems of the world identifiable by their similar types of soil, climate, plants and animals no matter where they are located.

**Biomonitoring** – The use of actual plant and animal species as reference points for determining environmental quality.

**Bioremediation** – Any process that uses microorganisms to transform harmful substances to nontoxic or less harmful compounds.

**Biosphere** – The part of the earth that supports life.

**Blood Borne Pathogens** – Disease-causing organisms that can be transmitted through blood.

**BOD** – See Biochemical Oxygen Demand.

**Boiling Point** – The point at which a liquid changes into its gaseous state.

**Bonded** – The electrical forces between the various parts of a compound that holds it together. The two major types of chemical bonding are ionic and covalent.

**Boyles Law** – A scientific law that states that the volume of a gas is inversely proportional to its pressure at constant temperature (PV=k).

**By-product** – Any material, other than the principal product, that is generated as a consequence of an industrial process.

**C** – See Ceiling Limit.

**CAA** – See Clean Air Act.

**CAER** – See Community Awareness and Emergency Response

**Calibration** – The checking and adjusting of an instrument to ensure it is giving accurate readings by comparing its readings with a known standard.

**Cancer** – A disease characterized by the unregulated overgrowth of cells (tumors).

**Carbon Adsorption** – A process in which substances adhere to the surface of specially treated (activated) carbon. This method is particularly effective in removing organic compounds from liquid or gaseous waste streams.

**Carbon Cycle** The term used to describe how carbon circulates through the air, plants, animals, and soil.

**Carcinogen** – An agent that can cause cancer.

**Carcinogenicity** – The tendency for cancer to occur.

**Carrying Capacity** – The ability of a region to support the people who live there without degrading the resources that are available.

**CAS#** – See Chemical Abstracts Service Number.

**Case Law** – The written decisions made by courts on similar cases.

**Cathodic Protection** – A technique to prevent corrosion of a metal surface by making that surface the cathode of an electrochemical cell. Cathodic protection is generally known as a method of applying suppressed current or sacrificial anodes to underground tanks.

**Ceiling Limit** – C, the concentration of airborne contaminant in the workplace that should not be exceeded during any part of the work day, even for an instant. The ceiling is an upper limit, not a time weighted average.

**CEMS** – See Continuous Emission Monitoring Systems

**CERCLA** – See Comprehensive Environmental Response, Compensation, and Liability Act

**CFC** – See Chlorofluorocarbons.

**CFR** – See Code of Federal Regulations

**Characteristic Waste** – A waste which is not on the hazardous waste lists, but which is regulated as hazardous because it displays one or more of the hazardous characteristics, ignitability, corrosivity, toxicity, or reactivity.

**Charles Law** – A scientific law that states that the volume of a gas is directly proportional to its absolute temperature, at constant pressure (V/T = k).

**Chemical Abstracts Service Number** – CAS#; the specific identification number for each chemical. A product with more than one component, will have a CAS# for each component.

**Chemical Manufacturers Association** – Founded in 1863, it is a nonprofit trade association of chemical manufacturers, which represents more than 90 percent of the U.S. and Canadian chemical industries. Through Chem Trec it provides emergency telephone information for accidents involving chemicals.

**Chemical Properties** – A unique set of characteristics that describes how the substance reacts with other substances.

**Chemical Substances** – Under TSCA, defined as any organic or inorganic substance of a particular molecular identity, including: any combination of such substances occurring in whole or part as a result of a chemical reaction or occurring in nature, and any element or uncombined radical. The definition further excludes: any mixture; any pesticide as defined in FIFRA, as long as it is used as a pesticide; tobacco or tobacco products; certain nuclear materials; firearms; and foods, food additives, drugs, cosmetics when covered under the provisions of the Federal Food, Drug, and Cosmetic Act.

**Chemical Treatment** – A process that results in the formation of a new substance or substances. The most common chemical wastewater treatments include coagulation, disinfection, water softening, and oxidation.

**Chlorine Demand** – A measure of the amount of chlorine that will combine with impurities and therefore will not be available to act as a disinfectant.

**Chlorofluorocarbons** – A group of gaseous chlorinated hydrocarbons that are used as coolants, refrigerants, propellants for aerosol cans, and in the production of foam products. The concern over their usage stems from their implication in the heating of the planet (global warming) and their affect on ozone depletion in the upper atmosphere.

**Chronic Toxicity** – A long-term exposure causing effects that appear after months or years of exposure.

**Clean Air Act** – CAA; this federal law was enacted in 1970 and amended most recently in 1990 with the fundamental objective of protecting public health and welfare from harmful effects of air pollution. To achieve this goal, the act sets maximum acceptable levels for pollution in ambient (outdoor) air.

**Clean Fuels** – Low-pollution fuels that can replace ordinary gasoline. These are alternative fuels including gasohol (gasoline-alcohol mixtures), natural gas, and LPG (liquefied petroleum gas).

**Clean Water Act** – CWA; federal law dating to 1972 with the objective to restore and maintain the chemical, physical, and biological integrity of the nations waters. Its long range goal is to eliminate the discharge of pollutants into navigable waters and to make national waters fishable and swimmable.

**Cleanup** – Actions taken to deal with a release or threat of release of a hazardous substance that could affect human health and/or the environment. The term cleanup is sometimes used interchangeably with the terms remedial action, removal action, response action, or corrective action.

**Closed-Loop Recycling** – Reclaiming or reusing wastewater for non-potable purposes in an enclosed process. Also used in hazardous waste context as a process by which wastes are returned for use in the production process.

**CMA** – See Chemical Manufacturers Association

**Coagulant** – A chemical that causes small particles to stick together to form larger particles.

**Coagulation** – A chemical treatment method. An irreversible combination or aggregation of semi-solid particles. Often brought about by the addition of a coagulating agent.

**Code of Federal Regulations** – CFR; the federal governments official publication of federal regulations. CFRs are numbered to correspond to a particular agencys or departments rules, for example, the Environmental Protection Agency has most of its regulations in 40 CFR.

**Commercial Nuclear Reactor** – A civilian nuclear power plant, owned by an electric utility or utilities, and operated for generating electricity for commercial sale. It is required to be licensed by the NRC.

**Community** – A small group of living things within an ecosystem.

**Community Awareness and Emergency Response** – CAER Program; a voluntary program established by the Chemical Manufacturers Association to encourages plant managers to listen to community concerns, participate in planning, and explain their plants operations and policies to the public. By working with the community to ensure safe handling, storage, transportation, and disposal of dangerous chemicals, industry can protect itself as well as the public from the high costs of chemical accidents.

**Community Water System** – A public water system that serves at least 15 service connections used by year-round residents or regularly serves at least 25 year-round residents.

**Compliance Safety Health Officer** – CSHO; an OSHA inspector.

**Compounds** – A pure substances composed of two or more elements, e.g. NaCl or KOH.

**Comprehensive Environmental Response, Compensation, and Liability Act** – CERCLA; also known as Superfund, was passed by Congress in 1980 and most recently amended in 1986 by the Superfund Amendments and Reauthorization Act (SARA). CERCLA provides a system for identifying and cleaning up chemical and hazardous materials releases into any part of the environment.

**Confined Aquifer** – Also known as the artesian aquifer, it is the area above the saturated zone of a confining bed, where the water is free to rise and fall.

**Confining Bed** – A rock unit that restricts the movement of ground water either in or out of adjacent aquifers.

**Consumer** – Also known as a heterotroph, it is any organism that is unable to manufacture or produce its own food, therefore, must rely on producers for food energy.

**Container** – Any portable device in which material is stored, transported, treated, disposed of, or otherwise handled.

**Contaminant** – Any physical, chemical, biological, or radiological substance or matter that has an adverse affect on air, water, or soil.

**Contingency Plan** – A document detailing organized, planned, and coordinated actions to be followed in case of a fire, explosion, or other accident that releases toxic chemicals, hazardous wastes, or radioactive materials that threaten human health or the environment.

**Continuous Emission Monitoring Systems** – Machines that measure, on a continuous basis, pollutants releases by a source.

**Control Techniques** – The equipment, processes or actions used to reduce air pollution.

**Conversion Factor** – An equality that can be used to convert a measured amount in one set of units to another set of units, e.g. 12 inches = 1 foot.

**Corrosive** – A material that can corrode standard steel containers or burn the skin, eyes or other body tissue.

**Criteria Pollutants** – As identified by the NAAQS, they include carbon monoxide, lead, nitrogen oxides, ozone, particulate matter and sulfur dioxide.

**Cross Connection** – A connection between safe drinking water and any source of contamination.

**CSHO** – See Compliance Safety Health Officer.

**Curtailment Programs** – Restrictions on operation of fireplaces and wood stoves in areas where they make major contributions to air pollution.

**CWA** – See Clean Water Act.

**Dechlorination** – A process that removes chlorine from a substance by chemically replacing it with hydrogen or hydroxide ions.

**Decomposer** – An organism that nourishes itself by breaking down dead organic matter for energy and nutrients.

**Degradation** – The process by which a chemical is reduced to a less complex form.

**Delist** – Use of the petition process to have a particular waste excluded from regulation at a particular generating facility.

**Department of Energy** – The U.S. Department of Energy.

**Department of Transportation** – DOT. The federal department that develops and enforces regulations dealing with all aspects of transportation, including transportation of hazardous materials.

**Diffusion** – The passage of one gas through another.

**Disinfection** – A chemical treatment method. The addition of a substance, e.g., chlorine, ozone or hydrogen peroxide, which destroys harmful microorganisms or inhibits their activity.

**Disintegration Series** – The steps followed by a radioactive nucleus as it is converted into a stable nucleus.

**Dispersion** – A term used to describe the distribution of the pollutant into the atmosphere.

**Disposal** – Final placement or destruction of toxic, radioactive, or other wastes. Disposal may be accomplished through use of approved secure landfills, surface impoundments, land farming, deep well injection, ocean dumping, or incineration.

**Dissociate** – The process of ion separation that occur when an ionic solid is dissolved in water.

**Distillation** – A separation process that makes use of the different boiling points of two or more liquids.

**DOE** – See Department of Energy.

**Dose** – Refers to the actual amount of chemical that enters and reacts with body systems.

**Dose-Response Relationships** – Describe the relationships between known exposure levels and the appearance of toxic effects in exposed populations.

**DOT** – See Department of Transportation.

**Dredging** – Removal of mud from the bottom of water bodies using a scooping machine. Dredging disturbs the ecosystem and causes silting that can kill aquatic life. Dredging of contaminated muds can expose aquatic life to heavy metals and other toxics. Dredging activities may be subject to regulation under the Clean Water Act.

**Drinking Water Standards** – Water quality standards measured in terms of suspended solids, unpleasant taste, and microbes harmful to human health. Drinking water standards are included in state water quality rules.

**Ecology** – The study of the relationship between living things and their environment. The word is derived from Greek words meaning the study of the home.

**Ecosystem** – A group of organisms interacting with each other and with their nonliving surroundings.

**Effluent** – The wastewater that is discharged as the result of a process.

**Effluent Limitations** – Standards developed by EPA to define the levels of pollutants that could be discharged into surface waters.

**EHS** – See Extremely Hazardous Substances.

**EIS** – See Environmental Impact Statement

**Electrochemistry** – The branch of chemistry concerned with the interaction between electrical forces and chemical phenomena.

**Electrodialysis** – An adaptation of the dialysis process that employs an alternating stack of cation and anion permeable membranes. As water is fed into the spaces between the membranes, a direct-current field is applied to the stack, causing the ions to migrate toward the opposite charged electrodes. This results in the concentration of ions in alternate spaces between membranes, and the water to become depleted of ions, or demineralized.

**Electronegativity** – The tendency for atoms that do not have a complete octet of electrons in their outer shell to become negatively charged.

**Electrons** – Particles with a negative electrical charge that orbit the nucleus of atoms.

**Elements** – Substances that cannot be further broken down by chemical reactions.

**Emission** – Release of pollutants into the air from a source.

**Emission Allowances** – An allowance made by the EPA to power plants in the acid rain program. Each allowance is worth one ton of sulfur dioxide released from the smokestack.

**Emission Standard** – The maximum amount of air polluting discharge legally allowed from a single source, mobile or stationary.

**Emissions Trading** – EPA policy that allows a plant complex with several facilities to decrease pollution from some facilities while increasing it from others, as long as total results are equal to or better than previous limits. Facilities where this is done are treated as if they exist in a bubble in which total emissions are averaged. Complexes that reduce emissions substantially may bank their credits or sell them to other industries. Emissions trading may be required to receive a permit to operate a new source in a nonattainment area.

**Encapsulate** – Total enclosure of a waste in another material that isolates the waste material so it cannot find its way out.

**Endangered Species** – Animals, birds, fish, plants, or other living organisms threatened with extinction by changes in their environment. Requirements for declaring a species endangered are contained in the Endangered Species Act.

**Enforcement** – Legal actions by regulatory agencies to obtain compliance with environmental laws, rules, regulations, or agreements and/or obtain penalties or criminal sanctions for violations. Enforcement procedures may vary, depending on the specific requirements for different environmental laws and related regulatory requirements.

**English System** – The measuring system in common usage in the USA, it is based on the foot as a unit length, the pound as a unit weight, the gallon as a liquid unit of volume and the second as a unit time.

**Environmental Assessment** – The cornerstone of NEPA. It is a requirement that the environmental impact of governmental actions be assessed for its impact to the environment.

**Environmental Audit** – An independent assessment of the current status of a facilitys compliance with applicable environmental requirements.

**Environmental Impact Statement** – EIS; a document required by NEPA that details the potential environmental impact of a proposed action.

**Environmental Medium or Compartment** – The part of the environment, typically air, water, soil, and biota that contaminants are carried by or transmitted through.

**Environmental Protection Agency** – EPA. The federal agency founded in 1970 with the responsibility of developing and enforcing regulations pertaining to the protection of environmental quality.

**Environmental Response Team** – EPA experts who can provide around-the-clock technical assistance to EPA regional offices and states during all types of emergencies involving hazardous waste sites and spills of hazardous substances.

**Environmental Technology** – The environmental field that uses applied science principles to manage environmental concerns while maintaining regulatory compliance.

**EPA** – See Environmental Protection Agency

**Epidemiology** – The field of science dealing with the study of disease occurrence in human populations.

**Episode (Pollution)** – An air pollution incident in a given area caused by concentration of atmospheric pollution reacting with meteorological conditions that may result in a significant increase in illnesses and deaths. Although most commonly used for air pollution, the term also is used with other kinds of environmental events such as a massive water pollution situation.

**Estuaries** – A coastal body of water that is partly enclosed.

**ET** – See Environmental Technology.

**Explosive** – A material that releases pressure, gas, and heat suddenly when subjected to shock, heat, or high pressure.

**Exposure** – The amount of toxic chemical our body comes into contact with in the air we breath, the food we eat, or our skin.

**Exposure Assessment** – Involves estimating the dosage of contaminants received by exposed populations, identifying the exposed population, and identifying the body sites at which toxic effects are produced.

**Extremely Hazardous Substances** – Any of the chemicals identified by EPA based on toxicity, and listed under SARA Title III. The list is periodically revised and published in the Federal Register. EPA has identified nearly 300 extremely hazardous substances.

**Facultative Bacteria** – A type of anaerobic bacteria that can metabolize its food either aerobically or anaerobically.

**Feasibility Study** – FS; used in site remediation and required by the EPA to identify and evaluate possible site cleanup alternatives.

**Federal Emergency Management Agency** – FEMA; federal agency with the lead role for coordinating civil emergency response planning and disaster response.

**Federal Insecticide, Fungicide, and Rodenticide Act** – FIFRA requires all pesticides to be registered with the EPA before being marketed.

**Federal Reference Methods** – The EPA standards for sampling and analysis of criteria pollutants in ambient air. The methods provide specific procedures for sampling, analysis, calibration, and calculation that must be followed for any monitoring activity related to compliance with the Clean Air Act.

**Felony** – A serious crime, punishable by imprisonment for more than one year. Felonies such as murder or treason are capital offenses and may be punishable by death.

**FEMA** – See Federal Emergency Management Agency.

**FIFRA** – See Federal Insecticide, Fungicide, and Rodenticide Act

**Filtration** – A physical treatment method for removing solid (particulate) matter from water by passing the water through porous media such as sand or a man-made filter.

**Finding of No Significant Impact** – FONSI; under NEPA, a FONSI is prepared if it is determined that a particular action presents no significant impact to the environment.

**Fire Hazard** – The ability of a material to burn easily when exposed to a heat source.

**Fire Triangle** – A concept used by safety trainers to help people understand that it requires the three components, fuel, heat and oxygen to start a fire.

**Flammable** – A material that catches on fire easily or spontaneously under conditions of standard temperature and pressure.

**Flammable Range** – Also known as flammable limits, it is the range of fuel percentages that will burn in air in the presence of an ignition source (heat). It can is also defined as the range between the LEL and UEL for a particular substance.

**Flash Point** – The temperature at which a liquid or solid gives off sufficient vapors to form an ignitable mixture with air. For official shipping regulations, it is determined by the Tagliabue open cup method.

**Flocculation** – A physical treatment method involving the clumping of particles as the result of coagulation. For this reason, the two terms are often used interchangeably.

**Flotation** – A process utilizing the difference in densities. Accomplished by allowing air bubbles to attach themselves to particles or droplets and buoy them to the surface.

**FOIA** – See Freedom of Information Act of 1966.

**FONSI** – See Finding of No Significant Impact.

**Food Chain** – The organisms in a line, where each is the food for the next in line.

**Food Web** – A food web is an intricate network of food chains. Because few organisms eat only one kind of food, a food web is a more accurate depiction of the complex nutritional cycles that exist in nature.

**Free Radicals** – Highly reactive uncharged molecular fragments, possessing an unpaired electron.

**Freedom of Information Act of 1966** – A federal law that entitles the public to access information gathered as a result of governmental agency functions.

**FS** – See Feasibility Study

**Fugitive Emissions** – Emissions not caught by a capture system.

**Fungicides** – A subgroup of pesticides, they are any materials that kill fungi.

**Gamma Rays** – A penetrating form of energy emitted during a fission reaction. They are designated by symbol $\gamma$, and are a form of electromagnetic radiation. Or short wave-length electromagnetic radiation emitted during the radioactive decay of certain nuclides.

**Gas** – One of the three states of matter. Gasses are characterized as having both an indefinite shape and volume.

**Gay-Lussacs Law** – A scientific law that states when gasses react their volumes, when measured at the same temperature and pressure, are simple whole number ratios.

**General Duty Clause** – Found in the Occupational Safety and Health Act, it is the overriding obligation of the employer to provide a workplace that is free from recognized hazards likely to cause death or serious physical harm.

**General Industry Safety Orders** – Found in 29 CFR 1910 Subparts B it is intended for those businesses that do not fall under a specific standard.

**Generation** – The act of originating or producing a waste.

**Geogenic Sources** – Sources of pollutants originating from earthly processes, such as volcanic eruptions.

**GISO** – See General Industry Safety Orders

**Gradient** – The direction and slope of a water table.

**Gravimetric** – A manual sampling method used for particulate matter that involves measurement by weight.

**Groundwater** – The fresh water found under the Earths surface, usually in aquifers. Groundwater is a major source of drinking water and there is a growing concern over areas where leaching agricultural or industrial pollutants, or substances from leaking underground storage tanks, are contaminating groundwater.

**Habitat** – The place where a specific plant and animal species are naturally found.

**Half-life** – The length of time required for one-half of the nucli of a radioactive substance to undergo discintegration.

**Halogen** – One of five nonmetal elements (astatine, bromine, chlorine, fluorine, or iodine), halogens are always part of a compound, such as a chlorinated hydrocarbon.

**Halogenated Organic Compounds** – Chemical compound made up of hydrogen, carbon, and one or more halogens.

**HAPs** – See Hazardous Air Pollutants.

**Hazard Classes** – There are nine hazard classes under the DOT system. They are: explosives, gases, flammable liquids, flammable solids, oxidizers, poisons, radioactive materials, corrosives, and miscellaneous hazards.

**Hazard Communication Standard** – HazCom; an OSHA workplace standard found in 29 CFR 1910.1200 that requires all employers to become aware of the chemical hazards in their workplace and relay that information to their employees. In addition, a contractor conducting work at the site of a client must provide chemical information to the client regarding the chemicals that are brought onto the work site.

**Hazard Evaluation** – One of the four steps in risk assessment. This step involves collecting and evaluating information about the toxic properties of contaminants.

**Hazard Ranking System** – The principle screening tool used by EPA to evaluate risks to public health and the environment associated with abandoned or uncontrolled hazardous waste sites. This screening tool is the primary factor in deciding if the site should be on the National Priorities List, and if so, what its ranking should be on that list.

**Hazardous Air Pollutants** – Chemicals that cause serious health and environmental effects and are released by sources such as chemical plants, dry cleaners, printing plants, and motor vehicles. These pollutants include asbestos, beryllium, mercury, and benzene.

**Hazardous and Solid Waste Amendments** – The 1984 Act (Public Law 98-616) that significantly expanded the earlier versions of RCRA.

**Hazardous Materials** – Substances that may either cause, or increase the occurrence of, death or irreversible illness in the population, or pose a hazard to human health or the environment when improperly treated, stored, transported, disposed of, or otherwise managed.

**Hazardous Materials Transportation Act** – HMTA; a federal law passed in 1975 to regulate commerce by improving the protections afforded the public against risks associated with the transportation of hazardous material.

**Hazardous Substance** – Is defined to include hazardous waste under RCRA, as well as substances regulated under the Clean Air Act, Clean Water Act, and Toxic Substances Control Act. DOT also has their own definition for a hazardous substance. It is any material, including mixtures and solutions, that is listed in Appendix A of the Hazardous Materials Table, is in a quantity, in one package, that equals or exceeds the RQ listed in Appendix A to the Hazardous Materials Table, and is in concentrations that meet or exceed those listed on a hazardous substance RQ table.

**Hazardous Waste Operations and Emergency Response Standard** – Hazwoper; an OSHA workplace standard that covers areas HazCom doesnt, including any clean-up operation required by a governmental body, whether Federal, State, or local agency.

**Hazardous Wastes** – Generally hazardous materials which have no further use. Legally, there are specific tests to determine whether a hazardous waste has been generated. Hazardous Wastes include solid wastes that are ignitable, corrosive, reactive, or toxic may pose a substantial or potential hazard to human health or the environment when improperly managed. It possesses at least one of four characteristics (ignitability, corrosivity, reactivity, or toxicity), or is listed by EPA or a state environmental agency.

**HazCom** – See Hazard Communication Standard.

**Hazwoper** – See Hazardous Waste Operations and Emergency Response Standard.

**Health Hazard** – The ability of a material to cause, either through direct or indirect exposure, temporary or permanent injury or incapacitation.

**Herbicides** – A subgroup of pesticides, they are materials that kill plants.

**High-Level Waste** – The highly radioactive material resulting from the reprocessing of spent nuclear fuel, including liquid waste produced directly in reprocessing and any solid material derived from such liquid waste, that contains fission products in sufficient concentrations; and other highly radioactive material that the NRC, consistent with existing law, determines by rule requires permanent isolation. In the US, spent nuclear fuel is considered to be high-level waste.

**HMTA** – See Hazardous Materials Transportation Act.

**HOC** – See Halogenated Organic Compounds.

**HRS** – See Hazard Ranking System.

**HSWA** – See Hazardous and Solid Waste Amendments.

**Hydrocarbon** – Any class of substances containing only hydrogen and carbon, such as methane or ethylene.

**Hydrogen Bonding** – The term used to describe the weak, but effective, attraction that occurs between polar covalent molecules.

**Hydrologic Cycle** – The term literally means the water-earth cycle. It refers to the movement of water, in all three of its physical forms, through the various environmental compartments.

**Hygroscopic** A substance that readily absorbs moisture.

**I/M Program** – See Inspection and Maintenance Program.

**ICS** – See Incident Command System.

**IDLH** – See Immediately Dangerous to Life or Health.

**Igneous Rocks** – One of the three types of rocks that comprise the crust of the earth. These rocks originate from the inner core of the earth and are therefore solidified molten material.

**Ignitable** – Capable of burning or causing a fire.

**IH** – See Industrial Hygienist.

**Immediately Dangerous to Life or Health** – IDLH; the concentration or condition that poses an immediate threat to life or health. The worker must be able to escape within thirty minutes without losing his or her life or suffering permanent health damage.

**Impoundment** – A body of water or sludge confined by a dam, dike, floodgate, or other barrier; see Surface Impoundment entry in this glossary.

**Impressed Current** – An AC to DC rectifier is used to place an electrical charge on a metal tank or pipe to resist the flow of electrons from the objects metal atoms into the surrounding soil.

**Incident Command System** – ICS; a Hazwoper requirement, it is a structured method of managing any emergency, the intent of which is to minimize the number of bosses and maximize the number of workers, which helps keep communications and information moving in the right directions.

**Incineration** – Treatment technology involving destruction of waste by controlled burning at high temperatures.

**Incinerator** – A furnace for burning wastes under controlled conditions.

**Indoor Air Pollution** – Chemical, physical, or biological contaminants in indoor air.

**Industrial Hygiene** – The application of scientific principles to reduce occupational exposures. The basic goals of industrial hygiene are the recognition, evaluation, and control of physical, chemical, and biological hazards in the work place.

**Industrial Hygienist** – IH; a person trained to practice industrial hygiene responsibilities.

**Infiltration** – The penetration of water through the ground surface into sub-surface soil.

**Influent** – Water, wastewater, or other liquid flowing into a reservoir, basin, or treatment plant.

**Initiative Process** – After obtaining the support of a minimum percentage of the voting public, it allows the public to place local or state measures directly on the ballot for a vote.

**Injection Well** – A well into which fluids are injected for waste disposal, to improve the recovery of crude oil, or for solution mining.

**Inorganic Compounds** – Compounds that do not contain carbon (with some exceptions, e.g. carbon dioxide, carbonates, and cyanides).

**Insecticides** – A subgroup of pesticides, they are substances that kill insects.

**Inspection and Maintenance Program** – Periodic auto inspection programs, usually done once a year or once every two years, that are required in some polluted areas.

**Inspection Warrant** – A legal document, obtained from a judge, that permits an enforcement agency reasonable access to observe the operations, permits, records, take samples and question employees of an uncooperative owner/operator.

**Interaction** – The reaction between the toxicant and other chemicals that may combine with or alter its behavior.

**Interim (Permit) Status** – Period during which treatment, storage, and disposal facilities (TSDFs) regulated under RCRA in 1980 are temporarily permitted to operate while awaiting denial or issuance of a permanent permit. Interim-status permits also are called Part A permits.

**Ion** – A form of an atom in which the number of protons in the nucleus no longer equals the number of electrons outside its nucleus. It it contains more electrons, it is negatively charged (anion) and if it has less electrons is positively charged (cation).

**Ionic Bond** – The attractive forces between oppositely charged ions. For example, the forces between the sodium and chloride ions in a sodium chloride crystal.

**Ionization** – The gain or loss of electrons from an atom. It results in the conversion of an atom into an ion.

**Ionizing radiation** – Any electromagnetic or particulate radiation capable of producing ions, directly or indirectly, in its passage through matter.

**ISO 14000** – Am emerging single environmental standard that will help a company meet its environmental goals and regulation regardless of its location. The standard is designed to eliminate potential non-tariff trade barriers resulting from multiple and sometimes conflicting environmental regulation and standards found throughout the world.

**Isotopes** – Two or more atoms of the same element that differ in the number of neutrons, e.g., carbon-12 and carbon-14.

**Joint and Several Liability** – Liability is shared among a group of people collectively and also individually.

**Kinetic Molecular Theory** – The theory that states that all matter is composed of particles that are in constant, chaotic motion.

**Labeling** – A generic term that includes labels, markings, or placarding.

**Labels** – Visual indications placed on a hazardous material container to identify the type of hazard posed by the hazardous material.

**Laboratory Safety Standard** – A more specific hazard communication program for laboratories found in 29 CFR 1910.1450. These regulations are essentially a blend of hazard communication and emergency response for laboratories. The cornerstone of the Lab Safety Standard is the requirement for a written Chemical Hygiene Plan.

**LAER** – See Lowest Achievable Emission Rate.

**Lagoon** A shallow pond where sunlight, bacterial action, and oxygen work to purify wastewater; also used to store wastewaters or spent nuclear fuel rods.

**Land Application** – Discharge of wastewater onto the ground for treatment or reuse.

**Land Ban** – The 1984 RCRA amendments mandated that by May 1990 all untreated hazardous waste must be banned from land disposal. The treatment standards and concentration levels were implemented in thirds beginning in November 1986.

**Landfarming of Waste** – A disposal process in which hazardous waste deposited on or in the soil is naturally degraded by microbes.

**Landfills** – Sanitary landfills are land disposal sites for non-hazardous solid wastes where waste is spread in layers, compacted to the smallest practical volume, and cover material is applied at the end of each operating day. Secure hazardous waste landfills are disposal sites permitted for hazardous waste. Both types of landfill are selected and designed to minimize the chance of hazardous substance releases into the environment.

**Latency Period** – The delay between the beginning of exposure and the resultant harmful effect(s).

**LC$_{50}$** – Lethal Concentration 50 per cent is the equivalent of LD50, but for toxicity from inhalation. It is the dose level at which 50% of the test animals died in a given amount of time.

**LD$_{50}$** – Lethal Dose 50 is a common way to report dose-response results. It is the dose level at which 50% of the test animals died in a given amount of time.

**Leachate** – The liquid that results when water moves through any non-water media and collects contaminants.

**LEL** – See Lower Explosive Limit.

**LEPC** – See Local Emergency Planning Committee.

**Liquid** – One of the three states of matter. Liquids are characterized as having a definite volume, but an indefinite shape.

**Listed Waste** – Waste listed as hazardous under RCRA or by equivalent state environmental regulations.

**LOAEL** – See Lowest Observable Adverse Effect Level.

**Lobbyist** – A person hired by and representing a special interest group that is attempting to affect the vote on a bill.

**Local Effect** – The effect of a toxicant occurs at the point where it first comes into contact with the body.

**Local Emergency Planning Committee** – A committee appointed by the state emergency response commission, as required by SARA Title III, to formulate a comprehensive emergency plan for its jurisdiction.

**Low-Level Waste** – Radioactive waste not classified as high-level radioactive waste, transuranic waste, spent nuclear fuel, or byproduct material as defined in the Atomic Energy Act.

**Lower Explosive Limit** LEL; the minimum concentration of a flammable gas in air required for ignition in the presence of an ignition source. It is listed as a percent by volume in air.

**Lowest Achievable Emission Rate** – Under the Clean Air Act, this is the rate of emissions that reflects the most stringent emission limit that is contained in the implementation plan of any state for the source unless the owner or operator of the proposed source demonstrates the limits are not achievable. It also means the most stringent emissions limit achieved in practice, whichever is more stringent.

**Lowest Observable Adverse Effect Level** – LOAEL; for a particular substance, the lowest amount of the substance that causes a harmful effect.

**MACT** – See Maximum Achievable Control Technology.

**Major Stationary Sources** – Used to determine applicability of Prevention of Significant Deterioration (PSD) and new source regulations. In a nonattainment area, any stationary pollutant source that has a potential to emit more than 100 tons per year is considered a major stationary source. In PSD areas, the cutoff level may be either 100 or 250 tons, depending on the type of source.

**Malignant Tumors** – A malignant tumor tends to invade surrounding tissue and spreads from its site of origin to additional distant sites elsewhere in the body.

**Marine Pollutant** – A material listed in Appendix B of the Hazardous Materials Table with a concentration that equals or exceeds 10% by weight, or 1% by weight for materials identified as severe marine pollutants.

**Marine Protection, Research, and Sanctuaries Act** – Also known as the Ocean Dumping Act, it regulates what can be dumped into the ocean in order to protect the marine environment.

**Market-based Approach** – The 1990 Clean Air Act incorporated program flexibility so that air pollution clean up could be accomplished as efficiently and inexpensively as possible.

**Markings** – Other informational items that are required on a container of hazardous materials such as the proper name, name of manufacturer, and instructions/cautions.

**Mass Number** – The sum of the protons and neutrons in the nucleus of an atom.

**Material Safety Data Sheet** – MSDS; chemical information sheets provided by the chemical manufacturer that include information such as: chemical and physical characteristics; long and short term health hazards; spill control procedures; personal protective equipment to be used when handling the chemical; reactivity with other chemicals; incompatibility with other chemicals; and manufacturers name, address and phone number. Employee access to and understanding of MSDSs are important parts of the HazCom Program.

**Maximum Achievable Control Technology** – MACT is the control technology which can be used effectively on a specific piece of equipment to reduce the maximum amount of emissions.

**Maximum Contaminant Levels** – MCLs; the maximum levels of certain contaminants allowed by the SDWA in drinking water from public systems. Under the 1986 amendments the EPA has set numerical standards or treatment techniques for an expanded number of contaminants.

**MCLs** – See Maximum Contaminant Levels.

**Measuring System** – An agreed upon set of units for expressing length, volume, weight and time.

**Media** – Specific environments, including air, water, or soil, that are subject to regulatory activities.

**Medical Monitoring** – An initial medical exam of a worker followed by periodic exams. The purpose of medical monitoring is to assess workers health, determine fitness to wear personal protective equipment, and maintain records of a persons health.

**Metamorphic Rocks** – One of the three types of rock that comprise the crust of the earth. These rocks were originally sedimentary or igneous rocks that were modified by temperature, pressure and chemically-active fluids.

**Metric System** – Introduced in the 1790s, it is a decimal system of units based on the meter as a unit length, the kilogram as a unit mass, and the second as a unit time.

**Metric ton** – 1,000 kilograms; about 2,200 pounds.

**Microgram per Gram** ($\mu$g/g) – One part per million (ppm)

**Microgram per Liter** ($\mu$g/l) – One part per billion (ppb)

**Milligram per Liter** (mg/l) – One per part million (ppm)

**Misdemeanor** – A less serious crime, punishable by a maximum sentence of a fine and/or up to one year in jail.

**Mitigation** – Measures taken to reduce adverse impacts on the environment.

**Mobile Sources** – Anthropogenic (man-made) sources of air pollution generated by moveable devices, such as cars, trucks, and buses.

**Moderator** – A substance that has little tendency to absorb, but are used to reduce the speed of neutrons given off by radioactive isotopes.

**Molecule** – The smallest part of a covalently bonded compound, e.g., $H_2O$ or $H_2SO_4$

**Monitor** – Another word for measure.

**Monitoring** – Periodic or continuous surveillance or testing to determine the level of compliance with statutory requirements and/or pollutant levels in various media or in humans, animals, or other living things.

**Monitoring Wells** – Wells drilled at a hazardous waste management facility or Superfund site to collect groundwater samples for the purpose of physical, chemical, or biological analysis to determine the amounts, types, and distribution of contaminants in the groundwater beneath the site.

**Montreal Protocol** – The agreement negotiated in September, 1987 by 24 nations to reduce the most widely used and most detrimental of the CFCs. The effective date was January 1, 1989.

**MSDS** – See Material Safety Data Sheet.

**Mutagen** – A chemical that has the ability to damage genetic material.

**Mutagenicity** – The tendency for genetic mutations to occur.

**Mutations** – Changes in the genetic information of a living cell that can be transmitted to offspring and reproduced in future generations.

**NAAQS** – See National Ambient Air Quality Standards.

**National Ambient Air Quality Standards** – NAAQS; a part of the Clean Air Act that sets standards for the criteria pollutants in outsice air throughout the country.

**National Contingency Plan** – NCP; required by regulations, this plan deals with how the EPA will use its authority and expend Superfund moneys. It actually details the process for investigating sites and cleaning them up.

**National Emissions Standards for Hazardous Air Pollutants** – NESHAPs; emissions standards set by EPA for an air pollutant not covered by NAAQS that may cause an increase in deaths, or in serious, irreversible, or incapacitating illness.

**National Environmental Policy Act** – NEPA; a federal law that states, as a national policy, our intent to achieve productive and enjoyable harmony between the activities of humans and the environment. It authorized a broad policy requiring that environmental consequences be considered whenever the federal government engaged in any activities that could have a negative impact to the environment.

**National Institute of Occupational Safety and Health** – NIOSH; the primary federal agency that conducts research to eliminate workplace hazards.

**National Pollutant Discharge Elimination System** – NPDES; a requirement of the CWA that discharges meet certain requirements prior to discharging waste to any water body. It sets the highest permissible effluent limits, by permit, prior to making any discharge.

**National Priorities List** – EPAs list of the most serious uncontrolled or abandoned hazardous waste sites identified for possible long-term remedial action under Superfund. The NPL is updated periodically and published in the Federal Register.

**National Priorities List** – NPL; currently a list of over 1,100 sites that have been evaluated and prioritized for the level of risk they present to public health and the environment. EPA can only take remedial actions at hazardous waste sites that are on the NPL.

**National Response Center** – The federal operations center that receives notifications of all releases of oil and hazardous substances into the environment. The center is operated by the U.S. Coast Guard.

**NCP** – See National Contingency Plan.

**Near Coastal Water Initiative** – This initiative was developed in 1985 to provide for management of specific problems in waters near coastlines that are not dealt with in other programs.

**Nematocides** – A subgroup of pesticides, they are substances that kill nematodes.

**NEPA** – See National Environmental Policy Act.

**NESHAP** – See National Emissions Standards for Hazardous Air Pollutants.

**Neutralization** – The process of combining acidic and basic substances to produce a nonhazardous (or reduced hazard) substance. This is accomplished by either adding alkaline substances to an acid or acidic substances to a base.

**Neutrons** – Particles having no charge that are found in the nucleus of an atom.

**New Source** – Any stationary source that is built or modified after a regulation has been published that prescribes a standard of performance intended to apply to that type of emissions source.

**New Source Performance Standards** – Uniform national EPA air emission and water effluent standards that limit the amount of pollution allowed from new sources or from existing sources that have been modified.

**Niche** – The effects an organism has on its surroundings and how the surroundings affect the organism.

**NIOSH** – See National Institute of Occupational Safety and Health.

**Nitrogen Cycle** – The circulation of nitrogen through plants and animals and back to the atmosphere.

**Nitrogen Oxides (NOx)** – One of the products of combustion of hydrocarbon fuels. Also one of the precursors, along with the volatile organic compounds that undergo photochemical reactions and produce ozone.

**No Observable Adverse Effect Level** – NOAEL; for a particular substance, the dose of the substance that produces no harmful effect.

**NOAEL** – See No Observable Adverse Effect Level.

**Nonattainment** – Areas, identified by the NAAQS, which fail to meet national standards for criteria pollutants. The federal government can impose sanctions on those areas to gain compliance.

**Nonattainment Pollutants** – Pollutants for which the California Air Resources Board has determined the air quality standard of having been reduced by five percent has not been met.

**Nonbiodegradable** – A term which means that it does not break down easily in the environment.

**Nonionizing Radiation** – Lower energy radiations, e.g., heat, light and radio waves, that do not cause ionization to occur.

**Nonpoint Sources** – Sources of water pollution that do not have a specific point of origin, such as rainwater carrying topsoil and chemicals from a field.

**Nonpolar Covalently Bonded** – A molecule composed of atoms that share their electrons equally, resulting in a molecule that does not have polarity.

**NPDES** – See National Pollutant Discharge Elimination System.

**NPL** – See National Priorities List.

**NPS** – See Nonpoint Sources.

**NRC** – See Nuclear Regulatory Commission.

**NSPS** – See New Source Performance Standards.

**Nuclear Fission** – A nuclear reaction in which an atomic nucleus, especially a heavy nucleus such as an isotope of uranium, splits into two fragments, several subatomic particles and a large amount of energy.

**Nuclear Regulatory Commission** – U.S. NRC; successor to the US Atomic Energy Commission.

**Nucleus** – The central part of an atom, made up of protons and neutrons.

**Occupational Safety and Health Act** – OSHA; a federal law passed in 1970 to assure, so far as possible, every working man and woman in the nation safe and healthful working conditions. To achieve this goal, the Act authorizes several functions such as encouraging safety and health programs in the workplace and encouraging labor-management cooperation in health and safety issues.

**Octet Theory** – A theory developed in 1916 suggesting that atoms were most stable when they had eight electrons in their outermost energy level.

**Offset** – A method used in the 1990 CAA to give companies that own or operate large sources in nonattainment areas flexibility in meeting overall pollution reduction requirements when changing production processes. If the owner or operator of the source wishes to increase release of a criteria air pollutant, an offset, which is a reduction of a somewhat greater amount of the same pollutant, must be obtained either at the same plant or by purchasing offsets from another company.

**Oil Spill** – An accidental or intentional oil discharge that reaches bodies of water. It may be controlled by chemical dispersion, combustion, mechanical containment, or adsorption.

**Ordinances** – Laws passed by local units of government.

**Organic Compounds** – Compounds that contain carbon (with some exceptions, e.g. carbon dioxide, carbonates, and cyanides).

**Organism** – Living things ranging from a bacteria in the soil, to a plant, to an animal.

**OSHA** – See Occupational Safety and Health Act.

**Oxidation** – When a substance either gains oxygen, loses hydrogen or electrons in a chemical reaction. One of the chemical treatment methods.

**Oxidizer** – Also known as an oxidizing agent, it is a substance that oxidizes another substance. Oxidizers are a category of hazardous materials that may assist in the production of fire by yielding oxygen readily.

**Oxygen Cycle** – The circulation of oxygen through the various environmental compartments. It is closely tied to the carbon cycle.

**Oxygen Deficient Atmospheres** – The legal definition of an atmosphere where the oxygen concentration is less than 19.5% by volume of air.

**Oxygenated Fuel** – A special type of gasoline, which burns more completely than regular gasoline in cold start conditions. More complete burning results in reduced production of carbon monoxide, a criteria air pollutant. In some parts of the country, carbon monoxide release from cars starting up in cold weather makes a major contribution to pollution. In these areas, gasoline refiners must market oxygenated fuels, which contain a higher oxygen content than regular gasoline.

**Paragraph** – In regulations, the division referring to the paragraphs within a section.

**Part** – A part is a body of regulations about the same topic within a title.

**Particulate Matter** – Substances such as diesel soot and combustion products resulting from the burning of wood that are released directly into the air. It can also be produced through various photochemical reactions.

**Pathogens** – Disease-causing organisms.

**PCBs** – See Polychlorinated Biphenyls.

**PEL** – See Permissible Exposure Limit.

**Performance Standards** – A form of OSHA regulation standard that lists the ultimate goal of compliance, but does not explain exactly how compliance is to be accomplished.

**Permeability** – The degree to which liquid can pass through a substance or mass.

**Permissible Exposure Limit** – PEL; a time weighted average concentration of an airborne contaminant that a healthy worker may be exposed to 8-hours per day or 40-hours per week without suffering any adverse health effects. It is established by legal means and is enforceable by OSHA.

**Permit Fees** – The fees a business are required to pay under the 1990 Clean Air Act to obtain air quality permits. The fee money collected helps to pay for administration and enforcement of the Act.

**Persistent** – When released, chemicals that tend to stay intact, possessing the same dangerous properties for long periods of time.

**Pesticides** – A substance defined in FIFRA as any substance or mixture of substances intended for preventing, destroying, repelling, or mitigating any pest or any substance or mixture of substances intended for use in a plant regulator, defoliant or desiccant.

**pH** – A measure of the acidity or alkalinity of a solution on a scale of 0 to 14 (low is acid; high is alkaline or base; and 7 is neutral).

**Photochemical Reactions** – A term used to describe the light driven chemical reactions that convert air pollutants into ozone.

**Physical Properties** – A unique set of properties that describes a substances boiling and melting points, color, odor, vapor pressure, density, etc.

**Physical States** – There are three physical states of matter, solids, liquids and gasses.

**Physical Treatment** – Any process that does not produce a new substance. The most common physical wastewater treatment methods are: screening, adsorption, aeration, flocculation, sedimentation, and filtration.

**Placards** – Similar to labels, except they are much larger (12) and are placed on the front, rear, and both sides of a vehicle transporting the hazardous material.

**Plume** – The shape of the contaminated area.

**Plutonium** – A heavy element (Atomic No. 94), which comprises about 1 percent of spent nuclear fuel from commercial reactors. One of its principle isotopes in fission is Pu-239 which has a half-life of 24,000 years.

**PM-10** – Particulate matter having a diameter of 10 microns or less.

**Point Sources** – Sources of water pollution that have identifiable single source origins, such as a discharge pipe.

**Polar Covalent Bond** – The shared pair of electrons between two atoms are not being equally held. This results in one of the atoms becoming slightly positively charged and the other atom becoming slightly negatively charged.

**Polar Covalent Molecule** – The result of one or more polar covalent bonds and result in a molecule that is polar covalent. Polar covalent molecules exhibit partial positive and negative poles, causing them to behave like tiny magnets. Water is the most common polar covalent substance.

**Pollutant** – Generally, any substance introduced into the environment that adversely affects the usefulness of the resource.

**Pollution** – Generally, the presence of matter or energy whose nature, location, or quantity produces undesired environmental effects. Under the Clean Water Act, for example, the term is defined as a man-made or man-induced alteration of the physical, biological, and radiological integrity of water.

**Polychlorinated Biphenyls** – PCBs; a group of organic compounds containing chlorine that were once widely used as liquid coolants and insulators in industrial equipment, especially in power transformers. They are particularly persistent in the environment.

**Population** – Members of the same species sharing a habitat.

**Porosity** – The percentage of the total volume of the material that is occupied by pores or other openings.

**Post Closure** – The time period following the shutdown of a waste management or manufacturing facility. For monitoring purposes, this is considered to be 30 years.

**Potentially Responsible Party** – Any individual or company, including owners, operators, transporters or generators, potentially responsible for, or contributing to, the contamination problems at a Superfund site. Whenever possible, EPA requires PRPs, through administrative and legal actions, to clean up hazardous waste sites they have contaminated.

**Potentiation** – This is where the presence of one substance increases the effect of another substance to a level more than it would have caused by itself.

**POTWs** – See Publicly Owned Treatment Works.

**PPM/PPB** – Parts per million/parts per billion, a way of expressing very small concentrations of pollutants in air, water, soil, human tissue, food, or other products.

**Precedents** – Earlier written decisions of courts that are used as patterns for making decisions on similar cases.

**Preliminary Assessment** – A quick analysis to determine how serious the situation is and to identify all potentially responsible parties. The preliminary assessment utilizes readily available information, for instance, from records, aerial photographs, and personnel interviews.

**Pretreatment** – The practice of industry removing toxic pollutants from their wastewaters before they are discharged into a municipal wastewater treatment plant.

**Primary Drinking Water Standards** – Those drinking water standards, outlined in the SDWA, that affect public health.

**Primary Pollutant** – Pollutants, such as the oxides of carbon and nitrogen and volatile organic compounds, that enter the atmosphere directly.

**Primary Standard** – A pollution limit for criteria pollutants that is based on their health effects.

**Primary Treatment** – The first step of treatment at a municipal wastewater treatment plant. It typically involves screening and sedimentation to remove the materials that float or settle.

**Privacy Act of 1974** – Requires that federal agencies provide individuals with information pertaining to them that is kept in governmental files. The government is required to amend or correct inaccurate files.

**Private Law** – One of the two main divisions of law; often referred to as civil law. The governments role in private law is to act as judge when one party claims injury by another party.

**Producer** – An organism, like a plant, that can make their own food by using a process like photosynthesis.

**Protons** – Particles having a positive charge that are found in the nucleus of an atom.

**PRP** – See Potentially Responsible Party.

**Public Law** – One of the two main divisions of law; often referred to as criminal law. Public Law includes laws describing the basic rules of our system of government, criminal acts against society, the operation and establishment of administrative agencies, and court cases.

**Publicly Owned Treatment Works** – A waste treatment works owned by a state, local government unit, or Indian tribe, usually designed to treat domestic wastewaters.

**RA** – See Removal Actions.

**Rad** – See Radiation Absorbed Dose.

**Radiation** – The Energy and/or particles released from radioactive decay.

**Radiation Absorbed Dose** – Measurement of the amount of energy deposited by ionizing radiation in any material including people, animals and objects.

**Radioactive Decay** – Spontaneous decay or disintegration of an unstable atomic nucleus, accompanied by the emission of ionizing radiation.

**Radioactive Material** – A material that emits harmful radiation.

**Radioisotopes** – See Radionuclides.

**Radionuclide** – An unstable radioactive isotope that decays toward a stable state at a characteristic rate by the emission of ionizing radiation.

**RAP** – See Remedial Action Plan.

**Raw Sewage** – Untreated wastewater.

**RCRA** – See Resource Conservation and Recovery Act.

**Reaction** – The interaction between the toxicant and other chemicals that may combine with or alter its behavior.

**Reactive** – A substance that reacts violently by catching on fire, exploding, or giving off fumes when it is exposed to water, air, or low heat.

**Reactivity Hazard** – The ability of a material to release energy when in contact with water. It can also be the tendency of a material, when in its pure state or as a commercially produced product, to vigorously polymerize, decompose or condense, or otherwise self-react and undergo violent chemical change.

**Receiving Waters** – A river, lake, ocean, stream, or other water source into which wastewater or treated effluent is discharged.

**Recharge** – The process by which water is added to a zone of saturation, usually by percolation from the soil surface.

**Reclamation** – Considered to be a part of recycling, it recovers valuable raw material for reuse.

**Recommended Exposure Limits** – RELs; workplace exposure limits recommended for use by NIOSH. RELs are reviewed by OSHA as input for revising the PELs.

**Recycle** – The process of using waste as raw material for new products that may or may not resemble the original material.

**Reduction** – The process of a substance gaining either hydrogen or electrons, or losing oxygen.

**Reference Dose** – RfD; an estimate of the amount of a chemical that a person can be exposed to on a daily basis that is not anticipated to cause adverse systemic health effects over the persons lifetime.

**Reformulated Gasoline** – Specially refined gasoline with low levels of smog-forming VOCs and low levels of hazardous air pollutants.

**Regulated Substances** – Substances defined as hazardous under the CERCLA of 1980 (with the exception of hazardous wastes) and petroleum. They are now found in Subtitle I of RCRA, as it applies to Federal underground storage tanks.

**RELs** – See Recommended Exposure Limits.

**Rem** – Radiation Equivalent Man; The unit of dose equivalence commonly used in the U.S. In most countries the rem has been replaced by the sievert, which is the unit of dose equivalent in the International System of Units. A sievert is equal to 100 rem.

**Remedial Action Plan** – RAP; a plan recommending the actual cleanup procedure at a CERCLA site. Usually, more than one method is used to perform cleanup. Before this plan is approved, it must be made available to other agencies, the responsible parties, and the public for **a minimum 30 day comment period.**

**Remedial Actions** – Authorized under CERCLA, these actions represent the final remedy for a site and generally are more expensive and longer in duration than removals. Remedial actions are intended to provide permanent solutions to hazardous substance threats.

**Remedial Investigation** – RI; a detailed investigation of a Superfund site. It entails collecting and analyzing data about the waste. Soil, ground water, surface water, or living organisms such as fish may be sampled in the contaminated area. The study provides all the information necessary to characterize a site.

**Remedial Investigation/Feasibility Study** – RI/FS; the investigative and sampling work at a Superfund site, results of which are compiled into a report that is used to recommend a final cleanup plan.

**Removal Actions** – RAs; actions at CERCLA cleanups that are short-term and usually address cleanup problems only at the surface of a site. These actions are limited to 12 months duration or $2 million in expenditures, although in certain cases these limits may be extended.

**Reportable Quantity** – RQ; the minimum amount of a hazardous material that, if spilled while in transport, must be reported immediately to the National Response Center. Minimum reportable quantities range from 1 pound to 5,000 pounds per 24 hour day.

**Repository** – A facility for the permanent deep geologic disposal of high-level radioactive waste and spent nuclear fuel. It includes both surface and subsurface areas where high-level radioactive waste and spent nuclear fuel handling activities are conducted.

**Reprocessing** – Recovery of fissile and/or fertile material from irradiated nuclear fuel by chemical separation from fission products and other radionuclides.

**Resource Conservation and Recovery Act** – RCRA; a federal law enacted in 1976 to deal with both municipal and hazardous waste problems and to encourage resource recovery and recycling.

**Reuse** – The process of minimizing the generation of waste by recovering usable products that might otherwise become waste.

**Reuse material** – A recovered material that is sold to another entity rather than being used in the facility in which it was generated.

**Reverse Osmosis** – Solutions of differing ion concentrations are separated by a semipermeable membrane. Typically, water flows from the chamber with lesser ion concentration into the chamber with the greater ion concentration, resulting in hydrostatic or osmotic pressure. In RO, enough external pressure is applied to overcome this hydrostatic pressure, thus reversing the flow of water. This results in the water on the other side of the membrane becoming depleted in ions or demineralized.

**RfD** – See Reference Dose.

**RI** – See Remedial Investigation.

**RI/FS** – See Remedial Investigation/Feasibility Study.

**Risk Assessment** – A process that uses scientific principles to determine the level of risk that actually exists in a contaminated area.

**Risk Characterization** – The final step in the risk assessment process, it involves determining a numerical risk factor. This step ensures that exposed populations are not at significant risk.

**Risk Management** – A political rather than scientific decision on how society chooses to minimize or control its risks.

**RO** – See Reverse Osmosis.

**Rodenticides** – A subgroup of pesticides, they are materials that kill rodents.

**RQ** – See Reportable Quantity.

**Runoff** – That part of precipitation, snow melt, or irrigation water that runs off the land into streams or other surface water. It may carry pollutants from the air and land into the receiving waters.

**Sacrificial Anode** – A metal anode, usually Mg or Zn, that is attached to a metal tank or pipe to resist the flow of electrons from the objects metal atoms into the surrounding soil.

**Safe Drinking Water Act** – The SDWA; a federal law passed in 1974 with the goal of establishing federal standards for drinking water quality, protecting underground sources of water, and setting up a system of state and federal cooperation to assure compliance with the law.

**Safety Factor** – Based on experimental data, it is the amount added, e.g., 1000-fold, to insure worker health and safety.

**SARA** – See Superfund Amendments and Reauthorization Act.

**Saturated Zone** – The layer below the vadose or unsaturated zone where all interconnected openings are full of water.

**Screening** – A physical treatment method. The use of a devise that has a predetermined hole size to selectively restrain larger particles and allow smaller particles to pass through.

**Scrubber** – A device installed in an incinerators stack to purify incinerator gases.

**SDWA** – See Safe Drinking Water Act.

**Secondary Containment** – A method using two containment systems so that if the first is breached, the second will contain all of the fluid in the first. For USTs, secondary containment consists of either a double-walled tank or a liner system.

**Secondary Drinking Water Standards** – Those drinking water standards, outlined in the SDWA, that affect public acceptance and aesthetic qualities of drinking water.

**Secondary Pollutant** – Pollutants, such as ozone, that are formed from other substances released into the atmosphere.

**Secondary Standard** – A pollution limit based on environmental effects such as damage to property, plants, visibility, etc. Secondary standards are set for criteria air pollutants.

**Secondary Treatment** – The second step of treatment at a municipal wastewater treatment plant. This step uses growing numbers of microorganisms to digest organic matter and reduce the amount of organic waste. Water leaving this process is chlorinated to destroy any disease-causing microorganisms before its release.

**Section** – Identified by the symbol, §, the basic designation for the next level below a part of a regulation.

**Sedimentary Rocks** – One of the three types of rock that comprise the crust of the earth. They are the result of the weathering of preexisting rocks.

**Sedimentation** – A physical treatment method. A process used to separate liquids from solids.

**Sensitizers** – Chemicals that in very low dose trigger an allergic response.

**SERC** – See State Emergency Response Commission.

**Sewage** – The waste and wastewater produced by residential and commercial establishments and discharged into sewers.

**Shipping Papers** – Documents that must accompany transported hazardous materials and contain a certification by the shipper that the material is offered for transport in accordance with applicable DOT regulations.

**Short Term Exposure Limit** – STEL; the time weighted average concentration to which workers can be exposed continuously for a short period of time (typically 15 minutes) without suffering: irritation, chronic or irreversible tissue damage, or impairment for self-rescue.

**SIP** – See State Implementation Plan.

**Site Investigation** – Part of the site cleanup process designed to provide additional site-specific information. The site investigation usually entails sampling. This information combined with the information from the preliminary assessment will allow ranking of the site using the Hazard Ranking System.

**Slip Law** – The first officially published, separately unbound pamphlet of a statute.

**Sludge** – The semi-solid to solid residue left from wastewater treatment.

**Small Quantity Generator** – A facility generating less than 1,000 kilograms of hazardous waste a month.

**Smog** – The origin of the word is credited to Dr. H.A. des Voeux who in the early 1900s blended the words smoke and fog to describe Londons foggy, dirty air.

**Soil** – The thin layer of the earths crust that we live on and that has been affected by weathering and decomposition of organisms.

**Soil Profile** – A recording of the depth and thickness of the various layers of the earth.

**Solid** – One of the three states of matter. A solid is characterized has having a definite shape and volume.

**Solid Waste** – As defined in RCRA the term means any garbage, refuse, sludge from a waste treatment plant, water supply treatment plant, or air pollution control facility and other discarded material, including solid, liquid, semisolid, or contained gaseous material resulting from indus-trial, commercial, mining, and agricultural operations, and from community activities, but does not include solid or dissolved material in domestic sewage, or solid or dissolved materials in irrigation return flows or industrial discharges which are point sources subject to permits under the Clean Water Act, or special nuclear or byproduct material as defined by the Atomic Energy Act of 1954.

**Solute** – The substance that is dissolved into another substance, and is usually present in smaller concentration.

**Solvated** – When either a positive or negative ion becomes completely surrounded by polar solvent molecules

**Solvent** – The substance in which another substance is dissolved; usually present in the greater concentration.

**Source or Waste Reduction** – The terms mean decreasing the generation of waste to avoid its handling, treatment, or disposal. Waste or source reduction includes in-plant practices that reduce or avoid the generation of waste.

**SPCC** – See Spill Prevention Control and Countermeasures Plan.

**Species** – A group of organisms able to interbreed freely with one another under natural conditions and produce offspring.

**Specific Gravity** – A ratio of the densities of a substance to water.

**Specific Hazard** – A special hazard concerning a particular product or chemical, and not covered by the other labeled hazard code items. The special hazard should be written in the white bottom diamond of the NFPA 704 system signs.

**Specific Standards** – OSHA regulatory standards that explain exactly how to comply.

**Spent Nuclear Fuel** – Irradiated fuel element not intended for further reactor service.

**Spill Prevention Control and Countermeasures Plan** – Plan covering responses to the release of hazardous substances as defined by the Clean Water Act.

**SQG** – See Small Quantity Generator.

**State Emergency Response Commission** – SERC; state commissions required under SARA, which generally includes representatives of public agencies and departments with expertise in environmental issues, natural resources, emergency services, public health, occupational safety, and transportation.

**State Implementation Plan** – A detailed description of the programs a state will use to carry out its responsibilities under the CAA.

**Stationary Sources** – Anthropogenic (man-made) air pollution sources generated by fixed facilities such as oil refineries or dry cleaning establishments.

**Statutes** – Also referred to as laws or acts. Statutes are the legal standards that govern society.

**Statutory Law** – Statutory law is composed of written laws, the exact wording of which has been approved by a federal, state or local legislative body.

**STEL** – See Short Term Exposure Limit.

**Succession** – The progressive changes in an ecosystem that eventually result in the establishment of a stable community.

**Sump** – A pit or tank that catches liquid runoff for drainage or disposal.

**Superfund** – Commonly used to refer to the program operated under the legislative authority of CERCLA that funds and carries out the EPA solid waste emergency and long-term removal remedial activities. These activities include establishing the National Priorities List, investigating sites for inclusion on the list, determining their priority level on the list, and conducting and/or supervising the ultimately determined cleanup and other remedial actions. More specifically, the Superfund is the remediation fund for these cleanups, funded by taxes on chemical feedstocks and petroleum products. Many states also have Superfund laws.

**Superfund Amendments and Reauthorization Act** — SARA of 1986; This amendment extended CERCLA for five years, increased by five-fold the amount of money in the fund, and added new community right to know requirements.

**Surface Tension** – The attractive forces exerted by the molecules below the surface upon those at the surface, resulting in them crowding together and forming a higher density.

**Surface Water** – All water naturally open to the atmosphere, and all springs, wells, or other collectors that are directly influenced by surface water.

**Sustainable Development** – The management of resources in a way that enable people living today to meet their needs without jeopardizing the ability of the earths future inhabitants to meet theirs.

**Synergism** – An interaction of two or more chemicals where their total effect is greater than the sum of their individual effects.

**Systemic Effect** – When the effect of a toxicant occurs only after it enters the body and travels via the bloodstream to reach vital organs, the reproductive system, brain and/or the nervous system.

**TC$_{50}$** – Toxic Concentration 50 per cent is the concentration in inhaled air needed to produce an observed toxic effect in 50 percent of the test animals in a given time period.

**TD$_{50}$** – Toxic Dosage 50 percent is the dosage by any route other than inhalation that produces an observed toxic effect in 50 percent of the test animals in a given time period.

**Technology** – The practical application of scientific laws and principles.

**Temperature Inversion** – A reversal in the normal air temperature gradient, where air temperature typically decreases about 5.4 degrees Fahrenheit for every 1000 foot rise in altitude. When warm air is above the cooler air, it creates a lid that prevents the components of air pollution from escaping.

**Teratogen** – An agent that causes abnormal development in the fetus (birth defects).

**Teratogenicity** – The tendency for specific interference with development of the unborn child.

**Tertiary Treatment** – The third step, sometimes employed, at a municipal wastewater treatment plant. It consists of advanced cleaning, which removes nutrients and most BOD.

**Thermal Treatment** – Treatment using heat, the most common of which for hazardous wastes is incineration, which destroys or makes the waste less hazardous through burning. It may also include pyrolysis and wet oxidative treatment processes.

**Threshold Limit Value** – TLV; the same concept as a PEL, except that TLVs do not have the force of governmental regulations behind them. They are based on recommended limits established and promoted by the American Conference of Governmental Industrial Hygienists.

**Threshold Planning Quantity** – A quantity designed for each chemical on the list of extremely hazardous substances that triggers notification by facilities to the state emergency response commission that such facilities are subject to emergency planning under SARA Title III.

**Time-Weighted Average** – The TWA; a mathematical average [(exposure in ppm × time in hours) ¼ time in hrs. = time weighted average in ppm] of exposure concentration over a specific time.

**Title** – When used in describing regulations, the title encompasses the whole area of one topic, e.g. 49 CFR is about transportation.

**TLV** – See Threshold Limit Value.

**Total Suspended Solids** – Solids present in wastewater.

**Toxic** – Another word for poisonous. Toxic effects can range from a minor irritation of mucus membranes, to damage of an internal organs, cancer or death.

**Toxic Chemical Release Form** – Information form required to be submitted by facilities that manufacture, process, or use, in quantities above a specific amount, chemicals listed under SARA Title III.

**Toxic Substance** – A chemical or mixture that may present an unreasonable risk of injury to health or the environment.

**Toxic Substances Control Act** – TSCA; a federal act enacted in 1976, the purpose of which is to control the risks of chemical substances that enter commerce. The Act deals with two major kinds of problems. First, are the newly created chemicals or chemicals entering into commerce that may do serious damage to humans and the environment before their potential danger is known; and second, existing chemicals that may require more stringent control.

**Toxicant** – The term used for toxic substances that are made by humans.

**Toxicity** – Toxicity is established by doing a long-term, low dosage concentration study of the affects of a substance on humans and other animals.

**Toxicology** – The study of poisons, which are substances that cause harmful effects to living things.

**Toxin** – The term used for poisons that are produced by living organisms.

**TPQ** – See Threshold Planning Quantity.

**Transuranic Waste** – Waste material contaminated with plutonium and other elements having atomic numbers higher than 92. In the commercial fuel cycle, transuranic waste is produced primarily from the reprocessing of spent fuel and the manufacture of mixed uranium-plutonium fuel.

**Treatment, Storage, and Disposal Facility** – Site where a hazardous waste is treated, stored, or disposed. TSDFs are regulated by EPA and states under RCRA.

**Trophic Level** – The term used to describe each level of energy consumption within a food web. Energy within a food web always flows in only one direction, starting with the producers.

**Troposphere** – The layer, approximately 10 miles thick, that is closest to the earth. It is the layer where weather forms and occurs.

**TSCA** – See Toxic Substances Control Act.

**TSD Facility** – Treatment, Storage, and Disposal Facility; a facility that is permitted to treat, store or dispose of hazardous wastes.

**TSS** – See Total Suspended Solids.

**Turbidity** – A measure of the cloudiness of water caused by the presence of suspended matter, which shelters harmful microorganisms and reduces the effectiveness of disinfecting compounds.

**TWA** – See Time-Weighted Average.

**UEL** – See Upper Explosive Limit.

**Underground Storage Tank** – A tank located all or partially underground that is designed to hold gasoline or other petroleum products or chemical solutions.

**Uniform Hazardous Waste Manifest** – A document required by RCRA when shipping hazardous wastes. It lists EPA identification numbers of the shipper, carrier, and the designated treatment, storage, and disposal facility, in addition to the standard information required by DOT.

**Unit** – As used in a measuring system, it is the label that expresses the smallest fixed value being used as the standard. For example, in the Metric Measuring System length units are meters, mass units are grams and volume units are liters.

**Unsaturated Zone** – Also known as the vadose zone, it is the layer found immediately below the land surface that contains both water and air.

**Upper Explosive Limit** – UEL; the maximum concentration of a flammable gas in air required for ignition in the presence of an ignition source.

**Uranium** – A naturally occurring radioactive element with the atomic number 92 that has become the basic raw material of nuclear energy.

**UST** – See Underground Storage Tank.

**Vadose Zone** – Also known as the unsaturated zone, it is the layer found immediately below the land surface and contains both water and air.

**Vapor Density** – Also called vapor specific gravity, it refers to the ratio of the densities of a gas and air.

**Vapor Pressure** – A measurement of the amount of pressure that exists, at a given temperature, above a liquid in a closed container.

**Vapor Recovery Nozzles** – Special gas pump nozzles that will reduce release of gasoline vapor into the air when people put gas in their cars.

**VOCs** – See Volatile Organic Compounds.

**Volatile** – Description of any substance that evaporates easily.

**Volatile Organic Compounds** – An organic compound that readily evaporates under normal conditions or temperature and pressure. Examples of VOCs include: benzene, trichloroethylene, and vinyl chloride. Also included are the precursors, along with the oxides of nitrogen that undergo photochemical reactions and produce ozone.

**Waste** – Unwanted materials left over from a manufacturing process or refuse from places of human or animal habitation.

**Waste Minimization** – A broad umbrella term that includes waste reduction, recycling, and waste treatment such as incineration.

**Waste Treatment Plant** – A facility containing a series of tanks, screens, filters, and other processes by which pollutants are removed from water.

**Wastewater** – The spent or used water from individual homes, a community, a farm, or an industry that contains dissolved or suspended matter.

**Water Quality Standards** – State-adopted or EPA-approved standards for water bodies. The standards cover the use of the water body and the water quality criteria that must be met to protect the designated use or uses.

**Water Softening** – A chemical treatment method. A process that uses either chemicals to precipitate or a zeolite to remove those metal ions (typically $Ca^{2+}$, $Mg^{2+}$, $Fe^{3+}$) responsible for hard water.

**Water Solubility** – The maximum concentration of a chemical compound which may result when it is dissolved in water. If a substance is water soluble, it may readily disperse through the environment.

**Watershed** – The land area that drains into a river, river system, or other body of water.

**Well** – A bored, drilled, or driven shaft, or a dug hole, whose depth is greater than the largest surface dimension and whose purpose is to reach underground water supplies or oil, or to store or bury fluids below ground.

**Wellhead Protection** – The protection of the surface and subsurface areas surrounding a water well or wellfield supplying a public water system that may be contaminated through human activity.

**Wetlands** – An area that is regularly saturated by surface or groundwater and subsequently is characterized by a prevalence of vegetation that is adapted for life in saturated soil conditions.

# Acknowledgments

## Chapter 1

**Figure 1-1:** Industrial Plant. Photograph courtesy of U.S. Department of the Army.

**Figure 1-3:** Henry David Thoreau. Photograph from CORBIS-BETTMANN.

**Figure 1-4:** Rachel Carson. Photograph from THE BETTMANN ARCHIVE.

**Figure 1-5:** Love Canal Emergency Declaration Area. Source: USEPA.

**Figure 1-6:** World Population Growth. Source: United Nations Population Division, Department for Economic and Social Information and Policy Analysis.

**Figure 1-7:** Environmental Technician taking samples of the soil. Photograph courtesy of U. S. Department of the Army.

**Figure 1-8:** Students "suit up" in a typical Environmental Technology classroom. Photograph courtesy of OSHA.

**Figure 1-9:** EPA regulations impact agricultural practices. Photograph courtesy of William Schaller / USDA—Soil Conservation Service.

## Chapter 2

**Figure 2-1:** Code of Hammurabi. Photograph from THE BETTMANN ARCHIVE.

**Figure 2-6:** Daily Federal Register. Photograph from Tom McCarthy.

**Figure 2-7:** Contents of Federal Register. Photograph from Tom McCarthy.

**Figure 2-8:** Code of Federal Regulations. Photograph from Tom McCarthy.

**Figure 2-9:** Page from Federal Register. Photograph from Tom McCarthy.

**Figure 2-11:** Compliance investigator. Photograph courtesy of U.S. Department of the Army.

**Figure 2-14:** Clean-up in progress. Photograph courtesy of U.S. Department of the Army.

## Chapter 3

**Figure 3-1:** Poison Hemlock. Photograph from Virginia P. Weinland / Photo Researchers, Inc.

**Figure 3-2:** Paracelsus. Photograph from CORBIS-BETTMANN.

**Figure 3-3:** Catherine de Medici. Photograph from CORBIS-BETTMANN.

**Figure 3-8:** Comparison of absorption by skin at various sites of the body. Source: Adaptation by OCAW/Labor Institute from E. Hodgson and P.E. Levi, *A Textbook of Modern Toxicology*, 1987, p. 34-35.

**Figure 3-10:** Prescription Bottle Labels. Photograph from Mark Antman/The Image Works.

**Figure 3-12:** $LD_{50}$ values. Source: Sax, 1984 and EPA Training Manual.

**Figure 3-13:** Toxicity ratings for pesticides. Source: EPA Pesticide Regulations.

**Figure 3-14:** Commonly used toxicity rating system. Source: Thomas J. Haley contributed chapter in Irving Sax, *Dangerous Properties of Industrial Materials*, Sixth Edition, 1984.

**Figure 3-15:** Average number of years after exposure for cancer first to appear. Source: Levy and Wegman, 1983.

**Figure 3-18:** Body systems affected by toxic agents. Source: Division of Agricultural Science, University of California, *Toxicology: The Science of Poisons*, 1981.

**Figure 3-20:** The respiratory system. Source: Hesis.

**Figure 3-22:** Lung with emphysema. Photograph courtesy of Massachusetts Audobon Society.

**Figure 3-25:** The Mad Hatter. Photograph from CORBIS-BETTMANN.

**Figure 3-28:** Some known human carcinogens. Source: U.S. Department of Health and Human Services, National Toxicology Program, *Sixth Annual Report on Carcinogens*, 1991.

**Figure 3-29:** How mutations occur. Source: Hesis.

**Figure 3-30:** Beer label warning of reproductive harm. Photograph from Jim West.

**Figure 3-35:**
  (a) Person parachuting from airplane. Photograph from Ellis Herwig/Stock, Boston.
  (b) Student at school. Photograph from Myrleen Ferguson/PhotoEdit.
  (c) Person sleeping. Photograph from D. Young-Wolff MR/PhotoEdit.
  (d) People crossing street. Photograph from Tony Freeman/PhotoEdit.

## ■ Chapter 4

**Figure 4-2:** Earth. Photograph courtesy of NASA.

**Figure 4-18:**
  (a) White rhino. Photograph from Leonard Lee Rue III / Photo Researchers, Inc.
  (b) *Trillium reliquum*. Photograph from Jeff Lepore / Photo Researchers, Inc.
  (c) Blunt nosed leopard lizard. Photograph from Dell O. Clark / Photo Researchers, Inc.

**Figure 4-19:** EIS table of contents. Photograph from Tom McCarthy.

**Figure 4-20:** An Environmental Impact Statement. Photograph from Tom McCarthy.

**Figure 4-21:** Incinerators. Photograph from John Griffin / The Image Works.

## ■ Chapter 5

**Figure 5-2:** Air pollution affecting different body systems. Source: ARB.

**Figure 5-4:** Number of days when ozone levels exceeded 0.12 ppm. Source: ARB.

**Figure 5-7:** Discolored Leaves. Photograph courtesy of U.S. Forest Service, USDA.

**Figure 5-8:** Air Pollution in the home. Source: USEPA.

**Figure 5-10:** Smoggy day in Los Angeles. Photograph from AP / WIDE WORLD PHOTOS.

**Figure 5-11:** Pollutants and their sources. Source: Modified from the 1986 305(b) National Report.

**Figure 5-13:** Major sources of ground water contamination. Source: USEPA.

**Figure 5-14:** Underground water. Source: USGS.

**Figure 5-15:** Aquifers and confining beds. Source: USGS.

**Figure 5-18:** Waste generated from manufacturing. Source: USEPA.

## ■ Chapter 6

**Figure 6-1:** George Bush signing CAA. Photograph from UPI / CORBIS-BETTMANN.

**Figure 6-2:** Areas in nonattainment for Particulate Matter. Source: USEPA.

**Figure 6-19:** ROG emissions per person. Source: ARB.

**Figure 6-20:** Hydrocarbon emissions by trip. Source: ARB.

## ■ Chapter 7

**Figure 7-1:**

(a) Wetlands. Photograph courtesy of U.S. Department of the Army.

(b) Surface water pollution. Photograph courtesy of Soil Conservation Service, USDA.

(c) Child at drinking fountain. Photograph from D. Young Wolff MR / PhotoEdit.

**Figure 7-3:** Example effluent limitations. Source: 40 CFR.

**Figure 7-4:** Examples of NPDES monitoring requirements. Source: 40 CFR.

**Figure 7-5:** Primary and secondary MCLs. Source: 40 CFR.

**Figure 7-6:** Comparison of MCLGs and MCLs. Source: USEPA.

**Figure 7-10:** Waterborne diseases. Source: Adapted from American Water Works Association, *Introduction to Water Treatment: Principles and Practices of Water Supply Operations*, Denver, Colorado, 1984.

**Figure 7-11:** Basic water treatment processes. Source: American Water Works Association, *Introduction to Water* Treatment, v.2, 1984.

**Figure 7-15:** Comparison of drinking water disinfectants. Source: EPA.

**Figure 7-18:** Readily and poorly adsorbed organics. Source: Drinking Water and Center for Environmental Research Information, *Technologies for Upgrading Existing or Designing New Water Treatment Facilities*, March 1990.

## ■ Chapter 8

**Figure 8-1:** Hazardous household products. Photograph from Aaron Haupt / Stock, Boston.

**Figure 8-2:** Hazardous materials contained in various vessels.

(a) Trains. Photograph from Tom McCarthy.

(b) Fuel Tank Truck. Photograph from Daudier/Jerrican, Photo Researchers, Inc.

(c) Barrels. Photograph from Tom McCarthy.

**Figure 8-3:** Gasoline fueling station. Photograph from Tom McCarthy.

**Figure 8-6:** Love Canal clean-up. Photograph from UPI / CORBIS-BETTMANN.

**Figure 8-10:** Emergency Response Guidebook. Photograph from Tom McCarthy.

**Figure 8-11:** Restricted pesticide. Photograph from Jack Dermid / Photo Researchers, Inc.

## ■ Chapter 9

**Figure 9-1:** Textile Factory. Photograph from CORBIS-BETTMANN.

**Figure 9-2:** Miner with black lung disease. Photograph from J. Jacobson / The Image Works.

**Figure 9-4:** Safety training class. Photograph from B. Daemmrich / The Image Works.

**Figure 9-5:** Compliance inspection. Photograph courtesy of OSHA.

**Figure 9-7:** Chimney Sweeps. Photograph from CORBIS-BETTMANN.

**Figure 9-8:** Alice Hamilton. Photograph from UPI / CORBIS-BETTMANN.

**Figure 9-10:** Lung function test. Photograph from Will/Deni McIntyre / Photo Researchers, Inc.

**Figure 9-12:** Asbestos workers. Photograph courtesy of U.S. Department of the Army.

**Figure 9-15:** Use of a respirator. Photograph courtesy of U.S. Department of the Army.

**Figure 9-17:** "Sharps" control. Photograph from Brian Yarvin / Photo Researchers, Inc.

## ■ Chapter 10

**Figure 10-1:** Pile of garbage. Photograph from Rafael Macia / Photo Researchers, Inc.

**Figure 10-3:** Dry cleaner. Photograph from Margot Granitsas / Photo Researchers, Inc.

**Figure 10-4:** Medical waste from Atlantic beach. Photograph from UPI/CORBIS-BETTMANN.

**Figure 10-7:** Sources of solid waste. Source: Council on Environmental Quality.

**Figure 10-8:** Mobro 4000. Photograph from UPI/CORBIS-BETTMANN.

**Figure 10-9:** Curbside recycling. Photograph from Barbara Rios / Photo Researchers, Inc.

**Figure 10-10:** Recycling glass. Photograph from David M. Grossman / Photo Researchers, Inc.

**Figure 10-12:** Landfill. Photograph from Spencer Grant / Photo Researchers, Inc.

**Figure 10-14:** Illegal dumping. Photograph from John Eastcott/VVA Momatiuk / Photo Researchers, Inc.

**Figure 10-15:** Hazardous waste drums. Photograph courtesy of U.S. Department of the Army.

**Figure 10-17:** Treatment-Storage-Disposal facility. U.S. Army Photograph by Visual Information Division/CRDEC.

**Figure 10-19:** Disposal of "sharps". Photograph from Larry Mulvehill / Photo Researchers, Inc.

**Figure 10-20:** Infectious waste labeling. Photograph from Blair Seitz / Photo Researchers, Inc.

**Figure 10-21:** Nuclear power plant. Photograph from Ulrike Welsch / Photo Researchers, Inc.

**Figure 10-27:** Biological effects of acute radiation. Source: DOE.

**Figure 10-28:** Biological effects of chronic radiation. Source: DOE.

## ■ Chapter 11

**Figure 11-1:** Protest sign. Photograph from Jim West.

**Figure 11-5:** Environmentally friendly products. Photograph from Ogust/The Image Works.

# Index